Chiral Organic Pollutants

Food Analysis & Properties
Series Editor
Leo M.L. Nollet
University College Ghent, Belgium

This CRC series **Food Analysis and Properties** is designed to provide a state-of-art coverage on topics to the understanding of physical, chemical, and functional properties of foods: including (1) recent analysis techniques of a choice of food components; (2) developments and evolutions in analysis techniques related to food; (3) recent trends in analysis techniques of specific food components and/or a group of related food components.

Spectroscopic Methods in Food Analysis
Edited by Adriana S. Franca, Leo M.L. Nollet

Phenolic Compounds in Food: Characterization and Analysis
Edited by Leo M.L. Nollet, Janet Alejandra Gutierrez-Uribe

Testing and Analysis of GMO-containing Foods and Feed
Edited by Salah E.O. Mahgoub, Leo M.L. Nollet

Fingerprinting Techniques in Food Authenticity and Traceability
Edited by K.S. Siddiqi and Leo M.L. Nollet

Hyperspectral Imaging Analysis and Applications for Food Quality
Edited by Nrusingha Charan Basantia, Leo M.L. Nollet, Mohammed Kamruzzaman

Ambient Mass Spectroscopy Techniques in Food and the Environment
Edited by Leo M.L. Nollet, Basil K. Munjanja

Food Aroma Evolution: During Food Processing, Cooking and Aging
Edited by Matteo Bordiga, Leo M.L. Nollet

Mass Spectrometry Imaging in Food Analysis
Edited by Leo M.L. Nollet

Proteomics for Food Authentication
Edited by Leo M.L. Nollet, Semih Ötleş

Analysis of Nanoplastics and Microplastics in Food
Edited by Leo M.L. Nollet, Khwaja Salahuddin Siddiqi

Chiral Organic Pollutants: Monitoring and Characterization in Food and the Environment
Edited by Edmond Sanganyado, Basil Munjanja, Leo M.L. Nollet

For more information, please visit the Series Page: https://www.crcpress.com/Food-Analysis–Properties/book-series/CRCFOODANPRO

Chiral Organic Pollutants

Monitoring and Characterization in Food and the Environment

Edited by

**Edmond Sanganyado
Basil K. Munjanja
Leo M.L. Nollet**

CRC Press
Taylor & Francis Group
Boca Raton London New York

CRC Press is an imprint of the
Taylor & Francis Group, an **informa** business

First edition published 2021
by CRC Press
6000 Broken Sound Parkway NW, Suite 300, Boca Raton, FL 33487-2742
and by CRC Press
2 Park Square, Milton Park, Abingdon, Oxon, OX14 4RN

© 2021 Taylor & Francis Group, LLC

CRC Press is an imprint of Taylor & Francis Group, LLC

Reasonable efforts have been made to publish reliable data and information, but the author and publisher cannot assume responsibility for the validity of all materials or the consequences of their use. The authors and publishers have attempted to trace the copyright holders of all material reproduced in this publication and apologize to copyright holders if permission to publish in this form has not been obtained. If any copyright material has not been acknowledged please write and let us know so we may rectify in any future reprint.

Except as permitted under U.S. Copyright Law, no part of this book may be reprinted, reproduced, transmitted, or utilized in any form by any electronic, mechanical, or other means, now known or hereafter invented, including photocopying, microfilming, and recording, or in any information storage or retrieval system, without written permission from the publishers.

For permission to photocopy or use material electronically from this work, access www.copyright.com or contact the Copyright Clearance Center, Inc. (CCC), 222 Rosewood Drive, Danvers, MA 01923, 978-750-8400. For works that are not available on CCC please contact mpkbookspermissions@tandf.co.uk

Trademark notice: Product or corporate names may be trademarks or registered trademarks, and are used only for identification and explanation without intent to infringe.

Library of Congress Control Number: 2020948723

ISBN: [978-0-367-42923-2] (hbk)
ISBN: [978-1-003-00016-7] (ebk)

Typeset in Times
by KnowledgeWorks Global Ltd.

Dedication

*To my wife, Surprise, for her cherished support and surprises.
And our children, Tinotenda and Akatendeka.*

Edmond Sanganyado

Contents

Series Preface ix
Preface xi
Contributors xiii
About the Editors xv

Chapter 1 Overview of Chiral Pollutants in the Environment and Food 1

    Ana Rita L. Ribeiro and Edmond Sanganyado

SECTION I Analysis, Fate, and Toxicity of Chiral Pollutants in the Environment

Chapter 2 Chiral Inversion of Organic Pollutants 27

    Simbarashe Nkomo and Edmond Sanganyado

Chapter 3 Chiral Pesticides 41

    Herbert Musarurwa, Mpelegeng Victoria Bvumbi, Nyeleti Bridget Mabaso, and Nikita Tawanda Tavengwa

Chapter 4 Analysis, Fate, and Toxicity of Chiral Pharmaceuticals in the Environment 77

    Maria Dolores Camacho-Muñoz, Malgorzata Baranowska, and Bruce Petrie

Chapter 5 Chiral Personal Care Products: Occurrence, Fate, and Toxicity 105

    Edmond Sanganyado

Chapter 6 Chiral Halogenated Organic Contaminants of Emerging Concern 131

    Edmond Sanganyado

Chapter 7 Persistent Organic Pollutants 153

    Leo M.L. Nollet

SECTION II Analysis, Fate, and Toxicity of Chiral Pollutants in Food

Chapter 8 Current Trends in Enantioselective Food Analysis 191

    Basil K. Munjanja

Chapter 9 Occurrence of Chiral Pesticides in Food and Their Effects on Food Processing and the Health of Humans 227

Herbert Musarurwa and Nikita Tawanda Tavengwa

Chapter 10 Chiral Pharmaceuticals: Source, Human Risk, and Future Studies 249

Christina Nannou and Dimitra Lambropoulou

Chapter 11 Food Authenticity and Adulteration 279

Leo M.L. Nollet

SECTION III Regulation and Remediation of Chiral Pollutants

Chapter 12 Regulatory Perspectives and Challenges in Risk Assessment of Chiral Pollutants 301

Edmond Sanganyado

SECTION IV Synthesis and Chiral Switching of Chiral Chemicals

Chapter 13 Chiral Pesticides: Synthesis, Chiral Switching, and Absolute Configuration 323

Herbert Musarurwa, Mpelegeng Victoria Bvumbi, and Nikita Tawanda Tavengwa

Chapter 14 Chiral Pharmaceuticals: Synthesis and Chiral Switching 347

Memory Zimuwandeyi, Buhlebenkosi Ndlovu, Qaphelani Ngulube, and Edmond Sanganyado

Index 363

Series Preface

There will always be a need for analyzing methods of food compounds and properties. Current trends in analyzing methods include automation, increasing the speed of analyses, and miniaturization. The unit of detection has evolved over the years from micrograms to pictograms.

A classical pathway of analysis is sampling, sample preparation, cleanup, derivatization, separation, and detection. At every step, researchers are working and developing new methodologies. A large number of papers are published every year on all facets of analysis. So, there is a need for books that gather information on one kind of analysis technique or on analysis methods of a specific group of food components.

The scope of the CRC Series on *Food Analysis & Properties* aims to present a range of books edited by distinguished scientists and researchers who have significant experience in scientific pursuits and critical analysis. This series is designed to provide state-of-the-art coverage on topics such as

1. Recent analysis techniques on a range of food components
2. Developments and evolution in analysis techniques related to food
3. Recent trends in analysis techniques of specific food components and/or a group of related food components
4. The understanding of physical, chemical, and functional properties of foods

The book *Chiral Organic Pollutants: Monitoring and Characterization in Food and the Environment* is volume number 13 of this series.

I am happy to be a series editor of such books for the following reasons:

- I am able to pass on my experience in editing high-quality books related to food.
- I get to know colleagues from all over the world more personally.
- I continue to learn about interesting developments in food analysis.

A lot of work is involved in the preparation of a book. I have been assisted and supported by a number of people, all of whom I would like to thank. I would especially like to thank the team at CRC Press/Taylor & Francis, with a special word of thanks to Steve Zollo, Senior Editor.

Many, many thanks to all the editors and authors of this volume and future volumes. I very much appreciate all their effort, time, and willingness to do a great job.

I dedicate this series to

- My wife, for her patience with me (and all the time I spend on my computer)
- All patients suffering from prostate cancer; knowing what this means, I am hoping they will have some relief

Dr. Leo M.L. Nollet (Retired)
University College Ghent
Ghent, Belgium

Preface

Neglect of stereochemistry in the study and evaluation of racemic products results in highly misleading data, the waste of researchers' time on "sophisticated nonsense" and the acceptance of inferior biological products.

Everhardus J. Ariëns
Chirality in Bioactive Agents and Its Pitfalls (1986)

In 1962, Rachel Carson published *Silent Spring* in which she exposed the harmful effects of organic pollutants on humans and wildlife. It is almost the diamond jubilee of *Silent Spring*, yet chemical pollution remains one of the leading causes of premature death and noncommunicable diseases worldwide. Industrial, agricultural, transport, medical, and domestic activities result in the discharge of thousands of chemicals, particularly organic pollutants, into the environment. Food products are contaminated by organic pollutants during the production of raw materials or products. They can also be contaminated during the packaging, transport, and storage of the product. A fraction of organic pollutants are chiral compounds. For example, more than 50%, 40%, and 35% of the pharmaceuticals, organic UV filters, and pesticides on the market are chiral compounds. The main organic pollutant Rachel Carson profiled in *Silent Spring* was DDT—a chiral compound. Recent studies have demonstrated that enantiomers of DDT and its metabolites exhibit different human and wildlife toxicities.

Two weeks after the publication of *Silent Spring*, the U.S. President John F. Kennedy signed into law the Kefauver-Harris Amendments in response to the Thalidomide Tragedy. Thalidomide, a drug introduced into the market globally for treating morning sickness, was shown to be responsible for congenital disabilities in 10,000 babies. Subsequent studies showed that the (S)-enantiomer of thalidomide was teratogenic, while the (R)-enantiomers were therapeutic. While *Silent Spring* revolutionized environmental risk assessment, the Kefauver-Harris Amendments introduced the concept of drug safety in discovery and development. However, both seminal works overlooked the role of stereochemistry in the observed adverse toxicological effects. The topic of this book offers ample evidence of the increasing importance of stereochemistry in chemical pollution. It took the US Food and Drug Administration 30 years to officially recognize that stereochemistry could affect the safety and efficacy of drugs. In contrast, the European Food Safety Authority issued the "*Guidance of EFSA on risk assessments for active substances of plant protection products that have stereoisomers as components or impurities and for transformation products of active substances that may have stereoisomers*" in 2019.

Pharmaceutical scientists have been concerned about the implications of chirality in pharmacodynamics and pharmacokinetics for over 70 years, while fragrance chemists have been aware of enantioselectivity in odor perception for over 140 years. In the past 30 years, sophisticated analytical techniques for establishing the enantiomeric composition of chiral organic pollutants at trace levels have been developed to ensure that environmental and food risk assessments are not mere "sophisticated nonsense" (Ariëns, 1986). Chiral compounds are available on the market as single enantiomers, but mostly as racemic mixtures. They undergo biotic and abiotic processes that can change their enantiomeric composition. This book provides critical information that will help environmental scientists and food scientists advance their knowledge on the fundamentals of stereochemistry and enantioselective analysis required to understand better the distribution, fate, and toxicity of chiral organic pollutants.

This book is divided into four sections: (I) Chiral pollutants in the environment; (II) Chiral pollutants in food; (III) Regulation of chiral pollutants, and (IV) Synthesis and chiral switching of chiral chemicals. Section I has six chapters focusing on the inversion of chiral pollutants in the environment (Chapter 2) as well as the enantioselective fate and toxicity of pesticides (Chapter 3), pharmaceuticals (Chapter 4), personal care products (Chapter 5), halogenated organic contaminants of emerging concern (Chapter 6), and persistent organic pollutants (Chapter 7). Research on chiral pollutants in food is still in its infancy; hence, Section II has three chapters focusing on enantioselective analysis in food (Chapter 8), chiral food contaminants (i.e., pesticides [Chapter 9] and pharmaceuticals [Chapter 10]), and food authentication and adulteration (Chapter 11). Section III has one chapter that discusses the challenges and recent developments in the regulation and prioritization of chiral pollutants (Chapter 12). In the concluding section, Section IV has two chapters on the enantioselective synthesis of chiral pesticides (Chapter 13) and pharmaceuticals (Chapter 14). Chiral switching attempts are also discussed in this section.

This book is intended to provide guidance on the enantioselective behavior of organic pollutants to environmental scientists, food scientists, pharmaceutical scientists, regulators, agriculture professionals, industrial chemists, process chemists, ecotoxicologists, and environmental engineers. Graduate students and researchers seeking to understand stereochemistry in crop science, toxicology, organic chemistry, analytical chemistry, pharmaceutical science, and legislative science will find the chapters compiled in this book valuable in their research endeavor. This book came to fruition through the unbridled enthusiasm and cooperation of the authors pooled from experts across the globe. Preparing a book chapter in the middle of a severe pandemic took heroic effort and commitment. Most of the authors had to develop material for online courses, engage in homeschooling, or endure separation from loved ones yet remained committed to ensuring this book reaches your hands. Their commitment and expertise, we think, will prove valuable to the student, researcher, regulator, and professor using this book.

Edmond Sanganyado
Shantou University
Shantou, Guangdong, China

REFERENCE

ADDIN Mendeley Bibliography CSL_BIBLIOGRAPHY Ariëns, E.J., 1986. Chirality in bioactive agents and its pitfalls. *Trends Pharmacol. Sci.* 7, 200–205. https://doi.org/10.1016/0165-6147(86)90313-5

Contributors

Malgorzata Baranowska
School of Pharmacy and Life Sciences
Robert Gordon University
Aberdeen, United Kingdom

Mpelegeng Victoria Bvumbi
Department of Chemistry
University of Venda
Thohoyandou, South Africa

Maria Dolores Camacho-Muñoz
School of Pharmacy and Life Sciences
Robert Gordon University
Aberdeen, United Kingdom

Dimitra Lambropoulou
Department of Chemistry
Aristotle University of Thessaloniki
Thessaloniki, Greece

Nyeleti Bridget Mabaso
Department of Chemistry
University of Venda
Thohoyandou, South Africa

Herbert Musarurwa
Department of Chemistry
University of Venda
Thohoyandou, South Africa

Christina Nannou
Department of Chemistry
Aristotle University of Thessaloniki
Thessaloniki, Greece

Buhlebenkosi Ndlovu
Department of Metallurgy
Gwanda State University
Gwanda, Zimbabwe

Qaphelani Ngulube
Department of Applied Chemistry
National University of Science and Technology
Bulawayo, Zimbabwe

Simbarashe Nkomo
Division of Natural Science and Mathematics
Oxford College
Emory University
Oxford, Georgia

Bruce Petrie
School of Pharmacy and Life Sciences
Robert Gordon University
Aberdeen, United Kingdom

Ana Rita L. Ribeiro
Laboratory of Separation and Reaction Engineering—Laboratory of Catalysis and Materials
Department of Chemical Engineering
Universidade do Porto
Porto, Portugal

Nikita Tawanda Tavengwa
Department of Chemistry
University of Venda
Thohoyandou, South Africa

Memory Zimuwandeyi
Molecular Sciences Institute
School of Chemistry
University of the Witwatersrand
Johannesburg, South Africa

About the Editors

Edmond Sanganyado is an associate professor in the College of Science at Shantou University, China. He obtained his BSc (Hons) in applied chemistry from the National University of Science and Technology, Zimbabwe, and his PhD in environmental toxicology from the University of California Riverside under the direction of Dr. Jay Gan. Dr. Sanganyado worked as an analytical chemist at Kutsaga Research Station, where he gained expertise in chromatographic analysis. He was a postdoctoral research fellow at Shantou University under Dr. Wenhua Liu before he was appointed an associate professor. Dr. Sanganyado has written over 30 peer-reviewed papers in journals such as the *Chemical Engineering Journal, Water Research, Environment International,* and *Environmental Pollution*. His research focuses on the application of chromatographic and spectroscopic techniques to understand the behavior of chemical pollutants in aquatic environments. He contributed to our understanding of the implications of stereochemistry on the analysis, fate, and toxicity of chiral pollutants. In recognition of his contribution to sustainability, the International Union of Pure and Applied Chemistry awarded him the Periodic Table of the Younger Chemists. Dr. Sanganyado is the recipient of numerous awards, including the Fulbright Fellowship (2011), Zhujiang Postdoctoral Fellowship (2017), Shantou University Excellent Postdoctoral Fellowship (2018), IUPAC Periodic Table of the Younger Chemists Award (2018), and Shantou University Outstanding Young Talent (2019).

Basil K. Munjanja received his BSc (Hons) in applied chemistry (2013) from the National University of Science and Technology, Zimbabwe. From 2013 to date, he has authored 13 book chapters on various topics such as biopesticides, spectroscopy, and mass spectrometry, for CRC Press/Taylor & Francis, in books namely *Handbook of Food Analysis* (3rd edition), *Biopesticides Handbook, Flow Injection Analysis of Food Additives, Chromatographic Analysis of the Environment,* and *Mass Spectrometry Based Approaches* (4th edition). All these books were under the editorship of Dr. Leo M.L. Nollet. He is an associate member of the South African Chemical Institute (SACI) and the Society of Environmental Toxicology and Chemistry (SETAC) Africa.

Leo M.L. Nollet earned an MS (1973) and PhD (1978) in biology from the Katholieke Universiteit Leuven, Belgium. He is an editor and associate editor of numerous books. He edited for M. Dekker, New York—now CRC Press/Taylor & Francis Group—the first, second, and third editions of *Food Analysis by HPLC* and *Handbook of Food Analysis*. The last edition is a two-volume book. Dr. Nollet also edited the *Handbook of Water Analysis* (first, second, and third editions) and *Chromatographic Analysis of the Environment* (third and fourth editions; CRC Press). With F. Toldrá, he coedited two books published in 2006, 2007, and 2017: *Advanced Technologies for Meat Processing* (CRC Press) and *Advances in Food Diagnostics* (Blackwell Publishing—now Wiley). With M. Poschl, he coedited the book *Radionuclide Concentrations in Foods and the Environment*, also published in 2006 (CRC Press). Dr. Nollet has also coedited with Y.H. Hui and other colleagues on several books: *Handbook of Food Product Manufacturing* (Wiley, 2007), *Handbook of Food Science, Technology, and Engineering* (CRC Press, 2005), *Food Biochemistry and Food Processing* (first and second editions; Blackwell Publishing—now Wiley, 2006 and 2012), and the *Handbook of Fruits and Vegetable Flavors* (Wiley, 2010). In addition, he edited the *Handbook of Meat, Poultry, and Seafood Quality* (first and second editions; Blackwell Publishing—now Wiley, 2007 and 2012). From 2008 to 2011, he published five volumes on animal product-related books with F. Toldrá: *Handbook of Muscle Foods Analysis, Handbook of Processed Meats and Poultry Analysis, Handbook of Seafood and Seafood Products Analysis, Handbook of Dairy Foods Analysis,* and *Handbook of Analysis of Edible Animal By-Products*. Also, in 2011, with F. Toldrá, he coedited two volumes for CRC Press:

Safety Analysis of Foods of Animal Origin and *Sensory Analysis of Foods of Animal Origin*. In 2012, they published the *Handbook of Analysis of Active Compounds in Functional Foods*. In a coedition with Hamir Rathore, *Handbook of Pesticides: Methods of Pesticides Residues Analysis* was marketed in 2009; *Pesticides: Evaluation of Environmental Pollution* in 2012; *Biopesticides Handbook* in 2015; and *Green Pesticides Handbook: Essential Oils for Pest Control* in 2017. Other finished book projects include *Food Allergens: Analysis, Instrumentation, and Methods* (with A. van Hengel; CRC Press, 2011) and *Analysis of Endocrine Compounds in Food* (Wiley-Blackwell, 2011). Dr. Nollet's recent projects include *Proteomics in Foods* with F. Toldrá (Springer, 2013) and *Transformation Products of Emerging Contaminants in the Environment: Analysis, Processes, Occurrence, Effects, and Risks* with D. Lambropoulou (Wiley, 2014). In the series Food Analysis & Properties, he edited (with C. Ruiz-Capillas) *Flow Injection Analysis of Food Additives* (CRC Press, 2015) and *Marine Microorganisms: Extraction and Analysis of Bioactive Compounds* (CRC Press, 2016). With A.S. Franca, he coedited *Spectroscopic Methods in Food Analysis* (CRC Press, 2017), and with Horacio Heinzen and Amadeo R. Fernandez-Alba he coedited *Multiresidue Methods for the Analysis of Pesticide Residues in Food* (CRC Press, 2017). Further volumes in the series Food Analysis & Properties are *Phenolic Compounds in Food: Characterization and Analysis* (with Janet Alejandra Gutierrez-Uribe, 2018), *Testing and Analysis of GMO-containing Foods and Feed* (with Salah E.O. Mahgoub, 2018), *Fingerprinting Techniques in Food Authentication and Traceability* (with K.S. Siddiqi, 2018), *Hyperspectral Imaging Analysis and Applications for Food Quality* (with N.C. Basantia, Mohammed Kamruzzaman, 2018), *Ambient Mass Spectroscopy Techniques in Food and the Environment* (with Basil K. Munjanja, 2019), *Food Aroma Evolution: During Food Processing, Cooking, and Aging* (with M. Bordiga, 2019), *Mass Spectrometry Imaging in Food Analysis* (2020), *Proteomics in Food Authentication* (with S. Ötleş, 2020), and *Analysis of Nanoplastics and Microplastics in Food* (with K.S. Siddiqi, 2020).

1 Overview of Chiral Pollutants in the Environment and Food

Ana Rita L. Ribeiro and Edmond Sanganyado

CONTENTS

1.1 Introduction 1
1.2 Chiral Environmental Contaminants 3
    1.2.1 Sources of Contaminants 3
    1.2.2 Pesticides 5
    1.2.3 Pharmaceuticals 6
    1.2.4 Personal Care Products 8
    1.2.5 Chiral Persistent Organic Pollutants 11
1.3 Chiral Chemicals in Food 12
    1.3.1 Food Authentication 12
    1.3.2 Food Additives and Contaminants 13
1.4 Risk Assessment of Chiral Pollutants 13
1.5 Objective and Contents of This Book 15
Acknowledgments 16
References 16

1.1 INTRODUCTION

Chirality was defined early in 1997 by the IUPAC (International Union of Pure and Applied Chemistry) as "the geometric property of a rigid object (or spatial arrangement of points or atoms) of being non-superposable on its mirror image." This definition is derived from the absence of any symmetry elements in the molecule, including a mirror plane, a center of inversion, or a rotation-reflection axis. Conversely, an achiral molecule is that which is superposable on its mirror image (IUPAC, 1997).

The most usual type of chirality is that originated by a center of asymmetry given by a stereogenic unit, i.e., an atom that is bonded to a set of ligands in a three-dimensional disposition not superposable to its mirror copy (Ribeiro et al., 2012b). Generally, the stereogenic element is generated by a tetrahedron carbon holding four different substituents, but other atoms as sulfur, phosphorous, and silicon may also generate a stereogenic unit (Testa et al., 2014). This stereogenic element will result in the so-called enantiomers that are two molecules with non-superposable left-handed and right-handed mirror images (Ribeiro et al., 2012b). Any exchange of two of the ligands in a molecule with a stereogenic center originates its enantiomer, being the resulting pair of molecules non-superposable and mirror images of each other (IUPAC, 1997). The molecule is chiral when only one unique chiral center is present, but the presence of more than one stereogenic center may lead to chiral or achiral molecules (Solomons, 2011). Moreover, axial, planar, and helical chirality are also possible (Allenmark, 1991; Allenmark and Gawronski, 2008). In an achiral context, enantiomers behave equally due to their identical thermodynamic properties, namely solubility, pKa, melting and boiling points, partition coefficient, etc. (Ribeiro et al., 2017). They only differ on their chiroptics by rotating the plane of polarized light in the opposite direction and thus circular dichroism (CD) (Siligardi and Hussain, 2017), optical rotatory dispersion (ORD), and polarimetry

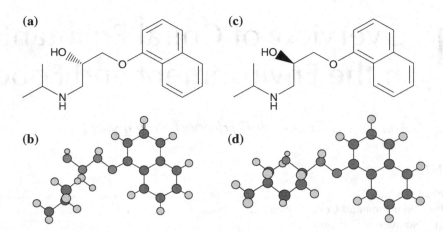

FIGURE 1.1 Chemical and three-dimensional structures of the pharmaceutical beta-blocker propranolol (1-(isopropylamino)-3-(1-naphthyloxy)-2-propanol): (a) chemical structure of (S)-(−)-propranolol; (b) three-dimensional structure of (S)-(−)-propranolol; (c) chemical structure of (R)-(+)-propranolol; (d) three-dimensional structure of (R)-(+)-propranolol.

(Brittain, 2017) are the available tools to distinguish them. Polarimetry is the traditional methodology to distinguish enantiomers which rotate differently the polarized light: to the right or clockwise (dextrorotatory, (d) or (+)-enantiomers) or to the left or counterclockwise (levorotatory, (l) or (−)-enantiomers) (Ribeiro et al., 2017). Depending on the three-dimensional arrangement of the substituents in relation to the stereogenic unit, enantiomers can be named as (R)- or (S)- from the Latin *rectus* and *sinister*, respectively (example in Figure 1.1).

Various proportions of enantiomers can occur leading to: (i) a racemate or racemic mixture, which corresponds to an equimolar mixture of enantiomers, thus resulting in the absence of rotation of the polarized light; (ii) an enantiomerically pure substance corresponding to a single enantiomer with consequent rotation of the polarized light; or (iii) a mixture of enantiomers in proportions different from 1:1, also resulting in the rotation of the polarized light (Eliel and Wilen, 1994). The similar physical and chemical properties of enantiomers are observed under achiral circumstances, whereas they can behave differently under chiral conditions that are usually found in biological media or alternatively when they react with other chiral compounds (Ribeiro et al., 2017). In fact, biological structures owe their frequent "intrinsic chirality" to their correspondent chiral units, namely proteins, glycoproteins, DNA, and RNA that are vital molecules consisting of important chiral molecules in their structure, respectively, amino acids, carbohydrates, deoxyribose, and ribose (Hühnerfuss and Shah, 2009; Müller and Kohler, 2004; Tiritan et al., 2016). This "intrinsic chirality" of biological entities (enzymes, receptors, membrane proteins, or other binding molecules) provides an enantioselective mechanism called chiral recognition, which originates dissimilar interactions between the enantiomers and the receptor and consequently different biological effects of enantiomers of a chiral compound, including different toxicity (Ribeiro et al., 2012b). As an example, the beta-blocker propranolol has one unique chiral center and thus two enantiomers (Figure 1.2), with the (S)-(−)-1-(isopropylamino)-3-(1-naphthyloxy)-2-propanol) having a pharmacological activity 100 times higher than the (R)-(+)-1-(isopropylamino)-3-(1-naphthyloxy)-2-propanol) (Pavlinov et al. 1990).

A lower activity or even any effect on a certain receptor is expected when the enantiomer binds to the receptor less intensively or does not bind, respectively. However, this "ineffective" enantiomer for the target receptor may be a ligand for another receptor responsible for other effects, which include unwanted toxic effects (Ribeiro et al., 2012b,c). Although chiral compounds naturally occurring in the environment (e.g., epinephrine, hyoscine, levodopa, levothyroxine,

Overview of Chiral Pollutants

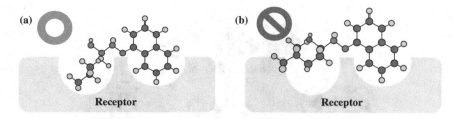

FIGURE 1.2 Illustrative interactions of the enantiomers of a chiral compound with its target receptor representing: (a) the interaction with the more potent enantiomer (*S*)-(−)-propranolol; and (b) the interaction with the less active (*R*)-(+)-propranolol.

morphine, etc.) (Gal, 2006; Tiritan et al., 2016) are typically pure enantiomers, anthropogenic chiral compounds (e.g., pharmaceuticals, illicit drugs, pesticides, etc.) are used either as enantiomerically pure or racemic mixtures, despite the desired biological effect is usually due to one enantiomer.

1.2 CHIRAL ENVIRONMENTAL CONTAMINANTS

This section discusses the main groups of chiral pollutants that have been detected in the environment, namely pesticides, pharmaceuticals and illicit drugs, personal care products, and other chiral persistent organic pollutants (POPs). Discussions in this section are focused on the use, source, and physicochemical properties of the group of compounds. An overall emphasis will be given on the nature of chirality and its implications on the use.

1.2.1 Sources of Contaminants

The immeasurable sources of production, usage, and disposal of thousands of substances frequently used in medicine (human and veterinary), industry, agriculture, household, among others, led to the extensive occurrence of organic pollutants in the environment (Sousa et al., 2018), including the chiral ones. Their release into the environment even at residual levels leads to their possible accumulation in the environmental compartments, which can originate potential harmful effects to ecosystems and to human health (Sousa et al., 2018). Contaminants can be originated from (i) domestic/hospital wastewater effluents (the main pathway in the case of pharmaceuticals/estrogens); (ii) livestock and aquaculture; (iii) agriculture runoff; (iv) landfill leachates; and (v) industrial wastewater (Barbosa et al., 2016). Figure 1.3 illustrates these sources of contamination that are common to chiral pollutants.

In this sense, their widespread occurrence in wastewater, surface and ground water, drinking water, sludge, soils, and air, as well as through the food web has led to a growing interest on the topic of chiral pollutants, and research on this subject has been increasing in the last two decades (Figure 1.4). Interestingly, approximately 65% of the papers retrieved from a Scopus search (up to May 29, 2020), using as keywords "chiral" and "pollutant" or "contaminant," were published in the last decade.

The enantiomeric fraction (EF) is the most used monitoring tool to assess the fate of chiral substances and their metabolites in the environment. EF is defined as the proportion between the concentration of one enantiomer and the sum of the concentrations of both enantiomers (Ribeiro et al., 2012c). An EF of 0.5 indicates that the composition is racemic, whereas an EF of 0 or 1 means that the compound is enantiomerically pure (Ribeiro et al., 2012c). In turn, a proportion of two enantiomers different than 1:1 gives an EF between 0 and 1 and different to 0.5, with an enrichment of one enantiomer. Tiritan et al. (2018) recently reviewed and discussed the many different notations that have been used to express the measured enantiomeric ratios and highlighted that

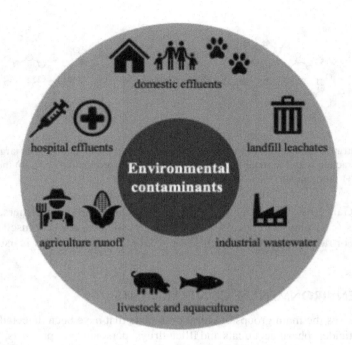

FIGURE 1.3 Sources of environmental chiral (and achiral) contaminants.

this issue is challenging for a proper data comparison. That critical review recommends the use of enantiomeric ratio (e.r. %) as obtained from the chromatograms (e.g., 80:20 (*R:S*), indicating 80% of (*R*)-enantiomer or conversely 20% of (*S*)-enantiomer, with enantiomeric enrichment of the (*R*)-form). The authors concluded that EF is also suitable since it is obtained by an equivalent approach but using a normalized scale (Tiritan et al., 2018). Anyway, the elution order can always be used when the absolute configuration of the enantiomers is not known (Tiritan et al., 2018).

The evaluation of EF variations in the environmental compartments allows a comprehensive understanding of the chemical transformations, the biotransformation, and the processes involved in the transfer between compartments, as well as the different ecotoxicological properties resulting from their enantioselective interaction with other natural chiral molecules (Ribeiro et al., 2020). The knowledge obtained by enantioselective analysis is nowadays considered a valuable tool for

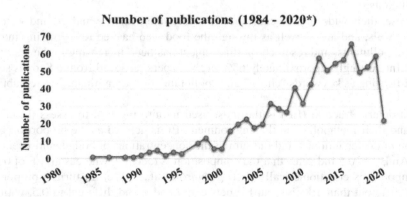

FIGURE 1.4 Evolution of the number of publications retrieved from Scopus search (May 29, 2020), using as keywords "chiral" and "pollutant" or "contaminant."

epidemiological and forensic purposes; namely, to track sources of contamination, to detect illicit discharges, and to study the usage patterns (Ribeiro et al., 2018).

1.2.2 Pesticides

A number of groups of chiral compounds used as racemic mixtures or enantiomerically pure compounds can contaminate the environment, namely pesticides, pharmaceuticals and illicit or abuse drugs, polycyclic aromatic hydrocarbons (PAHs), among others.

Agricultural activities are one of the most relevant sources of pesticides due to their intensive usage to protect plants and improve crop yields (Sousa et al., 2018). In fact, pesticides represent a large group of substances belonging to several classes, namely herbicides, insecticides, and fungicides (Maia et al., 2017). Other groups of pesticides include insect repellents, nematicides, molluscicides, rodenticides, and also plant growth regulators (Maia et al., 2017). Besides the vast number of pesticides that are widely applied worldwide, these compounds can easily achieve other environmental compartments rather than soils, namely air, surface water, and groundwater (Maia et al., 2017). In fact, this broad group of compounds is the most studied of environmental pollutants concerning enantiomeric composition, with a number of reports showing their occurrence, distribution (Faller et al., 1991), biodegradation, and toxicity in many matrices including biota (Carlsson et al., 2014; Hühnerfuss et al., 1993; Jantunen and Bidleman, 1996).

Most studies on chiral pesticides are focused on: (i) insecticides (including the banned organochlorine pesticides (OCPs) such as α-HCH, o,p'-DDT (dichlorodiphenyltrichloroethane), chlordane and toxaphene congers, but also the synthetic pyrethroids), (ii) fungicides (triazole), and (iii) herbicides (diclofop-methyl, lactonfen fluazifop-butyl, fluroxypyr methylheptyl ester, fenaxaprop-ethyl, etc.) (Xu et al., 2018).

As an example, the enantiomers of some phenylpyrazole chiral insecticides (e.g., fipronil, ethiprole, and flufiprole) were very recently reported as stereoselective endocrine disruptors (Hu et al., 2020). One of these phenylpyrazole insecticides, fipronil, was studied regarding its genomic mechanisms responsible for the enantiotoxicity and the authors found that the anxiety-like behavior enantioselectively induced in embryonic and larval zebrafish was ascribed to DNA methylation changes (Qian et al., 2019). Another study focused on the potential endocrine-disrupting effects of the chiral triazole fungicide prothioconazole and its metabolite suggested that the stereoselective effects of both the parent fungicide and its metabolite were partially attributed to enantiospecific receptor binding affinities (Zhang et al., 2019). The possible enantioselective agonistic/antagonistic effects on corticosteroids receptors and the influence on the production of corticosteroids were also assessed for eight chiral herbicides (Shen et al., 2020). Although neither the racemates or the enantiomers of all the studied chiral herbicides exhibited agonistic activity to glucocorticoid or mineralocorticoid receptors at non-cytotoxic concentrations, antagonism was found for rac-propisochlor and (S)-imazamox and (R)-napropamide (Shen et al., 2020). Rac-propisochlor and rac-/(R)-napropamide reduced the secretion of cortisol, while this glucocorticoid was induced by rac-diclofop-methyl and its two enantiomers. Exposure to rac-propisochlor, (S)-diclofop-methyl or rac-/(S)-/(R)-acetochlor, and rac-/(S)-/(R)-lactofen inhibited the aldosterone production (Shen et al., 2020). This enantioselective disruption of corticosteroid homeostasis was suggested for chiral herbicides (Shen et al., 2020).

Besides stereoselectivity in biodegradation pathways and ecotoxicity, chiral pesticides may enantioselectively bioaccumulate through the food web. This process can be followed by studying the chiral signatures, as reported by Zhou et al. (2018a). In that study evaluating the accumulation of the enantiomers in biota, it was shown that the majority of the target marine organisms had residues of non-racemic OCPs. The accumulation of pesticides through the food web was also demonstrated by the presence of banned pesticides in aquatic organisms, even many years after ban, which is ascribed to the persistence and lipophilic nature of many of them (Carlsson et al., 2014).

A very recent review collected information on another important phenomenon, which is the influence of local environmental conditions in the enantioselective behavior (Sanganyado et al., 2020). The authors highlighted that the possible interactions between the chiral pollutants and the external agents inducing homochirality is very complex and influenced by local environment conditions, including pH, redox conditions, humic acids, organic carbon, and organic nitrogen (Sanganyado et al., 2017). The pH and redox conditions of soils may affect the enantioselectivity degradation in this environmental compartment, as demonstrated for metalaxyl (Buerge et al., 2003). More recently, some reports have been published addressing other factors that may impact the enantioselectivity of pesticides. The EFs determined for the chiral o,p'- DDT, trans-chlordane, and cis-chlordane showed that these OCPs were mostly non-racemic in the soils (Zheng et al., 2020). The authors calculated the deviation of EFs from racemic and found that the deviations of cis-chlordane was highly correlated with organic carbon and the carbon-nitrogen ratio, suggesting that the enantioselective biodegradation of chiral compounds in soils may be affected by those factors by interfering with the microbial activity (Zheng et al., 2020). Another recent study demonstrated the impact of coexistent metals in aquatic environments in the environmental risk of pyrethroids, which are chiral insecticides (Yang et al., 2017). The authors studied the influence of metals on the known enantioselective toxicity and biotransformation of this class of pesticides in zebrafish (Yang et al., 2017). In that study, the effects of cadmium, copper, and lead on the enantioselective toxicity and metabolism of cis-bifenthrin were explored and the authors suggested that the simultaneous exposure to metals might affect the stereoselective biodegradation of pyrethroids and originate a stereoselective accumulation of the active enantiomer, which results in higher toxicity of pyrethroids to exposed fish.

Based on the massive use of chiral pesticides and nanomaterials, a study on the effect of carbon nanotubes on the toxicity of indoxacarb as co-existing pollutant showed that multi-walled carbon nanotubes could affect its enantioselective toxicity in zebrafish (Wang et al., 2020). The authors found a superior expression of some genes in the livers of males and brains of females, meaning a higher toxicity, after exposure to (S)-indoxacarb in comparison to those zebrafish treated with (R)-indoxacarb. These studies open a new research interest required to study the mechanisms underlying the effects of the interactions between achiral substances as metals or nanomaterials and the enantiomers in living organisms. Thus, these factors should be considered in future risk assessment exercises.

1.2.3 Pharmaceuticals

Pharmaceuticals are used for diagnosis, treatment, or prevention of health conditions in both humans and animals by interacting with the binding site of a receptor and the consequent formation of a drug-receptor complex, thus promoting the desired pharmacological activity (Stringer, 2006). Although the pharmacological activity is generally well known, their effects in non-target organisms are still not completely understood (Daughton and Ternes, 1999). In this context, there is a great environmental concern about chiral pharmaceuticals in particular due to the additional complexity in comparison to achiral ones, based on the possible different pharmacokinetic and pharmacodynamics of enantiomers and consequent dissimilar toxicological and ecotoxicological properties. Chiral illicit drugs have been receiving a great attention due to their high toxicity to aquatic organisms and in this topic, Kasprzyk-Hordern et al. have published several papers (Bagnall et al., 2013, 2012; Castrignanò et al., 2016; Evans et al., 2016, 2015; Evans and Kasprzyk-Hordern, 2014; Petrie et al., 2017, 2014; Vazquez-Roig et al., 2014). Interestingly, a very recent study described evidence on the effects of co-existing microplastics in the increasing toxic effects and enantioselectivity behavior of methylamphetamine to green alga and freshwater snail (Qu et al., 2020), as verified for many pesticides.

There are chiral pharmaceuticals commercialized as racemates, as single enantiomers, and as both racemic and enantiopure forms, depending on the interaction of the enantiomers with the receptors (Ribeiro et al., 2012b). There is a trend of commercialization of enantiopure pharmaceuticals and the chiral switching (re-evaluation of the license of the enantiomeric pure substances commercialized as racemate) has played an important role in this shift (Pavlinov et al., 1990). Other approach to obtain enantiopure substances is *de novo* development of pure enantiomers, which can be achieved by three processes: (i) screening enantiomerically pure natural drugs, also known as "chiral pool"; (ii) asymmetric synthesis; and (iii) chiral resolution that also complements the other two processes (Song et al., 2020). However, there are a number of pharmaceuticals that are commercialized both as racemic mixture and enantiopure forms, as example: bupivacaine/(S)-bupivacaine (levobupivacaine), cetirizine/(R)-cetirizine (levocetirizine), citalopram/(S)-citalopram (escitalopram), ibuprofen/(S)-ibuprofen (dexibuprofen), ketoprofen/(S)-ketoprofen (dexketoprofen), ofloxacin/(S)-ofloxacin (levofloxacin), omeprazole/(S)-omeprazole (esomeprazole), salbutamol/(S)-salbutamol (levalbuterol) (Tiritan et al., 2016). A less frequent alternative is the production of chiral pharmaceuticals with a proportion of enantiomers different from 1:1 (Ribeiro et al., 2012b). Another important aspect is that although the tendency to use enantiopure pharmaceuticals is driven by the lower therapeutic doses, less adverse effects, higher safety range, less interindividual variability, and less drug-drug interactions, the option of using enantiopure pharmaceuticals can also lead to unforeseen toxic effects (Ribeiro et al., 2012c). As an example, an unsuccessful chiral switching occurred with fluoxetine. The attempts of chiral switching for the development of a market license for (R)-fluoxetine led to an adverse cardiac effect, resulting in its abandonment (McConathy and Owens, 2003).

Chiral pharmaceuticals, and also chiral illicit drugs that follow the same pathways of pharmaceuticals, are metabolized and excreted by humans and animals and the parent compounds and the respective metabolites can reach the aquatic environment through their passage in wastewater treatment plants (WWTPs). Moreover, unused or expired drugs may be incorrectly dumped as solid waste or in toilets or sinks, reaching landfills and WWTPs, respectively (Sousa et al., 2018). The footprint of pharmaceutical industries cannot be disregarded even when industrial effluents are released into WWTPs, since the pharmaceuticals are not completely removed by the conventional treatment occurring in the WWTPs (Barbosa et al., 2016). Therefore, the release of effluents from municipal WWTPs is considered the most important pathway of pharmaceuticals in the aquatic environment (Ribeiro et al., 2012b, 2013). As the conventional WWTPs are not designed to completely remove residual organic compounds, those compounds not removed or partially degraded in WWTPs can reach surface waters that are interconnected to ground water and drinking water (Escuder-Gilabert et al., 2018; Ribeiro et al., 2013, 2012a). On the other hand, many industries are still not regulated in some countries and their non-treated effluents are often discharged into the environment (Sousa et al., 2018). The runoff from livestock areas and the release from aquaculture are other relevant sources of pharmaceuticals and estrogens (Barbosa et al., 2016). The use of reclaimed wastewater in agriculture can also be a source of chiral pharmaceuticals that may leach into receiving waters. Similarly, landfill leachates and septic tanks cannot be disregarded as a source of this type of chiral pollutants.

The general polar nature of pharmaceuticals as a wanted characteristic to promote their distribution and excretion within organisms makes them prone to be mostly removed in WWTP by adsorption, biodegradation, hydrolysis, and photodegradation, the same mechanisms occurring in the surface water (Ribeiro et al., 2020). In the case of chiral pollutants, biodegradation is an important degradation route since microorganisms are able to remove them enantioselectively (Ribeiro et al., 2012b). Despite the recent advances on the knowledge about environmental chirality and its recognition as an important research field, chiral pharmaceuticals and illicit drugs are mostly studied as unique molecular entities and EF is often neglected, therefore underestimating the enantioselective environmental behavior (e.g., occurrence, fate, distribution, degradation, uptake) and toxicological effects (Ribeiro et al., 2020). Many review papers have been published in this field, namely on their environmental determination (Dogan et al., 2020; Evans and Kasprzyk-Hordern, 2014; Ribeiro et al., 2020), occurrence and fate (Kasprzyk-Hordern, 2010;

Petrie et al., 2014; Sanganyado et al., 2017), biodegradation (Hashim et al., 2010; Maia et al., 2017; Wong, 2006; Xu et al., 2018), potential human health impacts and remediation (Sanganyado, 2019; Stanley and Brooks, 2009; Zhou et al., 2018b), and forensic application (Ribeiro et al., 2018; Sanganyado et al., 2020).

1.2.4 Personal Care Products

Humans use various chemicals such as cosmetics, disinfectants, preservatives, food additives, fragrances, and toothpaste to improve quality of life (Brausch and Rand, 2011). These personal care products contain a diverse group of bioactive ingredients that can potentially cause adverse effects to the environmental balance and human health (Hopkins and Blaney, 2016; Langdon et al., 2010; Nohynek et al., 2010). The main route of exposure for personal care products to humans is through dermal absorption (Tseng and Tsai, 2019). Thus, dermal exposure to personal care products may often result in acute effects such as dermatitis (Nohynek et al., 2010). Moreover, several studies have shown that personal care products can have carcinogenic, mutagenic, and estrogenic effects in humans (Jiménez-Díaz et al., 2014). Personal care products often enter the environment as parent products since they are often used externally, but their absorption also results in its metabolism and excretion as parent and/or metabolites (Kim and Choi, 2014; Schlumpf et al., 2010). Like pharmaceuticals, personal care products primarily enter the environment via wastewater effluent. Several studies have detected high concentrations of personal care products such as parabens, organic UV filters, synthetic musks, and disinfectants in wastewater effluent, rivers, and lakes (Brausch and Rand, 2011; Jiménez-Díaz et al., 2014; Li et al., 2018). Most personal care products are considered pseudo-persistent because they are continuously discharged into the environment, which is not overcome by their transformation. Several studies have shown that personal care products can cause adverse health effects on aquatic biota (Li et al., 2020). For that reason, personal care products are considered contaminants of emerging concern. Hence, understanding the sources, physicochemical properties, and environmental behavior of personal care products is important for improving the accuracy of environmental risk assessments.

Some personal care products such as synthetic musks and organic UV filters are chiral compounds. Chiral synthetic musks are extensively used as fragrances in several consumer products such as perfumes, shampoo, soaps, and cosmetics. Synthetic musks can be classified as macrocyclic, nitro, or polycyclic musks. Polycyclic musks are the most widely used synthetic musks. Examples of chiral polycyclic musks include galaxolide, tonalide, phantolide, traseolide and cashmeran. Galaxolide and traseolide have two chiral centers while tonalide, phantolide, and cashmeran have one unique stereogenic center (Table 1.1) (Lee et al., 2016). It is interesting to note that only two of the galaxolide stereoisomers can be sensed by the human nose. This suggests that the other stereoisomers are unnecessary contaminants that might need to be removed to reduce waste. However, the use of enantiomers can lead to more toxic effects as was observed in the failed chiral switching of fluoxetine (Agranat et al., 2002; Calcaterra and D'Acquarica, 2018). Chiral polycyclic musks have high lipophilicity (log K_{OW} = 4.5–6.3), hence they readily bioaccumulate in tissue as well as readily suffer partition into sediment and organic particles (Brausch and Rand, 2011). Previous studies have shown that chiral polycyclic musks can undergo enantioselective transformation in sewage sludge (Gao et al., 2019), surface water (Lee et al., 2016), sediments (Song et al., 2015), recycling water (Wang and Khan, 2014), and WWTPs (Lee et al., 2016). In biota, galaxolide, tonalide, phantolide, and traseolide were shown to undergo enantioselective metabolism in different fish species (Gatermann et al., 2002). A few studies have investigated the toxicity of chiral musks; for example, (R)-muscone was shown to cause higher toxicity to zebrafish embryo than (S)-muscone (Li et al., 2020). The enantiomeric composition of polycyclic musks in environmental matrices is widely determined using enantioselective capillary electrophoresis and gas chromatography coupled to mass spectrometry or tandem mass spectrometry (Martínez-Girón et al., 2010; Wang et al., 2013).

TABLE 1.1
A List of Five Polycyclic Musks and Their Molecular Structures

Chemical name	Trade name	Structure	
4,6,6,7,8,8-Hexamethyl-1,3,4,6,7,8-hexahydrocyclopenta[g]isocromene, HHCB	Galaxolide, Abbalide	(4R, 7R)-Galaxolide	(4S, 7S)-Galaxolide
		(4R, 7S)-Galaxolide	(4S, 7R)-Galaxolide
7-Acetyl-1,1,3,4,4,6-hexamethyl-tetraline, AHTN	Tonalide, Fixolide	(3R)-Tonalide	(3S)-Tonalide
5-Acetyl-1,1,2,3,3,6-hexamethylindane, AHDI	Phantolide	(2R)-Phantolide	(2S)-Phantolide
5-Acetyl-1,1,2,6-tetramethyl-3-isopropylindane, ATII	Traseolide	(2R, 3R)-Traseolide	(2S, 3S)-Traseolide
		(2S, 3R)-Traseolide	(2R, 3S)-Traseolide
1,1,2,3,3-Pentamethyl-1,2,3,5,6,7-hexahydro-4H-inden-4-one, DPMI	Cashmeran	(2S)-Cashmeran	(2R)-Cashmeran

Source: Chemical structures of the polycyclic musks are from Wang et al. (2013). Used with permission from Elsevier Ltd.

Organic UV filters are commonly used in sunscreens, lotions, hair products, and cosmetics for protecting the skin against UV radiation. UVA, UVB, and UVC light can cause radiation damage on the skin (Hopkins and Blaney, 2016). Organic UV filters are pseudo-persistent contaminants since they are continuously discharged into the environment due to their extensive usage. Several studies have detected organic UV filters in various environmental systems such as surface water, freshwater and marine sediments, and soil (Montes-Grajales et al., 2017). Organic UV filters can potentially bioaccumulate in tissue as they have been detected in mussels and fish (Yang et al., 2020). They can readily partition to organic matter in sediments or soil due to their high lipophilicity (log K_{OW} = 3.8–6.9) (Hopkins and Blaney, 2016; Rainieri et al., 2017). The exposure to organic UV filters has been shown to cause adverse effects in the reproduction, mobility, and development of fish, *Daphnia magna*, and coral larvae (Campos et al., 2017; Gilbert et al., 2013; Lozano et al., 2020; Park et al., 2017; Quintaneiro et al., 2019; Tsui et al., 2017). Organic UV filters have also been shown to cause anti-estrogenic, androgenic, and anti-androgenic activity in humans. Hence, organic UV filters are considered an important class of contaminants of emerging concern. However, some organic UV filters such as 4-methylbenzylidene camphor are chiral compounds that have stereoisomers which exhibit different fate and toxicity profiles in the environment (Buser et al., 2005). For example, the transformation of 2-ethylhexyl 4-dimethylaminobenzoate in rabbit liver and kidneys was shown to favor the (+)-enantiomer (Liang et al., 2017). Table 1.2 shows examples of chiral organic UV filters approved for use as sunscreens by the European Union. Therefore, the chiral nature of organic UV filters should not be overlooked when assessing their human health and environmental risks.

TABLE 1.2

Physicochemical Characteristics of Four Chiral Organic UV Filters Approved for Use by the European Union

Name	CAS no	Structure	Log K_{OW}	Solubility (g L^{-1})	λ_{max} (nm)
2-Ethylhexyl 4-(dimethylamino)benzoate	21245-02-3		6.15	2.1×10^{-3}	310
Homosalate	118-56-9		6.16	0.02	–
Ethylexyl methoxycinnamate	5466-77-3		5.8	0.15	306
4-methylbenzylidene camphor	36861-47-9		4.95	5.1×10^{-3}	300

Source: The log K_{OW}, solubility, and λ_{max} data for the selected chiral organic UV filters is from Díaz-Cruz et al. (2008). Used with permission from Elsevier Ltd.

1.2.5 CHIRAL PERSISTENT ORGANIC POLLUTANTS

The presence of POPs in the environment is an issue of major concern. POPs pose human health and environmental risks because they have high persistence, toxicity, and long-range transport, biomagnification, and bioaccumulation potential. The Stockholm Convention initially recognized 12 compounds as POPs, and these included halogenated organic contaminants such as OCPs, polychlorinated biphenyls (PCBs), and hexachlorobenzene (Liu et al., 2016; Raubenheimer and McIlgorm, 2018). New POPs were added to the Stockholm Convention and these included hexabromobiphenyl, hexabromocyclododecane (HBCD), α- and β-hexachlorocyclohexane, perfluorooctane sulfonic acid, perfluorooctanoic acid, tetrabromodiphenyl ether, and pentabromodiphenyl ether (English et al., 2016; Jiang et al., 2019). Hence, most POPs are halogenated organic contaminants, and this is probably because of the unique properties that the highly electronegative halogen atoms impart on molecules. Halogens are highly electronegative atoms, and carbon-halogen bonds often impart thermal and chemical stability to the molecule, hence their high recalcitrance in the environment (Tittlemier et al., 2002; Van Pée and Unversucht, 2003). In addition, aromatic and alkyl molecules with halogen atoms have high lipophilicity rendering them highly bioaccumulative. However, not all POPs are halogenated organic contaminants. For example, PAHs are non-halogenated compounds, but they originate halogenated transformation products.

Understanding the source and route of entry of POPs is important in developing their mitigation strategies. The primary source of PAHs of anthropogenic origin in the environment is the combustion of both biomass and fossil fuels (Guo et al., 2017; Tobiszewski and Namieśnik, 2012). They enter the environment through atmospheric deposition, surface runoff, and industrial effluents. Several studies have detected PAHs in surface water, sediment, soil, and even marine mammals (Sanganyado et al., 2018). Before their production was banned, PCBs were used in industry as flame retardants, plasticizers, coolants, and dielectrics. The main route of entry for PCBs was through industrial effluents. However, even after they were banned, this class of pollutants is present ubiquitously in the environment due to their high persistence. Likewise, PCBs, OCP use was globally banned, with only some countries still using OCPs, e.g., DDTs for malaria control. Currently, the primary source of OCPs is historically contaminated sediments and soils from which they can re-enter to the water column through different perturbation processes. In this way, these OCPs can subsequently bioaccumulate in aquatic biota such as fish, mussels, and marine mammals (Yu et al., 2020). Hence, OCPs and PCBs are classified as legacy contaminants.

OCPs, PAHs, and PCBs often contain a chiral element in their chemical structure. In fact, some OCPs such as chlordane, heptachlor, α-HCH, and o,p'-DDT contain at least one chiral center although they are marketed as racemic mixtures. A study of ambient air in an area without any immediate agricultural activities showed the enantiomeric composition of OCPs to be non-racemic (Leone et al., 2001). These results suggested the enantioselective transformation of the OCPs. A study in fish found that $(+)$-o,p'-DDT was dominant in fish suggesting that the $(-)$-enantiomer was preferentially degraded (Meng et al., 2009). Besides chiral centers, legacy POPs sometimes owe their chirality to axial chirality and exist as atropoisomers. PCBs and PAHs often contain an axially chiral aryl bond. Of the 209 possible PCBs, about 78 congeners exhibit axial chirality. About 19 PCB congeners have been detected in environmental matrices owing to their high rotational energy barriers. Although several studies have shown that the fate and toxicity of chiral PCBs is enantioselective, there are few studies that have explored the implications of chirality on the environmental behavior of PAHs.

Emerging contaminants include potentially toxic natural or synthetic compounds that are not regularly monitored (Bebianno and Gonzalez-Rey, 2015; Jiang et al., 2014; Reid et al., 2019). Examples of emerging contaminants that contain a chiral element include per- and polyfluoroalkyl substances (Rayne et al., 2008; Wang et al., 2011; Zhao et al., 2019), halogenated flame retardants (Ruan et al., 2018b, 2019; Zhang et al., 2013), and chlorinated paraffins. However, most studies on the effect of chirality on the distribution, fate, and toxicity of halogenated organic contaminants of emerging concern focused on halogenated flame retardants, particularly brominated flame retardants (Badea et al., 2016; Köppen et al., 2010; Ruan et al., 2019). Some brominated flame retardants such as HBCDs, polybrominated biphenyls, and tetrabromoethyl cyclohexane possess at least one chiral element. Brominated flame retardants are mainly used in textiles, electronics, plastics, firefighting foam, and food packaging (Birnbaum and Staskal, 2004; Brits et al., 2016). They enter the environment through landfills, atmospheric deposition, surface runoff, and wastewater effluents. In the marine food web, the trophic magnification potential of (−)-α-1,2,5,6,9,10- HBCD was significantly higher than that of the (+)-enantiomer with trophic magnification factors of 11.8 and 8.7, respectively (Ruan et al., 2018b). To corroborate these results, preferential enrichment of specific (−)-enantiomers of α-, β-, and γ-HBCD were observed in Indo-Pacific humpback dolphins and finless porpoises (Ruan et al., 2018a). Interestingly, *in vitro* studies found that (+)-HBCDs were more toxic than the (−)-enantiomers (Hamers et al., 2006; Khalaf et al., 2009). This suggests that the less toxic enantiomer is the one with the highest biomagnification potential. In contrast, exposing carp to HBCDs resulted in an enrichment of the more potent (+)-α- and -γ-enantiomers (Zhang et al., 2014). These results suggest that the enantioselectivity data of a specific compound cannot be extrapolated to different organisms.

1.3 CHIRAL CHEMICALS IN FOOD

Chemical analysis is important in the food industry as it helps in monitoring the quality, authenticity, and safety of the food. The chemical composition of food products is monitored regularly to ensure that the food product adheres to national and international regulations. Most countries have a standards association or a food and drug administration agency that monitors and/or regulates the quality and safety of foods.

1.3.1 Food Authentication

Chemicals in food can be endogenous, additives, or contaminants. Endogenous chemicals are those chemicals that naturally occur in the food products, and these may include amino acids and flavors. Chirality is an important property of food flavors and aroma. For example, *d*-asparagine has a sweet taste, while its antipode *l*-asparagine is tasteless (Engel, 2020) and (1R,2S,4R)-(+)-borneol and (1S,2R,4S)-(−)-borneol have a weak camphor-like odor and camphor- or turpentine-like odor, respectively (Bentley, 2006). The advances in enantioselective analysis supported the development of flavor and aroma analysis as powerful tools for food authentication. The determination of the enantiomeric compositions of naturally occurring food constituents at trace levels allowed to provide the essential data for establishing the pathways and enzymes involved in the biosynthesis of chiral substances. Moreover, enantioselective analysis of chiral substances is also important to understand the effect of enantiomeric composition on the odor and taste of foods that has been extensively studied (Engel, 2020). Considering that the enzymes involved in the synthesis of chiral substances are highly enantiospecific, the enantiomeric composition of the chiral substances is normally consistent, hence the application of EF values as an indicator of food authenticity. For example, raspberry flavor is primarily due to α-ionone that occurs naturally as the (R)-enantiomer (97–100%) as well as δ-octalactone, δ-decalactone, and terpenes-4-ol that occur naturally as the (S)-enantiomer (80–100%) (Hansen et al., 2016). Hence, a deviation of the EF values of these flavors from the natural enantiomeric composition suggests that the raspberry flavor is not natural. However, the application of EF values as indicators of food authenticity is challenging because flavor and aroma compounds may widely vary due to spatial, temporal, and local environmental conditions (Schäfer et al., 2015). In addition, food storage and processing can result in shifts in the enantiomeric composition of the chiral compounds due to an enzymatic action or physicochemical treatment (Schäfer et al., 2015). Hence, enantioselective analysis should be used in conjunction with other food authenticity techniques for improving the accuracy required.

1.3.2 FOOD ADDITIVES AND CONTAMINANTS

Food additives are a diverse group of chemicals that are added to food to enhance or maintain their sensorial, physicochemical, biological, and rheological properties (Martins et al., 2019). They can maintain or improve the color, flavor, quality, texture, and stability of foods. Food additives can be classified according to their industrial use into 25 classes such as acidity regulator, anticaking agent, antioxidant, colorant, emulsifier, foaming agent, preservative, and sweetener. However, previous studies found that the gut microbiota were altered by different food additives using *in vitro* studies and mammal models (Cao et al., 2020). Moreover, food additives have been shown to cause adverse reactions in some people (Wilson and Bahna, 2005). In fact, it was verified that azo dyes used as food colorants may cause proinflammatory responses *in vitro*, indicating a potential risk to human health (Leo et al., 2018). The European Food Safety Authority (EFSA) recommends that the enantiomeric stability of chiral food additives during storage should be investigated (Bura et al., 2019). Enantiomers can undergo chiral inversion whereby they change into their antipode. The EFSA further recommends that when the enantiomers of the food additives have different or unknown toxicological profiles, the enantiomers should be treated as different compounds (Bura et al., 2019).

Chemical food contaminants pose a risk to human health. The main route of entry of chemical contaminants to food is through packaging material, processing, soil, and water. Examples of chemicals that can contaminant food through packaging material include flame retardants, plasticizers, antioxidants, and thermal and light stabilizers. Several studies detected brominated flame retardants (Shaw et al., 2014), organic UV filters (Snedeker, 2014), and perfluorinated compounds (Rice et al., 2014) in food packaging and contact materials. However, as shown in the previous sections, some brominated flame retardants, organic UV filters, and perfluorinated compounds are potentially toxic chiral compounds. The second class of food contaminants is environmental contaminants that enter the raw material of the food due to contact, interaction, or exposure to contaminated environment. Examples of such contaminants include pharmaceuticals and personal care products which can enter plant products that are irrigated with contaminated water or grown in soil amended with biosolids (Fu et al., 2016). Additionally, pesticides are used in agriculture and can be taken up by plants. In fact, more than 35% of pesticides currently used possess at least one chiral center. Several studies have shown that the uptake of chiral pesticides in plants often results in the enrichment of one enantiomer over the other (Sanganyado et al., 2020; Zhang et al., 2018b, 2018a). In addition to land-based pathways, contaminant uptake in food can occur in aquatic ecosystems such as wild fisheries and aquaculture systems. Aquatic ecosystems are often the sink of most organic pollutants and hence, organic pollutants can bioaccumulate in fish (Meng et al., 2009). Interestingly, bioaccumulation of some chiral pollutants such as triazole fungicides (Konwick et al., 2006a; Wang et al., 2015), fipronil (Baird et al., 2013; Konwick et al., 2006b), and lactofen (Wang et al., 2018) has been shown to be enantioselective. These metabolism studies show that these chiral compounds should be treated as distinct compounds when assessing the safety of the fish for human consumption. The EFSA recommends that the distribution, behavior, and toxicity of the enantiomers shall be established individually to improve the robustness of the risk assessment (Bura et al., 2019).

1.4 RISK ASSESSMENT OF CHIRAL POLLUTANTS

Chemical risk assessment provides critical quantitative and qualitative data for determining environmental and human safety regarding chemical exposure (Hussain and Keçili, 2020). Risk assessment has to take into account whether exposing an organism to a specific chemical will result in adverse health effects. Moreover, the exposure to the hazardous chemical should be also considered. Therefore, in order to characterize risk, one must first establish the degree of exposure as well as the magnitude of the hazard. The processes of determining the degree of exposure and magnitude of hazard are called exposure and hazard assessment (i.e., hazard identification and hazard characterization), respectively. Hence, risk assessment comprises the highly interconnected stages of hazard evaluation, exposure assessment, and risk characterization (Benford, 2017). Chemical exposure occurs via multiple pathways such as inhalation, ingestion, and dermal contact. For example, humans can be exposed to chiral brominated flame retardants through dermal contact while using cosmetics, ingestion of contaminated milk, or inhalation of contaminated indoor dust. The type and degree of effects associated with exposure to a specific chemical are often dependent on the dose and length of exposure (Hussain and Keçili, 2020). However, the adverse effects are also influenced by the sex, age, weight, and life history of the target organism. Therefore, hazard assessment is not only related with the toxicological profile of the chemical, but also with the biological aspects of the target organism. Chiral pollutants complicate risk assessment due to possible enantioselectivity in toxicological and fate profiles. Table 1.3 shows the aims and tools used at various stages of risk assessment and the challenges posed by chirality at each stage.

TABLE 1.3
An Overview of the Steps Involved in Risk Assessment of Chiral Contaminants in Food and the Environment (Benford, 2017; Hussain and Keçili, 2020)

Risk Assessment Step	Aim	Tools	Implications of Chirality
Hazard identification	To determine the type and nature of biological response inherently caused by a contaminant to a biomolecule, cell, tissue, organ, organism, population, or community.	Short-term toxicity tests (e.g., genotoxicity, mutagenicity, reproductive toxicity, immunotoxicity, and developmental toxicity) and quantitative structure-activity relationship models.	Three scenarios are possible for compounds with a single chiral element: a. The enantiomers have equal potency. b. One enantiomer is more potent. c. Enantiomers target different toxicity pathways.
Hazard characterization	To qualitatively or quantitatively establish the inherent properties of a contaminant that have the potential to elicit adverse effects. To establish interspecies variations in toxicodynamics and toxicokinetics.	Dose-response relationship in various species.	Three scenarios are possible for compounds with a single chiral element: a. The enantiomers have similar dose-response curves. b. The enantiomers have different dose-response curves in one species. c. Interspecies differences in enantiomer toxicity.
Exposure assessment	To determine the occurrence of a chemical in food or in the environment through monitoring programs, total diet analyses, or targeted surveys.	Chemical analysis using various analytical tools such as gas chromatography, liquid chromatography, and capillary electrophoresis.	Enantiomeric composition is determined using enantioselective analysis, which can be a challenge. Four scenarios are possible: a. Chiral contaminant occurs in the food or in the environment as racemic mixture. b. Chiral contaminant occurs in the food or in the environment as a non-racemic mixture. c. An enantiopure contaminant converts to its antipode (e.g., naproxen in the environment). d. An enantiopure chemical is commercially available on the market.

TABLE 1.3 *(Continued)*
An Overview of the Steps Involved in Risk Assessment of Chiral Contaminants in Food and the Environment (Benford, 2017; Hussain and Keçili, 2020)

Risk Assessment Step	Aim	Tools	Implications of Chirality
Risk characterization	To quantitatively define the magnitude and nature of risk.	Health-based guidance value (e.g., acceptable daily intake and maximum acceptable concentration).	Uncertainties arising from the scenarios in hazard identification, hazard characterization, and exposure assessment are compounded into the risk characterization when stereochemistry is overlooked. Several scenarios are possible, but the main ones are: a. The risk is accurate if the enantiomers have the same fate and toxicity (very unlikely). b. The risk is overestimated when the most abundant enantiomer is the less toxic. c. The risk is underestimated when the most abundant enantiomer is the most toxic.

The implications of chirality on each step of the risk assessment are included. The simplest scenario whereby a contaminant has one chiral element is considered.

Despite the increase in research interest on stereochemistry in the food and in the environment, enantioselectivity in distribution, fate, and toxicity is often overlooked when estimating the risk of chiral pollutants (Stanley and Brooks, 2009). Analytical determinations in the food or in the environment are usually performed considering enantiomers as a unique entity, and racemic mixtures are often used to determine the health effects of chiral compounds. However, ignoring the enantiospecific biological effects and fate of chiral pollutants introduces uncertainties in the risk assessment. Stanley and Brooks (2009) proposed treating enantiomers as additive mixtures when they have different potencies but as racemate when they have similar potencies.

1.5 OBJECTIVE AND CONTENTS OF THIS BOOK

Chiral molecules are non-superimposable to their mirror images since they have one or more chiral centers. As a result, enantiomers often exhibit differences in their biological activity due to differences in adsorption, distribution, metabolism, and elimination in organisms. This book discusses the analysis, fate, and toxicity of chiral pollutants in food and environmental matrices. Understanding the behavior of chiral pollutants is a crucial step in risk assessment. Chapter 2 discusses chiral inversion focusing on the factors influencing configurational and conformational stability of enantiomers. It is important to understand chiral inversion because it does not only influence the fate and distribution of chiral pollutants, but has serious implications on enantioselective analysis. Several chemical pollutants are discussed in this book and these include pesticides (Chapter 3), pharmaceuticals (Chapter 4), personal care products (Chapter 5), halogenated organic contaminants of emerging concern (i.e., brominated flame retardants and per- and polyfluorinated alkyl substances) (Chapter 6), and POPs (Chapter 7). Besides determination of chiral composition of environmental samples, chiral analysis has found additional applications in discriminating abiotic and biotic fate processes of environmental contaminants, as well as a forensic tool.

The application of enantioselective analysis in the food industry is discussed in this book (Chapter 8). Chiral analysis is important in food analysis as it is a powerful tool for assessing food safety, quality, and authenticity. Chapters 9 and 10 discuss the contamination of food by chiral pesticides and chiral pharmaceuticals, respectively. In Chapter 11, the utility in enantioselective analysis as a tool for assessing food authenticity and adulteration is discussed.

Although chirality is often overlooked in environmental and food regulations, chiral pollutants make up more than one half and one third of pharmaceuticals and agrochemicals in current use, respectively. Chapter 12 discusses current regulations that comprise recommendations for chiral chemicals. Recommendations to consider stereochemistry in development of pesticides and pharmaceuticals have led to a growing interest of the scientific community on enantioselective synthesis of chiral pesticides and pharmaceuticals. When one enantiomer has the desirable effects while the antipode has lower or none of the effects, chiral switching is often recommended. Hence, Chapters 13 and 14 discuss enantioselective synthesis and chiral switching of pesticides and pharmaceuticals, respectively.

This book demonstrates the importance of incorporating the stereochemistry of a contaminant when conducting a risk assessment. By highlighting a diverse group of compounds, this book shows the ubiquity of chirality in potentially toxic chemicals. This book is expected to be useful as a reference book for undergraduate and graduate students majoring in food chemistry, environmental chemistry, pharmaceutical science, and environmental engineering. Agriculture and crop science professionals, government regulators (e.g., food and drug agencies), and environmental consultants will find this book valuable.

ACKNOWLEDGMENTS

Ana Rita L. Ribeiro acknowledges the support of: Project PTDC/QUI-QAN/30521/2017—POCI-01-0145-FEDER-030521—funded by FEDER funds through COMPETE2020—Programa Operacional Competitividade e Internacionalização (POCI), and by national funds (PIDDAC) through FCT/MCTES; Base Funding—UIDB/50020/ 2020 of the Associate Laboratory LSRE-LCM—funded by national funds through FCT/MCTES (PIDDAC).

REFERENCES

Agranat, I., Caner, H., Caldwell, J., 2002. Putting chirality to work: The strategy of chiral switches. *Nat. Rev. Drug Discov.* 1, 753–768. https://doi.org/10.1038/nrd915

Allenmark, S., 1991. *Chromatographic Enantioseparation: Methods and Applications*, 2nd ed. Ellis Horwood Ltd., Chichester, England.

Allenmark, S., Gawronski, J., 2008. Determination of absolute configuration—An overview related to this Special Issue. *Chirality*. 20, 606–608.

Badea, S. L., Niculescu, V. C., Ionete, R. E., Eljarrat, E., 2016. Advances in enantioselective analysis of chiral brominated flame retardants. Current status, limitations and future perspectives. *Sci. Total Environ.* 566–567, 1120–1130. https://doi.org/10.1016/j.scitotenv.2016.05.148

Bagnall, J., Malia, L., Lubben, A., Kasprzyk-Hordern, B., 2013. Stereoselective biodegradation of amphetamine and methamphetamine in river microcosms. *Water Res.* 47, 5708–5718. https://doi.org/10.1016/j.watres.2013.06.057

Bagnall, J. P., Evans, S. E., Wort, M. T., Lubben, A. T., Kasprzyk-Hordern, B., 2012. Using chiral liquid chromatography quadrupole time-of-flight mass spectrometry for the analysis of pharmaceuticals and illicit drugs in surface and wastewater at the enantiomeric level. *J. Chromatogr. A.* 1249, 115–29. https://doi.org/10.1016/j.chroma.2012.06.012

Baird, S., Garrison, A., Jones, J., Avants, J., Bringolf, R., Black, M., 2013. Enantioselective toxicity and bioaccumulation of fipronil in fathead minnows (*Pimephales promelas*) following water and sediment exposures. *Environ. Toxicol. Chem.* 32, 222–227. https://doi.org/10.1002/etc.2041

Barbosa, M. O., Moreira, N. F. F., Ribeiro, A. R., Pereira, M. F. R., Silva, A. M. T., 2016. Occurrence and removal of organic micropollutants: An overview of the watch list of EU Decision 2015/495. *Water Res.* 94, 257–279. https://doi.org/10.1016/j.watres.2016.02.047

Bebianno, M. J., Gonzalez-Rey, M., 2015. Ecotoxicological risk of personal care products and pharmaceuticals, aquatic ecotoxicology: Advancing tools for dealing with emerging risks. *Elsevier Inc.* https://doi.org/10.1016/B978-0-12-800949-9.00016-4

Benford, D. J., 2017. *Risk Assessment of Chemical Contaminants and Residues in Food*, Chemical Contaminants and Residues in Food, 2nd ed. Woodhead Publishing, Cambridge, MA. https://doi.org/10.1016/B978-0-08-100674-0.00001-1

Bentley, R., 2006. The nose as a stereochemist. Enantiomers and odor. *Chem. Rev.* 106, 4099–4112. https://doi.org/10.1021/cr050049t

Birnbaum, L. S., Staskal, D. F., 2004. Brominated flame retardants: Cause for concern? *Environ. Health Perspect.* 112, 9–17. https://doi.org/10.1289/ehp.6559

Brausch, J. M., Rand, G. M., 2011. A review of personal care products in the aquatic environment: Environmental concentrations and toxicity. *Chemosphere.* 82, 1518–1532. https://doi.org/10.1016/j.chemosphere.2010.11.018

Brits, M., de Vos, J., Weiss, J. M., Rohwer, E. R., de Boer, J., 2016. Critical review of the analysis of brominated flame retardants and their environmental levels in Africa. *Chemosphere.* 164, 174–189. https://doi.org/10.1016/j.chemosphere.2016.08.097

Brittain, H. G., 2017. ORD and Polarimetry Instruments, in: Lindon, J. C., Tranter, G. E., Koppenaal, D. W. (Eds.), *Encyclopedia of Spectroscopy and Spectrometry*. Elsevier, Oxford, pp. 501–506. https://doi.org/10.1016/B978-0-12-803224-4.00239-9

Buerge, I. J., Poiger, T., Müller, M. D., Buser, H.-R., 2003. Enantioselective degradation of metalaxyl in soils: Chiral preference changes with soil pH. *Environ. Sci. Technol.* 37, 2668–2674. https://doi.org/10.1021/es0202412

Bura, L., Friel, A., Magrans, J. O., Parra-Morte, J. M., Szentes, C., 2019. Guidance of EFSA on risk assessments for active substances of plant protection products that have stereoisomers as components or impurities and for transformation products of active substances that may have stereoisomers. *EFSA J.* 17, 5804. https://doi.org/10.2903/j.efsa.2019.5804

Buser, H.-R., Müller, M. D., Balmer, M. E., Poiger, T., Buerge, I. J., 2005. Stereoisomer composition of the chiral UV filter 4-methylbenzylidene camphor in environmental samples. *Environ. Sci. Technol.* 39, 3013–3019. https://doi.org/10.1021/es048265r

Calcaterra, A., D'Acquarica, I., 2018. The market of chiral drugs: Chiral switches versus de novo enantiomerically pure compounds. *J. Pharm. Biomed. Anal.* 147, 323–340. https://doi.org/10.1016/j.jpba.2017.07.008

Campos, D., Gravato, C., Quintaneiro, C., Golovko, O., Žlábek, V., Soares, A. M. V. M., Pestana, J. L. T., 2017. Toxicity of organic UV-filters to the aquatic midge *Chironomus riparius*. *Ecotoxicol. Environ. Saf.* 143, 210–216. https://doi.org/10.1016/j.ecoenv.2017.05.005

Cao, Y., Liu, H., Qin, N., Ren, X., Zhu, B., Xia, X., 2020. Impact of food additives on the composition and function of gut microbiota: A review. *Trends Food Sci. Technol.* 99, 295–310. https://doi.org/10.1016/j.tifs.2020.03.006

Carlsson, P., Warner, N. A., Hallanger, I. G., Herzke, D., Kallenborn, R., 2014. Spatial and temporal distribution of chiral pesticides in Calanus spp. from three Arctic fjords. *Environ. Pollut.* 192, 154–161. https://doi.org/10.1016/j.envpol.2014.05.021

Castrignanò, E., Lubben, A., Kasprzyk-Hordern, B., 2016. Enantiomeric profiling of chiral drug biomarkers in wastewater with the usage of chiral liquid chromatography coupled with tandem mass spectrometry. *J. Chromatogr. A.* 1438, 84–99. https://doi.org/10.1016/j.chroma.2016.02.015

Daughton, C. G., Ternes, T. A., 1999. Pharmaceuticals and personal care products in the environment: Agents of subtle change? *Environ. Health Perspect.* 107, 907–938.

Díaz-Cruz, M. S., Llorca, M., Barceló, D., 2008. Organic UV filters and their photodegradates, metabolites and disinfection by-products in the aquatic environment. *Trends Anal. Chem.* 27, 873–887. https://doi.org/10.1016/j.trac.2008.08.012

Dogan, A., Płotka-Wasylka, J., Kempińska-Kupczyk, D., Namieśnik, J., Kot-Wasik, A., 2020. Detection, identification and determination of chiral pharmaceutical residues in wastewater: Problems and challenges. *Trends Anal. Chem.* 122. https://doi.org/10.1016/j.trac.2019.115710

Eliel, E. L., Wilen, S. H., 1994. *Stereochemistry of Organic Compounds*. John Wiley & Sons, Inc., New York.

Engel, K.-H., 2020. Chirality: An important phenomenon regarding biosynthesis, perception, and authenticity of flavor compounds. *J. Agric. Food Chem.* https://doi.org/10.1021/acs.jafc.0c01512

English, K., Toms, L. M. L., Gallen, C., Mueller, J. F., 2016. BDE-209 in the Australian environment: Desktop review. *J. Hazard. Mater.* 320, 194–203. https://doi.org/10.1016/j.jhazmat.2016.08.032

Escuder-Gilabert, L., Martín-Biosca, Y., Perez-Baeza, M., Sagrado, S., Medina-Hernández, M. J., 2018. Trimeprazine is enantioselectively degraded by an activated sludge in ready biodegradability test conditions. *Water Res.* 141, 57–64. https://doi.org/10.1016/j.watres.2018.05.008

Evans, S. E., Bagnall, J., Kasprzyk-Hordern, B., 2016. Enantioselective degradation of amphetamine-like environmental micropollutants (amphetamine, methamphetamine, MDMA and MDA) in urban water. *Environ. Pollut.* 215, 154–163. https://doi.org/10.1016/j.envpol.2016.04.103

Evans, S. E., Davies, P., Lubben, A., Kasprzyk-Hordern, B., 2015. Determination of chiral pharmaceuticals and illicit drugs in wastewater and sludge using microwave assisted extraction, solid-phase extraction and chiral liquid chromatography coupled with tandem mass spectrometry. *Anal. Chim. Acta.* 882, 112–126. https://doi.org/10.1016/j.aca.2015.03.039

Evans, S. E., Kasprzyk-Hordern, B., 2014. Applications of chiral chromatography coupled with mass spectrometry in the analysis of chiral pharmaceuticals in the environment. *Trends Environ. Anal. Chem.* 1, e34–e51. https://doi.org/10.1016/j.teac.2013.11.005

Faller, J., Hühnerfuss, H., König, W. A., Ludwig, P., 1991. Gas chromatographic separation of the enantiomers of marine organic pollutants. *Mar. Pollut. Bull.* 22, 82–86. https://doi.org/10.1016/0025-326X(91)90142-F

Fu, Q., Sanganyado, E., Ye, Q., Gan, J., 2016. Meta-analysis of biosolid effects on persistence of triclosan and triclocarban in soil. *Environ. Pollut.* 210, 137–144. https://doi.org/10.1016/j.envpol.2015.12.003

Gal, J., 2006. Chiral Drugs from a Historical Point of View, in: Francotte, E., Lindner, W. (Eds.), *Chirality in Drug Research*. Wiley-VCH Verlag GmbH & Co. KGaA, Weinheim, Germany, pp. 1–26. https://doi.org/10.1002/9783527609437.ch1

Gao, S., Tian, B., Zeng, X., Yu, Z., 2019. Enantiomeric analysis of polycyclic musks AHTN and HHCB and HHCB-lactone in sewage sludge by gas chromatography/tandem mass spectrometry. *Rapid Commun. Mass Spectrom.* 33, 607–612. https://doi.org/10.1002/rcm.8390

Gatermann, R., Biselli, S., Hühnerfuss, H., Rimkus, G. G., Franke, S., Hecker, M., Kallenborn, R., Karbe, L., König, W. A., 2002. Synthetic musks in the environment. Part 2: Enantioselective transformation of the polycyclic musk fragrances HHCB, AHTN, AHDI, and ATII in freshwater fish. *Arch. Environ. Contam. Toxicol.* 42, 447–453. https://doi.org/10.1007/s00244-001-0042-1

Gilbert, E., Pirot, F., Bertholle, V., Roussel, L., Falson, F., Padois, K., 2013. Commonly used UV filter toxicity on biological functions: Review of last decade studies. *Int. J. Cosmet. Sci.* 35, 208–219. https://doi.org/10.1111/ics.12030

Guo, J., Chen, J., Wang, J., 2017. Sedimentary records of polycyclic aromatic hydrocarbons in China: A comparison to the worldwide. *Crit. Rev. Environ. Sci. Technol.* 47, 1612–1667. https://doi.org/10.1080/10643389.2017.1393262

Hamers, T., Kamstra, J. H., Sonneveld, E., Murk, A. J., Kester, M. H. A., Andersson, P. L., Legler, J., Brouwer, A., 2006. In vitro profiling of the endocrine-disrupting potency of brominated flame retardants. *Toxicol. Sci.* 92, 157–173. https://doi.org/10.1093/toxsci/kfj187

Hansen, A. M. S., Frandsen, H. L., Fromberg, A., 2016. Authenticity of raspberry flavor in food products using SPME-chiral-GC-MS. *Food Sci. Nutr.* 4, 348–354. https://doi.org/10.1002/fsn3.296

Hashim, N. H. H., Shafie, S., Khan, S. J. J., 2010. Enantiomeric fraction as an indicator of pharmaceutical biotransformation during wastewater treatment and in the environment—A review. *Environ. Technol.* 31, 1349–1370. https://doi.org/10.1080/09593331003728022

Hopkins, Z. R., Blaney, L., 2016. An aggregate analysis of personal care products in the environment: Identifying the distribution of environmentally-relevant concentrations. *Environ. Int.* 92–93, 301–316. https://doi.org/10.1016/j.envint.2016.04.026

Hu, K., Zhou, L., Gao, Y., Lai, Q., Shi, H., Wang, M., 2020. Enantioselective endocrine-disrupting effects of the phenylpyrazole chiral insecticides in vitro and in silico. *Chemosphere.* 252. https://doi.org/10.1016/j.chemosphere.2020.126572

Hühnerfuss, H., Faller, J., Kallenborn, R., König, W. A., Ludwig, P., Pfaffenberger, B., Oehme, M., Rimkus, G., 1993. Enantioselective and nonenantioselective degradation of organic pollutants in the marine ecosystem. *Chirality.* 5, 393–399. https://doi.org/10.1002/chir.530050522

Hühnerfuss, H., Shah, M. R., 2009. Enantioselective chromatography—A powerful tool for the discrimination of biotic and abiotic transformation processes of chiral environmental pollutants. *J. Chromatogr. A* 1216, 481–502. https://doi.org/10.1016/j.chroma.2008.09.043

Hussain, C. M., Keçili, R., 2020. Environmental Pollution and Environmental Analysis, *Modern Environmental Analysis Techniques for Pollutants*. Elsevier, Cambridge, MA. https://doi.org/10.1016/B978-0-12-816934-6.00001-1

IUPAC, 1997. *IUPAC Compendium of Chemical Terminology, "Gold Book,"* 2nd ed. IUPAC, Research Triagle Park, NC. https://doi.org/10.1351/goldbook

Jantunen, L. M., Bidleman, T., 1996. Air-water gas exchange of hexachlorocyclohexanes (HCHs) and the enantiomers of α-HCH in Arctic regions. *J. Geophys. Res. Atmos.* 101, 28837–28846. https://doi.org/10.1029/96JD02352

Jiang, J. J., Lee, C. L., Fang, M. Der, 2014. Emerging organic contaminants in coastal waters: Anthropogenic impact, environmental release and ecological risk. *Mar. Pollut. Bull.* 85, 391–399. https://doi.org/10.1016/j.marpolbul.2013.12.045

Jiang, Y., Yuan, L., Lin, Q., Ma, S., Yu, Y., 2019. Polybrominated diphenyl ethers in the environment and human external and internal exposure in China: A review. *Sci. Total Environ.* 696. https://doi.org/10.1016/j.scitotenv.2019.133902

Jiménez-Díaz, I., Zafra-Gómez, A., Ballesteros, O., Navalón, A., 2014. Analytical methods for the determination of personal care products in human samples: An overview. *Talanta* 129, 448–458. https://doi.org/10.1016/j.talanta.2014.05.052

Kasprzyk-Hordern, B., 2010. Pharmacologically active compounds in the environment and their chirality. *Chem. Soc. Rev.* 39, 4466. https://doi.org/10.1039/c000408c

Khalaf, H., Larsson, A., Berg, H., McCrindle, R., Arsenault, G., Olsson, P.-E., 2009. Diastereomers of the brominated flame retardant 1,2-Dibromo-4-(1,2 dibromoethyl) cyclohexane induce androgen receptor activation in the HepG2 hepatocellular carcinoma cell line and the LNCaP prostate cancer cell line. *Environ. Health Perspect.* 117, 1853–1859. https://doi.org/10.1289/ehp.0901065

Kim, S., Choi, K., 2014. Occurrences, toxicities, and ecological risks of benzophenone-3, a common component of organic sunscreen products: A mini-review. *Environ. Int.* 70, 143–157. https://doi.org/10.1016/j.envint.2014.05.015

Konwick, B. J., Garrison, A. W., Avants, J. K., Fisk, A. T., 2006a. Bioaccumulation and biotransformation of chiral triazole fungicides in rainbow trout (*Oncorhynchus mykiss*). *Aquat. Toxicol.* 80, 372–381. https://doi.org/10.1016/j.aquatox.2006.10.003

Konwick, B. J., Garrison, A. W., Black, M. C., Avants, J. K., Fisk, A. T., 2006b. Bioaccumulation, biotransformation, and metabolite formation of fipronil and chiral legacy pesticides in rainbow trout. *Environ. Sci. Technol.* 40, 2930–2936. https://doi.org/10.1021/es0600678

Köppen, R., Becker, R., Esslinger, S., Nehls, I., 2010. Enantiomer-specific analysis of hexabromocyclododecane in fish from Etnefjorden (Norway). *Chemosphere.* 80, 1241–1245. https://doi.org/10.1016/j.chemosphere.2010.06.019

Langdon, K., Warne, M. St. J., Kookana, R., 2010. Aquatic hazard assessment for pharmaceuticals, personal care products, and endocrine-disrupting compounds from biosolids-amended land. *Integr. Environ. Assess. Manag.* 6, 663–676. https://doi.org/10.1002/ieam.74

Lee, I., Gopalan, A. I., Lee, K. P., 2016. Enantioselective determination of polycyclic musks in river and wastewater by GC/MS/MS. *Int. J. Environ. Res. Public Health.* 13, 1–9. https://doi.org/10.3390/ijerph13030349

Leo, L., Loong, C., Ho, X. L., Raman, M. F. B., Suan, M. Y. T., Loke, W. M., 2018. Occurrence of azo food dyes and their effects on cellular inflammatory responses. *Nutrition.* 46, 36–40. https://doi.org/10.1016/j.nut.2017.08.010

Leone, A. D., Amato, S., Falconer, R. L., 2001. Emission of chiral organochlorine pesticides from agricultural soils in the Corn Belt region of the U.S. *Environ. Sci. Technol.* 35, 4592–4596. https://doi.org/10.1021/es010992o

Li, M., Yao, L., Chen, H., Ni, X., Xu, Y., Dong, Wengjing, Fang, M., Chen, D., Aowuliji, Xu, L., Zhao, B., Deng, J., Kwok, K. W., Yang, J., Dong, Wu, 2020. Chiral toxicity of muscone to embryonic zebrafish heart. *Aquat. Toxicol.* 222, 105451. https://doi.org/10.1016/j.aquatox.2020.105451

Li, X., Chu, Z., Yang, J., Li, M., Du, M., Zhao, X., Zhu, Z., Li, Y., 2018. Synthetic musks: A class of commercial fragrance additives in personal care products (PCPs) causing concern as emerging contaminants, in: Chen, B., Zhang, B., Zhu, Z., Lee, K. (Eds.), *Advances in Marine Biology.* Academic Press, London, UK, pp. 213–280. https://doi.org/10.1016/bs.amb.2018.09.008

Liang, Y., Zhan, J., Liu, X., Zhou, Z., Zhu, W., Liu, D., Wang, P., 2017. Stereoselective metabolism of the UV-filter 2-ethylhexyl 4-dimethylaminobenzoate and its metabolites in rabbits in vivo and vitro. *RSC Adv.* 7, 16991–16996. https://doi.org/10.1039/c7ra00431a

Liu, L.-Y., Ma, W.-L., Jia, H.-L., Zhang, Z.-F., Song, W.-W., Li, Y.-F., 2016. Research on persistent organic pollutants in China on a national scale: 10 years after the enforcement of the Stockholm Convention. *Environ. Pollut.* 217, 70–81. https://doi.org/10.1016/j.envpol.2015.12.056

Lozano, C., Matallana-Surget, S., Givens, J., Nouet, S., Arbuckle, L., Lambert, Z., Lebaron, P., 2020. Toxicity of UV filters on marine bacteria: Combined effects with damaging solar radiation. *Sci. Total Environ.* 722. https://doi.org/10.1016/j.scitotenv.2020.137803

Maia, A. S., Ribeiro, A. R., Castro, P. M. L. L., Tiritan, M. E., 2017. Chiral analysis of pesticides and drugs of environmental concern: Biodegradation and enantiomeric fraction. *Symmetry (Basel)*. 9, 196. https://doi.org/10.3390/sym9090196

Martínez-Girón, A. B., Crego, A. L., González, M. J., Marina, M. L., 2010. Enantiomeric separation of chiral polycyclic musks by capillary electrophoresis: Application to the analysis of cosmetic samples. *J. Chromatogr. A*. 1217, 1157–1165. https://doi.org/10.1016/j.chroma.2009.12.021

Martins, F. C. O. L., Sentanin, M. A., De Souza, D., 2019. Analytical methods in food additives determination: Compounds with functional applications. *Food Chem*. 272, 732–750. https://doi.org/10.1016/j.foodchem.2018.08.060

McConathy, J., Owens, M. J., 2003. Stereochemistry in drug action. *Prim. Care Companion J. Clin. Psychiatry*. 5, 70–73.

Meng, X. Z., Guo, Y., Mai, B. X., Zeng, E. Y., 2009. Enantiomeric signatures of chiral organochlorine pesticides in consumer fish from South China. *J. Agric. Food Chem*. 57, 4299–4304. https://doi.org/10.1021/jf900038u

Montes-Grajales, D., Fennix-Agudelo, M., Miranda-Castro, W., 2017. Occurrence of personal care products as emerging chemicals of concern in water resources: A review. *Sci. Total Environ*. 595, 601–614. https://doi.org/10.1016/j.scitotenv.2017.03.286

Müller, T. A., Kohler, H.-P. E., 2004. Chirality of pollutants—Effects on metabolism and fate. *Appl. Microbiol. Biotechnol*. 64, 300–16. https://doi.org/10.1007/s00253-003-1511-4

Nohynek, G. J., Antignac, E., Re, T., Toutain, H., 2010. Safety assessment of personal care products/cosmetics and their ingredients. *Toxicol. Appl. Pharmacol*. 243, 239–259. https://doi.org/10.1016/j.taap.2009.12.001

Park, C. B., Jang, J., Kim, S., Kim, Y. J., 2017. Single- and mixture toxicity of three organic UV-filters, ethylhexyl methoxycinnamate, octocrylene, and avobenzone on *Daphnia magna*. *Ecotoxicol. Environ. Saf*. 137, 57–63. https://doi.org/10.1016/j.ecoenv.2016.11.017

Pavlinov, S. A., Belolipetskaya, V. G., Piotrovskii, V. K., Metelitsa, V. I., Filatova, N. P., Bochkareva, E. V., 1990. Enantiomers of propranolol in the blood of patients with cardiovascular diseases during long-term therapy. *Pharm. Chem. J*. 24, 17–19. https://doi.org/10.1007/BF00769379

Petrie, B., Barden, R., Kasprzyk-Hordern, B., 2014. A review on emerging contaminants in wastewaters and the environment: Current knowledge, understudied areas and recommendations for future monitoring. *Water Res*. 72, 3–27. https://doi.org/10.1016/j.watres.2014.08.053

Petrie, B., Proctor, K., Youdan, J., Barden, R., Kasprzyk-Hordern, B., 2017. Critical evaluation of monitoring strategy for the multi-residue determination of 90 chiral and achiral micropollutants in effluent wastewater. *Sci. Total Environ*. 579, 569–578. https://doi.org/10.1016/j.scitotenv.2016.11.059

Qian, Y., Ji, C., Yue, S., Zhao, M., 2019. Exposure of low-dose fipronil enantioselectively induced anxiety-like behavior associated with DNA methylation changes in embryonic and larval zebrafish. *Environ. Pollut*. 249, 362–371. https://doi.org/10.1016/j.envpol.2019.03.038

Qu, H., Ma, R., Barrett, H., Wang, B., Han, J., Wang, F., Chen, P., Wang, W., Peng, G., Yu, G., 2020. How microplastics affect chiral illicit drug methamphetamine in aquatic food chain? From green alga (*Chlorella pyrenoidosa*) to freshwater snail (*Cipangopaludian cathayensis*). *Environ. Int*. 136. https://doi.org/10.1016/j.envint.2020.105480

Quintaneiro, C., Teixeira, B., Benedé, J. L., Chisvert, A., Soares, A. M. V. M., Monteiro, M. S., 2019. Toxicity effects of the organic UV-filter 4-methylbenzylidene camphor in zebrafish embryos. *Chemosphere*. 218, 273–281. https://doi.org/10.1016/j.chemosphere.2018.11.096

Rainieri, S., Barranco, A., Primec, M., Langerholc, T., 2017. Occurrence and toxicity of musks and UV filters in the marine environment. *Food Chem. Toxicol*. 104, 57–68. https://doi.org/10.1016/j.fct.2016.11.012

Raubenheimer, K., McIlgorm, A., 2018. Can the Basel and Stockholm conventions provide a global framework to reduce the impact of marine plastic litter? *Mar. Policy*. 96, 285–290. https://doi.org/10.1016/j.marpol.2018.01.013

Rayne, S., Forest, K., Friesen, K. J., 2008. Congener-specific numbering systems for the environmentally relevant C4 through C8 perfluorinated homologue groups of alkyl sulfonates, carboxylates, telomer alcohols, olefins, and acids, and their derivatives. *J. Environ. Sci. Heal. A Toxic/Hazard. Subst. Environ. Eng*. 43, 1391–1401. https://doi.org/10.1080/10934520802232030

Reid, A. J., Carlson, A. K., Creed, I. F., Eliason, E. J., Gell, P. A., Johnson, P. T. J., Kidd, K. A., MacCormack, T. J., Olden, J. D., Ormerod, S. J., Smol, J. P., Taylor, W. W., Tockner, K., Vermaire, J. C., Dudgeon, D., Cooke, S. J., 2019. Emerging threats and persistent conservation challenges for freshwater biodiversity. *Biol. Rev*. 94, 849–873. https://doi.org/10.1111/brv.12480

Ribeiro, A. R., Afonso, C. M., Castro, P. M. L., Tiritan, M. E., 2012a. Enantioselective HPLC analysis and biodegradation of atenolol, metoprolol and fluoxetine. *Environ. Chem. Lett.* 11, 83–90. https://doi.org/10.1007/s10311-012-0383-1

Ribeiro, A. R., Afonso, C. M., Castro, P. M. L. L., Tiritan, M. E., 2013. Enantioselective biodegradation of pharmaceuticals, alprenolol and propranolol, by an activated sludge inoculum. *Ecotoxicol. Environ. Saf.* 87, 108–14. https://doi.org/10.1016/j.ecoenv.2012.10.009

Ribeiro, A. R., Castro, P. M. L., Tiritan, M. E., 2012b. Chiral pharmaceuticals in the environment. *Environ. Chem. Lett.* 10, 239–253.

Ribeiro, A. R., Castro, P. M. L., Tiritan, M. E., 2012c. Environmental Fate of Chiral Pharmaceuticals: Determination, Degradation and Toxicity, in: Lichtfouse, E., Schwarzbauer, J., Robert, D. (Eds.), *Environmental Chemistry for a Sustainable World.* Springer Netherlands, Dordrecht, pp. 3–45. https://doi.org/10.1007/978-94-007-2439-6_1

Ribeiro, A. R. L., Maia, A. S., Ribeiro, C., Tiritan, M. E., 2020. Analysis of chiral drugs in environmental matrices: Current knowledge and trends in environmental, biodegradation and forensic fields. *TrAC Trends Anal. Chem.* 124. https://doi.org/10.1016/j.trac.2019.115783

Ribeiro, C., Ribeiro, A., Maia, A., Tiritan, M., 2017. Occurrence of chiral bioactive compounds in the aquatic environment: A review. *Symmetry (Basel).* 9, 215. https://doi.org/10.3390/sym9100215

Ribeiro, C., Santos, C., Gonçalves, V., Ramos, A., Afonso, C., Tiritan, M., 2018. Chiral drug analysis in forensic chemistry: An overview. *Molecules.* 23, 262. https://doi.org/10.3390/molecules23020262

Rice, P. A., Bandele, O. J., Honigfort, P., 2014. Perfluorinated compounds in food contact materials. *Mol. Integr. Toxicol.* 177–203. https://doi.org/10.1007/978-1-4471-6500-2_7

Ruan, Y., Lam, J. C. W., Zhang, X., Lam, P. K. S., 2018a. Temporal changes and stereoisomeric compositions of 1,2,5,6,9,10-hexabromocyclododecane and 1,2-dibromo-4-(1,2-dibromoethyl)cyclohexane in marine mammals from the South China Sea. *Environ. Sci. Technol.* 52, 2517–2526. https://doi.org/10.1021/acs.est.7b05387

Ruan, Y., Zhang, K., Lam, J. C. W., Wu, R., Lam, P. K. S., 2019. Stereoisomer-specific occurrence, distribution, and fate of chiral brominated flame retardants in different wastewater treatment systems in Hong Kong. *J. Hazard. Mater.* 374, 211–218. https://doi.org/10.1016/j.jhazmat.2019.04.041

Ruan, Y., Zhang, X., Qiu, J. W., Leung, K. M. Y., Lam, J. C. W., Lam, P. K. S., 2018b. Stereoisomer-specific trophodynamics of the chiral brominated flame retardants HBCD and TBECH in a marine food web, with implications for human exposure. *Environ. Sci. Technol.* 52, 8183–8193. https://doi.org/10.1021/acs.est.8b02206

Sanganyado, E., 2019. Comments on "Chiral pharmaceuticals: Environment sources, potential human health impacts, remediation technologies and future perspective." *Environ. Int.* 122, 412–415. https://doi.org/10.1016/j.envint.2018.11.032

Sanganyado, E., Lu, Z., Fu, Q., Schlenk, D., Gan, J., 2017. Chiral pharmaceuticals: A review on their environmental occurrence and fate processes. *Water Res.* 124, 527–542. https://doi.org/10.1016/j.watres.2017.08.003

Sanganyado, E., Lu, Z., Liu, W., 2020. Application of enantiomeric fractions in environmental forensics: Uncertainties and inconsistencies. *Environ. Res.* 184. https://doi.org/10.1016/j.envres.2020.109354

Sanganyado, E., Rajput, I. R., Liu, W., 2018. Bioaccumulation of organic pollutants in Indo-Pacific humpback dolphin: A review on current knowledge and future prospects. *Environ. Pollut.* 237, 111–125. https://doi.org/10.1016/j.envpol.2018.01.055

Schäfer, U., Kiefl, J., Zhu, W., Kempf, M., Eggers, M., Backes, M., Geissler, T., Wittlake, R., Reichelt, K. V., Ley, J. P., Krammer, G., 2015. Authenticity Control of Food Flavorings—Merits and Limitations of Chiral Analysis, in: Engel, K.-H., Takeoka, G. (Eds.), *Importance of Chirality to Flavor Compounds.* American Chemical Society, Washington, DC, pp. 3–12. https://doi.org/10.1021/bk-2015-1212.ch001

Schlumpf, M., Kypke, K., Wittassek, M., Angerer, J., Mascher, H., Mascher, D., Vökt, C., Birchler, M., Lichtensteiger, W., 2010. Exposure patterns of UV filters, fragrances, parabens, phthalates, organochlor pesticides, PBDEs, and PCBs in human milk: Correlation of UV filters with use of cosmetics. *Chemosphere.* 81, 1171–1183. https://doi.org/10.1016/j.chemosphere.2010.09.079

Shaw, S. D., Harris, J. H., Berger, M. L., Subedi, B., Kannan, K., 2014. Brominated Flame Retardants and Their Replacements in Food Packaging and Household Products: Uses, Human Exposure, and Health Effects, in: Snedeker, S. M. (Ed.), *Molecular and Integrative Toxicology, Molecular and Integrative Toxicology.* Springer London, London, pp. 61–93. https://doi.org/10.1007/978-1-4471-6500-2_3

Shen, Y., Zhang, J., Xie, J., Liu, J., 2020. In vitro assessment of corticosteroid effects of eight chiral herbicides. *J. Environ. Sci. Heal. B* 55, 91–102. https://doi.org/10.1080/03601234.2019.1665408

Siligardi, G., Hussain, R., 2017. Circular Dichroism, Applications, in: Lindon, J. C., Tranter, G. E., Koppenaal, D. W. (Eds.), *Encyclopedia of Spectroscopy and Spectrometry*. Elsevier, Oxford, pp. 293–298. https://doi.org/10.1016/B978-0-12-803224-4.00031-5

Snedeker, S. M., 2014. Benzophenone UV-Photoinitiators Used in Food Packaging: Potential for Human Exposure and Health Risk Considerations, in: Snedeker, S. M. (Ed.), *Molecular and Integrative Toxicology, Molecular and Integrative Toxicology*. Springer London, London, pp. 151–176. https://doi.org/10.1007/978-1-4471-6500-2_6

Solomons, T. W. G., 2011. *Organic Chemistry*, 10th ed. John Wiley & Sons, Inc., Hoboken, NJ.

Song, H., Zeng, X., Yu, Z., Zhang, D., Cao, S., Shao, W., Sheng, G., Fu, J., 2015. Enantiomeric composition of polycyclic musks in sediments from the Pearl River and Suzhou Creek. *Environ. Sci. Pollut. Res.* 22, 1679–1686. https://doi.org/10.1007/s11356-014-3687-9

Song, L., Pan, M., Zhao, R., Deng, J., Wu, Y., 2020. Recent advances, challenges and perspectives in enantioselective release. *J. Control. Release*. 324, 156–171. https://doi.org/10.1016/j.jconrel.2020.05.019

Sousa, J. C. G., Ribeiro, A. R., Barbosa, M. O., Pereira, M. F. R., Silva, A. M. T., 2018. A review on environmental monitoring of water organic pollutants identified by EU guidelines. *J. Hazard. Mater.* 344, 146–162. https://doi.org/10.1016/j.jhazmat.2017.09.058

Stanley, J. K., Brooks, B. W., 2009. Perspectives on ecological risk assessment of chiral compounds. *Integr. Environ. Assess. Manag.* 5, 364–373. https://doi.org/10.1897/IEAM_2008-076.1

Stringer, J. L., 2006. *Basic Concepts in Pharmacology: A Student's Survival Guide*, 3rd ed. McGraw-Hill, New York.

Testa, B., Caldwell, J., Kisakurek, M. V., 2014. *Organic Stereochemistry: Guiding Principles and Biomedicinal Relevance*. Wiley-VCH, Zürich, Switzerland.

Tiritan, M. E., Fernandes, C., Maia, A. S., Pinto, M., Cass, Q. B., 2018. Enantiomeric ratios: Why so many notations? *J. Chromatogr. A* 1569, 1–7. https://doi.org/10.1016/j.chroma.2018.07.039

Tiritan, M. E., Ribeiro, A. R., Fernandes, C., Pinto, M. M. M., 2016. Chiral Pharmaceuticals, in: Ley, C. (Ed.), *Kirk-Othmer Encyclopedia of Chemical Technology*. John Wiley & Sons, Inc., Hoboken, NJ, pp. 1–28. https://doi.org/10.1002/0471238961.1608011823092009.a01.pub2

Tittlemier, S. A., Blank, D. H., Gribble, G. W., Norstrom, R. J., 2002. Structure elucidation of four possible biogenic organohalogens using isotope exchange mass spectrometry. *Chemosphere*. 46, 511–517.

Tobiszewski, M., Namieśnik, J., 2012. PAH diagnostic ratios for the identification of pollution emission sources. *Environ. Pollut.* 162, 110–119. https://doi.org/10.1016/j.envpol.2011.10.025

Tseng, W.-J., Tsai, S.-W., 2019. Assessment of dermal exposures for synthetic musks from personal care products in Taiwan. *Sci. Total Environ.* 669, 160–167. https://doi.org/10.1016/j.scitotenv.2019.03.046

Tsui, M. M. P. P., Lam, J. C.W. W., Ng, T. Y. Y., Ang, P. O., Murphy, M. B., Lam, P. K. S. S., 2017. Occurrence, distribution, and fate of organic UV filters in coral communities. *Environ. Sci. Technol.* 51, 4182–4190. https://doi.org/10.1021/acs.est.6b05211

Van Pée, K. H., Unversucht, S., 2003. Biological dehalogenation and halogenation reactions. *Chemosphere*. 52, 299–312. https://doi.org/10.1016/S0045-6535(03)00204-2

Vazquez-Roig, P., Kasprzyk-Hordern, B., Blasco, C., Picó, Y., 2014. Stereoisomeric profiling of drugs of abuse and pharmaceuticals in wastewaters of Valencia (Spain). *Sci. Total Environ.* 494–495, 49–57. https://doi.org/10.1016/j.scitotenv.2014.06.098

Wang, F., Gao, J., Chen, L., Zhou, Z., Liu, D., Wang, P., 2018. Enantioselective bioaccumulation and metabolism of lactofen in zebrafish *Danio rerio* and combined effects with its metabolites. *Chemosphere*. 213, 443–452. https://doi.org/10.1016/j.chemosphere.2018.09.052

Wang, J., Jia, B., Li, Y., Ren, B., Liang, Hanlin, Yan, D., Xie, H., Zhang, X., Liang, Hongwu, 2020. Effects of multi-walled carbon nanotubes on the enantioselective toxicity of the chiral insecticide indoxacarb toward zebrafish (*Danio rerio*). *J. Hazard. Mater.* 397. https://doi.org/10.1016/j.jhazmat.2020.122724

Wang, L., Khan, S. J., 2014. Enantioselective analysis and fate of polycyclic musks in a water recycling plant in Sydney (Australia). *Water Sci. Technol.* 69, 1996–2003. https://doi.org/10.2166/wst.2014.095

Wang, L., McDonald, J. A., Khan, S. J., 2013. Enantiomeric analysis of polycyclic musks in water by chiral gas chromatography-tandem mass spectrometry. *J. Chromatogr. A* 1303, 66–75. https://doi.org/10.1016/j.chroma.2013.06.006

Wang, Y., Beesoon, S., Benskin, J. P., De Silva, A. O., Genuis, S. J., Martin, J. W., 2011. Enantiomer fractions of chiral perfluorooctanesulfonate (PFOS) in human sera. *Environ. Sci. Technol.* 45, 8907–8914. https://doi.org/10.1021/es2023434

Wang, Y., Xu, L., Li, D., Teng, M., Zhang, R., Zhou, Z., Zhu, W., 2015. Enantioselective bioaccumulation of hexaconazole and its toxic effects in adult zebrafish (*Danio rerio*). *Chemosphere*. 138, 798–805. https://doi.org/10.1016/j.chemosphere.2015.08.015

Wilson, B. G., Bahna, S. L., 2005. Adverse reactions to food additives. *Ann. Allergy, Asthma Immunol.* 95, 499–507. https://doi.org/10.1016/S1081-1206(10)61010-1

Wong, C. S., 2006. Environmental fate processes and biochemical transformations of chiral emerging organic pollutants. *Anal. Bioanal. Chem.* 386, 544–58.

Xu, C., Lin, X., Yin, S., Zhao, L., Liu, Y., Liu, K., Li, F., Yang, F., Liu, W., 2018. Enantioselectivity in biotransformation and bioaccumulation processes of typical chiral contaminants. *Environ. Pollut.* 243, 1274–1286. https://doi.org/10.1016/j.envpol.2018.09.095

Yang, H., Lu, G., Yan, Z., Liu, J., Dong, H., Bao, X., Zhang, X., Sun, Y., 2020. Residues, bioaccumulation, and trophic transfer of pharmaceuticals and personal care products in highly urbanized rivers affected by water diversion. *J. Hazard. Mater.* 391. https://doi.org/10.1016/j.jhazmat.2020.122245

Yang, Y., Ji, D., Huang, X., Zhang, J., Liu, J., 2017. Effects of metals on enantioselective toxicity and biotransformation of cis-bifenthrin in zebrafish. *Environ. Toxicol. Chem.* 36, 2139–2146. https://doi.org/10.1002/etc.3747

Yu, X., He, Q., Sanganyado, E., Liang, Y., Bi, R., Li, P., Liu, W., 2020. Chlorinated organic contaminants in fish from the South China Sea: Assessing risk to Indo-Pacific humpback dolphin. *Environ. Pollut.* 263. https://doi.org/10.1016/j.envpol.2020.114346

Zhang, Q., Xiong, W., Gao, B., Cryder, Z., Zhang, Z., Tian, M., Sanganyado, E., Shi, H., Wang, M., 2018a. Enantioselectivity in degradation and ecological risk of the chiral pesticide ethiprole. *L. Degrad. Dev.* 29, 4242–4251. https://doi.org/10.1002/ldr.3179

Zhang, Q., Zhang, Z., Tang, B., Gao, B., Tian, M., Sanganyado, E., Shi, H., Wang, M., 2018b. Mechanistic insights into stereospecific bioactivity and dissipation of chiral fungicide triticonazole in agricultural management. *J. Agric. Food Chem.* 66, 7286–7293. https://doi.org/10.1021/acs.jafc.8b01771

Zhang, Y., Ruan, Y., Sun, H., Zhao, L., Gan, Z., 2013. Hexabromocyclododecanes in surface sediments and a sediment core from rivers and harbor in the northern Chinese city of Tianjin. *Chemosphere.* 90, 1610–1616. https://doi.org/10.1016/j.chemosphere.2012.08.037

Zhang, Y., Sun, H., Ruan, Y., 2014. Enantiomer-specific accumulation, depuration, metabolization and isomerization of hexabromocyclododecane (HBCD) diastereomers in mirror carp from water. *J. Hazard. Mater.* 264, 8–15. https://doi.org/10.1016/j.jhazmat.2013.10.062

Zhang, Z., Du, G., Gao, B., Hu, K., Kaziem, A. E., Li, L., He, Z., Shi, H., Wang, M., 2019. Stereoselective endocrine-disrupting effects of the chiral triazole fungicide prothioconazole and its chiral metabolite. *Environ. Pollut.* 251, 30–36. https://doi.org/10.1016/j.envpol.2019.04.124

Zhao, L., Chen, F., Guo, F., Liu, W., Liu, K., 2019. Enantioseparation of chiral perfluorooctane sulfonate (PFOS) by supercritical fluid chromatography (SFC): Effects of the chromatographic conditions and separation mechanism. *Chirality.* 31, 870–878. https://doi.org/10.1002/chir.23120

Zheng, Q., Li, J., Wang, Y., Lin, T., Xu, Y., Zhong, G., Bing, H., Luo, C., Zhang, G., 2020. Levels and enantiomeric signatures of organochlorine pesticides in Chinese forest soils: Implications for sources and environmental behavior. *Environ. Pollut.* 262. https://doi.org/10.1016/j.envpol.2020.114139

Zhou, S., Pan, Y., Zhang, L., Xue, B., Zhang, A., Jin, M., 2018a. Biomagnification and enantiomeric profiles of organochlorine pesticides in food web components from Zhoushan Fishing Ground, China. *Mar. Pollut. Bull.* 131, 602–610. https://doi.org/10.1016/j.marpolbul.2018.04.055

Zhou, Yaoyu, Wu, S., Zhou, H., Huang, H., Zhao, J., Deng, Y., Wang, H., Yang, Y., Yang, J., Luo, L., 2018b. Chiral pharmaceuticals: Environment sources, potential human health impacts, remediation technologies and future perspective. *Environ. Int.* 121, 523–537. https://doi.org/10.1016/j.envint.2018.09.041

Section I

Analysis, Fate, and Toxicity of Chiral Pollutants in the Environment

Section 1

Analysis, Fate, and Toxicity of Chiral Pollutants in the Environment

2 Chiral Inversion of Organic Pollutants

Simbarashe Nkomo and Edmond Sanganyado

CONTENTS

2.1 Introduction 27
2.2 Chirality and Chiral Inversion 28
2.3 Implications of Chiral Inversion on Environmental Behavior 31
2.4 Factors Influencing Chiral Inversion in the Environment 31
    2.4.1 Solvent-Induced Chiral Inversion 32
    2.4.2 Thermal-Induced Chiral Inversion 34
    2.4.3 Photo-Induced Chiral Inversion 34
    2.4.4 Biologically Mediated Chiral Inversion 35
2.5 Chiral Inversion in the Environment 35
2.6 Determination of Chiral Inversion 36
2.7 Future Directions and Conclusion 37
References 38

2.1 INTRODUCTION

In the early 1950s, thalidomide was introduced on the market as a drug for treating nausea in pregnant women. Most women who took thalidomide ended up giving birth to babies with birth defects. Further studies confirmed that the birth defects were caused by consumption of thalidomide. By the time thalidomide was banned as a morning sickness drug, more than 10,000 children had been born with birth defects. Subsequent studies found that (*S*)-thalidomide is a teratogen while (*R*)-thalidomide is not (Eriksson et al., 1995). It was suggested that using (*R*)-thalidomide could have alleviated the tragedy. To administer only (*R*)-thalidomide required additional purification to ensure the absence of (*S*)-thalidomide. The enantiomeric separation would work only if the drug does not undergo chiral inversion in the body. In the case of thalidomide, enantiomeric separation is not useful because several studies found that (*R*)-thalidomide can undergo chiral inversion *in vivo* (Eriksson et al., 1995). Thalidomide was, therefore, banned for morning sickness treatment in pregnant women. Thalidomide is currently being used for treatment of leprosy and multiple myeloma. In recent years, there has been a surge in studies investigating its repurposing for treatment of other diseases such as cancer (Mori et al., 2018; Tokunaga et al., 2018). Despite the enantiomer-specificity tragedy, repurposed drugs continue to ignore the chirality of thalidomide.

In addition, understanding the configurational lability of enantiomers is important in assessing the environmental and human health risk posed by chiral pollutants. The chiral element (planar, helical, axis, or center) of a chiral molecule may become configurationally labile under certain conditions. When this happens, the stereoconfiguration of the chiral molecule may rapidly change reversibly or irreversibly depending on the conditions. The process by which an enantiomer changes to its antipode under certain conditions is called chiral inversion (Box 2.1). Chiral inversion can be classified as enantiomerization, racemization, epimerization, or diastereomization depending on the reversibility of the reaction and the presence of non-labile chiral

> **BOX 2.1 DEFINITIONS OF TERMS RELATED TO CHIRAL INVERSION (REIST ET AL., 1995)**
>
> A **chiral molecule** is a compound that has non-superimposable mirror images due to the absence of plane or center of symmetry. The plane and center are referred to as the axis of chirality and stereogenic center, respectively.
>
> **Chiral inversion** refers to the spatial rearrangement of atoms or groups of atoms in an asymmetric molecule, which produces a compound with a three-dimensional stereoconfiguration that cannot be superimposed on the parent molecule.
>
> **Enantiomerization** is a molecular process that occurs at microscopic level that involves a reversible interconversion of one enantiomer into the other.
>
> **Racemization** is a statistical process that occurs at macroscopic level that involves the irreversible conversion of one enantiomer into the other until a racemic mixture is produced.
>
> **Epimerization** occurs when a molecule has two or more chiral centers or axis, of which one of them is configurationally labile. In other words, epimerization is a type of chiral inversion that occurs at one chiral element but not the other(s).
>
> An **eutomer** is an enantiomer that exhibits a high affinity for a primary drug target, biological receptor, or plasma protein while a **distomer** has lower affinity. When chiral inversion in pharmaceuticals and pesticides results in formation of a distomer, the compound generally loses its efficacy.

elements in the molecule (see Box 2.1 for further details). We will start by discussing chirality and chiral inversions, and then move on to implications of chiral inversion on environmental behavior.

2.2 CHIRALITY AND CHIRAL INVERSION

Organic molecules can have different types of isomers (i.e., molecules with the same molecular formula but have different structures or spatial orientation of atoms) such as constitutional and configurational isomerism. Constitutional isomers have different connectivity of atoms, which means that conversion of one isomer into the other cannot happen without breaking a sigma bond. Figure 2.1 shows three constitutional isomers of pentane.

Constitutional isomers exhibit different physical and chemical properties. Configurational isomers have the same connectivity of atoms, but their spatial orientation is different. There are two types of configurational isomers, namely geometric and optical isomers. In this chapter, we will focus on optical isomers, which are capable of chiral inversion. Optical isomers rotate plane-polarized light either to the right or to the left, depending on the 3D orientation of atoms. Most of these molecules have at least one stereocenter. A stereocenter is an atom that is bonded to four different groups. For example, the carbon atom (C1) is a stereocenter because it is bonded to four different groups; namely, -CH_3, -Br, Cl, and H (Figure 2.2).

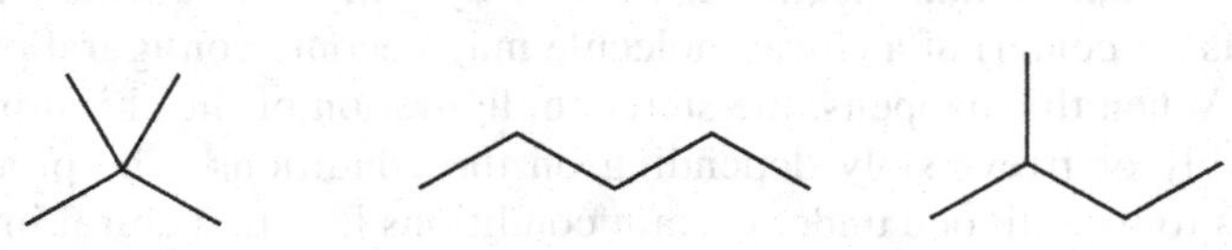

FIGURE 2.1 Constitutional isomers of pentane.

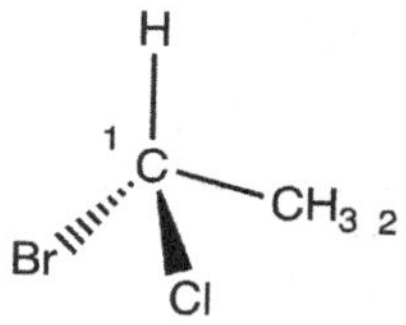

(*R*)-1-bromo-1-chloroethane

FIGURE 2.2 (R)-1-bromo-1-chloromethane.

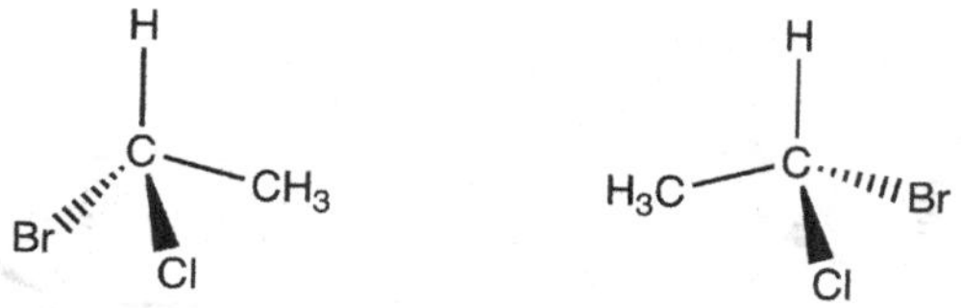

(*R*)-1-bromo-1-chloroethane (*S*)-1-bromo-1-chloroethane

FIGURE 2.3 Enantiomers of 1-bromo-1-chloromethane.

Carbon (C2) is not a stereocenter because it is bonded to at least two identical groups. 1-bromo-1-chloroethane is a chiral molecule and is optically active. The molecule exhibits chirality, which means it cannot be superimposed onto its mirror image, just like the left and the right hand. Figure 2.3 shows non-superimposable mirror images of 1-bromo-1-chloromethane.

Both molecules in Figure 2.3 rotate plane-polarized light but in opposite directions. A mixture containing equal amounts of these enantiomers is optically inactive because their effects cancel each other. Such a mixture is known as a **racemic mixture**. Enantiomers have the same physical properties except for the direction of rotation of plane-polarized light. Since the isomers have the same connectivity, the molecules have the same structural name. To distinguish enantiomers, scientists write a D for dextrorotatory or an L for levorotatory before the name, which indicates clockwise and counterclockwise rotation of plane-polarized light, respectively. One of the common examples is D-glucose. Chemists, mostly, use R and S configuration based on Cahn-Ingold-Prelog rules, which are outlined in most standard organic chemistry textbooks. The method involves assigning priorities to substituents; the heaviest is assigned the highest priority, number 1, and the lightest atom is the lowest priority (number 4). After assigning priorities, rotate the molecule so that the lowest priority substituent is pointing away from the viewer (Figure 2.4).

If tracing the direction of movement from highest to lowest priority (1-2-3) is in a clockwise direction, the molecule is designated 'R'. The molecule is designated S for a counterclockwise direction. Assigning R and S configuration is useful in distinguishing enantiomers and also identifying relationships between stereoisomers, especially those with more than one stereocenter. For example,

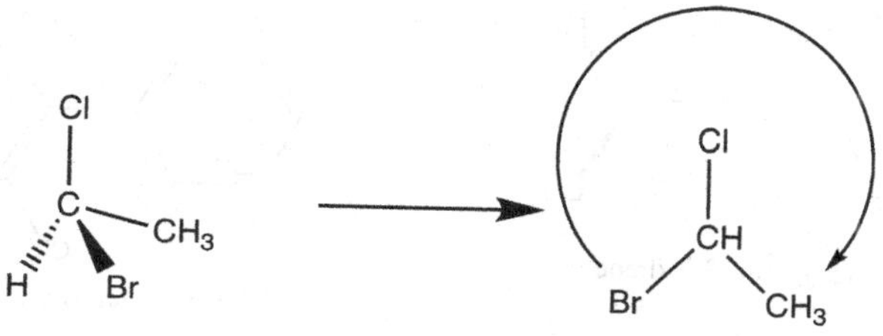

FIGURE 2.4 R configuration of 1-bromo-1-chloromethane.

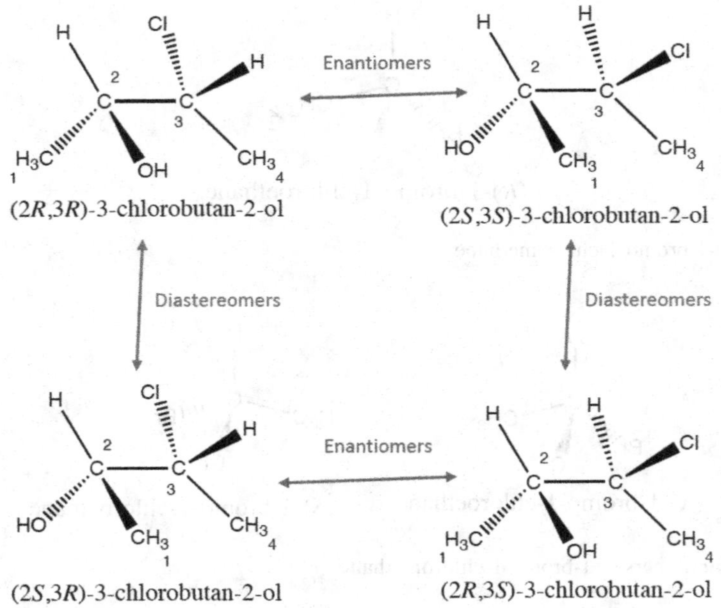

(2R,3R)-3-chlorobutan-2-ol   (2S,3S)-3-chlorobutan-2-ol

(2S,3R)-3-chlorobutan-2-ol   (2R,3S)-3-chlorobutan-2-ol

FIGURE 2.5 Stereoisomers of 3-chlorobutan-2-ol.

3-chlorobutan-2-ol has two stereocenters, C2 and C3, and has four possible stereoisomers. The number of possible stereoisomers is given by 2^n, where n is the number of stereocenters. Figure 2.5 shows stereoisomers of 3-chlorobutan-2-ol.

Isomers (1)-(2) and (3)-(4) are enantiomers, which means they are mirror images of one another. For enantiomers, the stereochemistry is flipped at each center as expected for mirror images. The relationship between isomers (1)-(3) and (2)-(4) is different because the pairs are not mirror images and they are non-superimposable on each other. These isomers are known as diastereomers. Unlike enantiomers, diastereomers have different physical properties and are, therefore, easier to separate compared to enantiomers. It is important to note that there are examples of allenes and biphenyls compounds that are optically active even though they do not have stereocenters. In both types of compounds, the optical activity is a consequence of restricted bond rotation. The rigidity of double bonds restricts rotation in allenes, whereas in biphenyls, it is due to the steric hindrance caused by substituent molecules (Figure 2.6). The presence of substituent molecules induces a rotational energy barrier, whose magnitude depends on the nature and size of substituent molecules (Nezel et al., 1997; Wong, 2006). Chiral inversion of biphenyls requires overcoming the rotational energy barrier.

a. (S)-penta-2,3-diene

b. 2,2',3,5',6-pentachloro-1,1'-biphenyl

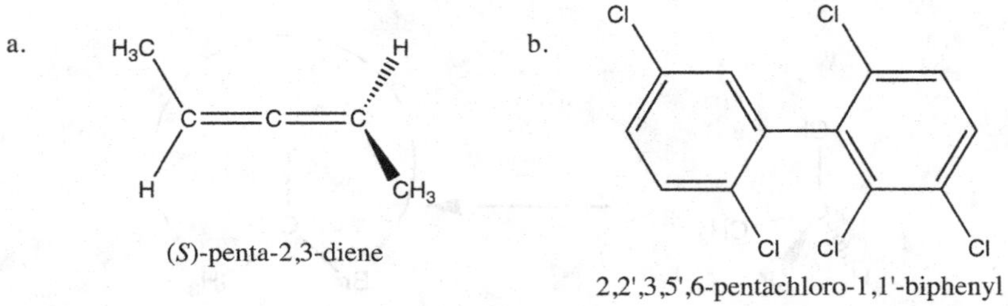

FIGURE 2.6 Chiral molecules without chiral carbons; (a) an allene (b) a biphenyl derivative (Wong, 2006).

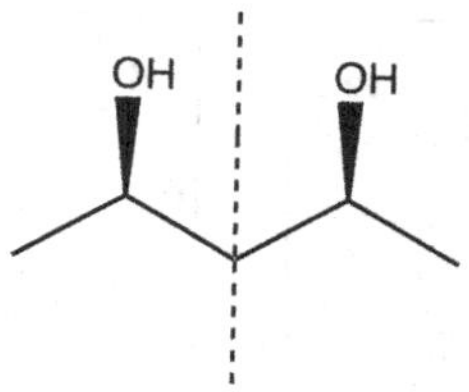

FIGURE 2.7 A meso compound has a plane of symmetry.

On the other hand, the presence of stereocenters does not necessarily mean optical activity. Some compounds may contain a plane or inversion center of symmetry that makes them optically inactive. An example of a molecule with a plane of symmetry (dotted line) is shown in Figure 2.7.

A *meso* compound has stereocenters, but it is optically inactive because it possesses a plane of symmetry or an inversion center. The presence of a plane of symmetry means that the two stereocenters on either side of the plane rotate plane-polarized light by the same magnitude but in opposite directions. Hence, optical inactivity of the compound.

We have discussed chirality and identification of compounds, and we will now discuss the implications of chiral inversion on the behavior of chiral molecules in the environment.

2.3 IMPLICATIONS OF CHIRAL INVERSION ON ENVIRONMENTAL BEHAVIOR

Enantiomers of chiral compounds often undergo enantioselective interactions in a chiral system. Chiral systems are common in the environment. They can be provided by various biotic and abiotic substances such as organic solvents, energy sources such as plane-polarized light, solid organic matter, solid surfaces, and biological building blocks such as proteins, oligosaccharides, phospholipids, and nucleic acids. Interactions in a chiral environment can be visualized using the three-point model (Figure 2.8). The moieties around a chiral element have high binding energy in the eutomer compared to the distomer. Hence, when a eutomer undergoes chiral inversion, it loses its affinity for the biological receptor. The stereoconfiguration of chiral compounds often elicit enantiomer-specific molecular binding, catalysis, and stabilization. Such enantiomer-specific interactions cause the enantioselectivity observed in the therapeutic and toxicological properties, environmental behavior, and human and environmental health effects. Hence, in the past three decades, research interest on the configurational stability of chiral pharmaceuticals and pesticides increased steadily.

Chiral inversion influences the biological properties and fate of configurationally labile pollutants. If the receptor was responsible for the biodegradation of the chiral pollutant in the environment, then the more persistent distomer will be enriched. However, if the distomer is the one that undergoes chiral inversion, then the chiral pollutant will be less persistent. In addition, if the more potent enantiomer undergoes chiral inversion, then the overall toxicity of the chiral pollutant decreases. Hence, chiral inversion has implications on the fate and toxicity of chiral pollutants in the environment and in food.

2.4 FACTORS INFLUENCING CHIRAL INVERSION IN THE ENVIRONMENT

Enantiomers undergo diverse types of chiral inversion depending on their form of chirality (helical, axial, or center) or the functional groups present around the chiral element. Table 2.1 shows the distinct types of chiral inversion that have been observed in pharmaceutical compounds. Chiral inversion can occur in humans and the environment due to solvent, heat, enzymatic, photo, and pH induction. Enzyme-mediated chiral inversion is the most observed pathway in humans and the

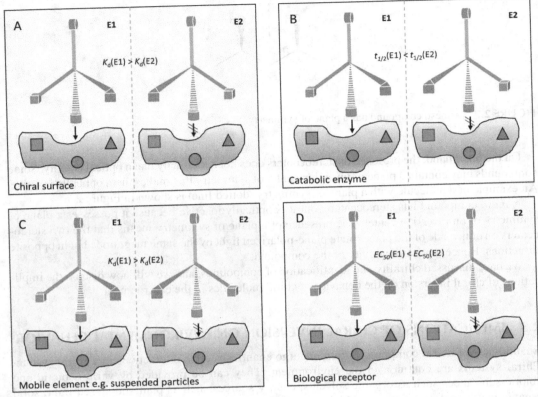

FIGURE 2.8 Chiral inversion of enantiomers and its implications on the environmental behavior of the contaminant. (From Sanganyado et al. (2020). Used with permission from Elsevier Ltd.)

environment. Solvent-, heat-, and photo-induced chiral inversion occurs in compounds with lower energy barriers.

2.4.1 Solvent-Induced Chiral Inversion

Understanding the stability of chiral pollutants in solvents is important for accurate quantitative human and environmental risk assessment. Solvents play a key role in the transport of chiral pollutants in the environment as well as in enantioselective analysis. For example, chiral pollutants are often found in aquatic environments where they can persist for short or lengthy periods depending on the stability of the compound. Previous studies have shown that some enantiomers that have low energy barriers such as helical compounds can undergo chiral inversion in water. A helical nanostructure formed by a L-glutamide amphiphile underwent helical inversion following exposure to water (Liu et al., 2019). It was proposed that the left-handed helices had hydrogen bonding predominantly while right-handed helices had π-π stacking (Liu et al., 2019). The addition of water affected the hydrogen bonding between L-glutamide amphiphile. In another study, water was shown to slow down the cyclization chiral α-diimines and promoted the rotation of the C-N bond (Aresu et al., 2013). These studies demonstrate that water-induced chiral inversion is important in synthesis of chiral compounds. Water-induced chiral inversion in compounds with a chiral center mainly occurs in enantiomers with low energy inversion barrier. Such compounds are normally less sterically hindered and often have a nitrogen rather than a carbon chiral center. Hence, the chiral inversion has been observed in amines (R_3N, R is an alkyl group) but not in aminium salts (R_3NH^+), which have a higher energy inversion

TABLE 2.1
Nature and Mechanisms of Chiral Inversion in Chiral Compounds of Pharmaceutical Concern

Compound	Nature of Chiral Inversion	Comment
Biphenyl and chlorobiphenyls	Exocyclic bond rotation	The enantiomers span a wide range of inversion barriers.
Gossypol	Exocyclic bond rotation	High inversion barrier warrants chiral stability to the enantiomers.
Peptides	Peptide bond rotation	The three-dimensional structure is stabilized by amino groups.
Lidocaine	N—CO bond rotation	Various conformational groups impact the pharmacological properties.
Cyclobutane, cyclopentane and cyclohexane	Rotation of the single endocyclic bonds	Partial rotation on the endocyclic single bond resulting in ring reversal.
Amfepramone and cathinone	Reversible chiral inversion at a configurationally labile asymmetric α-carbon	Metabolic racemization not catalyzed by an enzyme.
Hyoscyamine	Reversible chiral inversion at a configurationally labile asymmetric α-carbon	Racemization occurs slowly during storage.
Oxazepam	Reversible chiral inversion at a configurationally labile asymmetric α-carbon	Racemization occurs fast.
Piperidine and N-methylpiperidine	Ring reversal combined with nitrogen inversion	Conformational behavior about the nitrogen stereogenic center influenced by the presence of methyl group.

Source: From Testa et al. (2016). Used with permission from Elsevier Ltd.

barrier (Kaur and Vikas, 2017). Liu et al. (2005) found that synthetic pyrethroid insecticides underwent chiral inversion in water at the asymmetric α-carbon position. Further studies confirmed that water induced chiral inversion of synthetic pyrethroids such as cypermethrin and permethrin (Qin et al., 2006; Qin and Gan, 2007). However, the extent of chiral inversion caused by water is influenced by temperature and pH. A previous study found fenpropathrin, malathion, and phenthoate racemized more rapidly at pH 7.0 than at pH 5.8 (Li et al., 2010). This was probably because the chiral inversion in these chiral organophosphorus pesticides was via proton exchange at the asymmetric carbon. It was proposed that chiral inversion occurred following an exchange between an active proton at the asymmetric α-carbon and a protic solvent such as an alcohol or water. Briefly, the proton at the asymmetric α-carbon is comparatively acidic and can be donated resulting in an intermediate carbanion (Figure 2.9). The parent compound can be regenerated in protic solvents at different sides of the carbanion face. Chiral inversion will result depending on the face the proton is regenerated. Such a process can occur in the environment. Hence, these results suggest that abiotic processes such as dissolution in water may contribute to changes in enantiomeric composition of chiral pollutants in aquatic environments (Sanganyado et al., 2020). However, there are few studies that have investigated water-induced chiral inversion in the environment.

Chiral inversion has profound consequences on pharmacology and the accuracy of enantioselective analysis techniques. A previous study found that phosphate, hydroxyl ions, albumin, and amino acids catalyzed the chiral inversion of thalidomide in mammals (Reist et al., 1998). The yield of basic amino acids was higher than neutral and acidic amino acids, and this suggested that chiral inversion of thalidomide was base catalyzed. Hence, the catalysis activity observed in albumin was

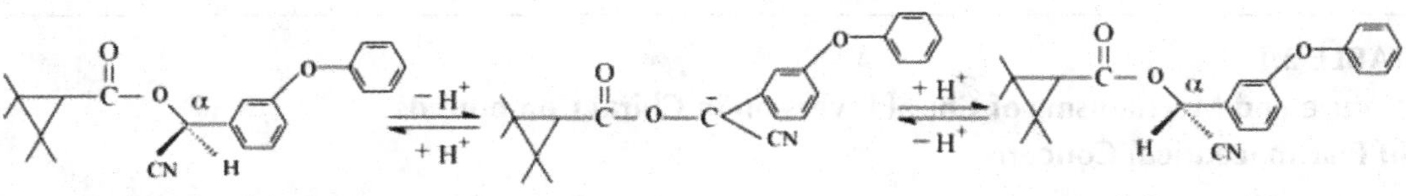

FIGURE 2.9 Proposed chiral inversion mechanism of fenpropathrin in protic solvents such as alcohols and water. (Used with permission from Li et al., 2010.)

ascribed to the presence of basic amino acid functional groups. It was proposed that chiral inversion in thalidomide occurred via electrophilic substitution. In analytical chemistry, chiral analytes are often stored in solvents prior to analysis and they can undergo chiral inversion during storage. Axially chiral 1,1'-binaphthyls were shown to undergo solvent-dependent chiral inversion in the presence of polar and nonpolar solvents (Takaishi et al., 2020). Chiral inversion occurred following the inversion of an excimer chirality due to intermolecular hydrogen bond interactions in the excited state. An excimer is a highly unstable dimeric or heterodimeric molecule that is formed by two smaller molecules. Cypermethrin was shown to undergo rapid enantiomerization at the asymmetric α-carbon position in isopropanol and methanol (Qin and Gan, 2007). However, no chiral inversion was observed when the cypermethrin was stored in methylene chloride, acetone, or hexane. Interestingly, the degree of enantiomerization was influenced by the temperature and presence of water as a co-solvent. These results show that the mobile phase conditions during enantioselective analysis can affect the accuracy of the analytical technique. Furthermore, chiral inversion in enantiopure commercial pesticides due to storage conditions can result in a decrease in the pesticide's efficacy.

2.4.2 Thermal-Induced Chiral Inversion

Enantiomers can undergo chiral inversion when the temperature of a system exceeds the inversion energy barrier. Thermally induced chiral inversion is common in compounds that are axially chiral and have relatively low steric hindrance about the chiral axis. Schurig and Reich (1998) found that increasing steric hindrance resulted in an increase in rotational energy barrier around the chiral axis. The estimated rotational energy $\Delta G^{\ddagger}$ (T) for PCB 95, 132, and 149 was 183–193 kJ/mol at 300°C while PCB 136 was 210 kJ/mol 320°C. PCB 136 had higher rotational energy barrier since it had four chlorines ortho-substituted about the chiral axis. In contrast, PCB 95, 132, and 149 only had three chlorines ortho-substituted. Liu et al. (2005) found cypermethrin underwent chiral inversion at the asymmetric α-carbon position at temperatures above 260°C during gas chromatography. However, no chiral inversion was observed in permethrin probably because unlike cypermethrin, cyfluthrin, deltamethrin, and fenvalerate, it does not have an asymmetric α-carbon. In gas chromatography, thermally induced chiral inversion can result in distortion, plateauing, and coalescence of the peaks depending on the enantioresolution and configurational lability of the chiral analyte (Trapp et al., 2002). These results suggest understanding the rotational energy barrier is imperative for differentiating environment changes in enantiomeric composition due to microbial action from thermal processes during enantioselective analysis.

2.4.3 Photo-Induced Chiral Inversion

When the irradiation from the sun exceeds the energy barriers in some chiral compounds, chiral inversion can occur. Photolysis can result in cleavage of bonds that are linked to the chiral element. The chiral compound is regenerated when the free radicals recombine. The stereoconfiguration of the chiral compound changes when the free radicals recombine from the face opposite

Chiral Inversion of Organic Pollutants

to where it cleaved. The cleavage of bonds on the chiral compound is not feasible due to high energy barriers. However, photolysis of a solvent might require less energy. The resultant free radicals can elicit chiral inversion on the chiral compound. For example, photolysis of water can produce hydroxyl radicals (OH•) which can abstract a hydrogen atom from a chiral center. Recently, reversible hydrogen atom transfer was observed in naproxen linked to tryptophan using dynamic NMR following irradiation with UV (Ageeva et al., 2019). Unidirectional chiral inversion occurred from (R)-naproxen to (S)-naproxen following recombination of the hydrogen atom to a biradical intermediate. Epimerization was previously reported in synthetic pyrethroids. Chiral inversion was observed at the 1-C and/or 3-C position in the cyclopropyl ring while the asymmetric α-carbon position remains unchanged (Li et al., 2015). The 3-C position was more labile than the 1-C position probably since it was attached to an electron-withdrawing alkyl group. There are few studies that explored the role of photolysis in changes in enantiomeric composition of chiral pollutants in the environment.

2.4.4 BIOLOGICALLY MEDIATED CHIRAL INVERSION

Several studies have investigated the mechanism of chiral inversion in biological systems. Enzymes such as epimerases and racemases play a key role in enzyme-mediated chiral inversion. Enzyme-mediated chiral inversion is more common than abiotic chiral inversion because enzymatic processes lower the energy barriers for inversion. There are two main catalytic pathways by which enzyme-mediated chiral inversion occur. First, chiral inversion often occurs after a hydrogen atom at the chiral center is abstracted to yield a carbonyl intermediate. The hydrogen atom is then transferred back to the carbonyl intermediate to regenerate the parent compound but with a different stereoconfiguration. Second, chiral inversion may involve oxidation of the chiral compound to give an intermediate followed by reduction to give the original compound but with a stereoconfiguration. For example, chiral inversion of 1-hydroxyethylpyrene was found to occur following the transfer of a sulfonyl functional group obtained from the cofactor 3′-phosphoadenosine-5′-phosphosulfate by sulfotransferases (Landsiedel et al., 1998). The sulfate group is then cleaved by hydrolysis through nucleophilic substitution. However, chiral inversion can also occur following successive reduction and oxidation. Chiral inversion of (R)-flosequinan at the sulphoxide position was shown to involve reduction of sulfoxide to sulfide via microbial activity followed by regeneration of the sulfoxide by oxidation of the sulfide to yield (S)-flosequinan (Eiji et al., 1994). A clinical study found proton pump inhibitors such as omeprazole, lansoprazole, pantoprazole, and rabeprazole underwent chiral inversion via reductive metabolism to thioethers by glutathione and cysteine followed by oxidative metabolism of the thioethers intermediates to regenerate the substrate (Tang et al., 2019). Chiral inversion potency was shown to be influenced by the substituent groups on the substrate since they affected the redox properties of the proton pump inhibitors. For 2-arylpropionic acids (also called profens, examples include ibuprofen, indoprofen, fenoprofen, flurbiprofen, ketoprofen, and suprofen), chiral inversion involves enantioselective formation of an (R)-profenyl-CoA thioester, followed by enzymatic epimerization to produce a diastereoisomeric (S)-profenyl-CoA thioester, and finally formation of the (S)-profen via hydrolysis (Thomason et al., 1997).

2.5 CHIRAL INVERSION IN THE ENVIRONMENT

Studies on the chiral inversion of chiral pollutants in the environment are scarce. Most studies focused on chiral inversion of pharmaceuticals in mammals with the goal of establishing the safety and efficacy of the compounds. At present, there are no studies on chiral inversion of chiral pollutants due to abiotic processes such as hydrolysis and photolysis. However, chiral inversion observed in environmental systems has been attributed to microbial action. A study on the biotransformation of 2-phenylbutyric acid, a transformation product of linear alkylbenzene sulfonates, was shown to undergo unidirectional chiral inversion mediated by *Xanthobacter flavus* strain PA1 (Liu et al., 2011). The *X. flavus* PA1 strain was isolated from sediments collected from a mangrove. This suggested that chiral inversion could occur in the environment via microbial mediation. Chiral inversion of (*R*)-fipronil to (*S*)-fipronil was observed in Chinese pond mussels (*Anodonta woodiana*) (Qu et al., 2016). Acute toxicity testing using LC50 at 72-h, (*S*)-fipronil (0.63 mg L^{-1}) was significantly more toxic than both the racemate (1.21 mg L^{-1}) and (*R*)-fipronil (3.27 mg L^{-1}) (Qu et al., 2016). These results suggest chiral inversion resulted in an increase in toxicity to *A. woodiana*. Hence, the need for risk assessments to consider enantiomers of chiral pollutants individually.

Chiral inversion of chiral pesticides has been observed in distinct types of soils. Incubation of (*S*)-(−)- and (*R*)-(+)-diclofop in soil under aerobic conditions resulted in bidirectional inversion with the (*S*) → (*R*) inversion proceeding with a higher rate (Diao et al., 2010). However, Diao et al. (2010) found that the degree of chiral inversion depended on the soil type. (*S*) → (*R*) inversion has been observed in herbicides such as haloxyfop (Poiger et al., 2015), fluazifop-butyl (Bewick, 1986), fenoxaprop (Wink and Luley, 1988), and quizalofop-ethyl (Li et al., 2012). Incubation of (*S*)-fluazifop-butyl in soils collected from Beijing and Harbin yielded the (*R*)-enantiomer with concentration increasing gradually with the decrease in (*S*)-fluazifop-butyl but not in soils from Anhui (Qi et al., 2016). The Beijing soils yielded more (*R*)-fluazifop-butyl up to 2.38 mg/kg. These results suggested that the biological and physicochemical properties of the soil may influence chiral inversion of chiral pesticides. However, chiral inversion in soil is not necessarily due to microbial activity. A study on chiral stability of phenthoate in soil found that there was higher chiral inversion of (+)-phenthoate to (−)-phenthoate in sterilized soils than in unsterilized soils suggesting microbial activity inhibited chiral inversion (Li et al., 2007).

Chiral inversion in pharmaceuticals has been reported in wastewater treatment plants and soil. A recent study found that naproxen and ibuprofen underwent bidirectional chiral inversion in soil matrices (Bertin et al., 2020). Pharmaceuticals enter soil through irrigation with recycled water, landfill disposal, or soil amendment using biosolids (Fu et al., 2016). Previous studies reported the chiral inversion of (*S*)-(+)-naproxen to (*R*)-(−)-naproxen in wastewater treatment plants (Hashim and Khan, 2011; Khan et al., 2014; Suzuki et al., 2014). (*S*)-naproxen is commercially marketed as a single enantiomer. Hashim et al. (2011) found that the enantiomeric fraction of (*S*)-naproxen in wastewater influent was approximately 1.0 but decreased significantly in the effluent suggesting the formation of (*R*)-naproxen. Similar results of chiral inversion of naproxen, ketoprofen, and ibuprofen were obtained in a membrane bioreactor (Hashim et al., 2011). As observed in mammalian studies, naproxen and profens can also undergo bidirectional chiral inversion in the environment. For example, bidirectional chiral inversion of naproxen, ketoprofen, and ibuprofen was observed in a membrane bioreactor (Nguyen et al., 2017). Hence, when chiral pharmaceuticals are discharged into the environment, their enantiomeric composition can be altered by various processes such as chiral inversion which can alter the toxicity of the pollutant.

2.6 DETERMINATION OF CHIRAL INVERSION

Dynamic chromatography and electrophoresis are powerful tools for assessing the changes in enantiomeric composition as a function of time (Wolf, 2005). Enantioselective analysis of chiral pollutants is often achieved using enantioselective gas chromatography and liquid chromatography. Following a successful separation of an analyte with one chiral center or axis, two peaks are observed on the chromatogram. However, when the analyte is conformationally or configurationally labile, the peaks coalesce (Krupcik et al., 2003). The degree of coalescence will depend on the rate of chiral inversion and the enantioresolution. With time, the peaks will move from being distinct to a plateau. Dynamic chromatography shows the changes in elution profile with time, and this is useful for determining the effect of pH, temperature, and solvents on the chiral inversion that occur on-column, on the injector, or in the detector (Wolf, 2005). Since chiral stationary phase is widely used in dynamic chromatography and electrophoresis, initial preparation of enantiopure

TABLE 2.2
Instruments and Experimental Approaches Used for Assessing Chiral Inversion of Conformationally or Configurationally Labile Enantiomers (Krupcik et al., 2003)

Instrument	Experimental Approaches
Dynamic NMR	Integrating enantioselective separation with classical kinetic studies
Dynamic gas chromatography	Continuous flow models
Dynamic supercritical fluid chromatography	Peak form analysis which involves experimentally obtained and simulated peaks
Dynamic liquid chromatography	Stopped-flow method
Dynamic capillary electrophoresis	Stochastic methods
Dynamic micellar electrokinetic chromatography	Deconvolution methods
Dynamic capillary electrochromatography	Approximation functions method
Chiroptical methods	

compounds prior to chiral inversion analysis is required (Krupcik et al., 2003). The inversion energy barrier range that can be investigated using dynamic chromatography is influenced by the run time, mobile phase, analysis temperature, and analyte stability. In addition, the inversion energy barriers obtained are also influenced by the type of chiral stationary phase used.

The choice for the instrument used in the analysis depends on the physicochemical characteristics of the chiral compound (i.e., the solubility, vapor pressure, thermal and solvent stability, and detection) (Krupcik et al., 2003). For example, an ionizable compound with high vapor pressure and high solubility in polar solvents can be analyzed using capillary electrophoresis or liquid chromatography (Sanganyado et al., 2017). A thermally stable compound with low vapor pressure can be determined using gas chromatography. Supercritical fluid chromatography has gained prominence in the analysis of chiral pollutants as it pairs well with mass spectrometers and offers fast analysis, high sensitivity, and a green approach compared to gas or liquid chromatography (Chen et al., 2019). In recent times, multidimensional approaches have been introduced to improve separation and detection. Table 2.2 shows the list of techniques and experimental approaches commonly used to determine chiral inversion. Any of the experimental approaches listed in Table 2.2 can then be used to determine the chiral inversion.

2.7 FUTURE DIRECTIONS AND CONCLUSION

Chiral inversion in chiral pollutants such as pesticides and pharmaceuticals can occur under environmental conditions through abiotic or biological processes. Examples of abiotic processes that can cause chiral inversion include heating, UV irradiation, solvolysis, and hydrolysis. Abiotic chiral inversion is limited to a few groups of chiral compounds that have low energy barriers. When such chiral pollutants enter the environment, they are exposed to UV irradiation from the sun and hydrolytic interactions, for example, in aqueous environments, and chiral inversion can result. This has only been observed in sterilized soils, but the study did not determine the exact mechanism of chiral inversion. Few studies have reported chiral inversion in environmental systems. Considering abiotic processes can change the enantiomeric composition of chiral pollutants due to chiral inversion, future studies should explore the mechanisms involved.

Biologically mediated chiral inversion often involves the action of enzymes that facilitate the oxidation, reduction, and hydrogen atom transfer about the chiral center or axis. Reductive and oxidative metabolism has been observed in microbial communities found in soil, sediments, and wastewater treatment plants. Enzymatic mediated chiral inversion has been widely reported in non-steroidal anti-inflammatory drugs such as ibuprofen, ketoprofen, and naproxen as well as in phenoxyphenoxypropionate herbicides such as diclofop, fluazifop-butyl, and haloxyfop. Physicochemical and biological properties of the environmental compartments may influence the rate or direction of chiral inversion. This implies that the results of chiral inversion in one soil type cannot be extrapolated to another. Therefore, there is a need for a systematic investigation of the factors that influence chiral inversion in different environmental compartments. Microbial communities can play an important part in determining the preferred direction of chiral inversion in a system. Therefore, future studies should explore the impact of microbial community, structure, and function on the chiral inversion of chiral pollutants.

Understanding the mechanism of chiral inversion should remain a priority in assessing the risk posed by chiral pollutants to environmental and human health. Chiral inversion can increase the toxicity of a pollutant when a less toxic enantiomer transforms into a more toxic and the inverse happens when the more toxic enantiomer turns to the less toxic. Therefore, more studies are required to determine chiral inversion in the environment.

REFERENCES

Ageeva, A. A., Babenko, S. V., Polyakov, N. E., Leshina, T. V., 2019. NMR investigation of photoinduced chiral inversion in (R)/(S)-naproxen–(S)-tryptophan linked system. *Mendeleev Commun.* 29, 260–262. https://doi.org/10.1016/j.mencom.2019.05.006

Aresu, E., Fioravanti, S., Pellacani, L., Sciubba, F., Trulli, L., 2013. Water-controlled chiral inversion of a nitrogen atom during the synthesis of diaziridines from α-branched N,N′-dialkyl α-diimines. *New J. Chem.* 37, 4125–4129. https://doi.org/10.1039/c3nj00780d

Bertin, S., Yates, K., Petrie, B., 2020. Enantiospecific behaviour of chiral drugs in soil. *Environ. Pollut.* 262. https://doi.org/10.1016/j.envpol.2020.114364

Bewick, D. W., 1986. Stereochemistry of fluazifop-butyl transformations in soil. *Pestic. Sci.* 17, 349–356. https://doi.org/10.1002/ps.2780170405

Chen, L., Dean, B., La, H., Chen, Y., Liang, X., 2019. Stereoselective supercritical fluidic chromatography–mass spectrometry (SFC-MS) as a fast bioanalytical tool to assess chiral inversion in vivo and in vitro. *Int. J. Mass Spectrom.* 444, 116–172. https://doi.org/10.1016/j.ijms.2019.06.008

Diao, J., Xu, P., Wang, P., Lu, Y., Lu, D., Zhou, Z., 2010. Environmental behavior of the chiral aryloxyphenoxypropionate herbicide diclofop-methyl and diclofop: Enantiomerization and enantioselective degradation in soil. *Environ. Sci. Technol.* 44, 2042–2047. https://doi.org/10.1021/es903755n

Eiji, K., Tsuyoshi, Y., Takashi, T., Masaaki, O., Tetsuya, K., 1994. Chiral inversion of drug: Role of intestinal bacteria in the stereoselective sulphoxide reduction of flosequinan. *Biochem. Pharmacol.* 48, 237–243. https://doi.org/10.1016/0006-2952(94)90093-0

Eriksson, T., Björkman, S., Roth, B., Fyge, Å., Höuglund, P., 1995. Stereospecific determination, chiral inversion in vitro and pharmacokinetics in humans of the enantiomers of thalidomide. *Chirality.* 7, 44–52. https://doi.org/10.1002/chir.530070109

Fu, Q., Sanganyado, E., Ye, Q., Gan, J., 2016. Meta-analysis of biosolid effects on persistence of triclosan and triclocarban in soil. *Environ. Pollut.* 210, 137–144. https://doi.org/10.1016/j.envpol.2015.12.003

Hashim, N. H., Khan, S. J., 2011. Enantioselective analysis of ibuprofen, ketoprofen and naproxen in wastewater and environmental water samples. *J. Chromatogr. A* 1218, 4746–54.

Hashim, N. H. H., Nghiem, L. D. D., Stuetz, R. M. M., Khan, S. J. J., 2011. Enantiospecific fate of ibuprofen, ketoprofen and naproxen in a laboratory-scale membrane bioreactor. *Water Res.* 45, 6249–58. https://doi.org/10.1016/j.watres.2011.09.020

Kaur, R., Vikas, 2017. From nitrogen inversion in amines to stereoinversion in aminium salts: Role of a single water molecule. *Theor. Chem. Acc.* 136, 1–16. https://doi.org/10.1007/s00214-017-2090-2

Khan, S. J., Wang, L., Hashim, N. H., Mcdonald, J. A., 2014. Distinct enantiomeric signals of ibuprofen and naproxen in treated wastewater and sewer overflow. *Chirality.* 26, 739–746. https://doi.org/10.1002/chir

Krupcik, J., Oswald, P., Májek, P., Sandra, P., Armstrong, D. W., 2003. Determination of the interconversion energy barrier of enantiomers by separation methods. *J. Chromatogr. A* 1000, 779–800. https://doi.org/10.1016/S0021-9673(03)00238-3

Landsiedel, R., Pabel, U., Engst, W., Ploschke, J., Seidel, A., Glatt, H., 1998. Chiral inversion of 1-hydroxyethylpyrene enantiomers mediated by enantioselective sulfotransferases. *Biochem. Biophys. Res. Commun.* 247, 181–185. https://doi.org/10.1006/bbrc.1998.8756

Li, Z., Li, Q., Cheng, F., Zhang, W., Wang, W., Li, J., 2012. Enantioselectivity in degradation and transformation of quizalofop-ethyl in soils. *Chirality.* 24, 552–557. https://doi.org/10.1002/chir.22053

Li, Z., Wu, T., Li, Q., Zhang, B., Wang, W., Li, J., 2010. Characterization of racemization of chiral pesticides in organic solvents and water. *J. Chromatogr. A* 1217, 5718–5723. https://doi.org/10.1016/j.chroma.2010.07.016

Li, Z. Y., Luo, X. N., Li, Q. L., Zhang, E. Q., Zhao, J. H., Zhang, W. S., 2015. Stereo and enantioselective separation and identification of synthetic pyrethroids, and photolytical isomerization analysis. *Bull. Environ. Contam. Toxicol.* 94, 254–259. https://doi.org/10.1007/s00128-014-1405-4

Li, Z. Y., Zhang, Z. C., Zhang, L., Leng, L., 2007. Enantioselective degradation and chiral stability of phenthoate in soil. *Bull. Environ. Contam. Toxicol.* 79, 153–157. https://doi.org/10.1007/s00128-007-9099-5

Liu, C., Yang, D., Zhang, L., Liu, M., 2019. Water inversed helicity of nanostructures from ionic self-assembly of a chiral gelator and an achiral component. *Soft Matter.* 15, 6557–6563. https://doi.org/10.1039/c9sm01176e

Liu, W., Qin, S., Gan, J., 2005. Chiral stability of synthetic pyrethroid insecticides. *J. Agric. Food Chem.* 53, 3814–3820. https://doi.org/10.1021/jf048425i

Liu, Y., Han, P., Li, X.-Y., Shih, K., Gu, J.-D., 2011. Enantioselective degradation and unidirectional chiral inversion of 2-phenylbutyric acid, an intermediate from linear alkylbenzene, by *Xanthobacter flavus* PA1. *J. Hazard. Mater.* 192, 1633–40. https://doi.org/10.1016/j.jhazmat.2011.06.088

Mori, T., Ito, T., Liu, S., Ando, H., Sakamoto, S., Yamaguchi, Y., Tokunaga, E., Shibata, N., Handa, H., Hakoshima, T., 2018. Structural basis of thalidomide enantiomer binding to cereblon. *Sci. Rep.* 8, 1–14. https://doi.org/10.1038/s41598-018-19202-7

Nezel, T., Müller-Plathe, F., Müller, M. D., Buser, H., 1997. Theoretical considerations about chiral PCBs and their methylthio and methylsulfonyl metabolites being possibly present as stable enantiomers. *Chemosphere.* 35, 1895–1906. https://doi.org/10.1016/S0045-6535(97)00229-4

Nguyen, L. N., Hai, F. I., McDonald, J. A., Khan, S. J., Price, W. E., Nghiem, L. D., 2017. Continuous transformation of chiral pharmaceuticals in enzymatic membrane bioreactors for advanced wastewater treatment. *Water Sci. Technol.* 76, 1816–1826. https://doi.org/10.2166/wst.2017.331

Poiger, T., Müller, M. D., Buser, H.-R. R., Buerge, I. J., 2015. Environmental behavior of the chiral herbicide haloxyfop. 1. Rapid and preferential interconversion of the enantiomers in soil. *J. Agric. Food Chem.* 63, 2583–2590. https://doi.org/10.1021/jf505241t

Qi, Y., Liu, D., Luo, M., Jing, X., Wang, P., Zhou, Z., 2016. Enantioselective degradation and chiral stability of the herbicide fluazifop-butyl in soil and water. *Chemosphere.* 146, 315–322. https://doi.org/10.1016/j.chemosphere.2015.12.040

Qin, S., Budd, R., Bondarenko, S., Liu, W., Gan, J., 2006. Enantioselective degradation and chiral stability of pyrethroids in soil and sediment. *J. Agric. Food Chem.* 54, 5040–5045. https://doi.org/10.1021/jf060329p

Qin, S., Gan, J., 2007. Abiotic enantiomerization of permethrin and cypermethrin: Effects of organic solvents. *J. Agric. Food Chem.* 55, 5734–5739. https://doi.org/10.1021/jf0708894

Qu, H., Ma, R., Liu, D., Jing, X., Wang, F., Zhou, Z., Wang, P., 2016. The toxicity, bioaccumulation, elimination, conversion of the enantiomers of fipronil in *Anodonta woodiana. J. Hazard. Mater.* 312, 169–174. https://doi.org/10.1016/j.jhazmat.2016.03.063

Reist, M., Carrupt, P. A., Francotte, E., Testa, B., 1998. Chiral inversion and hydrolysis of thalidomide: Mechanisms and catalysis by bases and serum albumin, and chiral stability of teratogenic metabolites. *Chem. Res. Toxicol.* 11, 1521–1528. https://doi.org/10.1021/tx9801817

Reist, M., Testa, B., Carrupt, P.-A., Jung, M., Schurig, V., 1995. Racemization, enantiomerization, diastereomerization, and epimerization: Their meaning and pharmacological significance. *Chirality.* 7, 396–400. https://doi.org/10.1002/chir.530070603

Sanganyado, E., Lu, Z., Fu, Q., Schlenk, D., Gan, J., 2017. Chiral pharmaceuticals: A review on their environmental occurrence and fate processes. *Water Res.* 124, 527–542. https://doi.org/10.1016/j.watres.2017.08.003

Sanganyado, E., Lu, Z., Liu, W., 2020. Application of enantiomeric fractions in environmental forensics: Uncertainties and inconsistencies. *Environ. Res.* 184. https://doi.org/10.1016/j.envres.2020.109354

Schurig, V., Reich, S., 1998. Determination of the rotational barriers of atropisomeric polychlorinated biphenyls (PCBs) by a novel stopped-flow multidimensional gas chromatographic technique. *Chirality.* 10, 316–320. https://doi.org/10.1002/(SICI)1520-636X(1998)10:4<316::AID-CHIR5>3.0.CO;2-5

Suzuki, T., Kosugi, Y., Hosaka, M., Nishimura, T., Nakae, D., 2014. Occurrence and behavior of the chiral anti-inflammatory drug naproxen in an aquatic environment. *Environ. Toxicol. Chem.* 33, 2671–8. https://doi.org/10.1002/etc.2741

Takaishi, K., Iwachido, K., Ema, T., 2020. Solvent-induced sign inversion of circularly polarized luminescence: Control of excimer chirality by hydrogen bonding. *J. Am. Chem. Soc.* 142, 1774–1779. https://doi.org/10.1021/jacs.9b13184

Tang, C., Chen, Z., Dai, X., Zhu, W., Zhong, D., Chen, X., 2019. Mechanism of reductive metabolism and chiral inversion of proton pump inhibitors. *Drug Metab. Dispos.* 47, 657–664. https://doi.org/10.1124/dmd.118.086090

Testa, B., Vistoli, G., Pedretti, A., 2016. Mechanisms and pharmaceutical consequences of processes of stereoisomerisation—A didactic excursion. *Eur. J. Pharm. Sci.* 88, 101–123. https://doi.org/10.1016/j.ejps.2016.04.007

Thomason, M. J., Rhys-Williams, W., Hung, Y. F., Baker, J. A., Hanlon, G. W., Lloyd, A. W., 1997. A mechanistic investigation of the microbial chiral inversion of 2-phenylpropionic acid by *Verticillium lecanii*. *Chirality*. 9, 254–260. https://doi.org/10.1002/(SICI)1520-636X(1997)9:3<254::AID-CHIR9>3.0.CO;2-F

Tokunaga, E., Yamamoto, T., Ito, E., Shibata, N., 2018. Understanding the thalidomide chirality in biological processes by the self-disproportionation of enantiomers. *Sci. Rep.* 8, 6–12. https://doi.org/10.1038/s41598-018-35457-6

Trapp, O., Schoetz, G., Schurig, V., 2002. Stereointegrity of thalidomide: Gas-chromatographic determination of the enantiomerization barrier. *J. Pharm. Biomed. Anal.* 27, 497–505. https://doi.org/10.1016/S0731-7085(01)00573-8

Wink, O., Luley, U., 1988. Enantioselective transformation of the herbicides diclofop-methyl and fenoxaprop-ethyl in soil. *Pestic. Sci.* 22, 31–40. https://doi.org/10.1002/ps.2780220104

Wolf, C., 2005. Stereolabile chiral compounds: Analysis by dynamic chromatography and stopped-flow methods. *Chem. Soc. Rev.* 34, 595–608. https://doi.org/10.1039/b502508g

Wong, C. S., 2006. Environmental fate processes and biochemical transformations of chiral emerging organic pollutants. *Anal. Bioanal. Chem.* 386, 544–558.

3 Chiral Pesticides

Herbert Musarurwa, Mpelegeng Victoria Bvumbi,
Nyeleti Bridget Mabaso, and Nikita Tawanda Tavengwa

CONTENTS

3.1 Introduction 41
3.2 Common Types of Chiral Pesticides 44
    3.2.1 Chiral Organophosphates 44
    3.2.2 Chiral Pyrethroids 45
    3.2.3 Chiral Herbicides 45
    3.2.4 Chiral Triazole Fungicides 45
3.3 Physicochemical Properties of Chiral Pesticides 45
3.4 Environmental Fate of Chiral Pesticides 49
    3.4.1 Enantioselective Biotransformation of Chiral Pesticides 49
    3.4.2 Enantioselective Biodegradation of Chiral Pesticides 50
    3.4.3 Enantioselective Bioaccumulation of Chiral Pesticides 52
    3.4.4 Enantiomerization 55
3.5 Enantioselective Action of Chiral Pesticides on Target Organisms 55
3.6 Enantioselective Toxicity of Chiral Pesticides on Non-Target Organisms 57
3.7 Pre-Concentration of Chiral Pesticides 60
3.8 Chiral Separation Techniques 60
    3.8.1 Chiral High-Performance Liquid Chromatography 60
    3.8.2 Chiral Gas Chromatography 65
    3.8.3 Chiral Supercritical Fluid Liquid Chromatography 66
3.9 Future Prospects and Challenges 66
3.10 Conclusion 67
Acknowledgment 68
Conflicts of Interest 68
References 68

3.1 INTRODUCTION

Pesticides have found widespread application in agriculture, industry, and homes to control or prevent invasive insects, diseases, and unwanted plant growth (Qu et al., 2016). About 30% of the known registered pesticides are chiral (Caballo et al., 2013; Chen et al., 2016; Cui et al., 2018). Such pesticides include the most frequently used herbicides (Buerge et al., 2019; Caballo et al., 2014), pyrethroid insecticides (Birolli et al., 2019), and organophosphorus insecticides (Chen et al., 2016; Gao et al., 2019). Chiral pesticides are chemicals consisting of one or two enantiomers that have identical physical and chemical properties in achiral environments (Asad et al., 2017; Zhang et al., 2016).

    Chirality is defined formally as the geometric property of a rigid object which is not superimposable with its mirror image. It is found throughout biological systems such as basic building blocks of life, such as amino acids, carbohydrates, and lipids as well as in synthetic compounds such as

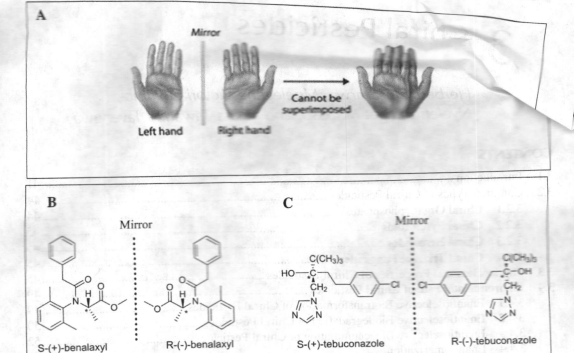

FIGURE 3.1 (A) Illustration of chirality using the right- and left-handedness, (B) enantiomers of benalaxyl pesticide, and (C) enantiomers of tebuconazole pesticide.

Note: *Asymmetric center.

chiral pesticides (Garrison, 2011). This chirality is in most cases illustrated with the idea of left- and right-handedness. The left hand and right hand are nonsuperimposable mirror images of each other (Figure 3.1A). Two mirror images of a chiral molecule, such as chiral pesticides, are called enantiomers. Just like with hands, enantiomers come in pairs. Chiral isomers have identical structures (same atoms and identical bonds) but they differ in their molecular conformation (+/−, D/L, or R/S). They have undistinguishable physicochemical properties in achiral environments. When enantiomers are placed in chiral environments, however, they show different behaviors. A chiral environment can either be physical such as plane-polarized light, or chemical such as a solvent, a reactant, or a catalyst (Mohan et al., 2009). The enantiomers of some of the chiral pesticides are shown in Figure 3.1B,C. The asymmetric center is usually a carbon bonded to four different substituents in a three-dimensional arrangement. However, other atoms such as sulfur and phosphorus can also be stereogenic centers (Figure 3.2). A mixture of enantiomers is called a racemate or a racemic mixture.

Most chiral pesticides have been commercialized in industry and agriculture as racemic forms (Asad et al., 2017; Maia et al., 2017). Although the enantiomers in the racemic mixture of a chiral pesticide have identical physical and chemical properties, as already alluded to, they usually display different physiochemical and biochemical properties in chiral environments (Qi et al., 2015; Xie et al., 2018). Most of the time, one of the enantiomers in the racemic mixture will have the desired effect on the target species and the other forms will be less active or even inactive (Cheng et al., 2018; Deng et al., 2019). Thus, the inactive enantiomers will be discharged into the environment where it may contaminate the soil and aquatic ecosystems and may be toxic to non-target species.

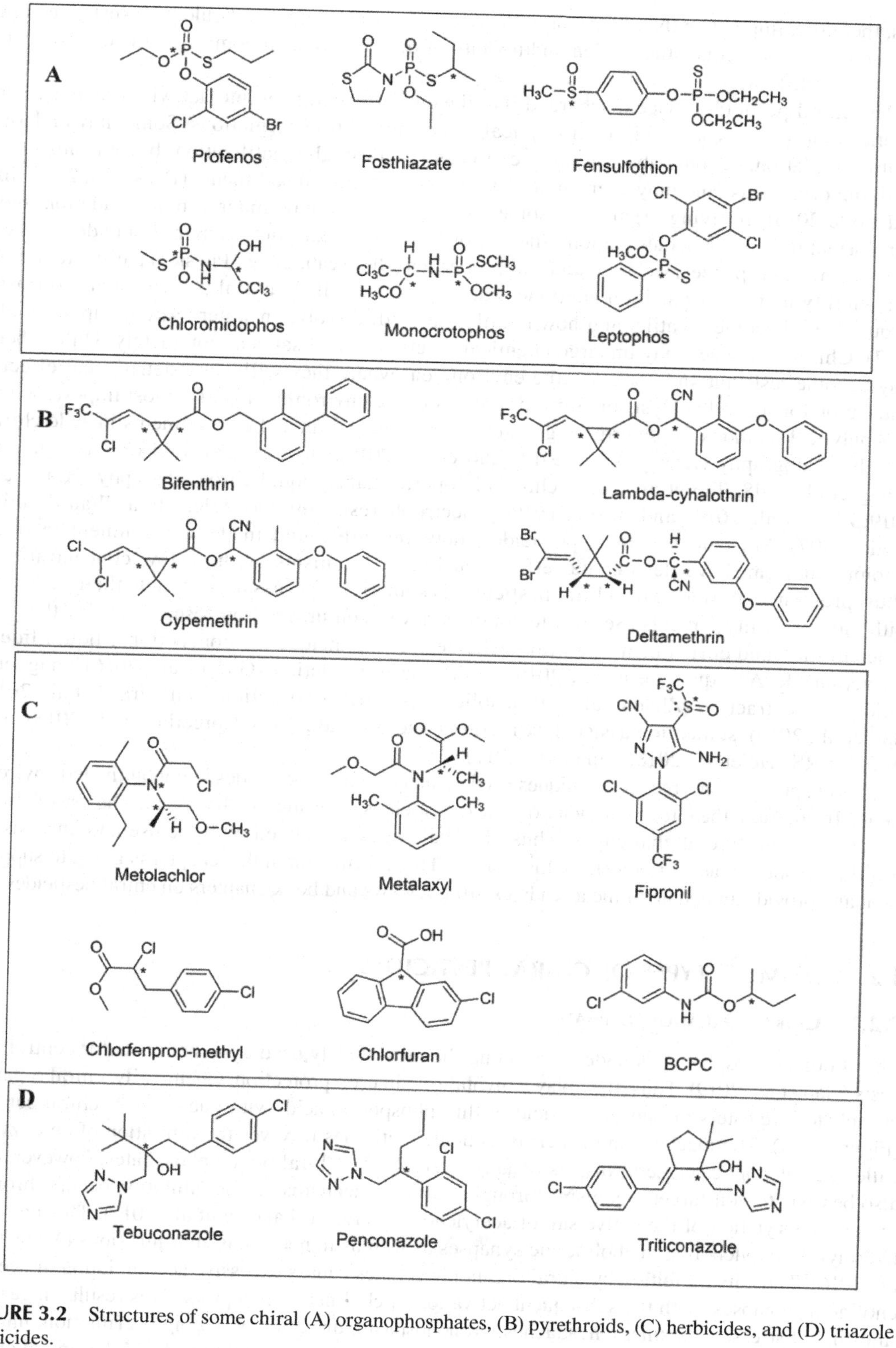

Chiral Pesticides

FIGURE 3.2 Structures of some chiral (A) organophosphates, (B) pyrethroids, (C) herbicides, and (D) triazole fungicides.

Note: *Asymmetric center.

It is, therefore, imperative that the enantioselectivity of chiral pesticides should be investigated and monitored so as to reduce the burden and toxicity of the inactive enantiomers on the environment (Qi et al., 2015).

The chiral pesticides, once discharged into the environment, may interact with the biotic and abiotic components resulting in their physical and chemical transformations. Some may undergo photodegradation, become absorbed in the atmosphere through volatilization, become absorbed by living organisms, and may even enter water sources, soil, and sediments (Brusseau, 2019; Jin and Rolle 2016). In living organisms, some chiral pesticides may undergo biodegradation, biotransformation, and bioaccumulation. The fate of chiral pesticides in the environment depends on their chemical properties, weather, and soil properties (Brusseau, 2019). Physical parameters such as solubility and water partition coefficient determine how likely chiral pesticides are to travel through soil, become volatile, and how readily they will dissolve in water (Saranjampour et al., 2017). Chiral pesticides may undergo chemical reactions and disappear completely while others may become resistant and persist in the environment where they will cause detrimental effects. Thus, monitoring and analysis of chiral pesticides in the environment is of utmost importance.

Analytical methods for determining enantiomers in contaminated environments include chiral gas chromatography (Xiang et al., 2019; Yao et al., 2019), supercritical fluid chromatography (Jiang et al., 2018; Tao et al., 2018), chiral high-performance liquid chromatography (Xia et al., 2019; Zhao et al., 2019), and chiral capillary electrophoresis–ultraviolet detection (Wan Ibrahim et al., 2007). Most of the chiral pesticides, however, are found in the environment in trace amounts and can have detrimental effects on living organisms at these low concentrations. Thus, pre-concentration of the chiral pesticides is sometimes necessary to enable their detection with analytical instruments. Several techniques have been utilized by researchers for the pre-concentration and enrichment of chiral pesticides. These include the convectional liquid-liquid extraction (de Albuquerque et al., 2016), solid phase extraction (Gao et al., 2016), magnetic solid phase extraction (Zhao et al., 2019), solid phase micro-extraction (Dallegrave et al., 2016; Gao et al., 2016), sonication assisted extraction (Buerge et al., 2015; Corcellas et al., 2015), and QuEChERS (He et al., 2018; Tan et al., 2017).

This chapter presents recent techniques used to analyze chiral pesticides in contaminated environments. In addition, the fate and toxicity of chiral pesticides are critically discussed. The use of chiral pesticides is not free of challenges. Thus, the challenges associated with the use and analysis of chiral pesticides is tackled as well in this chapter. The information in this chapter is going to supplement and provide an update of the already existing reviews and book chapters on chiral pesticides.

3.2 COMMON TYPES OF CHIRAL PESTICIDES

3.2.1 CHIRAL ORGANOPHOSPHATES

Chiral organophosphate pesticides are among the most widely used agrochemicals for controlling pests (Gao et al., 2019). Thus, they play a pivotal role in crop protection. Chemically, chiral organophosphates are esters of phosphoric acid or thio-phosphoric acid, with one or more chiral centers (Figure 3.2A). The mechanism of their insecticidal activities involves the inhibition of an enzyme called acetylcholinesterase in insects (Zhang et al., 2017). Chiral organophosphates, however, can also be toxic to non-target organisms through the same mechanism. The inhibition occurs through the phosphorylation of the active site of acetylcholinesterase (Shakoori et al., 2017). This enzyme hydrolyzes acetylcholine in cholinergic synapses as well as in neuromuscular junctions (Mangas et al., 2016). Thus, its inhibition by chiral organophosphates causes excessive accumulation of acetylcholine in synapses, with the subsequent activation of cholinergic receptors. This results in gastrointestinal distresses, dizziness, headaches, bronchospasm, urination, sweating, lacrimation, muscle weakness, hypertension, liver and kidney damage, coma, and ultimately death (Mohammadi et al., 2015; Shakoori et al., 2017).

3.2.2 CHIRAL PYRETHROIDS

Chiral pyrethroids are synthetic, xenobiotic compounds that have insecticidal properties and are related to the natural insecticides called pyrethrins (Corcellas et al., 2015). They have widespread use as insecticides in agriculture and forestry as well as for domestic purposes (Xiang et al., 2019; Zhang et al., 2019). Chiral pyrethroids are neurotoxins with one to three chiral centers (Figure 3.2B). They are usually sold as racemic mixtures, and their enantiomers usually have stereoselective toxicity to target and non-target organisms (Pérez-Fernández et al., 2010; Xiang et al., 2019). They affect the sodium channels in the nervous system resulting in the interference with the transmission of nerve impulses. This causes the stimulation of the nervous cells. Consequently, several electric shocks are produced that cause paralysis of the insect (Jiménez-Jiménez et al., 2019). Exposure to chiral pyrethroids causes neurotoxicity, immunity toxicity, development toxicity, and endocrine disrupting toxicity to living organisms (Xiang et al., 2019).

3.2.3 CHIRAL HERBICIDES

Chiral herbicides are chemicals manufactured by the agro-industry and they are commonly used to control weeds in farms (Hsiao et al., 2014; Xie et al., 2018; Zhao et al., 2019). Most of the chiral herbicides have one or more pairs of enantiomers and are therefore chiral (Asad et al., 2017; Buerge et al., 2015; Chen et al., 2019b) (Figure 3.2C). The enantiomers of the chiral herbicides have the same physical and chemical properties in achiral environments but their interaction with enzymes and other chiral biological molecules is different (Díaz Merino et al., 2019). In chiral environments, the asymmetric herbicides show enantioselectivity in their actions (Caballo et al., 2013; Ye et al., 2015).

3.2.4 CHIRAL TRIAZOLE FUNGICIDES

Chiral triazole fungicides are a type of broad-spectrum pesticides that contain the 1,2,4-triazole moiety (Wang et al., 2019; Zeng et al., 2019). Most of them have one or two chiral centers (Zeng et al., 2019) (Figure 3.2D). They are widely used in agriculture as racemic mixtures to prevent and treat a wide variety of fungal diseases in crops. The chiral triazole fungicides inhibit ergosterol biosynthesis in the fungi and this is an integral process during the formation of the fungal cell wall (Cui et al., 2018; Tong et al., 2019; Zhang et al., 2018). Thus, this forms the basis of their antifungal activities on fruit, vegetable, legume, and grain crops. In addition, racemates of triazole fungicides are also extensively used in antifouling agents, textiles, leather, wood preservatives, adhesives, and paints (Buerge et al., 2016; Jiang et al., 2018). The chiral triazole fungicides are moderately lipophilic and highly stable since they have long chemical and photochemical half-lives (Youness et al., 2018; Zhang et al., 2018). Thus, they can be easily transported to different environmental media where they accumulate and cause detrimental effects to fauna and flora. In addition, the chiral triazole fungicides have low biodegradability, and this makes them persistent in the environment (Wang et al., 2019).

3.3 PHYSICOCHEMICAL PROPERTIES OF CHIRAL PESTICIDES

The physicochemical properties of some of the chiral pesticides are shown in Table 3.1. In achiral environments, enantiomers have the same physicochemical properties such as water solubility, vapor pressure, and dissociation constant. The behavior of chiral pesticide residue in the environment is related to physicochemical properties (Lozowicka et al., 2015). Physicochemical properties determine the environmental fate of chiral pesticides such as degradation and overall pollution potential. N-octanol/water partition coefficient (K_{ow}) is used to estimate the transport and the exposure of chiral pesticides through a different compartment, bio-concentration factor, and acute toxicity in aquatic organisms (Saranjampour et al., 2017). Solubility parameter is important for migration potential and persistence in the environment and chiral pesticides having large solubility parameter

TABLE 3.1
Physicochemical Properties of Different Chiral Pesticides

Chiral pesticide	Structure	Water solubility (mg L⁻¹)	Behaviour of enantiomers in different environments	Log K_{ow}	pKa values
Fipronil	S-(+)-Fipronil / R-(−)-Fipronil	1.9–2.4	(S)-(+)-enantiomer shows higher level of genomic DNA methylation than (R)-(−)-fipronil in zebrafish embryos	4.0	—
Metalaxyl	R-metalaxyl / S-metalaxyl	8.40	(S)-metalaxyl degrade faster at pH < 4 in anaerobic and aerobic soil than (R)-metalaxyl (Celis et al. 2015)	1.65	Much less than zero
Benalaxyl	S-(+)-benalaxyl / R-(−)-Benalaxyl	28.6	There is a significant conversion of the R-enantiomer of benalaxyl to the S-form in the plasma of mice	1.65	No dissociation

TABLE 3.1 (Continued)
Physicochemical Properties of Different Chiral Pesticides

Chiral pesticide	Structure	Water solubility (mg L^{-1})	Behaviour of enantiomers in different environments	Log K$_{ow}$	pKa values
Tebuconazole	(+)-S-tebuconazole / (−)-R-tebuconazole	36	(R)-tebuconazole was found to have greater toxicity to non-target organisms as well as greater fungal activities	3.7	2.3
Isophenphos-methyl	(S)-(+)-isofenphos-methyl / (R)-(−)-isofenphos-methyl	22.1	(R)-isophenphos-methyl degrade faster than its S-form in river water	4.12	–
Myclobutanil	R-enantiomer / S-enantiomer	142	(R)-(−)-myclobutanil may stay longer in the body and cause harm while (S)-(+)-myclobutanil may not cause any harm as they are eliminated by the liver	2.94	–

(Continued)

TABLE 3.1 (Continued)
Physicochemical Properties of Different Chiral Pesticides

Chiral pesticide	Structure	Water solubility (mg L^{-1})	Behaviour of enantiomers in different environments	Log K$_{ow}$	pKa values
Triadimefon	R-(−)-triadimefon; S-(+)-triadimefon	71.5	RR-(+)-triadimefon produced by (R)-(−)-triadimefon gives highest fungal activities while RR-(+)-triadimefon produced by (R)-(−)-triadimefon mostly bioaccumulate in lizard (Shen et al. 2017)	2.77	-
Fluazifop-butyl	R-fluazifop-butyl; S-fluazifop-butyl	1.00	(R)-fluazifop-butyl degraded faster than its R form in Beijing soil and Anhui soil (Qi et al. 2016)	4.5	-
Indoxacarb	(−)-(R)-indoxacarb; (+)-(S)-indoxacarb	0.2	(S)-(+)-indoxacarb Dissipated fast in all three tea farms than (R)-(−)-indoxacarb (Zhang et al. 2015)	4.65	-

Note: - means no data.

tend to transfer into water and subject to water flow. For example, imazethapyr, which is a chiral herbicide, was found in the groundwater of Tibagi River, Brazil, due to it having relatively high-water solubility and moderate persistence in soil (Shen et al., 2019). Pyrethroids, on the other hand, have low water solubility and high octanol/water partition coefficient. Consequently, pyrethroids can partition into the organic compartment of sediment and can travel to surface water and sediments (Ulrich et al., 2017). Even though enantiomers have the same physicochemical properties in achiral environment, in chiral environments they are enantioselective and result in having different toxicity as well as degrading differently, as summarized in Table 3.1.

3.4 ENVIRONMENTAL FATE OF CHIRAL PESTICIDES

3.4.1 Enantioselective Biotransformation of Chiral Pesticides

Chiral pesticides can undergo chemical modifications due to the various metabolic processes that occur in living organisms (Asad et al., 2017; Buerge et al., 2016; de Albuquerque et al., 2017; Wang et al., 2020; Zhang et al., 2019). This phenomenon is called biotransformation, and it usually results in the formation of various metabolites in organisms that may have greater persistence and toxicity in the environment (Gao et al., 2019). Preferential metabolism of the enantiomers of chiral pesticides have been observed in living organisms by many researchers (Shen et al., 2017; Xie et al., 2019; Zhan et al., 2018). Thus, stereoselective metabolism of chiral pesticide enantiomers in different living organisms is a common occurrence (Wang et al., 2020).

Enantioselective biotransformation of chiral fungicides is common in animals (Xie et al., 2019; Zhang et al., 2019). For instance, Shen et al. (2017) studied metabolism of triadimefon and its chiral metabolite triadimenol in lizards. Their results indicated that triadimefon could undergo metabolic transformation in lizards resulting in the formation of triadimenol. The biotransformation of triadimefon to triadimenol in lizards involved the reduction of its carbonyl group to a hydroxyl group. Enantioselectivity was observed in the results and there was preferential metabolism of the S-enantiomer of triadimefon. In the same vein, Wang et al. (2020) studied the metabolism of tridiamenol in *Eremias argus* lizards. Tissue distribution studies revealed the occurrence of stereoselective metabolism of tridimenol. Their results showed that triadimenol had a greater fungicidal activity than triadimefon as well as greater toxicity on the environment. On the other hand, Xie et al. (2019) investigated the biotransformation of prothiocanazole fungicide in Chinese lizards. Their results showed that prothioconazol was bio-transformed to a chiral metabolite called prothioconazole-desthio in Chinese lizards. Prothioconazole-desthio was found to be more persistent and toxic to living organisms as compared to prothioconazole. In another study, Gao et al. (2019) investigated the biotransformation of ethiprole in algae. Figure 3.3 shows their proposed metabolic pathway of ethiprole in algae and water. Their results showed enantioselective biotransformation as (R)-ethiprole was preferentially metabolized in algae as compared to S-ethiprole. The results also indicated that most of the metabolites were more toxic than ethiprole.

Zhan et al. (2018) studied the biotransformation of an organochlorine pesticide called acetofenate in rabbits. The results revealed that acetofenate was rapidly bio-transformed into its chiral metabolite, acetofenate-alcohol, in rabbits. Enantioselective biotransformation was observed since there was preferential metabolism of (-)-acetofenate to (-)-acetofenate-alcohol. Wang et al. (2017), on the other hand, investigated enantioselective biotransformation of benalaxyl fungicide in mice. Biotransformation of benalaxyl in tissues and plasma of mice exhibited enantioselectivity with the enrichment of the (S)-(-)-enantiomer. In another study, Gao et al. (2020) studied the metabolic transformation of flufiprole pesticide in *Chlorella pyrenoidosa*. The enantiomers of flufiprole were mainly transformed to flufiprole amide and detrifluoromethylsulfinyl flufiprole in culture medium but various metabolites, which included amide derivatives and fipronil, were formed in *Chlorella pyrenoidosa*. The toxicity of fipronil, from the results obtained, was the highest and this metabolite poses serious problems to the environment.

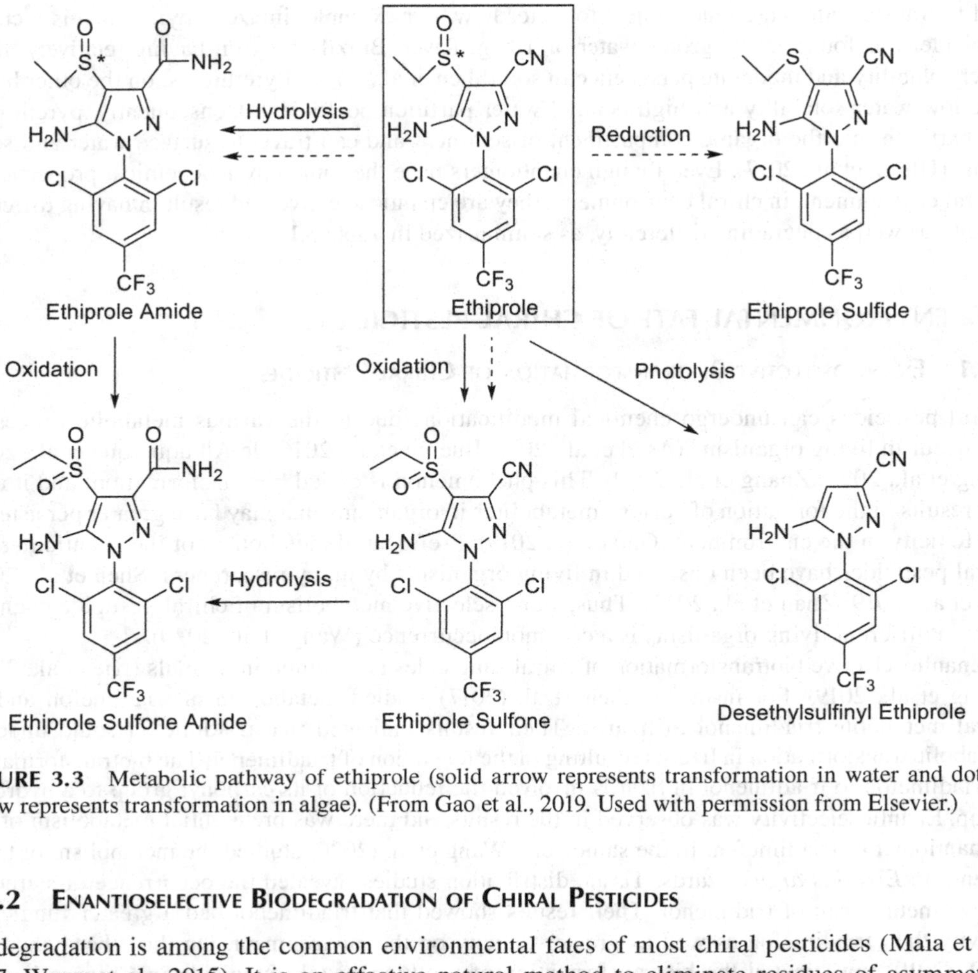

FIGURE 3.3 Metabolic pathway of ethiprole (solid arrow represents transformation in water and dotted arrow represents transformation in algae). (From Gao et al., 2019. Used with permission from Elsevier.)

3.4.2 Enantioselective Biodegradation of Chiral Pesticides

Biodegradation is among the common environmental fates of most chiral pesticides (Maia et al., 2017; Wang et al., 2015). It is an effective natural method to eliminate residues of asymmetric pesticides from contaminated environments. It involves the breakdown of the chiral pesticide residues into forms that are compatible with the environment (Gao et al., 2020; Talab et al., 2018). Enantioselectivity is a common occurrence during biodegradation of most of the enantiomers of chiral pesticides (Birolli et al., 2019; Nillos et al., 2009; Wang et al., 2015).

Cyhalothrin is a chiral pyrethroid that consists of four enantiomers, which are (1R,3R,αR) (1S,3S,αS), (1R,3R,αS) and (1S,3S,αR). The commercial formulation of cyhalothrin is called (±)-lambda-cyhalothrin. The enantiomer of cyhalothrin that has the highest insecticidal toxicity is the (1R,3R,αS)-isomer. Birolli et al. (2019) investigated the biodegradation of (±)-lambda-cyhalothrin with a consortium of bacteria. The biodegradation pathway they proposed is shown in Figure 3.4A. Their results showed that the (1R,3R,αS)-enantiomer, the most toxic one, was preferentially biodegraded by bacteria. This preferential biodegradation of (1R,3R,αS)-enantiomer in the environment reduces the risk of (±)-lambda-cyhalothrin to non-target organisms.

Chiral pesticides make their way into the soil through water leaching, field application and air deposition (Liu et al., 2016). Soil and sediments are the major compartments where chiral pesticides are usually deposited, and they are more likely to settle in these compartments affecting the overall dispersion of pesticides to the environment (Vryzas et al., 2018). Due to their interaction with matter, some chiral pesticides end up in contact with microorganisms in the soil resulting in their biodegradation. During the biodegradation of chiral pesticides, there is preferential interaction between different enzymes from the microorganisms with each enantiomer (Asad et al., 2017).

FIGURE 3.4 Biodegradation pathway for (A) lambda-cyhalothrin (Birolli et al., 2019 and (B) tebuconazole (Youness et al., 2018).

Thus, the biodegradation of chiral pesticides in the soil exhibit stereoselectivity. For instance, the enantiomers of metalaxyl degrade differently in soil with the S-enantiomer degrading rapidly at pH < 4 in anaerobic and aerobic soil as compared to the R-enantiomer. On the other hand, the R-enantiomer degrades faster than the S-form at pH > 5 in aerobic soils (Celis et al., 2015). In the same vein, Zhang et al. (2018) studied the biodegradation of the enantiomers of ethiprole pesticide in soil. The results also showed the enantioselective degradation of this chiral pesticide.

Chiral pesticides may also reach various water bodies through different mechanisms when applied in agriculture. They often bypass wastewater treatment and end up in natural aquatic systems. The reuse of wastewater saves water, but the nature of wastewater can have an impact on the discharge of pesticides to the environment. When wastewater is discharged into the environment, chiral pesticides may find their way into the environment resulting in enantioselectivity by chiral pesticides in the environment. For example, investigation on antifungal occurrence in untreated wastewater found imidazole was racemic to weakly non-racemic with weak enantioselectivity during treatment. Also, their enantiomeric composition and fractions of dissolved and suspended antifungals were different, suggesting a different behavior (Huang et al., 2013). The by-product of wastewater is sewage which when applied to the environment might influence the enantioselectivity and persistence of benalaxyl and can also increase the degradation of benalaxyl (Jing et al., 2017).

Some pesticides show little change in concentration of influents and effluents as they do not degrade via conventional wastewater treatment. For example, the influents and effluents concentration of mecoprop in wastewater were similar, ranging between 0.03 and 2 mg L^{-1}, which suggested that mecoprop was not degraded during wastewater treatment (Escolá Casas et al., 2017). Also, the enantiomeric fraction (EF) which quantifies stereoisomeric composition indicates enantioselectivity when it changes from influents and effluent. The EF values of miconazole in Table 3.2 indicate slight enantioselectivity as the EF value on the influent was 0.497–0.505 and 0.468 in the effluent (Hung et al., 2012). The influent and effluent show a small change suggesting that the biological action in the sewage system did not change the enantiomeric composition of the pesticide (Zhao et al., 2018).

The biodegradation of pesticides in wastewater depends on factors such as pH, dissolved organic matter, and electrolyte composition (Masbou et al., 2018). Some researchers have studied the biodegradation of chiral pesticides in water. Youness et al. (2018) investigated the biodegradation of the fungicide tebuconazole by *Baccillus* sp. bacteria in aqueous solutions. Their proposed biodegradation pathway of tebuconazole is shown in Figure 3.4B Their results exhibited stereoselectivity as there was preferential degradation of the S-enantiomers as compared to the R-enantiomer. This implied that there would be elevated concentrations of the R-enantiomer in the environment. This would pose serious environmental issues since the R-enantiomer of tebuconazole was found to have greater toxicity to non-target organisms as well as greater fungicidal activity.

3.4.3 Enantioselective Bioaccumulation of Chiral Pesticides

Bioaccumulation is the gradual increase of a chemical's concentration in an organism, and some chiral pesticides exhibit this phenomenon (Cui et al., 2018; Gao et al., 2014; Wang et al., 2018; Yin et al., 2017). During bioaccumulation, the rate of absorption of the chiral pesticides would be faster than its removal rate in the organism through processes such as catabolism and excretion. Chiral pesticides that are lipophilic and have longer half-lives have exhibited greater tendency of bioaccumulation (Cui et al., 2018; Zhou et al., 2018). Thus, chiral pesticides continue to accumulate in the food chain until they reach toxic levels, surpassing their maximum allowable limits (MALs). The concentrations of chiral pesticides are strictly controlled by many regulatory bodies in the world that set the MALs for toxic substances in the environment.

The potential negative long-term health impacts of chiral pesticides to humans is mostly unknown, which led to regulatory organizations stipulating different amounts of residue that a person should be allowed to take. The presence of pesticides in different compartments is regulated by national and international rules that establish the MAL concentration of pesticides (Table 3.3).

TABLE 3.2
Concentration of Different Chiral Pesticides in Influents and Effluents of Different Wastewater Treatment Plants

Name of Chiral Pesticide(s)	Location of the WWTP	Influent (ng L⁻¹)	Effluent (ng L⁻¹)	Removal Efficiency (%)	EF Values for Influent	EF Values of Effluent	Reference
Enantiomers of epoxiconazole, hexaconazole, myclobutanil, metalaxyl and napropamide	Sheyang	11.9–48.9	3.5–18.2	62–77	0.49–0.51[a]	No significant alteration	Zhao et al., 2018
Fipronil	North Carolina	131 ± 30	119 ± 70	-	-	-	Mcmahen et al., 2016
Fipronil	Southwestern U.S.	12–31	28 ± 6	44 ± 4	-	-	Supowit et al., 2016
Benalaxyl	Beijing, China	-	-	-	-	The EF value of (S)-benalaxyl was always Less than 0.5 indicating great abundance of S-enantiomer residues	Jing et al., 2017
Ketoconazole	Guangzhou	7–45	-	-	0.484–0.485	0.477	Huang et al., 2013
Miconazole	Guangzhou	0.5–9.0	-	-	0.497–0.527	0.468	Huang et al., 2013
Miconazole	Guangzhou	3.7–6.3	6.6–7.0	-	0.504–0.497	0.468	Huang et al., 2012
Ketoconazole	Guangzhou	42.7–45.5	0.54–0.61	-	0.484	0.485	Huang et al., 2012

Note: [a]The EF observed was of four pesticides; metalaxyl was not included. The four pesticides are still produced as racemates; - means no data.

TABLE 3.3
Maximum Allowable Limits of Different Groups of Chiral Pesticides by Several International Standards Regulating Bodies in Different Environmental Compartments

Regulatory Organization	Pesticide(s)	MAL	Environmental Compartment	Reference
EFSA	Benalaxyl	0.60 mg kg^{-1} 0.10 mg kg^{-1} 0.02 mg kg^{-1}	Table grapes Wine grapes Potatoes	European Food Safety Authority, 2013
Codex	Metalaxyl	0.50 mg kg^{-1} in 7 days	Tomato	Malhat, 2017
DOH	Myclobutanil	0.05–0.5 mg kg^{-1}	Cucurbits, dry beans, and pears	Quinn et al., 2011
Codex	Pebnconazole	0.1 mg kg^{-1}	peach, plum, plum apricot and mango	Abd-Alrahman and Ahmed (2012)
EU	Indoxacarb	0.1 mg kg^{-1}	brown rice	Shi et al., 2018
EU and Japan	Fipronil	5 and 2 μg kg^{-1}	Food and agricultural products	Chen et al., 2018
US	Metalaxyl and its metabolite	6.0 mg L^{-1} 20.0 mg L^{-1} 0.5 mg L^{-1}	Forge Hay Almond	Pohanish, 2015
EFSA	Metalaxyl and metalaxyl-M	Daily intake of 0.08 mg kg^{-1} body weight per day and an acute reference dose of 0.5 mg kg^{-1} bw	Human body	European Food Safety Authority, 2016
DOH	Penconazole	0.02–0.2 ng L^{-1}	Apples, cucurbits, grapes, pears, and peas	Quinn et al., 2011

Note: WHO = World Health Organization, EFSA = European Food Safety, DOH = Department of Health, EU = European Union.

These bodies include WHO, FAO, EU, and USEPA. The European Union stipulates a limit of 0.1 μg L^{-1} for individual pesticides in water that are allowed for human consumption. Chiral pesticides are assessed as racemates even though they have different toxicity in the environment. For example, the R-enantiomer of metalaxyl is sold as a single enantiomer but it is regulated by EFSA as a combination of R and S in the daily intake in a human body (European Food Safety Authority, 2016). Low concentrations of chiral pesticides are considered safe for human consumption in food and water (Table 3.3). Despite all these efforts to minimize exposure of humans to high amounts of chiral pesticides in food and water, their bioaccumulation in the food chains may cause an increase in their concentrations beyond MAL values.

Enantioselectivity have been observed during bioaccumulation of chiral pesticides in living organisms (Cui et al., 2018; Yin et al., 2017; Zhou et al., 2018). Many researchers have extensively studied the process of enantioselective bioaccumulation of chiral pesticides (Cui et al., 2018; Yin et al., 2017; Wang et al., 2019; Zhou et al., 2018). For instance, Wang et al. (2019) studied the bioaccumulation of triticonazole and prothioconazole in earthworms. The results showed that the bioaccumulation of triticonazole and prothioconazole in earthworms was enantioselective. The *R*-triticonazole and *S*-prothioconazole exhibited preferential bioaccumulation in the earthworms. In a similar study, Cui et al. (2018) investigated the bioaccumulation of tebuconazole in earthworms. Tebuconazole is a chiral fungicide that is a sterol demethylation inhibitor and is widely used to control pests in crops such as cereals, fruits, and vegetables. Cui et al. (2018), during their bioaccumulation study, exposed earthworms to rac-tebuconazole-contaminated soil and analyzed the concentrations of the enantiomers periodically. Their results showed that the enantiomers of tebuconazole were enantioselectively bio-accumulated in earthworms. The S-(+) enantiomer had priority as compared to the R-form during bioaccumulation of tebuconazole in earthworms. On the other hand, Yin et al. (2017) studied the bioaccumulation of furalaxyl pesticide in *Tenebrio molitor* larvae. The larvae were exposed to rac-furalaxyl and the concentration of the enantiomers were measured regularly using LC-MS/MS. It was observed that the bioaccumulation capacity of the two enantiomers of furalaxyl were generally very low. The results, however, showed a preferential bioaccumulation of (*S*)-furalaxyl as compared to (*R*)-furalaxyl. Thus, the bioaccumulation of furalaxyl also showed enantioselectivity.

3.4.4 ENANTIOMERIZATION

The enantiomers of chiral pesticides are not equally active towards the target organisms. In most cases, as already alluded to, one enantiomer has more activity and toxicity towards the target pest as compared to others. Thus, environmental risk of chiral pesticides can be reduced by using them in their enantiopure forms (Shi et al., 2018; Xie et al., 2018). This involves the manufacture of enantiopure formulations of the chiral pesticides so that the active enantiomer is the one applied to crops. In some cases, however, some asymmetric pesticides may be configurationally unstable under certain conditions and may undergo stereochemical inversion (Talab et al., 2018; Yin et al., 2017). This phenomenon is called enantiomerization and is one of the fates of many chiral pesticides in the environment.

Many researchers have investigated the enantiomerization of chiral pesticides under different environmental conditions (Gao et al., 2014; Talab et al., 2018; Wang et al., 2017; Yin et al., 2017). For instance, Talab et al. (2018) investigated the effect of soil pH on the enantiomerization of triadimefon pesticides using alkaline soil from Beijing, acidic soil from Wuhan, and neutral soil from Changchun, China. The results revealed significant interconversion of the enantiomers of triadimefon at different pH conditions. The alkaline soil from Beijing exhibited faster interconversion of the enantiomers of triadimefon than neutral soil from Changchun, while acidic soil from Wuhan showed no interconversion. In a similar study, Yin et al. (2017) investigated enantiomerization of furalaxyl pesticide in *Tenebrio molitor* larvae. Furalaxyl is a chiral fungicide that is manufactured in the agro-industry as a racemic mixture. Yin et al. (2017) results revealed that furalaxyl undergoes significant enantiomerization in *Tenebrio molitor* larvae under S- or R-exposure. In the same vein, Gao et al. (2014) studied the enantiomerization of a chiral fungicide called metalaxyl in *Tenebrio molitor* larvae. The larvae were exposed, in turn, to enantiopure (*R*)-metalaxyl and (*S*)-metalaxyl. It was observed that there was interconversion between the R- and S-enantiomers. Gao et al. (2014) attributed the enantiomerization to the presence of an enzyme in *Tenebrio molitor* larvae that catalyzed the interconversion.

Enantiomerization may have negative effects to the environment. This usually occurs when the active enantiomer is converted to the inactive form that has greater toxicity and persistence in the environment. For instance, Wang et al. (2017) studied the enantiomerization of benalaxyl in mice. Benalaxyl is a systemic chiral fungicide that is widely used on crops such as soybeans, grapes, tomatoes, potatoes, and onions to control diseases caused by *Oomycetes*. It has one asymmetric center and it, therefore, consists of two enantiomers. The R-enantiomer has been reported as the one with fungicidal activity. The results of Wang et al. (2017) revealed that there was a significant conversion of the R-enantiomer of benalaxyl to the S-form in the plasma of mice. Thus, enantiomerization, in this case, favored the formation of the inactive S-enantiomer. Consequently, this caused the excretion of large quantities of the S-enantiomer into the environment, putting non-target organisms at risk.

3.5 ENANTIOSELECTIVE ACTION OF CHIRAL PESTICIDES ON TARGET ORGANISMS

Stereoselectivity of chiral pesticides has received great attention from researchers in recent years (Duan et al., 2018; Li et al., 2018; Tong et al., 2019; Zhao et al., 2019). This was necessitated by the realization that the enantiomers of chiral pesticides exhibited significant differences in their bioactivity on target organisms (Cui et al., 2018; Qi et al., 2015). Thus, a lot of research efforts

TABLE 3.4
Enantioselective Bioactivity of the Enantiomers of Chiral Pesticides on Target Organisms

Class of Pesticide	Chiral Pesticide Studied	Target Organism Studied	Bioactivity on Target Organism	Active Enantiomer	Reference
Triazole fungicide	Prothioconazole	Various types of fungi	Inhibition of biosynthesis of ergosterol and deoxynivalenol	(R)-prothioconazole is more active than (S)-prothioconazole	Zhang et al., 2019
Triazole fungicide	Tetraconazole	Rhizoctonia cerealis and Fusarium graminearum	Inhibition of biosynthesis of ergosterol	(R)-(+)-tetraconazole had greater fungicidal activity than (S)-(−)-tetraconazole	Tong et al., 2019
Herbicide	Metolachlor	Echinochloa crusgali	Inhibition of gibberellic acid biosynthesis	(S)-metolachlor more active than the R-isomer	Zhao et al., 2019
Triazole fungicide	Tebuconazole	Botrytis cinerea	Inhibition of growth of Botrytis cinerea	(R)-tebuconazole more active than (S)-tebuconazole	Cui et al., 2018
Herbicide	Beflubutamid	Dicotyledonous weeds	Inhibition of carotenoid biosynthesis	(−)-beflubutamid more active than (+)-beflubutamid	Buerge et al., 2013
Triazole fungicide	Myclobutanil	Fusarium verticillioides	Inhibition of growth of Fusarium verticillioides	(R)-myclobutanil more active than (S)-myclobutanil	Li et al., 2018
Herbicide	Lactofen	Echinochloa crusgali	Inhibition of growth of Echinochloa crusgali	(R)-lactofen was more active than (S)-lactofen	Xie et al., 2018
Fungicide	Epoxiconazole	Chlorella vulgaris	Inhibition of growth of Chlorella vulgaris	(S,R)-epoxiconazole was more active than (R,S)-epoxiconazole	Kaziem et al., 2020
Herbicide	Imazapyr	Aradidopsis thalina	Inhibition of chlorophyll synthesis	(+)-imazapyr was more active than (−)-imazapyr	Hsiao et al., 2014
Herbicide	Napropamide	Poa annua	Inhibition of growth of Poa annua	(−)-napropamide was more active than (+)-napropamide	Qi et al., 2015
Triazole fungicide	Flutriafol	Rhizoctonia solani, Alternaria solani, Pyricularia grisea, Gibberella zeae and Botrytis cinerea	Inhibition of growth of fungi	(R)-flutriafol was more active than (S)-flutriafol	Zhang et al., 2015

were directed towards the establishment of the enantioselective activity of chiral pesticides on target organisms. Many researchers have found out that the enantiomers of chiral pesticides do not have same activity on the target organisms (Hsiao et al., 2014; Tong et al., 2019; Xie et al., 2018; Zhang et al., 2019) (Table 3.4). One enantiomer is usually more active than the others on the target organism.

Many researchers have studied the bioactivity of chiral fungicides in recent years (Cui et al., 2018; Kaziem et al., 2020; Li et al., 2018; Zhang et al., 2019) (Table 3.4). Chiral fungicides are chemical compounds that are used to kill parasitic fungi as well as their spores. They cause inhibition of the growth of the fungal cells. Research has shown that the bioactivity of chiral fungicides is enantioselective. For instance, Zhang et al. (2019) investigated the bioactivity of prothioconazole on fungi and they found that its action was stereoselective. The enantiomers had different bioactivities. (R)-prothioconazole was more potent on fungi as compared to S-prothioconazole. Tong et al. (2019) studied the bioactivity of tetraconazole on *Rhizoctonia cerealis* and *Fusarium graminearum*. Their results also established the enantioselective nature of the bioactivity of tetraconazole. The R-enantiomer had greater fungicidal activity than the S-enantiomer. In a similar study, Kaziem et al. (2020) investigated the activity of epoxiconazole on *Chlorella vulgaris*. Their results showed that (S,R)-epoxiconazole caused greater inhibition of the growth of *Chlorella vulgaris* as compared to (R,S)-epoxiconazole. Thus, chiral fungicides exhibit enantioselective bioactivity on their target organisms.

Some researchers studied the bioactivity of chiral herbicides (Hsiao et al., 2014; Qi et al., 2015; Xie et al., 2018; Zhao et al., 2019) (Table 3.4). Chiral herbicides are chemical compounds that are used to kill undesirable plants or weeds. Research work has established stereoselectivity in the bio-action of most of the chiral herbicides. For instance, Zhao et al. (2019) studied the bioactivity of the enantiomers of metolachlor on *Echinochloa crusgali*. Metolachlor retard growth of *Echinochloa crusgali* through inhibition of gibberellic acid biosynthesis. Their results showed that S-metolachlor had greater herbicidal activity on *Echinochloa crusgali* than (R)-metolachlor. In a similar study, Xie et al. (2018) investigated the activity of lactofen also on *Echinochloa crusgali* weed and enantioselective bioactivity was observed. They found out that (R)-lactofen was more potent on *Echinochloa crusgali* than (S)-lactofen. Thus, enantioselective bioactivity is a general occurrence when chiral herbicides are used in the agro-industry.

3.6 ENANTIOSELECTIVE TOXICITY OF CHIRAL PESTICIDES ON NON-TARGET ORGANISMS

Enantiomers of chiral pesticides are known to exhibit selective interaction with biological systems (Tong et al., 2019). Usually one of the enantiomers, as already alluded to, shows pronounced bioactivity to the target species in a chiral environment (Birolli et al., 2019). The other enantiomer will be inactive or less active to the target species and, therefore, is discharged into the environment in relatively large quantities where it may cause adverse effects to non-target fauna and flora. The toxicity of chiral pesticides to non-target organisms is, therefore, usually enantioselective (Birolli et al., 2019; Buerge et al., 2019; Duan et al., 2018; Fonseca et al., 2019; Kuhlmann et al., 2019; Qu et al., 2016; Tong et al., 2019; Xiang et al., 2019) (Table 3.5).

Toxicity of phenylpyrazole pesticides (fipronil, flufiprole, and ethiprol) on non-target organisms have been extensively studied by many researchers (Gao et al., 2020; Qian et al., 2019; Qu et al., 2016; S. Wang et al., 2019; Xu et al., 2019; Zhan et al., 2018) (Table 3.5). Their results showed that these pesticides have enantioselective toxicity to non-target organisms, in most cases. For instance, Qian et al. (2019) studied the toxicity of fipronil on zebrafish, a non-target organism and established that it was enantioselective. They reported that R-fipronil caused more severe anxiety-like behavior, such as increased swimming speed and dysregulated photoperiodic locomotion as well as increased neurotoxicity to embryonic and larval zebrafish as compared with (S)-fipronil. In a similar study, Qin et al. (2015) investigated the toxicity of fipronil on the earthworm, *Eisenia foetida*, and the results

TABLE 3.5
Enantioselective Toxicity of the Enantiomers of Chiral Pesticides on Non-Target Organisms

Class of Chiral Pesticide	Chiral Pesticide Studied	Non-Target Organism	Toxicity on Non-Target Organism	Toxic Enantiomer	Reference
Pyrethroid	Cis-bifenthrin	*Xenopus laevis*	Adverse effect on behavior and development of *Xenopus laevis*	(R)-cis-bifenthrin is more toxic than (S)-cis-bifenthrin	Zhang et al., 2019
Herbicide	Acetochlor	Zebrafish embryo-larvae	Disruption of thyroid gland activities	(S)-acetochlor is more potent than (R)-acetochlor	Xu et al., 2019
Pyrethroid	Cis-bifenthrin	Humans	Triggering development of obesity in humans	(S)-cis-bifenthrin triggers rapid lipid metabolism in human hepatoma cells than the R-form	Xiang et al., 2018
Herbicide	Glufosinate	Maize	Reduction in growth rate, shoot height and shoot weight	(S)-glufosinate is more toxic than (R)-glufosinate	Zhang et al., 2019
Pyrethroid	Cis-bifenthrin	Zebrafish	Damage of the reproductive system of zebrafish	(S)-cis-bifenthrin caused more severe damage to the reproductive system of zebrafish than (R)-cis-bifenthrin	Zhang et al., 2019
Herbicide	Dichlorprop	*Arabidopsis thaliana*	Disturbance of fatty acid synthesis	(R)-dichlorprop more toxic than (S)-dichlorprop	Chen et al., 2019a
Pyrethroid	Alpha-cypermethrin	Earthworm *Eisenia fetida*	Increased mortality rate of the earthworm	($1R$-cis-αS)-enantiomer more toxic	Diao et al., 2011
Herbicide	Lactofen	Zebrafish *Danio rerio*	Cause genetic defects in zebrafish *Danio rerio*	(S)-lactofen more potent than (R)-lactofen	Wang et al., 2018
Fungicide	Epoxiconazole	*Daphnia magna*	Increased death of *Daphnia magna*	(R,S)-epoxiconazole more toxic than S,R-epoxiconazole	Kaziem et al., 2020
Herbicide	Carfentrazone-ethyl	Maize	Inhibition of radicle growth	(S)-carfentrazone-ethyl more toxic than (R)-carfentrazone-ethyl	Duan et al., 2018
Fungicide	Imazalil	*Daphnia magna*	Increased death of *Daphnia magna*	(S)-imazalil was 2.25-fold more toxic than (R)-imazalil	Li et al., 2019
Herbicide	Lactofen	*Microcystis aeruginosa*	Inhibition of algal growth in aquatic environments	(S)-lactofen more potent than (R)-lactofen	Xie et al., 2018
Herbicide	Fenoxaprop-ethyl	Tadpoles	Increased mortality of tadpoles	(S)-fenoxaprop-ethyl more potent than (R)-fenoxaprop-ethyl	Jing et al., 2017
Fungicide	Prothioconazole	*Lemna minor* and *Chlorella pyrenoidosa*	Increased mortality of *Lemna minor* and *Chlorella pyrenoidosa*	(+)-prothioconazole more toxic than (−)-prothioconazole	Zhai et al., 2019
Fungicide	Prothioconazole	*Daphnia magna*	Increased mortality of *Daphnia magna*	(−)-prothioconazole more toxic than (+)-prothioconazole	Zhai et al., 2019

showed that the R-enantiomer was more toxic to these non-target organisms than the S-form. Qu et al. (2016) studied the enantioselective toxicity of fipronil to non-target organisms *Lemna minor* and *Anodonta woodiana* found in aquatic ecosystems. Their findings indicated that (*S*)-fipronil was more toxic than R-fipronil against *Anodonta woodiana*, while (*R*)-fipronil was more toxic for *Lemna minor*. Gao et al. (2020) investigated the toxicity of flufiprole to *Chlorella pyrenoidosa*. The results showed that flufiprole exhibited enantioselective toxicity towards *Chlorella pyrenoidosa*. (*R*)-flufiprole exhibited greater toxicity to *Chlorella pyrenoidosa* than the (*S*)-enantiomer. In a similar study, Gao et al. (2019) studied the enantioselective toxicity of the enantiomers of ethiprole to *Chlorella pyrenoidosa*. The results indicated that (*R*)-ethiprole was more toxic to non-target *Chlorella pyrenoidosa* than (*S*)-ethiprole.

Pyrethroid pesticides are among the most used agrochemicals for pest control (Brander et al., 2016). Most of them are chiral and highly lipophilic and have a high tendency to bio-accumulate in living organisms. These chiral pesticides have endocrine disrupting effects in organisms, and they exhibit enantioselective toxicity to non-target organisms. Some researchers have studied the enantioselective toxicity of chiral pyrethroids on non-target organisms. For instance, Xiang et al. (2019) investigated the toxicity of the enantiomers of *cis*-bifenthrin on the reproduction of non-target zebrafish. The results indicated that (*S*)-*cis*-bifenthrin caused more severe reproductive endocrine disturbance in zebrafish than (*R*)-*cis*-bifenthrin. In a similar study, Zhang et al. (2019) studied the enantioselective toxicity of *cis*-bifenthrin on non-target *Xenopus laevis*. Their findings indicated that R-cis-bifenthrin caused serious negative effects on the behavior and development of *Xenopus laevis* than S-cis-bifenthrin. Thus, cis-bifenthrin has enantioselective toxicity on *Xenopus laevis*. In the same vein, Xiang et al. (2018) investigated enantioselective effects of cis-bifenthrin on lipogenesis in human hepatoma cells. Their results indicated that (*S*)-cis-bifenthrin triggered rapid lipid metabolism in human hepatoma cells causing high prevalence of obesity. On the other hand, Diao et al. (2011) investigated the toxicity of the enantiomers of alpha-cypermethrin on non-target earthworm *Eisenia fetida*. They found out that (*1R-cis-αS*)-(+)-alpha-cypermethrin was 30 times more toxic than the (*1S-cis-αR*)-(−)-isomer as it caused increased mortality of the earthworm.

Many researchers have shown tremendous interest in the enantioselective toxicity of chiral herbicides (Chen et al., 2019a; Duan et al., 2018; Wang et al., 2018; Xu et al., 2019; Zhang et al., 2019) (Table 3.5). The general observation made by researchers was that the toxicity of most chiral herbicides on non-target organisms was enantioselective in nature. For instance, Xie et al. (2018) investigated the toxicity of the enantiomers of lactofen on non-target *Microcysts aeruginosa*. They found out that (*S*)-lactofen was more potent towards *Microcysts aeruginosa* than (*R*)-lactofen. In a similar study, Wang et al. (2018) studied enantioselective toxicity of lactofen on non-target zebrafish *Danio rerio*. They obtained similar results where the (*S*)-lactofen was more toxic towards zebrafish, *Danio rerio* than R-form. On the other hand, Jing et al. (2017) performed a research on the toxicity of the enantiomers of fenoxaprop-ethyl herbicide on non-target tadpoles. They reported that (*S*)-fenoxaprop-ethyl was more toxic to the tadpoles than (*R*)-fenoxaprop-ethyl. The investigation carried out by Zhang et al. (2019) on the toxicity of the enantiomers of glufosinate herbicide on non-target maize crop also showed enantioselectivity. They reported that (*S*)-glufosinate caused severe reduction in shoot growth of maize as compared to (*R*)-glufosinate.

Another class of chiral pesticides that is commonly used for both domestic and industrial purposes is the fungicide. Most fungicides have stereogenic centers and exhibit enantioselective toxicity to non-target organisms. For instance, Kaziem et al. (2020) investigated the toxicity of the enantiomers of epoxiconazole on non-target *Daphnia magna*. The results indicated that R,S-epoxiconazole was more potent than S,R-epoxiconazole. Thus, epoxiconazole had enantioselective toxicity towards *Daphnia magna*. Zhai et al. (2019) investigated the enantioselective toxicity of prothioconazole on non-target *Lemna minor*, *Chlorella pyrenoidosa*, and *Daphnia magna*. They reported that the toxicity of (+)-prothioconazole was higher than that of (−)-prothioconazole on *Lemna minor* and *Chlorella pyrenoidosa* while (−)-prothioconazole was more potent on *Daphnia magna* than (+)-prothioconazole.

3.7 PRE-CONCENTRATION OF CHIRAL PESTICIDES

Chiral pesticides can have detrimental effects on the environment even at trace concentrations. Thus, effective pre-concentration techniques are required to enable their monitoring in the environment at these trace levels. There are so many pre-concentration techniques that have been employed by researchers during chiral analysis by chromatographic techniques. These include QuEChERS (He et al., 2018; Jiang et al., 2018; Shi et al., 2018; Zhang et al., 2018), SPE (Gao et al., 2016; Zhang et al., 2019), SPME (Dallegrave et al., 2016), MSPE (Zhao et al., 2019), and liquid extraction (de Albuquerque et al., 2016; Xiang et al., 2019; Zhang et al., 2018) (Tables 3.6–3.8).

The QuEChERS technique is widely used during the pre-concentration of chiral pesticides prior to chromatographic analysis (Tables 3.6–3.8). Its popularity can be attributed to its simplicity, environmental friendliness, and robustness during pre-concentration of chiral pesticides. SPME and MSPE, just like QuEChERS, are miniaturized techniques with very low environmental footprints when used to pre-concentrate chiral pesticides. Thus, they are also used by environmentally conscious researchers for pre-concentration of chiral pesticides in different matrices. SPE and liquid-liquid extraction, although they use large volumes of toxic organic solvents, are equally popular techniques for pre-concentrating chiral pesticides. Their popularity probably stems from the fact that most laboratories have equipment used to perform these techniques. In addition, this may be due to resistance to change the old mind-set among those researchers who have used these techniques for a long time and inability to embrace new eco-friendly methods that can be used for pre-concentration with greater efficiency.

3.8 CHIRAL SEPARATION TECHNIQUES

3.8.1 Chiral High-Performance Liquid Chromatography

High-performance liquid chromatography (HPLC) is one of the most versatile and popular analytical techniques used during chiral analysis of pesticides (Díaz Merino et al., 2019; Gao et al., 2016; Zhang et al., 2019) (Table 3.6). The popularity of HPLC hinges on the fact that it is a rapid and a non-destructive technique. It is regarded as the best technique for the preparation and separation of enantiopure chemicals such chiral pesticides. For instance, Cui et al. (2018) successfully separated the enantiomers of tebuconazole using HPLC with a Chiralpak 1C column. The chromatogram they obtained is shown in Figure 3.5A. From the chromatogram, it can be observed that they managed to achieve baseline resolution of the enantiomers of tebuconazole. Gao et al. (2016), in the same vein, separated the enantiomers of isofenphos-methyl also using HPLC with a Lux Cellulose-3 column. Their chromatogram (Figure 3.5B) also shows good baseline resolution of the enantiomers of isofenphos-methyl. Thus, HPLC is a versatile and useful technique for the separation of the enantiomers of chiral pesticides.

Enantiomers of chiral pesticides cannot be separated in achiral environments. Thus, the success of the HPLC technique during chiral separation depends on the availability of efficient chiral stationary phases for effective chiral recognition. Many researchers have used chiral stationary phases derived from cellulose during analysis of chiral pesticides (Table 3.6). For instance, Zhang et al. (2019) successfully separated the enantiomers of a triazole fungicide called penconazole using ultra performance liquid chromatographic technique with a Lux Cellulose-2 column as a chiral stationary phase. A mixture of 0.1% formic acid in methanol and 10 mmol L^{-1} ammonium acetate in water (75/25, v/v) was used as the mobile phase. Enantioselective elution of the components of the racemic mixture was observed during this chromatographic technique. The enantiomer (+)-penconazole was first eluted followed by (−)-penconazole. The method showed reliable performances in linearity, recovery, and precision. In the same vein, Liu and Ding (2019) performed a simultaneous enantiomeric separation of prothioconazole and prothioconazole-destho using Lux Cellulose-1 column as a chiral stationary phase of HPLC. A mixture of acetonitrile and water was used as mobile phase and good linearity and recoveries were obtained. Wang et al. (2020) used the same column as a

TABLE 3.6
Analysis of Chiral Pesticides Using Enantioselective High-Performance Liquid Chromatography

Matrix	Class of Pesticide(s)	Chiral Pesticide(s)	Extraction Technique	Chiral Stationary Phase Column	Detector	Reference
Panax notogiseng, lyceum, barbarum, radixastragali, chrysanthemum, rose, and lily	Multi-class pesticides	Novaluron, triapenthenol, etoxazole, bromacil, surecide, and fenarimol	MSPE	Chiralpak 1G column	ng	Zhao et al., 2019
Cucumber	Carbamate	Fenobucarb	ng	Daicel IG-3 chiral column	UPLC-MS/MS	Xia et al., 2019
Grape, tea, lotus root, lotus leaf, lotus seeds and hulls	Triazole fungicide	Penconazole	Graphite carbon black SPE	Lux cellulose-2 column	MS/MS	Zhang et al., 2019
Eremias argus lizard	Fungicide	Triadimenol	Liquid extraction	Lux cellulose-1 column	MS/MS	Wang et al., 2019
Environmental matrices	Multi-class pesticides	30 chiral pesticides	ng	Chiralpak IG-3	DAD	Díaz Merino et al., 2019
Soil and water	Pyrethroids	Bifenthrin and lambda-cyhalothrin	Liquid-liquid extraction	Lux cellulose-1 column, luk-cellulose-3 column and chiralpak columns	DAD	Zhang et al., 2018
Water	Herbicides	Lactofan	ng	Chiralcel AD-H column	PDA	Xie et al., 2018
Food and environmental matrices	Triazole fungicide	Metconazole	QuEChERS	Enantiopak-OD column	UV	He et al., 2018
Corn and corn plant	α-Aminophosphate	Dufulin	QuEChERS	Chiralpak IC column	MS/MS	Shi et al., 2018
Soils	Pyrethroids	Fenpropathrin	ng	Lux cellulose-3 column	ng	P. Zhang et al., 2017
Water	Phenylpyrazole	Fipronil, flufiprole and ethiprole	ng	Chiralpak IB column	DAD	J. Gao et al., 2017
Water	Herbicides	Acetochlor, metolachlor, propisochlor and napropamide	ng	Chiralpak AS-H, OD-H and AY-H columns	DAD	J. Xie et al., 2016
Human liver microsomes	Nematocide	Fenamiphos	Liquid-liquid extraction	Chiralpak AS-H column	ng	de Albuquerque et al., 2016
Fruits, vegetables, and soils	Organophosphates	Isofenphos-methyl	SPE	Lux cellulose-3 column	UV	Gao et al., 2016
Orange pulp, orange peel, and kumquate	Organophosphates	Isocarbiphos	QuEChERS	Chiralpak AD-3R	MS/MS	Qi et al., 2015

Note: ng = not given.

TABLE 3.7
Analysis of Chiral Pesticides Using Enantioselective Gas Chromatography

Matrix	Class of Chiral Pesticide(s)	Chiral Pesticide(s)	Extraction Technique	Chiral Stationary Phase Column	Detector	Reference
Zebra fish	Pyrethroid	Bifenthrin	Liquid-liquid extraction	HP-5 column	ECD	Xiang et al., 2019
Tomato, cucumber, rape, cabbage, and pepper	Pyrethroid	α-Cypermethrin	Liquid-liquid extraction	HP-5 column	ECD	Yao et al., 2019
Beef, chicken, eggs, fish, and milk	Pyrethroid	17 Pyrethroids	Ultrasound extraction and SPME	DB-5MS capillary column	MS/MS	Dallegrave et al., 2016
Grapevines, sugar beets, and wheat	Fungicides	Fenpropidin, fenpropimorph, and spiroxamine	QuEChERS	BGB-5 column	MS/MS	Buerge et al., 2016
Blackgrass and garden cress	Herbicides	Haloxyfop-methyl	Sonication	BGB column coated with permethyl β-cyclodextrin	MS	Buerge et al., 2015
Soil	Fungicides	Metalaxyl	Ultrasonication	BGB-172 column	MS	Kalathoor et al., 2015
Human breast milk	Pyrethroids	Bifenthrin, cyhalothrin, cyfluthrin, cypermethrin, permethrin, and tetramethrin	Sonication	BGB-172 column	MS/MS	Corcellas et al., 2015
Soil and air	Organochlorine	DDT	ng	β-Cyclodextrin-based BGB-172 column	GC × GC-TOF-MS	Naudé and Rohwer, 2012

Note: ng – not given

TABLE 3.8
Analysis of Chiral Pesticides using Enantioselective Supercritical Fluid Chromatography

Matrix	Class of Pesticide(s)	Chiral Pesticide(s)	Extraction Technique	Mobile Phase	Chiral Stationary Phase Column	Detector	Reference
Tomato and soil	Triazole fungicide	Prothioconazole	QuEChERS	CO_2/ 2-propanol	EnantioPax OD column	VCD spectroscopy	Jiang et al., 2018
Tea, apple, and grapes	Triazole fungicide	Diniconazole	Liquid extraction	CO_2/ isopropanol	Chiral CCA column	Q-TOF/MS	Zhang et al., 2018
Environmental water	Organophosphates and phenylpyrazole pesticides	Isofenphos, isophenyl-methyl, isocarbophos, flufiprole, fipronil, and ethiprole	ng	CO_2	Chiralpax AD-3	ng	Zhang et al., 2018
Fruits, vegetables, cereals, and soil	Triazole fungicide	Fenbuconazole	QuEChERS	CO_2/ ethanol	ACQUITY UPC2 Trefoil AMY 1 column	Tandem mass spectrometer	Tao et al., 2018
Tomatoes and cucumber	Triazole fungicide	Triticonazole	QuEChERS	CO_2/ ethanol	EnantioPax OD column	UV	Tan et al., 2017
Wheat, grapes, and soil	Triazole fungicide	Propiconazole	QuEChERS	CO_2/ ethanol	Chiralpak AD-3 column	Tandem mass spectrometer	Cheng et al., 2017
Aqueous solutions	Triazole fungicide	Triticonazole	Ng	CO_2/ ethanol	EnantioPax OD column	PDA	He et al., 2016
Cucumber, tomato, and soil	Fungicide	Pyrisoxazole	QuEChERS	CO_2/ methanol	Chiralpak IA column	Tandem mass spectrometer	Pan et al., 2016
Soil	Acaricide	Cyflumetofen	QuEChERS	CO_2/ methanol	Trefoil AMY 1	Tandem mass spectrometer	Liu et al., 2016
Water and zebrafish	Triazole fungicide	Tebuconzole	QuEChERS	CO_2/ methanol	Chiralpak IA-3	Tandem mass spectrometer	Liu et al., 2015

Note: ng = not given; Q-TOF/MS = quadrupole time-of-flight mass spectrometry; VCD = vibrational circular dichroism.

stationary phase during the separation of the enantiomers of triadimenol from *Eremias argus* lizard tissues using LC-MS/MS. Tissue distribution experiments showed enantioselective biotransformation of the triadimenol in the lizards.

Other researchers used amylose-based chiral stationary phases during chiral separation using HPLC (Table 3.6). For instance, Díaz Merino et al. (2019) used silica-based column with immobilized amylose tris(3-chloro-5-methylphenycrbamates) (Chiralpak IG-3) as a stationary phase during chiral separation of thirty multi-class pesticides. Good resolution of the enantiomers of these pesticides was obtained. Qi et al. (2015) compared enantioseparation selectivity of isocarbophos enantiomers using four cellulose-based chiral columns and two amylose-based columns. The results showed that the fastest baseline separation of isocarbophos was obtained when amylose tris(3,5-dimethylphenylcarbamate) (Chiralpak AD-3R) was used as the stationary phase. In the same vein, Chai et al. (2013)

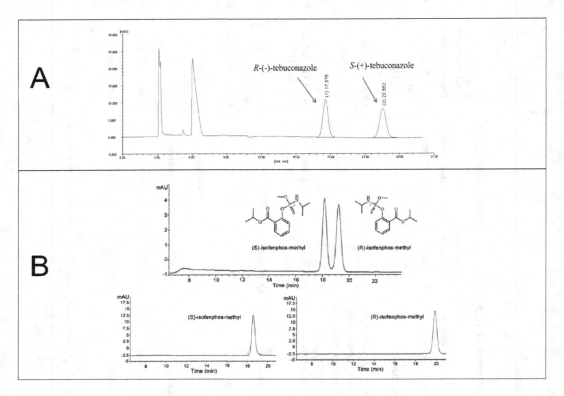

FIGURE 3.5 Chromatogram for the separation of (A) tebuconazole enantiomers by HPLC-DAD (Cui et al., 2018, used with permission from Elsevier); (B) isofenphos enantiomers by HPLC-UV (Gao et al., 2016); (C) enantiomers of fluazifop-methyl, fluazifop-ethyl, haloxyfop-methyl, and haloxyfop-ethyl by GC-MS (Buerge et al., 2015); (D) enantiomers of bromocyclen by GC-ECD (from Fidalgo-Used et al., 2008, used with permission of Elsevier); and (E) enantiomers of isofenphos by SFC-MS/MS (from Chen et al., 2016, used with permission of Elsevier). (Figure (A) was reprinted by permission from Springer Nature GmbH, Environmental Science and Pollution Research, Chiral triazole fungicide tebuconazole: Enantioselective bioaccumulation, bioactivity, acute toxicity, and dissipation in soils, Ning Cui et al., Copyright 2018. Figure (B) was reprinted by permission from Springer Nature GmbH, *Analytical and Bioanalytical Chemistry*, Enantioselective determination of the chiral pesticide isofenphos-methyl in vegetables, fruits, and soil and its enantioselective degradation in pakchoi using HPLC with UV detection, Beibei Gao et al., Copyright 2016. Figure (C) was reprinted with permission from Buerge, Ignaz J., Jürgen Krauss, Rocío López-Cabeza, Werner Siegfried, Michael Stüssi, Felix E. Wettstein, and Thomas Poiger, 2016. "Stereo-selective metabolism of the sterol biosynthesis inhibitor fungicides fenpropidin, fenpropimorph, and spiroxamine in grapes, sugar beets, and wheat." *Journal of Agricultural and Food Chemistry* 64, 5301–9. Copyright (2016) American Chemical Society.)

Chiral Pesticides

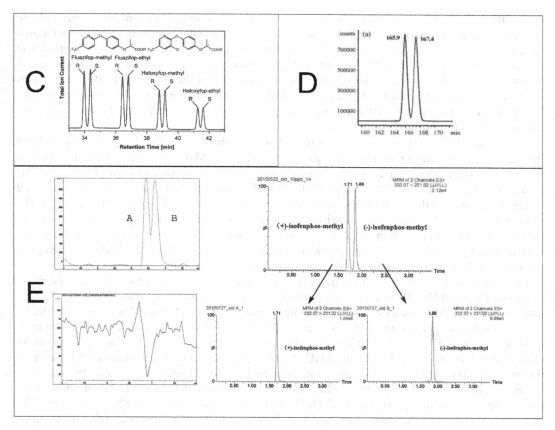

FIGURE 3.5 *(Continued)*

compared the resolution of cellulose-based stationary phase with the amylose-based stationary phase. It was observed that baseline separations were obtained with Phenomenex Lux Cellulose-1, Lux Cellulose-2, and Lux Amylose-2, using 2-propanol/hexane as the mobile phase. Li et al. (2013) used an amylose-based stationary phase (Chiralpak AD-RH) for the enantioseparation of nine multi-class pesticides from soil and water samples. The mean recoveries for all the enantiomers were above 77.8% and good linearity ($r^2 = 0.9986$) was obtained. Thus, amylose-based chiral stationary phases usually have effective chiral recognition sites for most chiral pesticides.

3.8.2 Chiral Gas Chromatography

Chiral gas chromatography is a chromatographic technique that can be used for the separation of the enantiomers of chiral compounds. Many researchers have used chiral GC for the enantioseparation of chiral pesticides from different complex matrices (Buerge et al., 2016; Dallegrave et al., 2016; Kalathoor et al., 2015; Xiang et al., 2019; Yao et al., 2019) (Table 3.7). Although chiral GC is a versatile analytical technique, it cannot be used to separate the enantiomers of all chiral pesticides. This technique can only separate the enantiomers of chiral pesticides that are volatile. The enantiomers of the chiral pesticides must easily vaporize within the temperature range used in the chiral GC oven without decomposing.

Researchers, using chiral GC, have managed to separate volatile enantiomers of many different classes of chiral pesticides. These include pyrethroids (Corcellas et al., 2015; Xiang et al., 2019; Yao et al., 2019), fungicides (Buerge et al., 2016; Kalathoor et al., 2015) and herbicides (Buerge et al., 2015) (Table 3.7). Generally, good resolution of the enantiomers was obtained when volatile chiral pesticides were subjected to chiral GC analysis. For instance, Buerge et al. (2015) separated the enantiomers of four herbicides, namely fluazifop methyl, fluazifop ethyl, haloxyfop methyl, and haloxyfop ethyl, using chiral GC. The chromatogram they obtained is shown in Figure 3.5C and it clearly shows that they managed to achieve baseline resolution of the enantiomers of the herbicides used. In the same vein, Fidalgo-Used et al. (2008) separated enantiomers of bromocyclen using chiral GC. Good resolution of the enantiomers was achieved as indicated in the chromatogram (Figure 3.5D) they obtained.

During chiral GC analysis, a carrier gas is used as the mobile phase. The carrier gas is usually an inert gas such as helium or an unreactive gas such as nitrogen. Many researchers have used helium as a carrier gas during the separation of the enantiomers of chiral pesticides by chiral GC (Corcellas et al., 2015; Kalathoor et al., 2015; Xiang et al., 2019). For instance, Kalathoor et al. (2015) achieved baseline separation of the enantiomers of metalaxyl using helium as a mobile phase. In a similar study, Corcellas et al. (2015) managed to effectively separate enantiomers of several pyrethroids also using helium as the mobile phase. Some researchers, however, used nitrogen as the mobile phase during the separation of the enantiomers of chiral pesticides by GC. For example, Yao et al. (2019) obtained good resolution of the enantiomers of alpha-cypermethrin using nitrogen as the mobile phase.

Effective chiral stationary phases are required during separation of the enantiomers of chiral pesticides by chiral GC. Many commercial chiral stationary phases are available on the market and many researchers have used them for the enantioseparation of chiral pesticides using GC (Buerge et al., 2016; Naudé and Rohwer, 2012; Xiang et al., 2019; Yao et al., 2019) (Table 3.7). For instance, Kalathoor et al. (2015) used the BGB-172 chiral column during the separation of the enantiomers of metalaxyl from soil samples by chiral GC. A good resolution of 0.95 was obtained for the enantiomers. Corcellas et al. (2015) used the same column, in a similar study, for the enantioseparation of several chiral pesticides in breast milk. Resolutions greater than 0.58 were obtained for all the enantiomers of the chiral pesticides used. On the other hand, Dallegrave et al. (2016) used DB-5MS chiral column during separation of the enantiomers of seventeen pyrethroids in different complex matrices with good resolution of the enantiomers. Another chiral stationary phase that has been used by some researchers during enantioseparation of asymmetric pesticides by GC is the HP-5 column. For instance, Xiang et al. (2019) successfully achieved baseline separation of the enantiomers of bifenthrin using the HP-5 chiral column.

3.8.3 Chiral Supercritical Fluid Liquid Chromatography

Supercritical fluid chromatography (SFC) is among the chromatographic techniques that are preferred during the enantioselective separations and purifications of chiral pesticides (Tao et al., 2018; Jiang et al., 2018) (Table 3.8). The mobile phase used during SFC is usually supercritical CO_2 (Liu et al., 2015; Pan et al., 2016). Polar organic modifiers or co-solvents are usually added to the supercritical CO_2 in order to enhance the elution strength of the chiral pesticides (Tan et al., 2017; Zhang et al., 2018). Alcohols such as propanol, ethanol, and methanol are the commonly used co-solvents during SFC (Table 3.8). The physicochemical properties of supercritical CO_2 such as lower viscosity and higher diffusivity give it an edge over the mobile phases used in other chromatographic techniques (Tao et al., 2018). Thus, SFC has higher throughput and higher flow rates as compared to HPLC and usually has good resolution of the enantiomers of chiral pesticides (Figure 3.5E). Consequently, SFC has shorter analysis time and faster column equilibration during analysis of chiral pesticides (Liu et al., 2015). SFC is also considered an environment-friendly method due to reduced consumption of toxic solvents and additives. Additional advantages are that CO_2 is characterized as cheap, non-flammable, and non-toxic as compared to organic solvents (Pan et al., 2016; Tao et al., 2018).

Chiral triazole fungicides are among the pesticides that have been extensively studied and analyzed by many researchers using SFC (He et al., 2016; Jiang et al., 2018; Pan et al., 2016; Tan et al., 2017; Tao et al., 2018) (Table 3.8). During the separation of the enantiomers of chiral triazole fungicides with SFC, some chemical modifiers are added to the supercritical CO_2. Triazole fungicides are polar compounds while the mobile phase, supercritical CO_2, is non-polar. Thus, polar chemical modifiers are added to the mobile phase, as already alluded to, for the enhancement of the elution of the polar triazole fungicide molecules. For instance, Jiang et al. (2018) added 2-propanol to the supercritical CO_2 mobile phase during chiral separation of the enantiomers of prothioconazole using SFC. Tan et al. (2017) used ethanol as a chemical modifier of supercritical CO_2 mobile phase during enantioseparation of triticonazole using SFC. In the same vein, Pan et al. (2016) combined supercritical CO_2 with methanol as a co-solvent during the SFC-based chiral separation of enantiomers of pyrisoxazole. Thus, alcohols are usually used as chemical modifiers during SFC and their popularity hinges largely on the fact that they are relatively inexpensive and easily accessible polar solvents.

SFC separation of the enantiomers of chiral pesticides often involves their direct resolution using columns packed with chiral stationary phases. So many commercial chiral stationary phases are on the market and have been used by researchers for enantioseparation of chiral pesticides (He et al., 2016; Jiang et al., 2018; Pan et al., 2016) (Table 3.7). For instance, Jiang et al. (2018) used enantiopak OD column for the separation of the enantiomers of prothioconazole in tomato and soil samples using SFC. The enantiomers were baseline separated with a resolution of 3.59. Tan et al. (2017) also successfully used enantiopak OD column for chiral separation of the enantiomers of triticonazole in tomato and cucumber using SFC. Baseline separation of the enantiomers of triticonazole was achieved using this column. In the same vein, He et al. (2016) also obtained good resolution of the enantiomers of triticonazole using SFC and the same column. Other researchers successfully separated the enantiomers of chiral pesticides using SFC with the Chiralpak AD-3 column. For instance, Cheng et al. (2017) successfully resolved enantiomers of propiconazole from wheat, grapes, and soil using SFC with a Chiralpak AD-3 column. In a similar study, Zhang et al. (2018) also separated the enantiomers of organophosphate pesticides using SFC with the Chiralpak AD-3 column. Other columns used by researchers for enantioseparation of chiral pesticides using SFC include Chiralpak IA column (Pan et al., 2016), trefoil AMY 1 column (Liu et al., 2016), and chiral CCA column (Zhang et al., 2018).

3.9 FUTURE PROSPECTS AND CHALLENGES

Chiral pesticides have contributed tremendously towards sustainable food security in the world. Their use, however, is not free of challenges. Most chiral pesticides that are on the market right now are sold as racemates and the enantiomers, as already alluded to, have enantioselective toxicity to the target organisms. One enantiomer, in most cases, will be active to the target organism and the other ones will be inactive. Thus, the accumulation of the inactive enantiomers in the environment is a serious environmental challenge. This entails that the analysis of the enantiomers of chiral pesticides in the environment is imperative. The pre-concentration and analysis of chiral pesticides, however, is hampered by the lack of enantiopure standards that are easy to synthesize and purify. This, therefore, has a negative impact on the quantification of the enantiomers of the chiral pesticides in different environmental matrices. Synthesis of enantiopure chiral pesticides seems to be the most attractive and plausible step in minimizing the choking of the environment with the inactive enantiomers of the chiral pesticides. This would ensure that only the active enantiomer would be applied to crops during pest control. Although this seems to be a better strategy, it is still far from providing a permanent solution to the environmental problems associated with chiral pesticides. This is because some chiral pesticides have labile asymmetric centers and enantiomers may undergo stereochemical inversion. This process of enantiomerization complicates control of pollution of the environment by chiral pesticides as well as their analysis. During chromatographic analysis of chiral pesticides, the stationary phases have chiral recognition sites that are specific to one of the enantiomers. Chiral inversion, therefore, makes it difficult to quantify accurately the target enantiomer through chromatographic analysis. Mitigation of pollution of the environment by chiral pesticides lies in minimization of their use during pest control and, where possible, to resort to the use of bio-based chiral pesticides that are not persistent in the environment and are biodegradable.

3.10 CONCLUSION

A significant number of the pesticides used for the mitigation of the detrimental effects of pests on crops are chiral. Some of these chiral pesticides are subsequently released into the environment where they exhibit enantioselective activity and toxicity to target and non-target organisms, respectively. Due to their toxicity to non-target organisms, the separation and analysis of the inactive enantiomers of chiral pesticides is mandatory. Thus, chiral analysis is a powerful and necessary tool for studying the behavior of chiral pesticides in the environment. Chiral analysis of pesticides, in recent years, has been dominated by chromatographic techniques. The commonly used enantioselective chromatographic techniques include gas chromatography, high-performance liquid chromatography, and supercritical fluid chromatography. Good resolution of the enantiomers during chromatography hinges on the correct choice of the chiral stationary phase. The most popular chiral stationary phases employed during separation of asymmetric pesticides are those that are polysaccharide-based. Cellulose and amylose are among the commonly used polysaccharides during the synthesis of chiral stationary phases. They have proved to have versatile enantioselective chiral recognition sites during chromatographic analysis of asymmetric pesticides.

ACKNOWLEDGMENT

The authors are grateful to the Research Center, University of Venda, for financial support.

CONFLICTS OF INTEREST

The authors declare no conflict of interest.

REFERENCES

Abd-Alrahman, Sherif H., and Nevin S. Ahmed. 2012. "Dissipation of penconazole in peach, plum, apricot, and mango by HPLC-DAD." *Bulletin of Environmental Contamination and Toxicology* 90(2): 260–263. doi:10.1007/s00128-012-0928-9.

Asad, Muhammad Asad Ullah, Michel Lavoie, Hao Song, Yujian Jin, Zhengwei Fu, and Haifeng Qian. 2017. "Interaction of chiral herbicides with soil microorganisms, algae and vascular plants." *Science of the Total Environment* 580: 1287–99. doi:10.1016/j.scitotenv.2016.12.092.

Birolli, Willian G., Marylyn S. Arai, Marcia Nitschke, and André L. M. Porto. 2019. "The pyrethroid (±)-lambda-cyhalothrin enantioselective biodegradation by a bacterial consortium." *Pesticide Biochemistry and Physiology* 156: 129–37. doi:10.1016/j.pestbp.2019.02.014.

Brander, Susanne M., Molly K. Gabler, Nicholas L. Fowler, Richard E. Connon, and Daniel Schlenk. 2016. "Pyrethroid pesticides as endocrine disruptors: Molecular mechanisms in vertebrates with a focus on fishes." *Environmental Science & Technology* 50(17): 8977–92. doi:10.1021/acs.est.6b02253.

Brusseau, Mark L., Ian L. Pepper, and Charles P. Gerba. (2019). *Environmental and Pollution Science*. Washinton, DC: Academic Press

Buerge, Ignaz J., Astrid Bächli, Jean-Pierre De Joffrey, Markus D. Müller, Simon Spycher, and Thomas Poiger. 2013. "The chiral herbicide beflubutamid (I): Isolation of pure enantiomers by HPLC, herbicidal activity of enantiomers, and analysis by enantioselective GC-MS." *Environmental Science & Technology* 47(13): 6806–11. doi:10.1021/es301876d.

Buerge, Ignaz J., Astrid Bächli, Roy Kasteel, Reto Portmann, Rocío López-Cabeza, Lars F. Schwab, and Thomas Poiger. 2019. "Behaviour of the chiral herbicide imazamox in soils: pH-dependent, enantioselective degradation, formation and degradation of several chiral metabolites." *Environmental Science & Technology* 53(10): 5725–32. doi:10.1021/acs.est.8b07209.

Buerge, Ignaz J., Astrid Bächli, Werner E. Heller, Martina Keller, and Thomas Poiger. 2015. "Environmental behaviour of the chiral herbicide haloxyfop: Unchanged enantiomer composition in blackgrass (*Alopecurus myosuroides*) and garden cress (*Lepidium sativum*)." *Journal of Agricultural and Food Chemistry* 63(10): 2591–96. doi:10.1021/jf505242f.

Buerge, Ignaz J., Jürgen Krauss, Rocío López-Cabeza, Werner Siegfried, Michael Stüssi, Felix E. Wettstein, and Thomas Poiger. 2016. "Stereo-selective metabolism of the sterol biosynthesis inhibitor fungicides fenpropidin, fenpropimorph, and spiroxamine in grapes, sugar beets, and wheat." *Journal of Agricultural and Food Chemistry* 64(26): 5301–9. doi:10.1021/acs.jafc.6b00919.

Caballo, C., M. D. Sicilia, and S. Rubio. 2013. "Stereo-selective quantitation of mecoprop and dichlorprop in natural waters by supramolecular solvent-based micro-extraction, chiral liquid chromatography and tandem mass spectrometry." *Analytica Chimica Acta* 761: 102–8. doi:10.1016/j.aca.2012.11.044.

Caballo, C., M. D. Sicilia, and S. Rubio. 2014. "Fast, simple and efficient supramolecular solvent-based micro-extraction of mecoprop and dichlorprop in soils prior to their enantioselective determination by liquid chromatography–tandem mass spectrometry." *Talanta* 119 (February): 46–52. doi:10.1016/j.talanta.2013.10.043.

Celis, Rafael, Beatriz Gámiz, Gracia Facenda, and Mafia C. Hermosín. 2015. "Enantioselective sorption of the chiral fungicide metalaxyl on soil from non-racemic aqueous solutions: Environmental implications." *Journal of Hazardous Materials* 300: 581–589. doi:10.1016/j.jhazmat.2015.07.059.

Chai, Tingting, Xueqing Wang, Mei Bie, Shouhui Dai, Hualin Zhao, Shuming Yang, and Jing Qiu. 2013. "Comparative enantioseparation of crufomate with cellulose- and amylose-based chiral stationary phases on reverse-phase and normal-phase high-performance liquid chromatography." *Asian Journal of Chemistry* 25(2): 797–802. doi:10.14233/ajchem.2013.12908.

Chen, Hongping, Guanwei Gao, Peng yin, Jinxia Dai, Yungfeng Chai, Xin Liu, and Chengyin Lu. 2018. "Enantioselectivity and residue analysis of fipronil in tea (*Camellia sinesis*) by ultra-performance liquid chromatography orbitrap mass spectrometry." *Food Additives & Contaminants: Part A* 35(10): 2000–2010. doi:10.1080/19440049.2018.1497306.

Chen, Siyu, Hui Chen, Zunwei Chen, Yuezhong Wen, and Weiping Liu. 2019a. "Enantioselective phytotoxic disturbances of fatty acids in *Arabidopsis thaliana* by dichlorprop." *Environmental Science & Technology* 53(15): 9252–59. doi:10.1021/acs.est.9b03744.

Chen, Siyu, Lijuan Zhang, Hui Chen, Zunwei Chen, and Yuezhong Wen. 2019b. "Enantioselective toxicity of chiral herbicide metolachlor to *Microcystis aeruginosa*." *Journal of Agricultural and Food Chemistry* 67(6): 1631–37. doi:10.1021/acs.jafc.8b04813.

Chen, Xixi, Fengshou Dong, Jun Xu, Xingang Liu, Zenglong Chen, Na Liu, and Yongquan Zheng. 2016. "Enantioseparation and determination of isofenphos-methyl enantiomers in wheat, corn, peanut and soil with supercritical fluid chromatography/tandem mass spectrometric method." *Journal of Chromatography B* 1015-1016: 13–21. doi:10.1016/j.jchromb.2016.02.003.

Cheng, Cheng, Rui Ma, Yuele Lu, Chunxiao Liu, Wenjun Zhang, Shanshan Di, Li Chen, Jinling Diao, Zhiqiang Zhou, and Yuxia Hou. 2018. "Enantioselective toxic effects and digestion of furalaxyl enantiomers in *Scenedesmus obliquus*." *Chirality* 30(12): 1269–76. doi:10.1002/chir.23020.

Cheng, Feng, Feifei Cheng, Jianyong Zheng, Guanzhong Wu, Yinjun Zhang, and Zhao Wang. 2018. "A novel esterase from *Pseudochrobactrum asaccharolyticum* WZZ003: Enzymatic properties toward model substrate and catalytic performance in chiral fungicide intermediate synthesis." *Process Biochemistry* 69: 92–98. doi:10.1016/j.procbio.2018.03.011.

Cheng, Youpu, Yongquan Zheng, Fengshou Dong, Jing Li, Yaofang Zhang, Shuhong Sun, Ning Li, et al. 2017. "Stereo-selective analysis and dissipation of propiconazole in wheat, grapes, and soil by supercritical fluid chromatography–tandem mass spectrometry." *Journal of Agricultural and Food Chemistry* 65(1): 234–43. doi:10.1021/acs.jafc.6b04623.

Corcellas, Cayo, Ethel Eljarrat, and Damià Barceló. 2015. "Enantiomeric-selective determination of pyrethroids: Application to human samples." *Analytical and Bioanalytical Chemistry* 407(3): 779–86. doi:10.1007/s00216-014-7905-6.

Cui, Ning, Haoyu Xu, Shijie Yao, Yiwen He, Hongchao Zhang, and Yunlong Yu. 2018. "Chiral triazole fungicide tebuconazole: Enantioselective bioaccumulation, bioactivity, acute toxicity, and dissipation in soils." *Environmental Science and Pollution Research* 25(25): 25468–75. doi:10.1007/s11356-018-2587-9.

Dallegrave, Alexsandro, Tânia Mara Pizzolato, Fabiano Barreto, Ethel Eljarrat, and Damià Barceló. 2016. "Methodology for trace analysis of 17 pyrethroids and chlorpyrifos in foodstuff by gas chromatography–tandem mass spectrometry." *Analytical and Bioanalytical Chemistry* 408(27): 7689–97. doi:10.1007/s00216-016-9865-5.

de Albuquerque, Nayara Cristina Perez de, Daniel Blascke Carrão, Maísa Daniela Habenschus, and Anderson Rodrigo Moraes de Oliveira. 2018. "Metabolism studies of chiral pesticides: A critical review." *Journal of Pharmaceutical and Biomedical Analysis* 2017, 147: 89–109. doi:10.1016/j.jpba.2017.08.011.

de Albuquerque, Nayara Cristina Perez de, Juliana Vicentin de Matos, and Anderson Rodrigo Moraes de Oliveira. 2016. "In-line coupling of an achiral-chiral column to investigate the enantioselective in vitro metabolism of the pesticide fenamiphos by human liver microsomes." *Journal of Chromatography A, Enantioseparations* 1467: 326–34. doi:10.1016/j.chroma.2016.08.039.

Deng, Yue, Wenjun Zhang, Yinan Qin, Rui Liu, Luyao Zhang, Zikang Wang, Zhiqiang Zhou, and Jinling Diao. 2019. "Stereo-selective toxicity of metconazole to the antioxidant defences and the photosynthesis system of *Chlorella pyrenoidosa*." *Aquatic Toxicology* 210: 129–38. doi:10.1016/j.aquatox.2019.02.017.

Diao, Jinling, Peng Xu, Donghui Liu, Yule Lu, and Zhiqiang Zhou. 2011. "Enantiomer-specific toxicity and bioaccumulation of alpha-cypermethrin to earthworm *Eisenia fetida*." *Journal of Hazardous Materials* 192(3): 1072–78. doi:10.1016/j.jhazmat.2011.06.010.

Díaz Merino, Matías E., Romina N. Echevarría, Ester Lubomirsky, Juan M. Padró, and Cecilia B. Castells. 2019. "Enantioseparation of the racemates of a number of pesticides on a silica-based column with immobilized amylose tris(3-chloro-5-methylphenylcarbamate)." *Microchemical Journal* 149: 103970. doi:10.1016/j.microc.2019.103970.

Duan, Jinsheng, Mingna Sun, Yang Shen, Beibei Gao, Zhaoxian Zhang, Tongchun Gao, and Minghua Wang. 2018. "Enantioselective acute toxicity and bioactivity of carfentrazone-ethyl enantiomers." *Bulletin of Environmental Contamination and Toxicology* 101(5): 651–56. doi:10.1007/s00128-018-2474-6.

Escolá Casas, Mònica, Tue kjaergaard Nielsen, Witold Kot, Lars Hestbjerg Hansen, Anders Johansen, and Kai Bester. 2017. "Degradation mocoprop in polluted landfill leachate and wastewater in a moving bed biofilm reactor." *Water Research* 121: 213–220. doi:10.1016/j.watres.2017.05.031

European Food Safety Authority (EFSA). 2013. "Reasoned opinion on the review of the existing maximum residue levels (MRLs) for benalaxyl according to Article 12 of Regulation (EC) No 396/2005." *EFSA Journal* 11(10): 43. https://doi.org/10.2903/j.efsa.2013.3405.

European Food Safety Authority (EFSA). 2015. "Reasoned opinion on combined review of the existing maximum residue levels (MRLs) for the active substances metalaxyl and metalaxyl-M." *EFSA Journal* 13(4): 56. https://doi.org/10.2903/j.efsa.2015.4076.

European Food Safety Authority (EFSA). 2016. "Modification of the existing maximum residue levels for metalaxyl in various crops. For benalaxyl according to Article 12 of Regulation (EC) No 396/2005. (2013)." *EFSA Journal* 14(7). doi:10.2903/j.efsa.2016.4521.

Fidalgo-Used, Natalia, Maria Montes-Bayón, Elisa Blanco-González, and Alfredo Sanz-Medel. 2008. "Enantioselective determination of the organochlorine pesticide bromocyclen in spiked fish tissue using solid-phase micro-extraction coupled to gas chromatography with ECD and ICP-MS detection." *Talanta, Special Section: Remote Sensing*, 75(3): 710–16. doi:10.1016/j.talanta.2007.12.006.

Fonseca, Franciele S., Daniel B. Carrão, Nayara C. P. de Albuquerque, Viviani Nardini, Luís G. Dias, Rodrigo M. da Silva, Norberto P. Lopes, and Anderson R. M. de Oliveira. 2019. "Myclobutanil enantioselective risk assessment in humans through in Vitro CYP450 reactions: Metabolism and inhibition studies." *Food and Chemical Toxicology* 128: 202–11. doi:10.1016/j.fct.2019.04.009.

Gao, Beibei, Qing Zhang, Mingming Tian, Zhaoxian Zhang, and Minghua Wang. 2016. "Enantioselective determination of the chiral pesticide isofenphos-methyl in vegetables, fruits, and soil and its enantioselective degradation in pakchoi using HPLC with UV detection." *Analytical and Bioanalytical Chemistry* 408(24): 6719–27. doi:10.1007/s00216-016-9790-7.

Gao, Beibei, Zhaoxian Zhang, Lianshan Li, Amir E. Kaziem, Zongzhe He, Qianwen Yang, Peiyang Qing, Qing Zhang, and Minghua Wang. 2019. "Stereo-selective environmental behaviour and biological effect of the chiral organophosphorus insecticide isofenphos-methyl." *Science of the Total Environment* 648: 703–10. doi:10.1016/j.scitotenv.2018.08.182.

Gao, Jing, Fang Wang, Peng Wang, Wenqi Jiang, Zhenhua Zhang, Donghui Liu, and Zhiqiang Zhou. 2019. "Enantioselective toxic effects and environmental behavior of ethiprole and its metabolites against *Chlorella pyrenoidosa*." *Environmental Pollution* 244: 757–65. doi:10.1016/j.envpol.2018.10.056.

Gao, Jing, Fang Wang, Wenqi Jiang, Jiajun Han, Peng Wang, Donghui Liu, and Zhiqiang Zhou. 2020. "Biodegradation of chiral flufiprole in *Chlorella pyrenoidosa*: Kinetics, transformation products, and toxicity evaluation." *Journal of Agricultural and Food Chemistry* 68(7): 1966–73. doi:10.1021/acs.jafc.9b05860.

Gao, Jing, Han Qu, Chuntao Zhang, Weijia Li, Peng Wang, and Zhiqiang Zhou. 2017. "Direct chiral separations of the enantiomers of phenylpyrazole pesticides and the metabolites by HPLC." *Chirality* 29(1): 19–25. doi:10.1002/chir.22661.

Gao, Yongxin, Huili Wang, Fang Qin, Peng Xu, Xiaotian Lv, Jianzhong Li, and Baoyuan Guo. 2014. "Enantiomerization and enantioselective bioaccumulation of metalaxyl in *Tenebrio molitor* larvae." *Chirality* 26(2): 88–94. doi:10.1002/chir.22269.

Garrison, A. Wayne. 2011. "An introduction to pesticide chirality and the consequences of stereoselectivity." *ACS Symposium Series* 1085: 1–7. https://doi.org/10.1021/bk-2011-1085.ch001.

He, Jianfeng, Jun Fan, Yilun Yan, Xiaodong Chen, Tai Wang, Yaomou Zhang, and Weiguang Zhang. 2016. "Triticonazole enantiomers: Separation by supercritical fluid chromatography and the effect of the chromatographic conditions." *Journal of Separation Science* 39(21): 4251–57. doi:10.1002/jssc.201600820.

He, Rujian, Jun Fan, Qi Tan, Yecai Lai, Xiaodong Chen, Tai Wang, Ying Jiang, Yaomou Zhang, and Weiguang Zhang. 2018. "Enantioselective determination of metconazole in multi-matrices by high-performance liquid chromatography." *Talanta* 178: 980–86. doi:10.1016/j.talanta.2017.09.045.

Huang, Qiuxin, Kun Zhang, Zhifang Wang, Chunwei Wang, and Xianzhi Peng. 2012. "Enantiomeric determination of azole antifungals in wastewater and sludge by liquid chromatography-tandem mass spectrometry." *Analytical and Bioanalytical Chemistry* 403(6): 1751–1760. doi:10.1007/s00216-012-5976-9.

Huang, Quiuxin, Zhifang Wang, Chunwei Wang, and Xianzhi Peng. 2013. "Chiral profiling of azole antifungals in municipal wastewater and recipient rivers of the Pearl River delta, China." *Environmental Science and Pollution Research* 20(12): 8890–8899. doi:10.1007/s11356-013-1862-z.

Hsiao, Yu-Ling, Yei-Shung Wang, and Jui-Hung Yen. 2014. "Enantioselective effects of herbicide imazapyr on *Arabidopsis thaliana*." *Journal of Environmental Science and Health, Part B* 49(9): 646–53. doi: 10.1080/03601234.2014.922404.

Jiang, Ying, Jun Fan, Rujian He, Dong Guo, Tai Wang, Hui Zhang, and Weiguang Zhang. 2018. "High-fast enantioselective determination of prothioconazole in different matrices by supercritical fluid chromatography and vibrational circular dichroism spectroscopic study." *Talanta* 187: 40–46. doi:10.1016/j.talanta.2018.04.097.

Jiménez-Jiménez, Sara, Natalia Casado, María Ángeles García, and María Luisa Marina. 2019. "Enantiomeric analysis of pyrethroids and organophosphorus insecticides." *Journal of Chromatography A* 1605: 360345. doi:10.1016/j.chroma.2019.06.066.

Jin, Biao, and Massimo Rolle. 2016. "Joint interpretation of enantiomer and stable isotope fractionation for chiral pesticides degradation." *Water Research* 105: 178–186. doi:10.1016/j.watres.2016.08.057.

Jing, Xu, Guojun Yao, Donghui Liu, Yiran Liang, Mai Luo, Zhiqiang Zhou, and Peng Wang. 2017. "Effects of wastewater irrigation and sewage sludge application on soil residues of chiral fungicide benalaxyl." *Environmental Pollution* 224: 1–6. doi:10.1016/j.envpol.2017.03.004.

Jing, Xu, Guojun Yao, Donghui Liu, Chang Liu, Fang Wang, Peng Wang, and Zhiqiang Zhou. 2017. "Exposure of frogs and tadpoles to chiral herbicide fenoxaprop-ethyl." *Chemosphere* 186: 832–38. doi:10.1016/j.chemosphere.2017.07.132.

Kalathoor, Roschni, Jens Botterweck, Andreas Schäffer, Burkhard Schmidt, and Jan Schwarzbauer. 2015. "Quantitative and enantioselective analyses of non-extractable residues of the fungicide metalaxyl in soil." *Journal of Soils and Sediments* 15(3): 659–70. doi:10.1007/s11368-014-1027-9.

Kaziem, Amir E., Beibei Gao, Lianshan Li, Zhaoxian Zhang, Zongzhe He, Yong Wen, and Ming-hua Wang. 2020. "Enantioselective bioactivity, toxicity, and degradation in different environmental mediums of chiral fungicide epoxiconazole." *Journal of Hazardous Materials* 386: 121951. doi:10.1016/j.jhazmat.2019.121951.

Kuhlmann, Janna, Andreas C. Kretschmann, Kai Bester, Ulla E. Bollmann, Kristoffer Dalhoff, and Nina Cedergreen. 2019. "Enantioselective mixture toxicity of the azole fungicide imazalil with the insecticide α-cypermethrin in *Chironomus riparius*: Investigating the importance of toxicokinetics and enzyme interactions." *Chemosphere* 225: 166–73. doi:10.1016/j.chemosphere.2019.03.023.

Li, Na, Luqing Deng, Jianfang Li, Zhengbing Wang, Yiye Han, and Chenglan Liu. 2018. "Selective effect of myclobutanil enantiomers on fungicidal activity and fumonisin production by *Fusarium verticillioides* under different environmental conditions." *Pesticide Biochemistry and Physiology, Special Issue: Fungicide Toxicology in China* 147: 102–9. doi:10.1016/j.pestbp.2017.12.010.

Li, Runan, Xinglu Pan, Yan Tao, Duoduo Jiang, Zenglong Chen, Fengshou Dong, Jun Xu, Xingang Liu, Xiaohu Wu, and Yongquan Zheng. 2019. "Systematic evaluation of chiral fungicide imazalil and its major metabolite R14821 (imazalil-M): Stability of enantiomers, enantioselective bioactivity, aquatic toxicity, and dissipation in greenhouse vegetables and soil." *Journal of Agricultural and Food Chemistry* 67(41): 11331–39. doi:10.1021/acs.jafc.9b03848.

Li, Yuanbo, Fengshou Dong, Xingang Liu, Jun Xu, Xiu Chen, Yongtao Han, Xuyang Liang, and Yongquan Zheng. 2013. "Development of a multi-residue enantiomeric analysis method for 9 pesticides in soil and water by chiral liquid chromatography/tandem mass spectrometry." *Journal of Hazardous Materials* 250–251: 9–18. doi:10.1016/j.jhazmat.2013.01.071.

Liu, Hui, and Wei Ding. 2019. "Enantiomeric separation of prothioconazole and prothioconazole-desthio on chiral stationary phases." *Chirality* 31(3): 219–29. doi:10.1002/chir.23050.

Liu, Na, Fengshou Dong, Jun Xu, Xingang Liu, Zenglong Chen, Xinglu Pan, Xixi Chen, and Yongquan Zheng. 2016. "Enantioselective separation and pharmacokinetic dissipation of cyflumetofen in field soil by ultra-performance convergence chromatography with tandem mass spectrometry." *Journal of Separation Science* 39(7): 1363–70. doi:10.1002/jssc.201501123.

Liu, Na, Fengshou Dong, Jun Xu, Xingang Liu, Zenglong Chen, Yan Tao, Xinglu Pan, XiXi Chen, and Yongquan Zheng. 2015. "Stereo-selective determination of tebuconazole in water and zebrafish by supercritical fluid chromatography tandem mass spectrometry." *Journal of Agricultural and Food Chemistry* 63(28): 6297–6303. doi:10.1021/acs.jafc.5b02450.

Lozowicka, Bozena, Magdalena Jankowska, Izabela Hrynko, and Piotr Kaczynski. 2015. "Removal of 16 pesticide residues from strawberries by washing with tap and ozone water, ultrasonic cleaning and boiling." *Environmental Monitoring and Assessment* 188(1). doi:10.1007/s10661-015-4850-6.

Maia, Alexandra S., Ana R. Ribeiro, Paula M. L. Castro, and Maria Elizabeth Tiritan. 2017. "Chiral analysis of pesticides and drugs of environmental concern: Biodegradation and enantiomeric fraction." *Symmetry-Basel* 9(9): 196–225. doi:10.3390/sym9090196.

Malhat, Farag Mahmoud. 2017. Persistence of metalaxyl residues on tomato fruit using high-performance liquid chromatography and QuEChERS methodology. *Arabian Journal of Chemistry* 10: S765–S768. doi:10.1016/j.arabjc.2012.12.002.

Mangas, Iris, Eugenio Vilanova, Jorge Estévez, and Tanos C. C. França. 2016. "Neurotoxic effects associated with current uses of organophosphorus compounds." *Journal of the Brazilian Chemical Society* 27(5): 809–25. doi:10.5935/0103-5053.20160084.

Masbou, Jèrèmy, Fatima Meite, Benoit Guyot, and Gwanaël Imfled. 2018. Enantiomer-specific stable carbon isotope analysis (ESIA) to evaluate degradation of the chiral fungicide metalaxyl in soils. *Journal of Hazardous Materials* 353: 99–107. doi:10.1016/j.jhazmat.2018.03.047.

Mcmahen, Rebecca L., Mark J. Strynar, Larry McMillan, Eugene Derose, and Andrew B. Lindstrom. 2016. Comparison of fipronil sources in North Carolina surface water and identification of a novel fipronil transformation product in recycled wastewater. *Science of the Total Environment* 569–570: 880–887. doi:10/1016/j.scitotenv.2016.05.085.

Mohammadi, Mehrnoush, Hamed Tavakoli, Yaser Abdollahzadeh, Amir Khosravi, Rezvan Torkaman, and Ashkan Mashayekhi. 2015. "Ultra-pre-concentration and determination of organophosphorus pesticides in soil samples by a combination of ultrasound assisted leaching-solid phase extraction and low-density solvent based dispersive liquid–liquid micro-extraction." *RSC Advances* 5(92): 75174–81. doi:10.1039/C5RA11959F.

Mohan, Somagoni Jagan, Eaga Chandra Mohan, and Madhsudan Rao Yamsani. 2009. "Chirality and its importance in pharmaceutical field—an overview." *International Journal of Pharmaceutical Sciences and Nanotechnology* 1: 309–16.

Naudé, Yvette, and Egmont R. Rohwer. 2012. "Two multidimensional chromatographic methods for enantiomeric analysis of o,p'-DDT and o,p'-DDD in contaminated soil and air in a malaria area of South Africa." *Analytica Chimica Acta*, Africa 730: 120–26. doi:10.1016/j.aca.2012.03.028.

Nillos, Mae G., Kunde Lin, Jay Gan, Svetlana Bondarenko, and Daniel Schlenk. 2009. "Enantioselectivity in fipronil aquatic toxicity and degradation." *Environmental Toxicology and Chemistry* 28(9): 1825–33. doi:10.1897/08-658.1.

Pan, Xinglu, Fengshou Dong, Jun Xu, Xingang Liu, Zenglong Chen, and Yongquan Zheng. 2016. "Stereoselective analysis of novel chiral fungicide pyrisoxazole in cucumber, tomato and soil under different application methods with supercritical fluid chromatography/tandem mass spectrometry." *Journal of Hazardous Materials* 311: 115–24. doi:10.1016/j.jhazmat.2016.03.005.

Pérez-Fernández, Virginia, Maria Ángeles García, and Maria Luisa Marina. 2010. "Characteristics and enantiomeric analysis of chiral pyrethroids." *Journal of Chromatography A, Chiral Separations* 1217(7): 968–89. doi:10.1016/j.chroma.2009.10.069.

Ponash, Richard P. (Ed.). 2015. *Sittig's handbook of pesticides and agricultural chemicals* (2nd edition). William Andrew Publishing, 518–597. ISBN 9781455731480, https://doi.org/10.1016/B978-1-4557-3148-0.00013-3.

Qi, Peipei, Xiangyun Wang, Hu Zhang, Xinquan Wang, Hao Xu, and Qiang Wang. 2015. "Rapid enantioseparation and determination of isocarbophos enantiomers in orange pulp, peel, and kumquat by chiral HPLC-MS/MS." *Food Analytical Methods* 8(2): 531–38. doi:10.1007/s12161-014-9922-7.

Qi, Yanli, Donghui Liu, Mai Luo, Xu Jing, Peng Wang, and Zhiqiang Zhou. 2016. "Enantioselective degradation and chiral stability of the herbicide fluazifop-butyl in soil and water." *Chemosphere* 146: 315–322. doi:10.1016/j.chemosphere. 2015.12.040.

Qi, Yanli, Donghui Liu, Wenting Zhao, Chang Liu, Zhiqiang Zhou, and Peng Wang. 2015. "Enantioselective phytotoxicity and bioactivity of the enantiomers of the herbicide napropamide." *Pesticide Biochemistry and Physiology* 125: 38–44. doi:10.1016/j.pestbp.2015.06.004.

Qin, F., Y. Gao, P. Xu, B. Guo, J. Li, and H. Wang. (2015). "Enantioselective bioaccumulation and toxic effects of fipronil in the earthworm Eisenia foetida following soil exposure." *Pest Management Science* 71: 553–561. doi:10.1002/ps.3841.

Qian, Yi, Chenyang Ji, Siqing Yue, and Meirong Zhao. 2019. "Exposure of low-dose fipronil enantioselectively induced anxiety-like behaviour associated with DNA methylation changes in embryonic and larval zebrafish." *Environmental Pollution* 249: 362–71. doi:10.1016/j.envpol.2019.03.038.

Qu, Han, Ruixue Ma, Dong-hui Liu, Jing Gao, Fang Wang, Zhi-qiang Zhou, and Peng Wang. 2016. "Environmental behaviour of the chiral insecticide fipronil: Enantioselective toxicity, distribution and transformation in aquatic ecosystem." *Water Research* 105: 138–46. doi:10.1016/j.watres.2016.08.063.

Quinn, L. P., B. J. de Vos, M. Fernandes-Whaley, C. Roos, H. Bouwman, H. Kylin, R. Pieters, and J. van den Berg. 2011. "Pesticide Use in South Africa: One of the Largest Importers of Pesticides in Africa, Pesticides in the Modern World - Pesticides Use and Management." Ed. Margarita Stoytcheva. *Pesticides in the Modern World - Pesticides Use and Management*. London, UK: IntechOpen, DOI: 10.5772/16995. Available from: https://www.intechopen.com/books/pesticides-in-the-modern-world-pesticides-use-and-management/pesticide-use-in-south-africa-one-of-the-largest-importers-of-pesticides-in-africa

Saranjampour, Parichehr, Emily N. Vebrosky, and Kevin L. Armbrust. 2017. "Salinity impacts on water solubility and n-octanol/water partition coefficients of selected pesticides and oil constituent." *Environmental Toxicology and Chemistry* 36(9): 2274–2280. doi:10.1002/etc.3784.

Shakoori, Attaollah, Iran, Peyman Mahasti, and Vahideh Moradi. 2017. "Determination of twenty organophosphorus pesticides in wheat samples from different regions of Iran." *Iranian Journal of Toxicology* 11(5): 37–44. doi:10.29252/arakmu.11.5.37.

Shen, Qiuxuan, Jitong Li, Peng Xu, Wei Li, Guoqiang Zhuang, and Yinghuan Wang. 2017. "Enantioselective metabolism of triadimefon and its chiral metabolite triadimenol in lizards." *Ecotoxicology and Environmental Safety* 143: 159–65. doi:10.1016/j.ecoenv.2017.05.024.

Shen, Yuqing, Jianyun Zhang, Jingqian Xie, and Jing Liu. 2019. "In vitro assessment of corticosteroid effects of eight chiral herbicides." *Journal of Environmental Science and Health, Part B* 55(2): 91–102. doi: 10.1080/03601234.2019.1665408.

Shi, Li H., Shan Zhao, Ting Gui, Jin Xu, Fei Wang, Yu P. Zhang, and De Y. Hu. 2018. "Degradation dynamics, residues and risk assessment of Dufulin enantiomers in corn plants and corn by LC/MS/MS." *Journal of Environmental Science and Health, Part B* 53(11): 761–69. doi:10.1080/03601234.2018.1480165.

Shi, Lihong, Ting Gui, Shan Zhao, Jin Xu, Fei Wang, Changling Sui, Yuping Zhang, and Deyu Hu. 2018. "Degradation and residues of indoxacarb enantiomers in rice plants, rice hulls and brown rice using enriched S-indoxacarb formulation and enantiopure formulation." *Biomedical Chromatography* 32(10): 4301. doi:10.1002/bmc.4301.

Supowit, Samuel D., Akash M. Sadaria, Edward J. Reyes, and Rolf U. Halden. 2016. "Mass balance of fipronil and total toxicity of fipronil-related compounds in process streams during conventional wastewater and wetland treatment." *Environmental Science & Technology* 50(3): 1519–1526. doi:10.1021/acs.est.5b04516.

Talab, Khaled Mohamed Ahmed, Zhong-Hua Yang, Jian-Hong Li, Yue Zhao, Sara Alrasheed Mohamed Omer, and Ya-Bing Xiong. 2018. "The influence of microbial communities for triadimefon enantiomerization in soils with different pH values." *Chirality* 30(3): 293–301. doi:10.1002/chir.22796.

Tan, Qi, Jun Fan, Ruiqi Gao, Rujian He, Tai Wang, Yaomou Zhang, and Weiguang Zhang. 2017. "Stereoselective quantification of triticonazole in vegetables by supercritical fluid chromatography." *Talanta* 164: 362–67. doi:10.1016/j.talanta.2016.08.077.

Tao, Yan, Zuntao Zheng, Yang Yu, Jun Xu, Xingang Liu, Xiaohu Wu, Fengshou Dong, and Yongquan Zheng. 2018. "Supercritical fluid chromatography–tandem mass spectrometry-assisted methodology for rapid enantiomeric analysis of fenbuconazole and its chiral metabolites in fruits, vegetables, cereals, and soil." *Food Chemistry* 241: 32–39. doi:10.1016/j.foodchem.2017.08.038.

Tong, Zhou, Xu Dong, Shasha Yang, Mingna Sun, Tongchun Gao, Jinsheng Duan, and Haiqun Cao. 2019. "Enantioselective effects of the chiral fungicide tetraconazole in wheat: Fungicidal activity and degradation behaviour." *Environmental Pollution* 247: 1–8. doi:10.1016/j.envpol.2019.01.013.

Ulrich, Elin M., Patti L. Tenbrook, Larry M. Mcmillan, Qianheng Wang, and Wenjian Lao. 2017. "Enantiomer-specific measurements of current-use pesticides in aquatic systems." *Environmental Toxicology and Chemistry* 37: 99–106. doi:10.1002/etc.3938.

Vryzas, Zisis. 2018. "Pesticide fate in soil-sediment-water environmental in relation to contamination preventing actions." *Current Opinion in Environmental Science & Health* 4: 5–9. doi:10.1016/j.coesh.2018.03.001.

Wan Ibrahim W. A., D. Hermawan, M. M. Sanagi. (2007). "On-line preconcentration and chiral separation of propiconazole by cyclodextrin-modified micellar electrokinetic chromatography." *Journal of Chromatography A* 1170:107–113.

Wang, Fang, Jing Gao, Li Chen, Zhiqiang Zhou, Donghui Liu, and Peng Wang. 2018. "Enantioselective bioaccumulation and metabolism of lactofen in zebrafish *Danio rerio* and combined effects with its metabolites." *Chemosphere* 213: 443–52. doi:10.1016/j.chemosphere.2018.09.052.

Wang, Meiyun, Xiude Hua, Qing Zhang, Yu Yang, Haiyan Shi, and Minghua Wang. 2015. "Enantioselective degradation of metalaxyl in grape, tomato, and rice plants." *Chirality* 27(2): 109–14. doi:10.1002/chir.22397.

Wang, Shunhui, Huizhen Li, and Jing You. 2019. "Enantioselective degradation and bioaccumulation of sediment-associated fipronil in *Lumbriculus variegatus*: Toxicokinetic analysis." *Science of the Total Environment* 672: 335–41. doi:10.1016/j.scitotenv.2019.03.490.

Wang, Xinru, Wentao Zhu, Jing Qiu, Dezhen Wang, and Zhiqiang Zhou. 2017. "Enantioselective metabolism and enantiomerization of benalaxyl in mice." *Chemosphere* 169: 308–15. doi:10.1016/j.chemosphere.2016.11.091.

Wang, Zhaokun, Xia Wang, Shuang Li, Zhen Jiang, and Xingjie Guo. 2019. "Magnetic solid-phase extraction based on carbon nanosphere@Fe_3O_4 for enantioselective determination of eight triazole fungicides in water samples." *Electrophoresis* 40(9): 1306–13. doi:10.1002/elps.201800530.

Wang, Zikang, Zhongnan Tian, Li Chen, Wenjun Zhang, Luyao Zhang, Yao Li, Jinling Diao, and Zhiqiang Zhou. 2020. "Stereo-selective metabolism and potential adverse effects of chiral fungicide triadimenol on *Eremias argus*." *Environmental Science and Pollution Research* 27(8): 7823–34. doi:10.1007/s11356-019-07205-4.

Xia, Weitong, Zongzhe He, Kunming Hu, Beibei Gao, Zhaoxian Zhang, Minghua Wang, and Qiang Wang. 2019. "Simultaneous separation and detection chiral fenobucarb enantiomers using UPLC-MS/MS." *Applied Sciences* 1(7): 795. doi:10.1007/s42452-019-0822-8.

Xiang, Dandan, Linxi Zhong, Shuyuan Shen, Zhuoying Song, Guonian Zhu, Mengcen Wang, Qiangwei Wang, and Bingsheng Zhou. 2019. "Chronic exposure to environmental levels of cis-bifenthrin: enantioselectivity and reproductive effects on zebrafish *(Danio rerio)*." *Environmental Pollution* 251: 175–84. doi:10.1016/j.envpol.2019.04.089.

Xiang, Dandan, Tianyi Chu, Meng Li, Qiangwei Wang, and Guonian Zhu. 2018. "Effects of pyrethroid pesticide cis-bifenthrin on lipogenesis in hepatic cell line." *Chemosphere* 201: 840–49. doi:10.1016/j.chemosphere.2018.03.009.

Xie, Jingqian, Lijuan Zhang, Lu Zhao, Qiaozhi Tang, Kai Liu, and Weiping Liu. 2016. "Metolachlor stereoisomers: Enantioseparation, identification and chiral stability." *Journal of Chromatography A* 1463: 42–48. https://doi.org/10.1016/j.chroma.2016.07.045.

Xie, Jingqian, Lu Zhao, Kai Liu, Fangjie Guo, and Weiping Liu. 2016. "Enantioseparation of four amide herbicide stereoisomers using high-performance liquid chromatography." *Journal of Chromatography A* 1471: 145–54. doi:10.1016/j.chroma.2016.10.029.

Xie, Jingqian, Lu Zhao, Kai Liu, Fangjie Guo, and Weiping Liu. 2018. "Enantioselective effects of chiral amide herbicides napropamide, acetochlor and propisochlor: The more efficient R-enantiomer and its environmental friendly." *Science of the Total Environment* 626: 860–66. doi:10.1016/j.scitotenv.2018.01.140.

Xie, Jingqian, Lu Zhao, Kai Liu, Fangjie Guo, Zunwei Chen, and Weiping Liu. 2018. "Enantiomeric characterization of herbicide lactofen: Enantioseparation, absolute configuration assignment and enantioselective activity and toxicity." *Chemosphere* 193: 351–57. doi:10.1016/j.chemosphere.2017.10.168.

Xie, Yun, Leon Yu, Zheng Li, Weiyu Hao, Jing Chang, Peng Xu, Baoyuan Guo, Jianzhong Li, and Huili Wang. 2019. "Comparative toxicokinetics and tissue distribution of prothioconazole and prothioconazole desthio in Chinese lizards (*Eremias argus*) and transcriptional responses of metabolic-related genes." *Environmental Pollution* 247: 524–33. doi:10.1016/j.envpol.2019.01.055.

Xu, Chao, Xiaohui Sun, Lili Niu, Wenjing Yang, Wenqing Tu, Liping Lu, Shuang Song, and Weiping Liu. 2019. "Enantioselective thyroid disruption in zebrafish embryo-larvae via exposure to environmental concentrations of the chloroacetamide herbicide acetochlor." *Science of the Total Environment* 653: 1140–48. doi:10.1016/j.scitotenv.2018.11.037.

Yao, Guojun, Jing Gao, Chuntao Zhang, Wenqi Jiang, Peng Wang, Xueke Liu, Donghui Liu, and Zhiqiang Zhou. 2019. "Enantioselective degradation of the chiral alpha-cypermethrin and detection of its metabolites in five plants." *Environmental Science and Pollution Research* 26(2): 1558–64. doi:10.1007/s11356-018-3594-6.

Ye, Jing, Meirong Zhao, Lili Niu, and Weiping Liu. 2015. "Enantioselective environmental toxicology of chiral pesticides." *Chemical Research in Toxicology* 28(3): 325–38. doi:10.1021/tx500481n.

Yin, Jing, Yongxin Gao, Feilong Zhu, Weiyu Hao, Qi Xu, Huili Wang, and Baoyuan Guo. 2017. "Enantiomerization and stereo-selectivity in bioaccumulation of furalaxyl in Tenebrio molitor larvae." *Ecotoxicology and Environmental Safety* 145: 244–49. doi:10.1016/j.ecoenv.2017.07.041.

Youness, Mohamed, Martine Sancelme, Bruno Combourieu, and Pascale Besse-Hoggan. 2018. "Identification of new metabolic pathways in the enantioselective fungicide tebuconazole biodegradation by *Bacillus* sp. 3B6." *Journal of Hazardous Materials* 351: 160–68. doi:10.1016/j.jhazmat.2018.02.048.

Zeng, Huiyun, Xiujuan Xie, Yejing Huang, Jieming Chen, Yingtao Liu, Yi Zhang, Xiaoman Mai, Jianchao Deng, Huajun Fan, and Wei Zhang. 2019. "Enantioseparation and determination of triazole fungicides in vegetables and fruits by aqueous two-phase extraction coupled with online heart-cutting two-dimensional liquid chromatography." *Food Chemistry* 301: 125265. doi:10.1016/j.foodchem.2019.125265.

Zhai, Wangjing, Linlin Zhang, Jingna Cui, Yimu Wei, Peng Wang, Donghui Liu, and Zhiqiang Zhou. 2019. "The biological activities of prothioconazole enantiomers and their toxicity assessment on aquatic organisms." *Chirality* 31(6): 468–75. doi:10.1002/chir.23075.

Zhan, Jing, Yiran Liang, Donghui Liu, Chang Liu, Hui Liu, Peng Wang, and Zhiqiang Zhou. 2018. "Organochlorine pesticide acetofenate and its hydrolytic metabolite in rabbits: Enantioselective metabolism and cytotoxicity." *Pesticide Biochemistry and Physiology* 145: 76–83. doi:10.1016/j.pestbp.2018.01.007.

Zhang, Huili, Jianhua Wang, Li Li, and Ying Wang. 2017. "Determination of 103 pesticides and their main metabolites in animal origin food by QuEChERS and liquid chromatography–tandem mass spectrometry." *Food Analytical Methods* 10(6): 1826–43. doi:10.1007/s12161-016-0736-7.

Zhang, Lijun, Yelong Miao, and Chunmian Lin. 2018. "Enantiomeric separation of six chiral pesticides that contain chiral sulfur/phosphorus atoms by supercritical fluid chromatography." *Journal of Separation Science* 41(6): 1460–70. doi:10.1002/jssc.201701039.

Zhang, Ping, Qian Yu, Xiulong He, Kun Qian, Wei Xiao, Zhifeng Xu, Tian Li, and Lin He. 2018. "Enantiomeric separation of type I and type II pyrethroid insecticides with different chiral stationary phases by reversed-phase high-performance liquid chromatography." *Chirality* 30(4): 420–31. doi:10.1002/chir.22801.

Zhang, Ping, Qian Yu, Yuhan He, Wentao Zhu, Zhiqiang Zhou, and Lin He. 2017. "Chiral pyrethroid insecticide fenpropathrin and its metabolite: Enantiomeric separation and pharmacokinetic degradation in soils by reverse-phase high-performance liquid chromatography." *Analytical Methods* 9(30): 4439–46. doi:10.1039/C7AY01124E.

Zhang, Qing, Beibei Gao, Mingming Tian, Haiyan Shi, Xiude Hua, and Minghua Wang. 2016. "Enantioseparation and determination of triticonazole enantiomers in fruits, vegetables, and soil using efficient extraction and clean-up methods." *Journal of Chromatography B* 1009–1010: 130–37. doi:10.1016/j.jchromb.2015.12.018.

Zhang, Qing, Wu Xiong, Beibei Gao, Zachary Cryder, Zhaoxian Zhang, Mingming Tian, Edmond Sanganyado, Haiyan Shi, and Minghua Wang. 2018. "Enantioselectivity in degradation and ecological risk of the chiral pesticide ethiprole." *Land Degradation & Development* 29(12): 4242–51. doi:10.1002/ldr.3179.

Zhang, Qing, Xiu-de Hua, Hai-yan Shi, Ji-song Liu, Ming-ming Tian, and Ming-hua Wang. 2015. "Enantioselective bioactivity, acute toxicity and dissipation in vegetables of the chiral triazole fungicide flutriafol." *Journal of Hazardous Materials* 284: 65–72. doi:10.1016/j.jhazmat.2014.10.033.

Zhang, Quan, Qingmiao Cui, Siqing Yue, Zhengbiao Lu, and Meirong Zhao. 2019. "Enantioselective effect of glufosinate on the growth of maize seedlings." *Environmental Science and Pollution Research* 26(1): 171–78. doi:10.1007/s11356-018-3576-8.

Zhang, Wenjun, Li Chen, Jinling Diao, and Zhiqiang Zhou. 2019. "Effects of cis-bifenthrin enantiomers on the growth, behavioural, biomarkers of oxidative damage and bioaccumulation in *Xenopus laevis*." *Aquatic Toxicology* 214: 105237. doi:10.1016/j.aquatox.2019.105237.

Zhang, Xinzhong, Xinru Wang, Fengjian Luo, Huishan Sheng, Li Zhou, Qing Zhong, Zhengyun Lou, et al. 2019. "Application and enantioselective residue determination of chiral pesticide penconazole in grape, tea, aquatic vegetables and soil by ultra-performance liquid chromatography-tandem mass spectrometry." *Ecotoxicology and Environmental Safety* 172: 530–37. doi:10.1016/j.ecoenv.2019.01.103.

Zhang, Xinzhong, Yuechen Zhao, Xinyi Cui, Xinru Wang, Huishan Shen, Zongmao Chen, Chun Huang, et al. 2018. "Application and enantiomeric residue determination of diniconazole in tea and grape and apple by supercritical fluid chromatography coupled with quadrupole-time-of-flight mass spectrometry." *Journal of Chromatography A* 1581–1582: 144–55. doi:10.1016/j.chroma.2018.10.051.

Zhang, Yuping, Deyu Hu, Xingang Meng, Qingcai Shi, Pei Li, Linhong Jin, Kankan Zhang, and Baoan Song. 2015. "Enantiosective degradation of indoxacarb from different commercial formulations applied to tea." *Chirality* 27: 262–267. doi:10.1002/chir.22422.

Zhang, Zhaoxian, Beibei Gao, Zongzhe He, Lianshan Li, Haiyan Shi, and Minghua Wang. 2019. "Enantioselective metabolism of four chiral triazole fungicides in rat liver microsomes." *Chemosphere* 224: 77–84. doi:10.1016/j.chemosphere.2019.02.119.

Zhang, Zhaoxian, Beibei Gao, Zongzhe He, Lianshan Li, Qing Zhang, Amir E. Kaziem, and Minghua Wang. 2019. "Stereo-selective bioactivity of the chiral triazole fungicide prothioconazole and its metabolite." *Pesticide Biochemistry and Physiology* 160: 112–18. doi:10.1016/j.pestbp.2019.07.012.

Zhao, Lu, Yue Gao, Jingqian Xie, Qiong Zhang, Fangjie Guo, Shuren Liu, and Weiping Liu. 2019. "A strategy to reduce the dose of multi-chiral agricultural chemicals: The herbicidal activity of metolachlor against *Echinochloa crusgalli*." *Science of the Total Environment* 690: 181–88. doi:10.1016/j.scitotenv.2019.06.521.

Zhao, Pengfei, Shuang Li, Xiaoming Chen, Xingjie Guo, and Longshan Zhao. 2019. "Simultaneous enantiomeric analysis of six chiral pesticides in functional foods using magnetic solid-phase extraction based on carbon nanospheres as adsorbent and chiral liquid chromatography coupled with tandem mass spectrometry." *Journal of Pharmaceutical and Biomedical Analysis* 175: 112784. doi:10.1016/j.jpba.2019.112784.

Zhao, Pengfei, Shuo Lei, Mingming Xing, Shihang Xiong, and Xingjie Guo. 2018. "Simultaneous enantioselective determination of six pesticides in aqueous environmental samples by chiral liquid chromatography with tandem mass spectrometry." *Journal of Separation Science* 41: 1287–1297. doi:10.1002/jssc.201701259.

Zhou, Shanshan, Yongqiang Pan, Lina Zhang, Bin Xue, Anping Zhang, and Meiqing Jin. 2018. "Biomagnification and enantiomeric profiles of organochlorine pesticides in food web components from Zhoushan fishing ground, China." *Marine Pollution Bulletin* 131: 602–10. doi:10.1016/j.marpolbul.2018.04.055.

4 Analysis, Fate, and Toxicity of Chiral Pharmaceuticals in the Environment

Maria Dolores Camacho-Muñoz, Malgorzata Baranowska, and Bruce Petrie

CONTENTS

4.1 Introduction 77
4.2 Analysis of Chiral Pharmaceuticals in the Environment 78
    4.2.1 Sampling and Sample Storage 78
    4.2.2 Sample Preparation..79
    4.2.3 Instrumental Method and Chiral Stationary Phases 84
    4.2.4 Signal Suppression and Method of Quantitation 88
4.3 Fate of Chiral Pharmaceuticals in the Environment 88
    4.3.1 Antidepressants..88
    4.3.2 Beta-Blockers..90
    4.3.3 Stimulants 90
    4.3.4 Analgesics 91
    4.3.5 Antibiotics...93
4.4 Toxicity of Chiral Pharmaceuticals in the Environment 93
4.5 Future Perspectives..96
References..96

4.1 INTRODUCTION

Pharmaceuticals are compounds of environmental concern due to their ubiquity and design to biologically affect specific biochemical and metabolic pathways. They have been detected in different environmental matrices including river waters, sediments, and soils (Dogan et al., 2020; Petrie et al., 2015; Sanganyado et al., 2017; Zhou et al., 2018). The main release pathway of pharmaceuticals and their metabolites into aquatic ecosystems is via discharge of effluents from wastewater treatment plants (WWTPs) to water bodies, due to their incomplete removal during treatment (Fatta-Kassinos et al., 2011; Luo et al., 2014).

Most research to date has investigated pharmaceuticals in river waters with concentrations typically reported in the ng/L to µg/L range observed (Camacho-Muñoz et al., 2010; Lindim et al., 2016; Muñoz et al., 2009; Pereira et al., 2017). Such concentrations pose a threat to the ecology of the receiving environment (Liu et al., 2018; Minguez et al., 2016; Osorio et al., 2016; Watanabe et al., 2016). Sludge disposal, direct animal excretion, or manure application to agricultural soils are the main pathways to enter the terrestrial environment (Ghirardini and Verlicchi, 2019; He et al., 2020; Martín et al., 2012). Pharmaceutical concentrations reported in soils are typically ng/g (Kumirska et al., 2019; Martín et al., 2012; Picó et al., 2020). Whilst there is considerable research on the fate and toxicological effects of pharmaceuticals in the environment, there is a lack of studies which consider pharmaceutical stereochemistry.

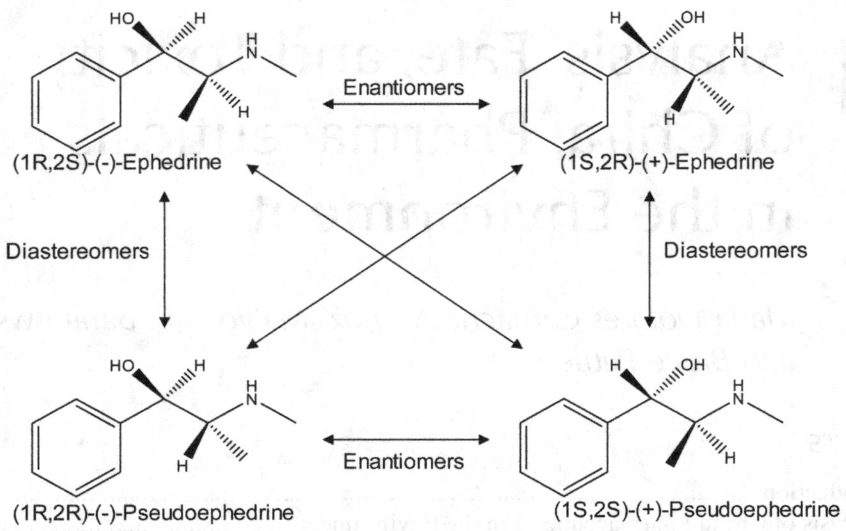

FIGURE 4.1 The four stereoisomers of ephedrine.

Approximately 50% of pharmaceuticals on the market are chiral compounds whereby they have one or more stereogenic center (Mane, 2016). A molecule containing one stereogenic center has two enantiomers which share the same chemical structure but different spatial arrangement of atoms around the stereogenic center (e.g., see Figure 4.1). Enantiomeric fraction (EF) is commonly used to describe the enantiomeric composition of chiral compounds:

$$EF = \frac{(+)}{[(+)+(-)]} \qquad (4.1)$$

where (+) is the concentration of the (+)-enantiomer and (−) is the concentration of the corresponding (−)-enantiomer. The EF can vary between 0 and 1, and 0.5 denotes a racemic composition. An EF of 0 or 1 represents the presence of a single enantiomer only. Compounds with two stereogenic centers will have two pairs of enantiomers and so on. Diastereomers are those stereoisomers that are not mirror images (e.g., see Figure 4.1). Chiral pharmaceuticals are often produced as racemates (i.e., equal concentrations of enantiomers). Less often, pharmaceuticals are dispensed as the racemate and the enantiomerically pure form (e.g., $R/S(\pm)$-ofloxacin and $S(-)$-ofloxacin or levofloxacin), or just in enantiomerically pure form ($S(+)$-naproxen).

Chiral pharmaceuticals exhibit stereoselectivity in their metabolism within the body and degradation during wastewater treatment (Camacho-Muñoz et al., 2019; Duan et al., 2018; Vanessa et al., 2009), environmental fate (Camacho-Muñoz et al., 2019; Ma et al., 2016), and toxicity (Andrés-Costa et al., 2017). Therefore, to better appreciate the environmental risk and impact of chiral pharmaceuticals it is essential to undertake investigations at the enantiomeric level. However, a limiting factor has been the lack of analytical methodologies capable of enantiomeric determinations in complex environmental matrices.

4.2 ANALYSIS OF CHIRAL PHARMACEUTICALS IN THE ENVIRONMENT

4.2.1 Sampling and Sample Storage

Sample collection is an essential step in any analytical protocol. Appropriate sampling protocols as well as proper sample handling are necessary to ensure representative data is obtained. However, it is often under-investigated and could lead to erroneous data and misleading interpretations.

Ort et al. (2010) demonstrated that different active sampling approaches (grab and composite sampling) can provide biased results for pharmaceuticals in wastewater unless sampling is undertaken in a flow-proportional manner with adequate sub-sampling collection frequencies. However, similar studies in environmental matrices such as river water are lacking. It can be considered that temporal changes to pharmaceuticals in river water will be comparatively less than those observed in wastewaters. Indeed, the majority of studies undertaken on chiral pharmaceuticals in river waters adopt grab sampling (Table 4.1). An important consideration for such sampling is enantiomer stability in collected samples during transport and storage prior to sample preparation.

Ramage et al. (2019) investigated the stability of pharmaceutical enantiomers in collected stream water samples. $S(+)$-amphetamine, $S(+)$-fluoxetine and $R(-)$-fluoxetine degraded by >25% in samples stored at 18°C over 48 h (Ramage et al., 2019). Other than the loss of total pharmaceutical concentration (i.e., sum of both enantiomers), the degradation of $S(+)$-amphetamine resulted in a change in EF from 0.5 to 0.1. In samples stored at 4°C no significant enantiomer degradation was observed over 48 h (Ramage et al., 2019).

Alternatively, the addition of a biocide (e.g., sodium azide) or pH adjustment can be used to limit microbial activity. However, care is needed if such approaches are to be considered. For example, the addition of sodium azide can be detrimental to subsequent enantiomeric separations with loss of chiral recognition previously noted (e.g., salbutamol and chlorpheniramine in river water treated with sodium azide at 1 g/L) (Baranowska, 2018). Sodium azide is also reported to be unsuitable for the stabilization of some pharmaceuticals including fluoxetine (Vanderford et al., 2011). Alternatively, sample adjustment to acidic pH can influence the partitioning of pharmaceuticals between the liquid phase and solid phase (suspended matter) of environmental matrices (Petrovic, 2014). Thus filtration of samples prior to pH adjustment is needed which may not be possible in the field. Nevertheless, analyte stability within samples needs to be incorporated into analytical method validation to ensure collected data is representative of what was collected.

Passive sampling is proposed as an alternative method of sampling river water (Pinasseau et al., 2019; Wong and MacLeod, 2009). Passive sampling relies on the transport of a pharmaceutical from the sample matrix to a receiving phase or sorbent contained within a passive sampling device. Both the Chemcatcher® and polar organic integrative sampler have been applied to determine pharmaceuticals in river waters (Rimayi et al., 2019; Toušová et al., 2019). They can be used to estimate time-weighted average concentrations for seven days or longer. However, there is a paucity of information on their ability to maintain the enantiomeric composition of chiral pharmaceuticals during deployment. Nevertheless, the Chemcatcher with a disk receiving phase containing Oasis HLB successfully maintained the EF of atenolol, tramadol, and venlafaxine in wastewater effluent (Petrie et al., 2016). Further studies are needed to fully assess the suitability of passive samplers to provide quantitative information on chiral pharmaceuticals at the enantiomeric level.

4.2.2 Sample Preparation

Pharmaceuticals are often present in liquid environmental matrices at concentrations below the detection capabilities of analytical instrumentation. Thus, sample pre-concentration and clean up is employed. This is normally achieved by solid phase extraction (SPE) and ensures the extracted sample can be reconstituted in the mobile phase so not to disrupt the equilibrium of the chiral chromatography. For chiral analysis, procedures applied are achiral to avoid any stereoselectivity during the process. The most popular SPE cartridge is Oasis HLB which contains a copolymer of divinylbenzene and vinylpyrrolidone (Table 4.1). Relying on reversed phase and ionic interactions, it is well suited for multi residue analysis of pharmaceuticals without the need of pH sample adjustment. However, due to its non-selective nature many matrix interferences remain in extracted samples.

Mixed mode sorbents based on Oasis HLB and containing cation-exchange (Oasis MCX) and anion-exchange groups (Oasis MAX) are also available. Such sorbents are more selective and produce cleaner extracts (Kasprzyk-Hordern et al., 2010). However, they require acidic or basic

TABLE 4.1
Recently Reported Enantioselective LC-MS/MS Methods for Chiral Pharmaceuticals in Environmental Matrices

Chiral Pharmaceutical	Sample Type + Preparation	Chromatographic Column	Mobile Phase Conditions	Run Time (min)	MS Detector	Enantiomer R_S	Recovery (%)	MDL (ng L^{-1} or ng g^{-1})	Reference
Liquid Environmental Matrices									
Ibuprofen, naproxen, flurbiprofen, ketoprofen, and warfarin.	Estuary water (500 mL), acidified to pH 2 and Oasis MCX SPE. Reconstituted in 0.25 mL mobile phase.	Whelk-O®1 (1-(3,5-dinitrobenzamido)-1,2,3,4-tetrahydrophenanthrene) 250 × 4.6 mm, 5 μm @ room temp.	Methanol and water (60/40) + 0.1 % formic acid @ 1 mL min^{-1}	90	QqQ	0.7–32	96–118	0.01–2.66	Coelho et al. (2019)
O-desmethyltramadol, salbutamol, alprenolol, bisoprolol, metoprolol, propranolol, fluoxetine, mirtazapine, venlafaxine, and O-desmethylvenlafaxine.	Estuary water (500 mL), acidified to pH 2 and Oasis MCX SPE. Reconstituted in 0.25 mL mobile phase.	Chirobiotic V® (vancomycin) 150 × 2.1 mm, 5 μm @ 25 °C	10 mM ammonium acetate + 92.5 % ethanol (pH 6.8) @ 0.32 mL min^{-1}	30	QqQ	0.6–1.7	55–81	0.01–2.66	Coelho et al. (2019)
Amphetamine, atenolol, chlorpheniramine, citalopram, fluoxetine, propranolol, and salbutamol	River water (250 mL), filtered (0.7 μm) and Oasis HLB SPE.	Chirobiotic V2® (vancomycin) 250 × 2.1 mm, 5 μm @ 15 °C	1 mM ammonium acetate + 0.01 % acetic acid in methanol @ 0.17 mL min^{-1}	60	QqQ	1–2.3	20–118	0.1–870	Ramage, Camacho-Muñoz, and Petrie (2019)
Ibuprofen, naproxen, and flurbiprofen	River water (500 mL), filtered (0.7 μm) and Oasis HLB SPE.	Chiralpak AD-RH® (amylose-tris(3,5-dimethylphenylcarbamate)) 150 × 2.1 mm, 5 μm @ 30 °C	10 mM ammonium acetate (pH 5) + 35 % acetonitrile @ 0.4 mL min^{-1}	25	QqQ	1–2.3	74–101	0.35–11.1	Ma et al. (2019)

TABLE 4.1 (Continued)
Recently Reported Enantioselective LC-MS/MS Methods for Chiral Pharmaceuticals in Environmental Matrices

Chiral Pharmaceutical	Sample Type + Preparation	Chromatographic Column	Mobile Phase Conditions	Run Time (min)	MS Detector	Enantiomer R_S	Recovery (%)	MDL (ng L^{-1} or ng g^{-1})	Reference
Ofloxacin, ofloxacin-N-oxide, desmethylofloxacin, flumequine, moxifloxacin, nadifloxacin, and prulifloxacin	River water (100 mL), filtered (0.7 µm) and Oasis HLB SPE. Reconstituted in 0.5 mL mobile phase.	Chiralpak OZ-RH® (cellulose-tris(3-chloro-4-methylphenylcarbamate)) 150 × 4.6 mm, 5 µm @ 25 °C	10 mM ammonium formate and 0.05 % formic acid + 99 % methanol @ 0.1 mL min^{-1}	40	QqQ	0.7–2.9	72–119	0.3–33.9	Castrignanò et al. (2018)
Carboxyibuprofen, chloramphenicol, 2-hydroxyibuprofen, ibuprofen, indoprofen, ifosfamide, ketoprofen, naproxen, 2-phenylpropionic acid, and praziquantel	River water (200 mL), filtered (0.7 µm) and Oasis HLB-MAX SPE. Reconstituted in 0.5 mL mobile phase.	Chirobiotic T® (teicoplanin) 250 × 4.6 mm, 5 µm	10 mM ammonium acetate (pH 4.3) + 30 % methanol @ 0.08 mL min^{-1}	90	QqQ	0.4–0.9	8–127	0.12–263	Camacho-Muñoz and Kasprzyk-Hordern (2017)
Omeprazole, lansoprazole, pantoprazole, and rabeprazole	River water (100 mL) adjusted to pH 10 and Cleanert PEP-2 SPE and DLLME.	Chiralpak IC® (cellulose-tris(3,5-dichlorophenylcarbamate) 250 × 4.6 mm, 5 µm	Acetonitrile:5 mM ammonium acetate in water (40:60, v/v) @ 0.6 mL min^{-1}	30	QqQ	>1.5	90–107	0.67–2.29	Zhao et al. (2016)
Aminorex, chloramphenicol, dihydroketoprofen, fexofenadine, ibuprofen, ifosfamide, naproxen, ketoprofen, praziquantel, 3-N-dechloroethylifosfamide and 10,11-dihydro-10-hydroxycarbamazepine, and tetramisole	River water (500 mL) filtered (0.7 µm) and Oasis HLB-MAX SPE. Reconstituted in 0.5 mL mobile phase	Chiral AGP (α1-acid glycoprotein) 100 × 2 mm, 5 µm @ 25 °C	10 mM ammonium acetate in water with 1 % acetonitrile (pH 6.7)	100	QqQ	≥0.7	2–158	0.04–34.7	Camacho-Muñoz and Kasprzyk-Hordern (2015)

(Continued)

TABLE 4.1 *(Continued)*
Recently Reported Enantioselective LC-MS/MS Methods for Chiral Pharmaceuticals in Environmental Matrices

Chiral Pharmaceutical	Sample Type + Preparation	Chromatographic Column	Mobile Phase Conditions	Run Time (min)	MS Detector	Enantiomer R_S	Recovery (%)	MDL (ng L^{-1} or ng g^{-1})	Reference
Ketoconazole, econazole, miconazole, and tebunconazole	River water filtered (0.7 μm) and Oasis HLB SPE.	Chiral AGP (α1-acid glycoprotein) 100 × 4 mm, 5 μm @ 25 °C	10 mM ammonium acetate in water (pH 7) with 15-30 % acetonitrile (gradient elution)	50	QqQ	0.8–2.1	75–103	0.5–7[a]	Huang et al. (2013)
Albuterol, pindolol, propranolol, atenolol, metoprolol, clenbuterol, sotalol, timolol, and fluoxetine	River water (500 mL) filtered (0.7 μm) and Oasis HLB SPE. Reconstituted in 0.5 mL mobile phase	Chirobiotic V® (vancomycin) 250 × 4.6 mm, 5 μm @ 25 °C	4 mM ammonium acetate + 0.005 % formic acid in methanol @ 0.1 mL min^{-1}	65	QqQ	≥0.4–1.1	56–116	0.1–11	López-Serna et al. (2013)
Amphetamine, methamphetamine, MDMA, propranolol, atenolol, metoprolol, venlafaxine, and fluoxetine	River water (250 mL) filtered (0.7 μm) and Oasis HLB SPE. Reconstituted in 0.5 mL mobile phase	Chirobiotic V® (vancomycin) 250 × 4.6 mm, 5 μm @ 25 °C	4 mM ammonium acetate + 0.005 % formic acid in methanol @ 0.1 mL min^{-1}	40	QTOF-MS	0.9–4.7	61–126	0.2–23	Bagnall et al. (2012)
Solid Environmental Matrices									
Salbutamol, propranolol, atenolol, amphetamine, chlorpheniramine, and fluoxetine	Soil (5 g) extracted by ASE and Oasis HLB SPE. Reconstituted in 0.5 mL mobile phase	Chirobiotic V2® (vancomycin) 250 × 2.1 mm, 5 μm @ 20 °C	1 mM ammonium acetate + 0.01 % acetic acid in methanol @ 0.2 mL min^{-1}	65	QqQ	1.1–3.0	76–121	0.02–0.24	Petrie et al. (2018)

TABLE 4.1 (Continued)
Recently Reported Enantioselective LC-MS/MS Methods for Chiral Pharmaceuticals in Environmental Matrices

Chiral Pharmaceutical	Sample Type + Preparation	Chromatographic Column	Mobile Phase Conditions	Run Time (min)	MS Detector	Enantiomer R_S	Recovery (%)	MDL (ng L^{-1} or ng g^{-1})	Reference
Ketoconazole, econazole, miconazole, and tebunconazole	Sediment extracted by USE and Oasis HLB SPE.	Chiral AGP (α1-acid glycoprotein) 100 × 4 mm, 5 μm @ 25 °C	10 mM ammonium acetate in water (pH 7) with 15-30 % acetonitrile (gradient elution)	50	QqQ	0.8–2.1	75–103	0.3–3	Huang et al. (2013)

Key: [a]limit of quantification; MS/MS, tandem mass spectrometry; MDL, method detection limit; QqQ, triple quadrupole; SPE, solid phase extraction; ASE, accelerated solvent extraction; USE, ultrasonic solvent extraction; NH$_4$OAc, ammonium acetate; MeOH, methanol; ACN, acetonitrile; HCOOH, formic acid; CH$_3$COOH, acetic acid; QTOF, quadrupole time of flight; MDMA, 3,4-methylenedioxy-methamphetamine modifiers in the elution solvent which can influence subsequent chromatographic separations (Evans et al., 2015). For example, the elution of basic compounds from Oasis MCX cartridges with 7% ammonium hydroxide in methanol, drastically reduced the chromatographic resolution (R_S) of enantiomers of almost all compounds trialed (amphetamines, beta-blockers, antidepressants, analgesics, bronchodilators) (Kasprzyk-Hordern et al., 2010). On the other hand, the use of Oasis MAX cartridges with an eluting agent of methanol containing 2% formic acid did not report any detrimental effect to chiral separation (Camacho-Muñoz and Kasprzyk-Hordern, 2015). Nevertheless, care is needed when selecting sample preparation methods as the SPE method which provides the greatest analyte recovery may be detrimental to subsequent chiral separations.

The analysis of solid environmental matrices such as soils and sediments requires an additional sample extraction method. This involves the transfer of the analyte from the solid phase to a liquid phase. Both accelerated solvent extraction and ultrasonic solvent extraction have been reported prior to chiral analysis (Petrie et al., 2018; Pérez-Lemus et al., 2019). Microwave assisted extraction can also be used for the analysis of pharmaceuticals from solid matrices (Azzouz and Ballesteros, 2016; Evans et al., 2015; Sanchez-Prado et al., 2015). Once samples have been extracted using a suitable solvent they can be diluted using water and treated as a liquid sample (i.e., undergo SPE).

4.2.3 INSTRUMENTAL METHOD AND CHIRAL STATIONARY PHASES

Broadly, there are two means of separating enantiomers so they can be measured independently. Indirect methods involve the derivatization of enantiomers to diastereomers. Diastereomers have different physicochemical properties (whereas enantiomers do not) enabling their separation using achiral methods. Direct methods involve the use of a chiral selector added to the mobile phase, or more commonly, to the stationary phase. In both instances, one enantiomer has a higher affinity to the chiral selector than the other which facilitates separation. A broad range of commercially available enantioselective columns are available for gas chromatography (GC) and in particular, high-performance liquid chromatography (HPLC).

Due to the polar nature of pharmaceuticals, HPLC is the preferred separation system as pre-column derivatization is needed to make analytes GC amenable. Furthermore, coupling the separation to mass spectrometry (MS) detection and more specifically tandem MS (MS/MS) is essential to provide the desired specificity and sensitivity for the determination of pharmaceuticals in environmental matrices. However, traditionally enantioselective methods have been developed on HPLC systems utilizing ultraviolet detectors for quality control purposes in the pharmaceutical industry. Such methods typically utilize normal phase conditions with mobile phases not compatible with MS. Therefore, new chromatography methods have been developed using volatile solvents and additives making them compatible with HPLC-MS/MS (Camacho-Muñoz and Kasprzyk-Hordern, 2015, 2017; Caballo et al., 2015; Petrie et al., 2015; Ribeiro et al., 2014).

Macrolide glycoprotein antibiotic stationary phases (vancomycin, teicoplanin, etc.) are popular for the enantioselective separation of pharmaceuticals (Table 4.1). The features of these phases include a basket-shaped aglycon framework formed by either three or four fused macrocyclic rings and several functional groups attached (e.g., hydroxyl, amine, aromatic, and carboxylic acid moieties, amide linkages, etc.). They separate enantiomers via hydrogen bonding, dipole interactions (ion dipole, dipole-dipole, dipole-induced dipole), π-π interactions, inclusion complexation, steric interactions, Van der Waals forces, and anionic and cationic binding (Berthod et al., 2010). Their wide versatility enables operation in reversed phase, polar organic, and polar ionic modes which are all compatible with MS.

Methods in the literature report the use of vancomycin stationary phases mainly for the analysis of basic pharmaceuticals (Guo et al., 2018; López-Serna et al., 2013) whereas teicoplanin stationary phases have shown a preference for the analysis of acidic pharmaceuticals, such as profens (Camacho-Muñoz and Kasprzyk-Hordern, 2017). Electrostatic interactions are considered one of the dominant interactions occurring between these stationary phases and analytes. Vancomycin has an isoelectric point of ~7.2 and teicoplanin ~3.5. The chiral selector in macrocyclic antibiotic-based phases is ionizable, therefore changes in the acid/base ratio will affect the degree of ionization of the chiral selector itself (Ilisz et al., 2013). Nevertheless, it should be noted that chiral recognition is a sum of interactions not the result of just one (Beesley and Lee, 2009; Ilisz et al., 2019). The three-point interaction model has traditionally been used to describe the chiral recognition process. Here, three functional groups of one of the enantiomers undergo molecular interactions with three sites on the chiral stationary phase forming a diastereomeric complex (Sanganyado et al., 2017). Therefore, this enantiomer has a stronger affinity to the stationary phase than the other enantiomer, which only has one point of interaction.

The vancomycin and teicoplanin stationary phases have collectively resolved a considerable number of pharmaceuticals in environmental samples, including profens, anticancer drugs, β-blockers, β-agonists, antihistamines, antidepressants, antibiotics, and anthelmintics among others (Bagnall et al., 2012; Coelho et al., 2019; Camacho-Muñoz and Kasprzyk-Hordern, 2017; López-Serna et al., 2013; Ramage et al., 2019) (Table 4.1). Most methods report mobile phases comprising a high percentage of organic solvents with small percentages of acid and/or basic modifiers and buffer concentrations which improve ionization as well as chromatographic separation (López-Serna et al., 2013). Polar ionic mode (e.g., methanol containing small concentrations of volatile additives) is ideally suited for MS and provides enhanced sensitivity over reversed phase mode (Bagnall et al., 2012). Previous work on vancomycin stationary phases has found that the buffer concentration (typically ammonium acetate) has the greatest influence on enantioresolution (Petrie et al., 2018; Sanganyado et al., 2014).

Protein-based chiral stationary phases are also popular for the separation of pharmaceutical enantiomers due to the inherent chiral distinction capability of enzymes and plasma proteins (Lämmerhofer, 2010). The chiral recognition ability of proteins is related to the formation of a three-dimensional structure which plays an essential physiological role. Due to the complexity of proteins, hydrophobic, ionic, π-π, and steric interactions and hydrogen bonding are assumed to be the main retention mechanisms (Haginaka, 2001). Factors such as the nature and content of the organic modifier, the mobile phase pH, the ionic strength, and the nature and content of charged organic modifiers, can influence chiral recognition on these stationary phases (Haginaka, 2001).

Among a series of commercially available protein-based stationary phases include α_1-acid glycoprotein (AGP), cellobiohydrolase (CBH), and human serum albumin (HSA). AGP shows the broadest enantiomer separation capabilities (Lämmerhofer, 2010), whereas CBH is preferred for solely basic chiral analytes and HAS for acidic chiral analytes. Their preferential operational mode with aqueous or aqueous-organic eluents makes them easily coupled with MS. However, these stationary phases are particularly sensitive to pH (4–7 for AGP, 3–7 for CBH, and 5–7 for HSA), organic content in the mobile phase (<20%), and temperature (max. 40°C, normal operating range is 20–25°C), with denaturing likely when operated outside these conditions (Sigma-Aldrich, 2020).

In the case of the CBH stationary phase, a cellobiohydrolase enzyme is immobilized onto spherical 5 μm silica particles. According to Henriksson et al. (1999), three main active chiral-recognition areas are apparent: an enzymatically active core, a cellobiohydrolase domain with 36 amino acids forming two disulphide-bridged loops and a connected area. The dominating enantioselective site is located on the core. The major chiral recognition mechanisms have been attributed to the electrostatic and hydrophobic interactions between CBH and the chiral analytes (Henriksson et al., 1997). CBH has been successfully applied, using an isocratic mobile phase of water with 10% 2-propanol and 1 mM ammonium acetate to separate amphetamine-like compounds (such as amphetamine, methamphetamine, 3,4-methylenedioxy-methamphetamine, 3,4-methylenedioxyamphetamine, 3,4-methylenedioxy-N-ethyl-amphetamine, ephedrine/pseudoephedrine, norphedrine) as well as venlafaxine and atenolol in environmental samples within one analytical run (Bagnall et al., 2012; Kasprzyk-Hordern and Baker, 2012).

Developed initially by Hermansson (1984), AGP consists of a single peptide chain with 181 amino acids and five heteropolysaccharide units, containing 14 residues of sialic acid. The mechanism of chiral recognition by AGP is not clearly stated but a combination of hydrophobic and electrostatic interactions and hydrogen bonding are considered important. To date, the most comprehensive method for the environmental analysis of pharmaceuticals was developed by Camacho-Muñoz and Kasprzyk-Hordern (2015). In this study, excellent chromatographic separation of enantiomer (Rs ≥ 1.0) was achieved for chloramphenicol, fexofenadine, ifosfamide, naproxen, tetramisole, ibuprofen, and their metabolites: aminorex and dihydroketoprofen (three of four enantiomers), an partial separation (Rs = 0.7–1.0) was achieved for ketoprofen, praziquantel, and the metabolite 3-N-dechloroethylifosfamide and 10,11-dihydro-10-hydroxycarbamazepine (Figure 4.2). This was achieved using a mobile phase comprising 10 mM ammonium acetate in water with 1% acetonitrile (pH 6.7) (Camacho-Muñoz and Kasprzyk-Hordern, 2015). However, chromatographic run time is the limitation of enantioselective methods for multi residue analysis. Run times of ≥55 minutes are typical (Table 4.1).

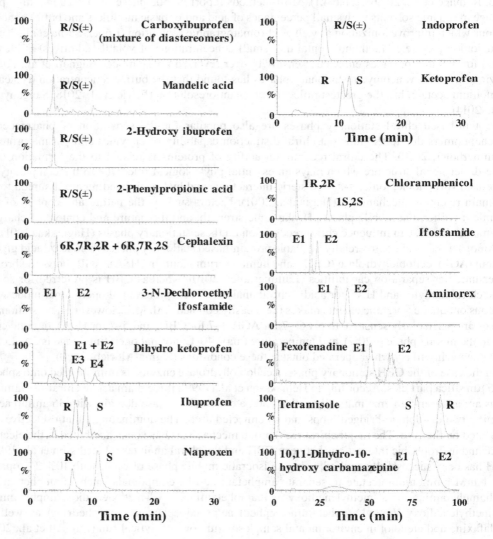

FIGURE 4.2 Enantioselective separation of human and veterinary pharmaceuticals in extracted river water using an α_1-acid glycoprotein stationary phase. (Adapted from Camacho-Muñoz and Kasprzyk-Hordern, 2015.)

A further group of chiral stationary phases popular for environmental analysis are polysaccharide derivatives. Traditionally, these phases were prepared by coating polysaccharide derivatives on silica beds (Okamoto et al., 1984). More recent immobilization versions facilitate expanded solvent resistance resulting in greater robustness and a wider range of applications. A large variety of polysaccharide-based derivatives are now available (Shen et al., 2014). These are amylose or cellulose derivatives. The chains of glucose units on the amylose lie side by side in a helical way, whereas this conformation is linear in the case of the cellulose. The most widely used stationary phases are amylose tris-(3,5-dimethylphenylcarbamate), cellulose tris-(3,5-dimethylphenylcarbamate) and cellulose tris-(3,5-dichlorophenylcarbamate) (Camacho-Muñoz et al., 2016; Karakka Kal et al., 2019). It has been reported that 95% of chiral compounds have been separated on polysaccharide-based phases (Zhang et al., 2005). The exceptional chiral recognition ability of polysaccharide-based phases not only depends on the cellulose or amylose backbone but also the nature and location of the substituents (Pan et al., 2017).

Castrignanò et al. (2018) used a cellulose-tris(3-chloro-4-methylphenylcarbamate) phase to separate several fluoroquinolone antibiotics including ofloxacin and its metabolites desmethylofloxacin and ofloxacin-N-oxide. Separation was achieved using 10 mM ammonium formate and 0.05% formic acid in 99:1 methanol:water for this analytically challenging group of pharmaceuticals (Table 4.1). Ma et al. (2019) used a tris(3,5-dimethylphenylcarbamate) phase for the separation of ibuproxen, naproxen, and flurbiprofen. In this study a more conventional reversed phase system was utilized with a mobile phase of 10 mM ammonium acetate (pH 5) and 35% acetonitrile. A tris(3,5-dichlorophenylcarbamate) stationary phase has been used to separate the proton pump inhibitors omeprazole, lansoprazole, pantoprazole, and rabeprazole (Zhao et al., 2016). Again, reversed phase conditions (5 mM ammonium acetate and acetonitrile (60:40)) were utilized.

An emerging technology for enantioselective separation is supercritical fluid chromatography (SFC). The benefits of SFC over conventional HPLC systems are fast analysis speeds, wide polarity compatibility, high column efficiency, and environmentally friendly mobile phases. Similar to HPLC chiral stationary phases are utilized for separation (3–5 µm particle size). To date, the polysaccharide-based phases are the favored choice for fast enantioselective separations (Camacho-Muñoz et al., 2016; Svan et al., 2015).

In SFC the main mobile phase component is carbon dioxide. Carbon dioxide offers interesting features as it is non-toxic, non-flammable, low viscosity, available as a by-product of many industries, inexpensive, low values of temperature (31°C) and pressure (74 bar), and environmentally friendly when compared with most commonly used organic solvents in HPLC. However, pure carbon dioxide has a low solvent strength for elution of highly polar compounds, such as pharmaceuticals. In this case, the use of various co-solvents (organic modifiers) in proportions typically varying from 1 to 50%, under both isocratic (to achieve the highest possible enantioresolution) and gradient (for initial screening conditions) elution modes are used. The addition of organic modifiers is used to manipulate retention, selectivity, and efficiency. The most widely used co-solvents are alcohols, especially methanol due to its superior hydrogen bonding properties, followed by ethanol and isopropanol, as well as mixtures (De Klerck et al., 2012). The use of acetonitrile in binary and/or ternary mixtures with alcohols has been reported only on a few occasions (Camacho-Muñoz et al., 2016).

Similar to other chromatographic techniques, small quantities of additives (0.1–1%) can be employed to improve resolution and peak shapes. MS-compatible basic additives (diethylamine, ammonia, or ammonium hydroxide) are added when analyzing basic analytes, while acidic additives (acetic, formic, or trifluoroacetic acid) when analyzing acidic analytes (Mosiashvili et al., 2013). They act by modifying the stationary phase surface or as ion pairs affecting selectivity. Volatile salts, including ammonium acetate or formate, can also be used to manipulate the separation.

Although SFC has been extensively applied for purification and chiral separation of chiral pharmaceuticals in the pharmaceutical industry, its application for environmental research using MS detection is limited. Recently, several veterinary and human pharmaceuticals were separated using ultrahigh-performance supercritical fluid chromatography tandem mass spectrometry (UHPSFC-MS/MS), albeit for wastewater matrices (Camacho-Muñoz et al., 2016). The use of two coated modified 2.5 μm polysaccharide-based stationary phases (amylose tris-(3,5,dimethylphenylcarbamate) and cellulose tris-(3-chloro-4-methylphenylcarbamate)) allowed the baseline enantioresolution of 13 pharmaceuticals (aminorex, carprofen, chloramphenicol, 3-N-dechloroethylifosfamide, flurbiprofen, 2-hydroxyibuprofen, ifosfamide, imazalil, naproxen, ofloxacin, omeprazole, praziquantel, and tetramisole) and partial resolution of two pharmaceuticals (ibuprofen and indoprofen) in under 10 minutes.

4.2.4 SIGNAL SUPPRESSION AND METHOD OF QUANTITATION

The most popular ionization source for both HPLC-MS/MS and SFC-MS/MS is electro-spray ionization (ESI) (Laaniste et al., 2019). However, a well-known drawback of ESI for environmental analysis is the loss of analyte signal response during ionization (Kruve et al., 2009; Laaniste et al., 2019). This is a result of co-extracted matrix components suppressing analyte ionization. On the other hand, signal enhancement can also be observed (Ramage et al., 2019). However, a further consideration for chiral analysis is the possibility of enantioselective suppression. For example, in wastewater $S(-)$-propranolol-d_7 and $R(+)$-propranolol-d_7 had signal suppressions of 44% and 94%, respectively (López-Serna et al., 2013). It is essential that such interferences are accounted for, such that erroneous results are not reported. This can be achieved in two ways: (i) preparing an external calibration curve in the matrix to be analyzed, and (ii) using deuterated surrogates or deuterated internal standards added to both the sample and calibration standards at the same concentration.

By using matrix matched calibrations both the calibrations standards and samples are subject to the same extent of suppression and in theory will be accounted for. However, such an approach is not popular because it is time consuming. It is also not always possible to obtain the same matrix which is analyte free to prepare the calibration. Sample specific suppression has previously been observed in environmental analysis (Kruve et al., 2008). Therefore, differences in suppression could be observed from different samples collected from the same river, for example. The internal standard method is the preferred method of quantitation. Deuterated standards are typically added prior to sample extraction as a surrogate standard such that all analyte losses are accounted for (not exclusively signal suppression during ESI). Despite the loss in analyte signal that can be observed, method detection limits can still reach sub-ng/L concentrations (Table 4.1). The development of these robust enantioselective methods capable of trace analysis in environmental matrices has helped improve our understanding of chiral drug fate.

4.3 FATE OF CHIRAL PHARMACEUTICALS IN THE ENVIRONMENT

4.3.1 ANTIDEPRESSANTS

Fluoxetine, citalopram, and venlafaxine are highly prescribed chiral antidepressants. Due to their high usage and incomplete removal during wastewater treatment, they have been widely reported in the aquatic environment (Boogaerts et al., 2019; Golovko et al., 2014). Fluoxetine is typically reported with an EF > 0.5 due to an enrichment of $S(+)$-fluoxetine (Evans et al., 2017; López-Serna et al., 2013) (Table 4.2). Wastewater effluents are typically enriched with $S(+)$-fluoxetine due to enantioselective metabolism in the human body and during wastewater treatment (Evans et al., 2017; López-Serna et al., 2013). However, once in the environment enantioselective degradation can take place. Enantioselective processes that occur in the environment are typically investigated in laboratory scale microcosms. Here the environmental matrix of interest (e.g., river water) is kept under controlled conditions (temperature, light exposure, etc.) and spiked with the pharmaceutical of interest. Samples are then collected at different time intervals to investigate exposure conditions to pharmaceutical concentration and enantiomeric composition.

TABLE 4.2
Example Enantiomeric Fractions of Pharmaceuticals in Freshwaters

Therapeutic Group	Chiral Pharmaceutical	Location	Matrix	Enantiomeric Fraction	Reference
Anti-inflammatory drugs	Ibuprofen	Spain	River	0.46	Caballo et al. (2015)
		Spain	Rivers	0.30–0.50	López-Serna et al. (2013)
		Switzerland	Lakes and rivers	0.41–0.47	Buser et al. (1999)
		Australia	River	0.60–0.80	Khan et al. (2014)
		China	Canal and river	0.13–0.33	Wang et al. (2013)
		China	River	0.69–0.71	Ma et al. (2019)
		China	River	0.38	Wang et al. (2018)
	Naproxen	Australia	River	1.00	Khan et al. (2014)
		China	River	0.92–0.94	Ma et al. (2019)
		UK	River	0.92–0.97	Camacho-Muñoz et al. (2019)
	Flurbiprofen	China	River	0.45	Wang et al. (2018)
Beta-blockers	Propranolol	USA	Rivers	0.21–0.53	Fono and Sedlak (2005)
		UK	River	0.45	Bagnall et al. (2012)
		Spain	Rivers	0.38–0.60	López-Serna et al. (2013)
		UK	River	0.38–0.41	Evans et al. (2017)
	Atenolol	UK	River	0.38–0.56	Kasprzyk-Hordern and Baker (2012)
		UK	River	0.47	Bagnall et al. (2012)
		Spain	Rivers	0.38–0.50	López-Serna et al. (2013)
		UK	River	0.47	Evans et al. (2017)
		UK	River	0.45–0.50	Ramage et al. (2019)
	Sotalol	Spain	Rivers	0.41–0.65	López-Serna et al. (2013)
	Metoprolol	Spain	Rivers	0.42–0.51	López-Serna et al. (2013)
		Germany	River	0.43–0.46	Kunkel and Radke (2012)
Antidepressant drugs	Fluoxetine	Spain	Rivers	0.64–0.68	López-Serna et al. (2013)
		UK	River	0.65	Evans et al. (2017)
	Norfluoxetine	Spain	Rivers	0.79	López-Serna et al. (2013)
	Venlafaxine	UK	River	0.50	Evans et al. (2017)
		UK	River	0.58	Bagnall et al. (2012)
	Amphetamine	UK	Small stream	0.43	Ramage et al. (2019)
	Ofloxacin	UK	River	0.51	Castrignanò et al. (2018)

Andrés-Costa et al. (2017) investigated the fate of fluoxetine in river water microcosms under different biotic and light conditions. Over 16 days of study, little enantioselectivity was observed. Nevertheless, degradation was observed under light (biotic and abiotic) and dark (biotic only) conditions confirming both microbial processes and photolysis contribute to fluoxetine degradation (Andrés-Costa et al., 2017). Interestingly, a similar study by Evans et al. (2017) reported no photolysis of fluoxetine in river water microcosms. However, mild enantioselectivity was observed due to microbial processes leading to enrichment of $S(+)$-fluoxetine and an EF > 0.5. Therefore, the fate of fluoxetine will vary between rivers (and temporally) and needs consideration when assessing the environmental risk.

In wastewater effluents, citalopram is typically enriched with $S(-)$-citalopram and EFs are <0.5 (Ramage et al., 2019). In river water microcosms, little degradation was observed with no enantioselectivity over 14 days (Evans et al., 2017). Interestingly, its metabolite desmethylcitalopram degraded by 70% during this time with considerable enantioselectivity. This resulted in an enrichment of the $S(+)$-enantiomer (Evans et al., 2017). Similar observations were noted for venlafaxine and desmethylvenlafaxine. Venlafaxine showed little degradation whereas desmethylvenlafaxine degraded considerably during 14 days (93%) and enantioselectively (Evans et al., 2017). The lack of change to the enantiomeric composition of citalopram and venlafaxine in river water suggest they could be used as markers for identifying their source (e.g., between treated and untreated wastewater). This would require changes to their enantiomeric composition during wastewater treatment. However, care must be taken as there are limited studies on the environmental fate, and enantiospecific fate of fluoxetine was found to differ between river waters (Andrés-Costa et al., 2017; Evans et al., 2017).

Only limited studies have investigated the enantiospecific fate of chiral pharmaceuticals in soil. Antidepressants such as fluoxetine have moderate hydrophobicity (log K_{OW} 4.1) and partition into sludge during wastewater treatment (Radjenović et al., 2009). Sludge then undergoes anaerobic digestion and many pharmaceuticals including fluoxetine persist during this treatment (Radjenović et al., 2009). Digested sludge (or biosolids) is then applied to agricultural land as fertilizer. However, no data exists on the enantiomeric composition of background chiral pharmaceutical concentrations found in soil.

Nevertheless, in soil microcosms fluoxetine was found to be resistant to degradation over 56 days incubation (Bertin et al., 2020; Petrie et al., 2018). In similar studies, citalopram showed some degradation over 56 days (up to 25%) but no enantioselectivity was observed (Bertin et al., 2020). Therefore, the enantiomeric composition of fluoxetine and citalopram may not change once present in soil. However, digested sludge used for land application typically has non-racemic compositions of citalopram and fluoxetine (Evans et al., 2015). For example, the EF of fluoxetine in digested sludge was found to be 0.7±0.1 (Evans et al., 2015).

4.3.2 BETA-BLOCKERS

Beta-blockers are another group of high use chiral pharmaceuticals which are prescribed for the treatment of cardiovascular diseases. The most well studied beta-blockers and most commonly reported in the environment are atenolol and propranolol (Ma et al., 2016; Ribeiro et al., 2012; Zhou et al., 2018). Both atenolol and propranolol have been reported in river waters with EFs typically being <0.5 (Table 4.2). In river water microcosms, atenolol and propranolol were susceptible to biodegradation with little change to their enantiomeric composition (Evans et al., 2017). The mild enantioselectivity observed favored enrichment of the $S(-)$-enantiomer and EFs <0.5. Fono and Sedlak (2005) reported little or no change to propranolol EF in river water microcosms over 20 days.

In soil microcosms, propranolol showed moderate degradation (<40%) over 56 days incubation (Bertin et al., 2020; Petrie et al., 2018). In both studies, no enantioselectivity was observed. On the other hand, atenolol was found to degrade rapidly reaching concentration method detection limits within 14 days. In this case, enantiospecific degradation was observed whereby elimination of $R(+)$-atenolol was favored (Bertin et al., 2020; Petrie et al., 2018). Microcosms spiked with individual enantiomers confirmed enantiospecific degradation took place over other potential enantioselective mechanisms (e.g., chiral inversion) (Bertin et al., 2020). This study also reported that the enantiomer degradation rate was concentration dependent. In this case, higher initial concentrations (100 ng/g versus 10 ng/g) had greater rates of degradation.

4.3.3 STIMULANTS

Stereoselectivity in the environment has been reported for amphetamine-like compounds (Kasprzyk-Hordern and Baker, 2012). In particular, amphetamine has received the greatest attention. In river water microcosms amphetamine degrades enantioselectively, with degradation favoring the removal of $S(+)$-amphetamine (Baranowska, 2018; Bagnall et al., 2012; Evans et al., 2016). Amphetamine has been found to be photo-stable with degradation only occurring under biotic

Chiral Pharmaceuticals in the Environment

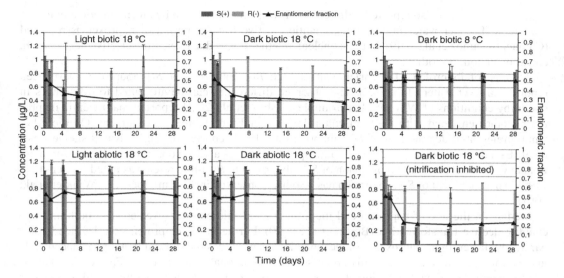

FIGURE 4.3 Enantioselective degradation of R/S(±)-amphetamine in river water microcosms (River Dee, Aberdeenshire) under different biotic and abiotic treatment conditions (Baranowska, 2018).

conditions (Bagnall et al., 2012). Degradation of S(+)-amphetamine resulted in EFs close to 0 before complete removal of amphetamine was observed (Evans et al., 2016).

Further studies where river water was maintained at 8°C (against 18 or 20°C used in other studies) resulted in the cessation of any amphetamine biodegradation (Baranowska, 2018) (Figure 4.3). This indicates a seasonal dependency on chiral drug fate particularly under changing environmental temperatures typical of temperate climates. Further studies were conducted with river water treated with allylthiourea (ATU) to inhibit nitrifying activity. It was found that in ATU-treated river water no effect to amphetamine degradation was observed against untreated river water (Figure 4.3). Although ATU is non-selective and inhibits the activity of other mono-oxygenases it demonstrates nitrifying bacteria do not contribute to amphetamine biodegradation. Therefore, other bacteria such as heterotrophs may be responsible for amphetamine biodegradation (Baranowska, 2018). However, additional studies are needed to investigate this further.

In soil microcosms amphetamine was completely degraded within 28 d (Bertin et al., 2020; Petrie et al., 2018). Within this time degradation of S(+)-amphetamine was comparatively faster than R(−)-amphetamine. This resulted in EFs close to 0 before complete degradation was observed. The enantioselective degradation behavior was similar to that observed in river waters (Baranowska, 2018; Bagnall et al., 2012; Evans et al., 2016). Furthermore, single enantiomer microcosms revealed that chiral inversion did not contribute to the changes in enantiomeric composition (Bertin et al., 2020).

4.3.4 Analgesics

Analgesics such as ibuprofen and naproxen are some of the most studied in the environment (da Silva et al., 2012; Li, 2014). Nevertheless, many studies do not consider their stereochemistry. Naproxen is sold in enantiomerically pure form (S(+)-naproxen) and ibuprofen in racemic form. River waters in the UK have been found to be enriched with S(+)-naproxen due to its dispensed form (Camacho-Muñoz et al., 2019). This is similar to river waters in Australia and China (Khan et al., 2014; Ma et al., 2019) (Table 4.2). However, the presence of R(−)-naproxen was also noted and EFs <1. Chiral inversion has been reported for naproxen whereby S(+)-naproxen coverts to its antipode during wastewater treatment (Hashim and Khan, 2011; Suzuki et al., 2014).

Nguyen et al. (2017) also reported bidirectional inversion of naproxen by an enzymatic membrane bioreactor. In river water microcosms, enantioselective transformation of naproxen is reported (Camacho-Muñoz et al., 2019), but it is unclear whether this is due to enantioselective degradation, chiral inversion, or both.

A range of ibuprofen EFs (enrichment of both $R(-)$-ibuprofen or $S(+)$-ibuprofen) have been reported in river water (Table 4.2). The observed differences can be attributed to differences in operating conditions of WWTPs. For example, Hashim et al. (2011) reported that racemic ibuprofen becomes enriched with $R(-)$-ibuprofen during membrane bioreactor treatment. On the other hand, preferential degradation of $R(-)$-ibuprofen has been reported during activated sludge treatment (Escuder-Gilabert et al., 2018). Enantiospecific changes are also observed in the environment. In lake water microcosms, the inversion of $S(+)$-ibuprofen to $R(-)$-ibuprofen was noted (Buser et al., 1999). An enrichment of $S(+)$-ibuprofen has also been observed in river microcosms (Winkler et al., 2001).

Studies on the fate of naproxen and ibuprofen in soil microcosms revealed they both undergo enantioselective transformation (Bertin et al., 2020). Furthermore, ibuprofen was degraded comparatively faster than naproxen. Single enantiomer microcosms showed both pharmaceuticals underwent chiral inversion, and this process was bidirectional (see naproxen; Figure 4.4). This is significant because the introduction of a less or non-toxic enantiomer to agricultural soils could lead to the formation of the toxic antipode. In this case enantiomer enrichment favored $S(+)$-naproxen and $R(-)$-ibuprofen, respectively (Bertin et al., 2020).

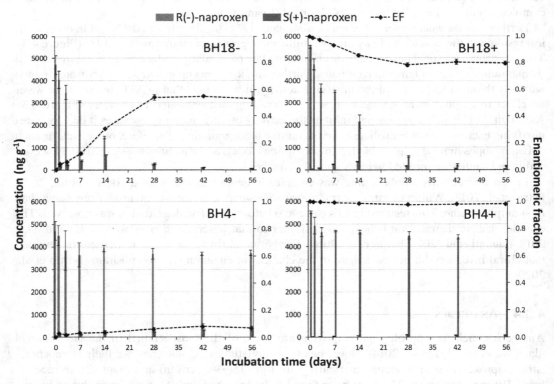

FIGURE 4.4 Concentration of $R(-)$-naproxen and $S(+)$-naproxen and the corresponding enantiomeric fraction in soil microcosms spiked with individual naproxen enantiomers (Bertin et al., 2020).
KEY: BH18-, biotic high spike level of (−)-enantiomer 18°C microcosm; BH18+, biotic high spike level of (+)-enantiomer 18°C microcosm; BH4-, biotic high spike level of (−)-enantiomer 4°C microcosm; BH4+, biotic high spike level of (+)-enantiomer 4°C microcosm.

4.3.5 ANTIBIOTICS

Antibiotics are ubiquitous in the aquatic environment (Michael et al., 2013). There are concerns due to their ability to select for antimicrobial resistance (Sharma et al., 2016). The enantiomeric composition of a chiral antibiotic will play an important role in its ability to select for resistance if pharmacological activity varies between enantiomers. For example, $S(-)$-ofloxacin is considered to be two orders of magnitude more potent than $R(+)$-ofloxacin (Hayakawa et al., 1986; Morrissey et al., 1996). However, the enantiospecific fate and behavior of chiral antibiotics has received little attention. Castrignanò et al. (2018) reported a racemic ofloxacin composition in river water upstream of a WWTP discharge point. Downstream an enrichment of the active $S(-)$-ofloxacin was found.

Another antibiotic studied at the enantiomeric level is chloramphenicol. Chloramphenicol has two asymmetric carbons and four stereoisomers, or two pairs of enantiomers. However, it is prescribed in an enantiomerically pure form as $R,R(-)$-chloramphenicol. Interestingly, all four chloramphenicol enantiomers have been detected in wastewater in China, with $R,R(-)$-chloramphenicol being predominant (Ruan et al., 2019). Further research is needed to determine the source of those enantiomers that are not dispensed. No data exists on the enantiomeric composition of background chloramphenicol concentrations in the environment. River water microcosms have revealed that under biotic conditions $R,R(-)$-chloramphenicol degraded faster than $S,S(+)$-chloramphenicol (Camacho-Muñoz et al., 2019).

4.4 TOXICITY OF CHIRAL PHARMACEUTICALS IN THE ENVIRONMENT

Numerous studies have investigated the toxicity of chiral pharmaceuticals to aquatic organisms (Santos et al., 2010). However, most studies do not account for the possibility of enantiospecific toxicity, or report the enantiomeric composition of the pharmaceutical reference standard used in the test. This is concerning considering the enantiomeric composition of pharmaceuticals observed in the environment (Table 4.2). To date, only a few studies have conducted toxicity testing on individual pharmaceutical enantiomers (Andrés-Costa et al., 2017; De Andrés et al., 2009; Stanley et al., 2006; Stanley et al., 2007). Furthermore, these have been limited to a few pharmaceuticals (Table 4.3).

The beta-blocker propranolol has shown enantiospecific toxicity towards both *Daphnia magna* and *Pimephales promelas* (Stanley et al., 2006). In both instances, $R(+)$-propranolol was more toxic than $S(-)$-propranolol. The LOEC(7d) (lowest observed effect concentration) to *P. promelas* growth was 0.13 mg/L and 0.46 mg/L for $R(+)$-propranolol and $S(-)$-propranolol, respectively. Interestingly, environmental data shows an enrichment of $S(-)$-propranolol as EFs are typically <0.5 (Table 4.2). This suggests that without considering enantioselectivity in both toxicity and composition of propranolol in the environment, results in the potential risk being overestimated. However, the enantiospecific toxicity data is very limited and conducted at concentrations considerably greater than those found in the environment. This highlights the need to conduct further toxicity testing at the enantiomeric level to better appreciate the environmental risks of chiral pharmaceuticals. Similarly, the enantiospecific toxicity of atenolol is limited to one study (De Andrés et al., 2009). In this case, a maximum four-fold difference in enantiomer toxicity was observed.

The pharmaceutical which has received the greatest attention at the enantiomeric level is the antidepressant fluoxetine (Table 4.3). Three studies have investigated toxicity at the enantiomeric level. Andrés-Costa et al. (2017) recorded the greatest difference in enantiomer toxicity to *Tetrahymena thermophila*. Here, growth EC_{50}(24h) were 35.2 mg/L and 1.3 mg/L for $S(+)$-fluoxetine and $R(-)$-fluoxetine, respectively (Andrés-Costa et al., 2017). However, De Andrés et al. (2009) report an opposite trend towards *T. thermophilia*. It is essential to further understand the enantiospecific toxicity of fluoxetine considering river waters are typically enriched with $S(+)$-fluoxetine (Table 4.3).

Similarly, Rice et al. (2018) investigated the toxicity of ephedrine stereoisomers ($1S,2R(+)$-ephedrine, $1R,2S(-)$-ephedrine, $1S,2S(+)$-pseudoephedrine, and $1R,2R(-)$-pseudoephedrine) to

TABLE 4.3
Aquatic Toxicity Tests of Pharmaceutical Enantiomers

Chiral Pharmaceutical	Test Organism/ Test Cells	Test	Endpoint	Toxicity (mg L^{-1}) (S)-	(R)-	Difference[e]	Reference
Propranolol	*Daphnia magna*	LOEC(21d)	Immobility	0.87	0.41	2.1	Stanley et al. (2006)
	Pimephales promelas	LOEC(7d)	Growth	0.46	0.13	3.5	Stanley et al. (2006)
Atenolol	*Daphnia magna*	LC$_{50}$(48h)	Immobility	755	1450	1.9	De Andrés et al. (2009)
	Tetrahymena thermophila	EC$_{50}$(24h)	Growth	55.7	13.7	4.1	De Andrés et al. (2009)
	Pseudokirchneriella subcapitata	EC$_{50}$(72h)	Growth	143	190	1.3	De Andrés et al. (2009)
Fluoxetine	*Daphnia magna*	LOEC(21d)	Immobility	0.44	0.43	—	Stanley et al. (2007)
		LOEC(21d)	Grazing	0.20	none	—	Stanley et al. (2007)
		LOEC(21d)	Reproduction	0.44	0.43	—	Stanley et al. (2007)
		LC$_{50}$(48h)	Immobility	6.9	8.1	1.2	De Andrés et al. (2009)
		EC$_{50}$(48h)	Immobility	3.6	4.1	1.1	Andrés-Costa et al. (2017)
	Pimephales promelas	LOEC(7d)	Survival	0.10	0.17	1.7	Stanley et al. (2007)
		LOEC(7d)	Feeding	0.05	0.17	3.4	Stanley et al. (2007)
		LOEC(7d)	Growth	0.05	0.17	3.4	Stanley et al. (2007)
		LC$_{50}$(48h)	Survival	0.22	0.21	—	Stanley et al. (2007)
	Tetrahymena thermophila	EC$_{50}$(24h)	Growth	35.2	1.3	27.1	Andrés-Costa et al. (2017)
		EC$_{50}$(24h)	Growth	3.2	30.5	9.5	De Andrés et al. (2009)
Norfluoxetine	*Daphnia magna*	EC$_{50}$(48h)	Immobility	2.8	2.9	—	Andrés-Costa et al. (2017)
	Tetrahymena thermophila	EC$_{50}$(24h)	Growth	3.8	5.8	1.5	Andrés-Costa et al. (2017)
Naproxen	*Photobacterium leiognathi*	EC$_{50}$(30min)	Inhibition	0.9	0.8	1.2	Neale et al. (2019)
	Pseudokirchneriella subcapitata	EC$_{10}$(24h)	Inhibition	23.8	19.8	1.2	Neale et al. (2019)
	PLHC-1 cells	EC$_{IR1.5}$(6h)	EROD activity	5.1	12.6	2.5	Neale et al. (2019)
Flurbiprofen	*Photobacterium leiognathi*	EC$_{50}$(30min)	Inhibition	1.2	2.1	1.7	Neale et al. (2019)
	Pseudokirchneriella subcapitata	EC$_{10}$(24h)	Inhibition	9.1	5.5	1.7	Neale et al. (2019)
	PLHC-1 cells	EC$_{IR1.5}$(6h)	EROD activity	>12.5	8.4	1.5	Neale et al. (2019)
Ibuprofen	*Photobacterium leiognathi*	EC$_{50}$(30min)	Inhibition	3.1	2.8	1.1	Neale et al. (2019)
Ketoprofen	*Photobacterium leiognathi*	EC$_{50}$(30min)	Inhibition	4.6	4.2	1.1	Neale et al. (2019)
	PLHC-1 cells	EC$_{IR1.5}$(6h)	EROD activity	>27.2	4.3	>6.3	Neale et al. (2019)

TABLE 4.3 (Continued)
Aquatic Toxicity Tests of Pharmaceutical Enantiomers

Chiral Pharmaceutical	Test Organism/ Test Cells	Test	Endpoint	Toxicity (mg L^{-1})		Difference[e]	Reference
Ephedrine	*Daphnia magna*	EC$_{50}$(48h)	Immobility	170.8[a]	253.7[b]	1.5	Rice et al. (2018)
	Pseudokirchneriella subcapitata	EC$_{50}$(72h)	Inhibition	754.5[a]	259.1[b]	2.9	Rice et al. (2018)
	Tetrahymena thermophila	EC$_{50}$(24h)	Inhibition	42.6[a]	36.0[b]	1.2	Rice et al. (2018)
Pseudoephedrine	*Daphnia magna*	EC$_{50}$(48h)	Immobility	274.3[c]	107.2[d]	2.6	Rice et al. (2018)
	Pseudokirchneriella subcapitata	EC$_{50}$(72h)	Inhibition	417.9[c]	44.8[d]	9.3	Rice et al. (2018)
	Tetrahymena thermophila	EC$_{50}$(24h)	Inhibition	99.3[c]	4.6[d]	21.6	Rice et al. (2018)

Key: LOEC, lowest observed effect concentration; LC$_{50}$, concentration which is lethal to 50% of test organisms; EC$_{50}$, concentration which induces an effect halfway between the baseline and maximum response; EC$_{10}$, concentration which induces an effect to 10% of the test organisms; EC$_{IR1.5}$, concentration which induces an induction ratio of 1.5; EROD, ethoxyresorufin-O-deethylase.

[a] 1S,2R(+)-ephedrine
[b] 1R,2S(−)-ephedrine
[c] 1S,2S(+)-pseudoephedrine
[d] 1R,2R(−)-pseudoephedrine
[e] Difference = $\dfrac{\text{Higher effect or lethal enantiomer concentration}}{\text{Lower effect or lethal enantiomer concentration}}$

aquatic organisms. Toxicity was investigated towards *D. magna*, *T. thermophilia*, and *Pseudokirchneriella subcapitata*. Enantiospecific toxicity was observed towards all organisms with *T. thermophilia* the most sensitive. In this case the EC$_{50}$(24h) ranged from 4.6 mg/L for *1R,2R(−)-*pseudoephedrine and 99.3 mg/L for *1S,2S(+)-*pseudoephedrine (Table 4.3).

Active pharmaceutical metabolites are also excreted from the human body and have been widely reported in the environment. Furthermore, microbial metabolites can be formed during degradation during wastewater treatment or in the environment itself (Fent et al., 2006). The environmental toxicity of chiral pharmaceutical metabolites is poorly understood. This is attributed to a lack of analytical reference standards available for individual metabolite enantiomers. Nevertheless, Andrés-Costa et al. (2017) synthesized the enantiomers of fluoxetine and its main metabolite norfluoxetine to determine their fate and ecotoxicity. Interestingly, *S*-norfluoxetine was 10 times more toxic to *T. thermophilia* than *S*(+)-fluoxetine (Andrés-Costa et al., 2017). This is concerning as metabolites are not always monitored or included in risk assessments (Fent et al., 2006).

Despite the ubiquity of analgesics reported in the environment worldwide (Fent et al., 2006; Ziylan and Ince, 2011), there has only been limited study of their ecotoxicity at the enantiomeric level. Neale et al. (2019) investigated the toxicity of individual flurbiprofen, ibuprofen, ketoprofen, and naproxen enantiomers to bacteria, algae, and fish cells. Flurbiprofen showed greatest enantiospecific toxicity towards the bacteria *Photobacterium leiognathi* albeit only 1.7 times difference between *S*(+)-flurbiprofen and *R*(−)-flurbiprofen (EC$_{50}$(30 min) % inhibition; 1.2 *versus* 2.1 mg L^{-1}) (Neale et al., 2019) (Table 4.3). In the case of naproxen, *S*(+)-naproxen was 2.5 times more active than *R*(−)-naproxen in the assay measuring ethoxyresorufin-O-deethylase activity in fish cells (this measures the induction of cytochrome P450 1A enzymes which are important for xenobiotic metabolism (Fent, 2001).

At the studied concentrations, ibuprofen enantiomers only exhibited toxicity towards *P. leiognathi*. Toxicity was not enantiospecific with EC_{50}(30min) (% inhibition) of 3.1 and 2.8 mg L^{-1} for S(−)-ibuprofen and R(+)-ibuprofen, respectively (Neale et al., 2019). Finally, greatest enantiospecific toxicity of ketoprofen was observed towards ethoxyresorufin-O-deethylase activity. The effect concentration caused an induction ratio of 1.5 ($EC_{IR1.5}$(6 h)). It was 4.3 mg L^{-1} for R(−)-ketoprofen and >27.2 mg L^{-1} for S(+)-ketoprofen demonstrating >6.3 times potency between enantiomers (Neale et al., 2019). Here, the more potent enantiomer (R(−)-ketoprofen) was the less prevalent reported in wastewater effluents (Caballo et al., 2015; Camacho-Muñoz et al., 2019). However, all the studied analgesics showed that enantiospecific differences in toxicity were less than an order of magnitude for the studied bioassays (Neale et al., 2019).

Although chiral pharmaceuticals are known to persist in amended soils and some are likely to be present in non-racemic form, no toxicity testing has been undertaken at the enantiomeric level towards terrestrial organisms. Such investigations are needed to assess the risk of chiral pharmaceuticals in soil.

4.5 FUTURE PERSPECTIVES

A considerable limitation of existing enantioselective methods is chromatographic run time. Methods capable of multi residue separation are typically ≥60 minutes (Table 4.1). This limits sample throughput and advancement in this area (Sanganyado et al., 2017). Chiral separations have been limited to HPLC mode with columns of 5 μm stationary phase particle sizes (Table 4.1). Nevertheless, enantioselective columns with sub-2 μm particle sizes are becoming available. Such columns will enable UPLC performance and reduced chromatographic run times. This should make chiral pharmaceutical analysis more desirable for environmental researchers in the future.

Several studies have been conducted on the fate of chiral pharmaceuticals in the environment. This has been undertaken using both river water and soil in controlled laboratory microcosm studies (Bertin et al., 2020; Camacho-Muñoz et al., 2019; Evans et al., 2016; Petrie et al., 2018). Such studies are essential to understand the enantiospecific fate of chiral pharmaceuticals in environmental compartments. However, to date only limited research has been undertaken on single enantiomers. Using racemic analytical standard mixtures in such studies could result in overlooking important fate mechanisms (e.g., chiral inversion) and associated risks of such processes (Bertin et al., 2020). A contributor of this has been the lack of enantiomerically pure analytical standards. However, they are becoming more readily available for use in such studies.

The majority of enantiospecific studies to date have focused on biodegradation in the environment. However, enantioselective sorption could take place and be an important process in the environment. For example, Sanganyado et al. (2016) reported that sorption of acebutolol, atenolol, and metoprolol to wastewater sludge was enantioselective in nature. Chiral drugs are known to partition into river sediments (Díaz-Cruz et al., 2003; Kunkel and Radke, 2012). Therefore, enantioselective sorption to sediments as well as suspended particulate matter in rivers needs to be investigated.

Overall, there is a lack of acute and chronic ecotoxicity testing of pharmaceuticals and their metabolites at the enantiomer level (Table 4.3). Such tests at different trophic levels are essential to enable the incorporation of stereochemistry into environmental risk assessments. However, care is needed because adverse effects can be observed at non-traditional endpoints (Brooks et al., 2003). Furthermore, it is important to consider mixtures as molecular interactions of enantiomers can be additive, antagonistic, and synergistic (Stanley and Brooks, 2009).

REFERENCES

Andrés-Costa, M. J., K. Proctor, M. T. Sabatini, A. P. Gee, S. E. Lewis, Y. Pico, and B. Kasprzyk-Hordern. 2017. "Enantioselective transformation of fluoxetine in water and its ecotoxicological relevance." *Scientific Reports* 7 (1):15777. doi: 10.1038/s41598-017-15585-1.

Azzouz, A., and E. Ballesteros. 2016. "Determination of 13 endocrine disrupting chemicals in environmental solid samples using microwave-assisted solvent extraction and continuous solid-phase extraction followed by gas chromatography–mass spectrometry." *Analytical and Bioanalytical Chemistry* 408 (1):231–241. doi: 10.1007/s00216-015-9096-1.

Bagnall, J. P., S. E. Evans, M. T. Wort, A. T. Lubben, and B. Kasprzyk-Hordern. 2012. "Using chiral liquid chromatography quadrupole time-of-flight mass spectrometry for the analysis of pharmaceuticals and illicit drugs in surface and wastewater at the enantiomeric level." *Journal of Chromatography A* 1249:115–129. doi: 10.1016/j.chroma.2012.06.012.

Baranowska, M. 2018. "Fate of chiral pharmacologically active compounds in river water microcosms." MSc Thesis, Robert Gordon University, UK.

Beesley, T. E., and J.-T. Lee. 2009. "Method development strategy and applications update for CHIROBIOTIC chiral stationary phases." *Journal of Liquid Chromatography & Related Technologies* 32 (11-12):1733–1767. doi: 10.1080/10826070902959489.

Berthod, A., H. X. Qiu, S. M. Staroverov, M. A. Kuznestov, and D. W. Armstrong. 2010. "Chiral Recognition with Macrocyclic Glycopeptides: Mechanisms and Applications." In *Chiral Recognition in Separation Methods: Mechanisms and Applications*, edited by Alain Berthod, 203–222. Berlin, Heidelberg: Springer Berlin Heidelberg.

Bertin, S., K. Yates, and B. Petrie. 2020. "Enantiospecific behaviour of chiral drugs in soil." *Environmental Pollution* 262:114364. doi: https://doi.org/10.1016/j.envpol.2020.114364.

Boogaerts, T., M. Degreef, A. Covaci, and A. L. N. van Nuijs. 2019. "Development and validation of an analytical procedure to detect spatio-temporal differences in antidepressant use through a wastewater-based approach." *Talanta* 200:340–349. doi: https://doi.org/10.1016/j.talanta.2019.03.052.

Brooks, B. W., C. M. Foran, S. M. Richards, J. Weston, P. K. Turner, J. K. Stanley, K. R. Solomon, M. Slattery, and T. W. La Point. 2003. "Aquatic ecotoxicology of fluoxetine." *Toxicology Letters* 142 (3):169–183. doi: https://doi.org/10.1016/S0378-4274(03)00066-3.

Buser, H.-R., T. Poiger, and M. D. Müller. 1999. "Occurrence and environmental behavior of the chiral pharmaceutical drug ibuprofen in surface waters and in wastewater." *Environmental Science & Technology* 33 (15):2529–2535. doi: 10.1021/es981014w.

Caballo, C., M. D. Sicilia, and S. Rubio. 2015. "Enantioselective determination of representative profens in wastewater by a single-step sample treatment and chiral liquid chromatography–tandem mass spectrometry." *Talanta* 134:325–332. doi: https://doi.org/10.1016/j.talanta.2014.11.016.

Camacho-Muñoz, D., and B. Kasprzyk-Hordern. 2015. "Multi-residue enantiomeric analysis of human and veterinary pharmaceuticals and their metabolites in environmental samples by chiral liquid chromatography coupled with tandem mass spectrometry detection." *Analytical and Bioanalytical Chemistry* 407 (30):9085–9104. doi: 10.1007/s00216-015-9075-6.

Camacho-Muñoz, D., and B. Kasprzyk-Hordern. 2017. "Simultaneous enantiomeric analysis of pharmacologically active compounds in environmental samples by chiral LC–MS/MS with a macrocyclic antibiotic stationary phase." *Journal of Mass Spectrometry* 52 (2):94–108. doi: 10.1002/jms.3904.

Camacho-Muñoz, D., B. Kasprzyk-Hordern, and K. V. Thomas. 2016. "Enantioselective simultaneous analysis of selected pharmaceuticals in environmental samples by ultrahigh-performance supercritical fluid based chromatography tandem mass spectrometry." *Analytica Chimica Acta* 934:239–251. doi: https://doi.org/10.1016/j.aca.2016.05.051.

Camacho-Muñoz, D., J. Martín, J. L. Santos, I. Aparicio, and E. Alonso. 2010. "Occurrence, temporal evolution and risk assessment of pharmaceutically active compounds in Doñana Park (Spain)." *Journal of Hazardous Materials* 183 (1):602–608. doi: https://doi.org/10.1016/j.jhazmat.2010.07.067.

Camacho-Muñoz, D., B. Petrie, L. Lopardo, K. Proctor, J. Rice, J. Youdan, R. Barden, and B. Kasprzyk-Hordern. 2019. "Stereoisomeric profiling of chiral pharmaceutically active compounds in wastewaters and the receiving environment – A catchment-scale and a laboratory study." *Environment International* 127:558–572. doi: https://doi.org/10.1016/j.envint.2019.03.050.

Castrignanò, E., A. M. Kannan, E. J. Feil, and B. Kasprzyk-Hordern. 2018. "Enantioselective fractionation of fluoroquinolones in the aqueous environment using chiral liquid chromatography coupled with tandem mass spectrometry." *Chemosphere* 206:376–386. doi: 10.1016/j.chemosphere.2018.05.005.

Coelho, M. M., A. R. Lado Ribeiro, J. C. G. Sousa, C. Ribeiro, C. Fernandes, A. M. T. Silva, and M. E. Tiritan. 2019. "Dual enantioselective LC–MS/MS method to analyse chiral drugs in surface water: Monitoring in Douro River estuary." *Journal of Pharmaceutical and Biomedical Analysis* 170:89–101. doi: https://doi.org/10.1016/j.jpba.2019.03.032.

da Silva, A. K., M. J. M. Wells, A. N. Morse, M.-L. Pellegrin, S. M. Miller, J. Peccia, and L. C. Sima. 2012. "Emerging pollutants – Part I: Occurrence, fate and transport." *Water Environment Research* 84 (10):1878–1908. doi: 10.2175/106143012x13407275695878.

De Andrés, F., G. Castañeda, and Á. Ríos. 2009. "Use of toxicity assays for enantiomeric discrimination of pharmaceutical substances." *Chirality* 21 (8):751–759. doi: 10.1002/chir.20675.

De Klerck, K., D. Mangelings, and Y. Vander Heyden. 2012. "Supercritical fluid chromatography for the enantioseparation of pharmaceuticals." *Journal of Pharmaceutical and Biomedical Analysis* 69:77–92. doi: https://doi.org/10.1016/j.jpba.2012.01.021.

Díaz-Cruz, M. S., M. a. J. López de Alda, and D. Barceló. 2003. "Environmental behavior and analysis of veterinary and human drugs in soils, sediments and sludge." *TrAC Trends in Analytical Chemistry* 22 (6):340–351. doi: https://doi.org/10.1016/S0165-9936(03)00603-4.

Dogan, A., J. Płotka-Wasylka, D. Kempińska-Kupczyk, J. Namieśnik, and A. Kot-Wasik. 2020. "Detection, identification and determination of chiral pharmaceutical residues in wastewater: Problems and challenges." *TrAC Trends in Analytical Chemistry* 122:115710. doi: https://doi.org/10.1016/j.trac.2019.115710.

Duan, L., Y. Zhang, B. Wang, S. Deng, J. Huang, Y. Wang, and G. Yu. 2018. "Occurrence, elimination, enantiomeric distribution and intra-day variations of chiral pharmaceuticals in major wastewater treatment plants in Beijing, China." *Environmental Pollution* 239:473–482. doi: https://doi.org/10.1016/j.envpol.2018.04.014.

Escuder-Gilabert, L., Y. Martín-Biosca, M. Perez-Baeza, S. Sagrado, and M. J. Medina-Hernández. 2018. "Direct chromatographic study of the enantioselective biodegradation of ibuprofen and ketoprofen by an activated sludge." *Journal of Chromatography A* 1568:140–148. doi: https://doi.org/10.1016/j.chroma.2018.07.034.

Evans, S., J. Bagnall, and B. Kasprzyk-Hordern. 2017. "Enantiomeric profiling of a chemically diverse mixture of chiral pharmaceuticals in urban water." *Environmental Pollution* 230:368–377. doi: https://doi.org/10.1016/j.envpol.2017.06.070.

Evans, S. E., J. Bagnall, and B. Kasprzyk-Hordern. 2016. "Enantioselective degradation of amphetamine-like environmental micropollutants (amphetamine, methamphetamine, MDMA and MDA) in urban water." *Environmental Pollution* 215:154–163. doi: https://doi.org/10.1016/j.envpol.2016.04.103.

Evans, S. E., P. Davies, A. Lubben, and B. Kasprzyk-Hordern. 2015. "Determination of chiral pharmaceuticals and illicit drugs in wastewater and sludge using microwave assisted extraction, solid-phase extraction and chiral liquid chromatography coupled with tandem mass spectrometry." *Analytica Chimica Acta* 882:112–126. doi: https://doi.org/10.1016/j.aca.2015.03.039.

Fatta-Kassinos, D., S. Meric, and A. Nikolaou. 2011. "Pharmaceutical residues in environmental waters and wastewater: Current state of knowledge and future research." *Analytical and Bioanalytical Chemistry* 399 (1):251–275. doi: 10.1007/s00216-010-4300-9.

Fent, K. 2001. "Fish cell lines as versatile tools in ecotoxicology: Assessment of cytotoxicity, cytochrome P4501A induction potential and estrogenic activity of chemicals and environmental samples." *Toxicology in Vitro* 15 (4):477–488. doi: https://doi.org/10.1016/S0887-2333(01)00053-4.

Fent, K., A. A. Weston, and D. Caminada. 2006. "Ecotoxicology of human pharmaceuticals." *Aquatic Toxicology* 76 (2):122–159. doi: https://doi.org/10.1016/j.aquatox.2005.09.009.

Fono, L. J., and D. L. Sedlak. 2005. "Use of the chiral pharmaceutical propranolol to identify sewage discharges into surface waters." *Environmental Science & Technology* 39 (23):9244–9252. doi: 10.1021/es047965t.

Ghirardini, A., and P. Verlicchi. 2019. "A review of selected microcontaminants and microorganisms in land runoff and tile drainage in treated sludge-amended soils." *Science of the Total Environment* 655:939–957. doi: https://doi.org/10.1016/j.scitotenv.2018.11.249.

Golovko, O., V. Kumar, G. Fedorova, T. Randak, and R. Grabic. 2014. "Seasonal changes in antibiotics, antidepressants/psychiatric drugs, antihistamines and lipid regulators in a wastewater treatment plant." *Chemosphere* 111:418–426. doi: https://doi.org/10.1016/j.chemosphere.2014.03.132.

Guo, H., M. F. Wahab, A. Berthod, and D. W. Armstrong. 2018. "Mass spectrometry detection of basic drugs in fast chiral analyses with vancomycin stationary phases." *Journal of Pharmaceutical Analysis* 8 (5):324–332. doi: https://doi.org/10.1016/j.jpha.2018.08.001.

Haginaka, J. 2001. "Protein-based chiral stationary phases for high-performance liquid chromatography enantioseparations." *Journal of Chromatography A* 906 (1):253–273. doi: https://doi.org/10.1016/S0021-9673(00)00504-5.

Hashim, N. H., and S. J. Khan. 2011. "Enantioselective analysis of ibuprofen, ketoprofen and naproxen in wastewater and environmental water samples." *Journal of Chromatography A* 1218 (29):4746–4754. doi: https://doi.org/10.1016/j.chroma.2011.05.046.

Hashim, N. H., L. D. Nghiem, R. M. Stuetz, and S. J. Khan. 2011. "Enantiospecific fate of ibuprofen, ketoprofen and naproxen in a laboratory-scale membrane bioreactor." *Water Research* 45 (18):6249–6258. doi: https://doi.org/10.1016/j.watres.2011.09.020.

Hayakawa, I., S. Atarashi, S. Yokohama, M. Imamura, K. Sakano, and M. Furukawa. 1986. "Synthesis and antibacterial activities of optically active ofloxacin." *Antimicrobial Agents and Chemotherapy* 29 (1):163–164. doi: 10.1128/AAC.29.1.163.

He, K., Y. Asada, S. Echigo, and S. Itoh. 2020. "Biodegradation of pharmaceuticals and personal care products in the sequential combination of activated sludge treatment and soil aquifer treatment." *Environmental Technology* 41 (3):378–388. doi: 10.1080/09593330.2018.1499810.

Henriksson, H., G. Pettersson, and G. Johansson. 1999. "Discrimination between enantioselective and non-selective binding sites on cellobiohydrolase-based stationary phases by site specific competing ligands." *Journal of Chromatography A* 857 (1):107–115. doi: https://doi.org/10.1016/S0021-9673(99)00776-1.

Henriksson, H., J. Ståhlberg, A. Koivula, G. Pettersson, C. Divne, L. Valtcheva, and R. Isaksson. 1997. "The catalytic amino-acid residues in the active site of cellobiohydrolase 1 are involved in chiral recognition." *Journal of Biotechnology* 57 (1):115–125. doi: https://doi.org/10.1016/S0168-1656(97)00094-1.

Hermansson, J. 1984. "Direct liquid chromatographic resolution of racemic drugs by means of α1-acid glycoprotein as the chiral complexing agent in the mobile phase." *Journal of Chromatography A* 316:537–546. doi: https://doi.org/10.1016/S0021-9673(00)96181-8.

Huang, Q., Z. Wang, C. Wang, and X. Peng. 2013. "Chiral profiling of azole antifungals in municipal wastewater and recipient rivers of the Pearl River delta, China." *Environmental Science and Pollution Research* 20 (12):8890–8899. doi: 10.1007/s11356-013-1862-z.

Ilisz, I., A. Aranyi, Z. Pataj, and A. Péter. 2013. "Enantioseparations by High-Performance Liquid Chromatography Using Macrocyclic Glycopeptide-Based Chiral Stationary Phases: An Overview." In *Chiral Separations: Methods and Protocols*, edited by Gerhard K. E. Scriba, 137–163. Totowa, NJ: Humana Press.

Ilisz, I., T. Orosz, and A. Péter. 2019. "High-Performance Liquid Chromatography Enantioseparations Using Macrocyclic Glycopeptide-Based Chiral Stationary Phases: An Overview." In *Chiral Separations: Methods and Protocols*, edited by Gerhard K. E. Scriba, 201–237. New York, NY: Springer New York.

Karakka Kal, A. K., T. K. Karatt, R. Sayed, M. Philip, S. Meissir, and J. Nalakath. 2019. "Separation of ephedrine and pseudoephedrine enantiomers using a polysaccharide-based chiral column: A normal phase liquid chromatography–high-resolution mass spectrometry approach." *Chirality* 31 (8):568–574. doi: 10.1002/chir.23104.

Kasprzyk-Hordern, B., and D. R. Baker. 2012. "Enantiomeric profiling of chiral drugs in wastewater and receiving waters." *Environmental Science and Technology* 46 (3):1681–1691. doi: 10.1021/es203113y.

Kasprzyk-Hordern, B., V. V. R. Kondakal, and D. R. Baker. 2010. "Enantiomeric analysis of drugs of abuse in wastewater by chiral liquid chromatography coupled with tandem mass spectrometry." *Journal of Chromatography A* 1217 (27):4575–4586. doi: https://doi.org/10.1016/j.chroma.2010.04.073.

Khan, S. J., L. Wang, N. H. Hashim, and J. A. Mcdonald. 2014. "Distinct enantiomeric signals of ibuprofen and naproxen in treated wastewater and sewer overflow." *Chirality* 26 (11):739–746. doi: 10.1002/chir.22258.

Kruve, A., A. Künnapas, K. Herodes, and I. Leito. 2008. "Matrix effects in pesticide multi-residue analysis by liquid chromatography–mass spectrometry." *Journal of Chromatography A* 1187 (1):58–66. doi: https://doi.org/10.1016/j.chroma.2008.01.077.

Kruve, A., I. Leito, and K. Herodes. 2009. "Combating matrix effects in LC/ESI/MS: The extrapolative dilution approach." *Analytica Chimica Acta* 651 (1):75–80. doi: https://doi.org/10.1016/j.aca.2009.07.060.

Kumirska, J., P. Łukaszewicz, M. Caban, N. Migowska, A. Plenis, A. Białk-Bielińska, M. Czerwicka, F. Qi, and S. Piotr. 2019. "Determination of twenty pharmaceutical contaminants in soil using ultrasound-assisted extraction with gas chromatography-mass spectrometric detection." *Chemosphere* 232:232–242. doi: https://doi.org/10.1016/j.chemosphere.2019.05.164.

Kunkel, U., and M. Radke. 2012. "Fate of pharmaceuticals in rivers: Deriving a benchmark dataset at favorable attenuation conditions." *Water Research* 46 (17):5551–5565. doi: https://doi.org/10.1016/j.watres.2012.07.033.

Laaniste, A., I. Leito, and A. Kruve. 2019. "ESI outcompetes other ion sources in LC/MS trace analysis." *Analytical and Bioanalytical Chemistry* 411 (16):3533–3542. doi: 10.1007/s00216-019-01832-z.

Lämmerhofer, M. 2010. "Chiral recognition by enantioselective liquid chromatography: Mechanisms and modern chiral stationary phases." *Journal of Chromatography A* 1217 (6):814–856. doi: https://doi.org/10.1016/j.chroma.2009.10.022.

Li, W. C. 2014. "Occurrence, sources, and fate of pharmaceuticals in aquatic environment and soil." *Environmental Pollution* 187:193–201. doi: https://doi.org/10.1016/j.envpol.2014.01.015.

Lindim, C., J. van Gils, D. Georgieva, O. Mekenyan, and I. T. Cousins. 2016. "Evaluation of human pharmaceutical emissions and concentrations in Swedish river basins." *Science of the Total Environment* 572:508–519. doi: https://doi.org/10.1016/j.scitotenv.2016.08.074.

Liu, J., X. Dan, G. Lu, J. Shen, D. Wu, and Z. Yan. 2018. "Investigation of pharmaceutically active compounds in an urban receiving water: Occurrence, fate and environmental risk assessment." *Ecotoxicology and Environmental Safety* 154:214–220. doi: https://doi.org/10.1016/j.ecoenv.2018.02.052.

López-Serna, R., B. Kasprzyk-Hordern, M. Petrović, and D. Barceló. 2013. "Multi-residue enantiomeric analysis of pharmaceuticals and their active metabolites in the Guadalquivir River basin (south Spain) by chiral liquid chromatography coupled with tandem mass spectrometry." *Analytical and Bioanalytical Chemistry* 405 (18):5859–5873. doi: 10.1007/s00216-013-6900-7.

Luo, Y., W. Guo, H. H. Ngo, L. D. Nghiem, F. I. Hai, J. Zhang, S. Liang, and X. C. Wang. 2014. "A review on the occurrence of micropollutants in the aquatic environment and their fate and removal during wastewater treatment." *Science of the Total Environment* 473-474:619–641. doi: https://doi.org/10.1016/j.scitotenv.2013.12.065.

Ma, R., H. Qu, B. Wang, F. Wang, Y. Yu, and G. Yu. 2019. "Simultaneous enantiomeric analysis of non-steroidal anti-inflammatory drugs in environment by chiral LC-MS/MS: A pilot study in Beijing, China." *Ecotoxicology and Environmental Safety* 174:83–91. doi: https://doi.org/10.1016/j.ecoenv.2019.01.122.

Ma, R., B. Wang, S. Lu, Y. Zhang, L. Yin, J. Huang, S. Deng, Y. Wang, and G. Yu. 2016. "Characterization of pharmaceutically active compounds in Dongting Lake, China: Occurrence, chiral profiling and environmental risk." *Science of the Total Environment* 557-558:268–275. doi: https://doi.org/10.1016/j.scitotenv.2016.03.053.

Mane, S. 2016. "Racemic drug resolution: a comprehensive guide." *Analytical Methods* 8 (42):7567–7586. doi: 10.1039/C6AY02015A.

Martín, J., D. Camacho-Muñoz, J. L. Santos, I. Aparicio, and E. Alonso. 2012. "Occurrence of pharmaceutical compounds in wastewater and sludge from wastewater treatment plants: Removal and ecotoxicological impact of wastewater discharges and sludge disposal." *Journal of Hazardous Materials* 239-240:40–47. doi: https://doi.org/10.1016/j.jhazmat.2012.04.068.

Michael, I., L. Rizzo, C. S. McArdell, C. M. Manaia, C. Merlin, T. Schwartz, C. Dagot, and D. Fatta-Kassinos. 2013. "Urban wastewater treatment plants as hotspots for the release of antibiotics in the environment: A review." *Water Research* 47 (3):957–995. doi: https://doi.org/10.1016/j.watres.2012.11.027.

Minguez, L., J. Pedelucq, E. Farcy, C. Ballandonne, H. Budzinski, and M.-P. Halm-Lemeille. 2016. "Toxicities of 48 pharmaceuticals and their freshwater and marine environmental assessment in northwestern France." *Environmental Science and Pollution Research* 23 (6):4992–5001. doi: 10.1007/s11356-014-3662-5.

Morrissey, I., K. Hoshino, K. Sato, A. Yoshida, I. Hayakawa, M. G. Bures, and L. L. Shen. 1996. "Mechanism of differential activities of ofloxacin enantiomers?" *Antimicrobial Agents and Chemotherapy* 40 (8):1775–1784.

Mosiashvili, L., L. Chankvetadze, T. Farkas, and B. Chankvetadze. 2013. "On the effect of basic and acidic additives on the separation of the enantiomers of some basic drugs with polysaccharide-based chiral selectors and polar organic mobile phases." *Journal of Chromatography A* 1317:167–174. doi: https://doi.org/10.1016/j.chroma.2013.08.029.

Muñoz, I., J. C. López-Doval, M. Ricart, M. Villagrasa, R. Brix, A. Geiszinger, A. Ginebreda, H. Guasch, M. J. L. de Alda, A. M. Romaní, S. Sabater, and D. Barceló. 2009. "Bridging levels of pharmaceuticals in river water with biological community structure in the Llobregat River basin (northeast Spain)." *Environmental Toxicology and Chemistry* 28 (12):2706–2714. doi: 10.1897/08-486.1.

Neale, P. A., A. Branch, S. J. Khan, and F. D. L. Leusch. 2019. "Evaluating the enantiospecific differences of non-steroidal anti-inflammatory drugs (NSAIDs) using an ecotoxicity bioassay test battery." *Science of the Total Environment* 694:133659. doi: https://doi.org/10.1016/j.scitotenv.2019.133659.

Nguyen, L. N., F. I. Hai, J. A. McDonald, S. J. Khan, W. E. Price, and L. D. Nghiem. 2017. "Continuous transformation of chiral pharmaceuticals in enzymatic membrane bioreactors for advanced wastewater treatment." *Water Science and Technology* 76 (7):1816–1826. doi: 10.2166/wst.2017.331.

Okamoto, Y., M. Kawashima, and K. Hatada. 1984. "Chromatographic resolution. 7. Useful chiral packing materials for high-performance liquid chromatographic resolution of enantiomers: phenylcarbamates of polysaccharides coated on silica gel." *Journal of the American Chemical Society* 106 (18):5357–5359. doi: 10.1021/ja00330a057.

Ort, C., M. G. Lawrence, J. Rieckermann, and A. Joss. 2010. "Sampling for pharmaceuticals and personal care products (PPCPs) and illicit drugs in wastewater systems: Are your conclusions valid? A critical review." *Environmental Science & Technology* 44 (16):6024–6035. doi: 10.1021/es100779n.

Osorio, V., A. Larrañaga, J. Aceña, S. Pérez, and D. Barceló. 2016. "Concentration and risk of pharmaceuticals in freshwater systems are related to the population density and the livestock units in Iberian rivers." *Science of the Total Environment* 540:267–277. doi: https://doi.org/10.1016/j.scitotenv.2015.06.143.

Pan, X., F. Dong, Z. Chen, J. Xu, X. Liu, X. Wu, and Y. Zheng. 2017. "The application of chiral ultra-high-performance liquid chromatography tandem mass spectrometry to the separation of the zoxamide enantiomers and the study of enantioselective degradation process in agricultural plants." *Journal of Chromatography A* 1525:87–95. doi: https://doi.org/10.1016/j.chroma.2017.10.016.

Pereira, A. M. P. T., L. J. G. Silva, C. S. M. Laranjeiro, L. M. Meisel, C. M. Lino, and A. Pena. 2017. "Human pharmaceuticals in Portuguese rivers: The impact of water scarcity in the environmental risk." *Science of the Total Environment* 609:1182–1191. doi: https://doi.org/10.1016/j.scitotenv.2017.07.200.

Pérez-Lemus, N., R. López-Serna, S. I. Pérez-Elvira, and E. Barrado. 2019. "Analytical methodologies for the determination of pharmaceuticals and personal care products (PPCPs) in sewage sludge: A critical review." *Analytica Chimica Acta* 1083:19–40. doi: https://doi.org/10.1016/j.aca.2019.06.044.

Petrie, B., R. Barden, and B. Kasprzyk-Hordern. 2015. "A review on emerging contaminants in wastewaters and the environment: Current knowledge, understudied areas and recommendations for future monitoring." *Water Research* 72:3–27. doi: https://doi.org/10.1016/j.watres.2014.08.053.

Petrie, B., D. Camacho-Muñoz, E. Castrignanò, S. Evans, and B. Kasprzyk-Hordern. 2015. "Chiral liquid chromatography coupled with tandem mass spectrometry for environmental analysis of pharmacologically active compounds." *LC-GC Europe* 28 (3).

Petrie, B., A. Gravell, G. A. Mills, J. Youdan, R. Barden, and B. Kasprzyk-Hordern. 2016. "In situ calibration of a new Chemcatcher configuration for the determination of polar organic micropollutants in wastewater effluent." *Environmental Science & Technology* 50 (17):9469–9478. doi: 10.1021/acs.est.6b02216.

Petrie, B., J. Mrazova, B. Kasprzyk-Hordern, and K. Yates. 2018. "Multi-residue analysis of chiral and achiral trace organic contaminants in soil by accelerated solvent extraction and enantioselective liquid chromatography tandem–mass spectrometry." *Journal of Chromatography A* 1572:62–71. doi: https://doi.org/10.1016/j.chroma.2018.08.034.

Petrovic, M. 2014. "Methodological challenges of multi-residue analysis of pharmaceuticals in environmental samples." *Trends in Environmental Analytical Chemistry* 1:e25–e33. doi: https://doi.org/10.1016/j.teac.2013.11.004.

Picó, Y., R. Alvarez-Ruiz, A. H. Alfarhan, M. A. El-Sheikh, H. O. Alshahrani, and D. Barceló. 2020. "Pharmaceuticals, pesticides, personal care products and microplastics contamination assessment of Al-Hassa irrigation network (Saudi Arabia) and its shallow lakes." *Science of the Total Environment* 701:135021. doi: https://doi.org/10.1016/j.scitotenv.2019.135021.

Pinasseau, L., L. Wiest, A. Fildier, L. Volatier, G. R. Fones, G. A. Mills, F. Mermillod-Blondin, and E. Vulliet. 2019. "Use of passive sampling and high resolution mass spectrometry using a suspect screening approach to characterise emerging pollutants in contaminated groundwater and runoff." *Science of the Total Environment* 672:253–263. doi: https://doi.org/10.1016/j.scitotenv.2019.03.489.

Radjenović, J., A. Jelić, M. Petrović, and D. Barceló. 2009. "Determination of pharmaceuticals in sewage sludge by pressurized liquid extraction (PLE) coupled to liquid chromatography-tandem mass spectrometry (LC-MS/MS)." *Analytical and Bioanalytical Chemistry* 393 (6):1685–1695. doi: 10.1007/s00216-009-2604-4.

Ramage, S., D. Camacho-Muñoz, and B. Petrie. 2019. "Enantioselective LC-MS/MS for anthropogenic markers of septic tank discharge." *Chemosphere* 219:191–201. doi: https://doi.org/10.1016/j.chemosphere.2018.12.007.

Ribeiro, A. R., P. M. L. Castro, and M. E. Tiritan. 2012. "Environmental Fate of Chiral Pharmaceuticals: Determination, Degradation and Toxicity." In *Environmental Chemistry for a Sustainable World: Volume 2: Remediation of Air and Water Pollution*, edited by Eric Lichtfouse, Jan Schwarzbauer, and Didier Robert, 3–45. Dordrecht: Springer Netherlands.

Ribeiro, A. R., L. H. M. L. M. Santos, A. S. Maia, C. Delerue-Matos, P. M. L. Castro, and M. E. Tiritan. 2014. "Enantiomeric fraction evaluation of pharmaceuticals in environmental matrices by liquid chromatography-tandem mass spectrometry." *Journal of Chromatography A* 1363:226–235. doi: https://doi.org/10.1016/j.chroma.2014.06.099.

Rice, J., K. Proctor, L. Lopardo, S. Evans, and B. Kasprzyk-Hordern. 2018. "Stereochemistry of ephedrine and its environmental significance: Exposure and effects directed approach." *Journal of Hazardous Materials* 348:39–46. doi: https://doi.org/10.1016/j.jhazmat.2018.01.020.

Rimayi, C., L. Chimuka, A. Gravell, G. R. Fones, and G. A. Mills. 2019. "Use of the Chemcatcher® passive sampler and time-of-flight mass spectrometry to screen for emerging pollutants in rivers in Gauteng Province of South Africa." *Environmental Monitoring and Assessment* 191 (6):388. doi: 10.1007/s10661-019-7515-z.

Ruan, Y., R. Wu, J. C. W. Lam, K. Zhang, and P. K. S. Lam. 2019. "Seasonal occurrence and fate of chiral pharmaceuticals in different sewage treatment systems in Hong Kong: Mass balance, enantiomeric profiling, and risk assessment." *Water Research* 149:607–616. doi: https://doi.org/10.1016/j.watres.2018.11.010.

Sanchez-Prado, L., C. Garcia-Jares, T. Dagnac, and M. Llompart. 2015. "Microwave-assisted extraction of emerging pollutants in environmental and biological samples before chromatographic determination." *TrAC Trends in Analytical Chemistry* 71:119–143. doi: https://doi.org/10.1016/j.trac.2015.03.014.

Sanganyado, E., Q. Fu, and J. Gan. 2016. "Enantiomeric selectivity in adsorption of chiral β-blockers on sludge." *Environmental Pollution* 214:787–794. doi: https://doi.org/10.1016/j.envpol.2016.04.091.

Sanganyado, E., Z. Lu, Q. Fu, D. Schlenk, and J. Gan. 2017. "Chiral pharmaceuticals: A review on their environmental occurrence and fate processes." *Water Research* 124:527–542. doi: https://doi.org/10.1016/j.watres.2017.08.003.

Sanganyado, E., Z. Lu, and J. Gan. 2014. "Mechanistic insights on chaotropic interactions of liophilic ions with basic pharmaceuticals in polar ionic mode liquid chromatography." *Journal of Chromatography A* 1368:82–88. doi: https://doi.org/10.1016/j.chroma.2014.09.054.

Santos, L. H. M. L. M., A. N. Araújo, A. Fachini, A. Pena, C. Delerue-Matos, and M. C. B. S. M. Montenegro. 2010. "Ecotoxicological aspects related to the presence of pharmaceuticals in the aquatic environment." *Journal of Hazardous Materials* 175 (1):45–95. doi: https://doi.org/10.1016/j.jhazmat.2009.10.100.

Sharma, V. K., N. Johnson, L. Cizmas, T. J. McDonald, and H. Kim. 2016. "A review of the influence of treatment strategies on antibiotic resistant bacteria and antibiotic resistance genes." *Chemosphere* 150:702–714. doi: https://doi.org/10.1016/j.chemosphere.2015.12.084.

Shen, J., T. Ikai, and Y. Okamoto. 2014. "Synthesis and application of immobilized polysaccharide-based chiral stationary phases for enantioseparation by high-performance liquid chromatography." *Journal of Chromatography A* 1363:51–61. doi: https://doi.org/10.1016/j.chroma.2014.06.042.

Sigma-Aldrich. 2020. "Operating Guidelines for ChromTech CHIRAL-AGP, CHIRAL-HSA, and CHIRAL-CBH HPLC Columns." accessed 08/05/2020. https://www.sigmaaldrich.com/content/dam/sigma-aldrich/docs/Supelco/Product_Information_Sheet/t709074.pdf.

Stanley, J. K., and B. W. Brooks. 2009. "Perspectives on ecological risk assessment of chiral compounds." *Integrated Environmental Assessment and Management* 5 (3):364–373. doi: 10.1897/ieam_2008-076.1.

Stanley, J. K., A. J. Ramirez, C. K. Chambliss, and B. W. Brooks. 2007. "Enantiospecific sublethal effects of the antidepressant fluoxetine to a model aquatic vertebrate and invertebrate." *Chemosphere* 69 (1):9–16. doi: https://doi.org/10.1016/j.chemosphere.2007.04.080.

Stanley, J. K., A. J. Ramirez, M. Mottaleb, C. K. Chambliss, and B. W. Brooks. 2006. "Enantiospecific toxicity of the β-blocker propranolol to *Daphnia magna* and *Pimephales promelas*." *Environmental Toxicology and Chemistry* 25 (7):1780–1786. doi: 10.1897/05-298r1.1.

Suzuki, T., Y. Kosugi, M. Hosaka, T. Nishimura, and D. Nakae. 2014. "Occurrence and behavior of the chiral anti-inflammatory drug naproxen in an aquatic environment." *Environmental Toxicology and Chemistry* 33 (12):2671–2678. doi: 10.1002/etc.2741.

Svan, A., M. Hedeland, T. Arvidsson, J. T. Jasper, D. L. Sedlak, and C. E. Pettersson. 2015. "Rapid chiral separation of atenolol, metoprolol, propranolol and the zwitterionic metoprolol acid using supercritical fluid chromatography–tandem mass spectrometry – Application to wetland microcosms." *Journal of Chromatography A* 1409:251–258. doi: https://doi.org/10.1016/j.chroma.2015.07.075.

Toušová, Z., B. Vrana, M. Smutná, J. Novák, V. Klučárová, R. Grabic, J. Slobodník, J. P. Giesy, and K. Hilscherová. 2019. "Analytical and bioanalytical assessments of organic micropollutants in the Bosna River using a combination of passive sampling, bioassays and multi-residue analysis." *Science of the Total Environment* 650:1599–1612. doi: https://doi.org/10.1016/j.scitotenv.2018.08.336.

Vanderford, B. J., D. B. Mawhinney, R. A. Trenholm, J. C. Zeigler-Holady, and S. A. Snyder. 2011. "Assessment of sample preservation techniques for pharmaceuticals, personal care products, and steroids in surface and drinking water." *Analytical and Bioanalytical Chemistry* 399 (6):2227–2234. doi: 10.1007/s00216-010-4608-5.

Vanessa, L. C., S. C. B. Lilian, and C. Ivone. 2009. "Stereoselectivity in drug metabolism: Molecular mechanisms and analytical methods." *Current Drug Metabolism* 10 (2):188–205. doi: http://dx.doi.org/10.2174/138920009787522188.

Wang, Z., Q. Huang, Y. Yu, C. Wang, W. Ou, and X. Peng. 2013. "Stereoisomeric profiling of pharmaceuticals ibuprofen and iopromide in wastewater and river water, China." *Environmental Geochemistry and Health* 35 (5):683–691. doi: 10.1007/s10653-013-9551-x.

Wang, Z., P. Zhao, B. Zhu, Z. Jiang, and X. Guo. 2018. "Magnetic solid-phase extraction based on Fe3O4/graphene nanocomposites for enantioselective determination of representative profens in the environmental water samples and molecular docking study on adsorption mechanism of graphene." *Journal of Pharmaceutical and Biomedical Analysis* 156:88–96. doi: https://doi.org/10.1016/j.jpba.2018.04.023.

Watanabe, H., I. Tamura, R. Abe, H. Takanobu, A. Nakamura, T. Suzuki, A. Hirose, T. Nishimura, and N. Tatarazako. 2016. "Chronic toxicity of an environmentally relevant mixture of pharmaceuticals to three aquatic organisms (alga, daphnid, and fish)." *Environmental Toxicology and Chemistry* 35 (4):996–1006. doi: 10.1002/etc.3285.

Winkler, M., J. R. Lawrence, and T. R. Neu. 2001. "Selective degradation of ibuprofen and clofibric acid in two model river biofilm systems." *Water Research* 35 (13):3197–3205. doi: https://doi.org/10.1016/S0043-1354(01)00026-4.

Wong, C. S., and S. L. MacLeod. 2009. "JEM Spotlight: Recent advances in analysis of pharmaceuticals in the aquatic environment." *Journal of Environmental Monitoring* 11 (5):923–936. doi: 10.1039/B819464E.

Zhang, T., C. Kientzy, P. Franco, A. Ohnishi, Y. Kagamihara, and H. Kurosawa. 2005. "Solvent versatility of immobilized 3,5-dimethylphenylcarbamate of amylose in enantiomeric separations by HPLC." *Journal of Chromatography A* 1075 (1):65–75. doi: https://doi.org/10.1016/j.chroma.2005.03.116.

Zhao, P., M. Deng, P. Huang, J. Yu, X. Guo, and L. Zhao. 2016. "Solid-phase extraction combined with dispersive liquid-liquid microextraction and chiral liquid chromatography-tandem mass spectrometry for the simultaneous enantioselective determination of representative proton-pump inhibitors in water samples." *Analytical and Bioanalytical Chemistry* 408 (23):6381–6392. doi: 10.1007/s00216-016-9753-z.

Zhou, Y., S. Wu, H. Zhou, H. Huang, J. Zhao, Y. Deng, H. Wang, Y. Yang, J. Yang, and L. Luo. 2018. "Chiral pharmaceuticals: Environment sources, potential human health impacts, remediation technologies and future perspective." *Environment International* 121:523–537. doi: https://doi.org/10.1016/j.envint.2018.09.041.

Ziylan, A., and N. H. Ince. 2011. "The occurrence and fate of anti-inflammatory and analgesic pharmaceuticals in sewage and fresh water: Treatability by conventional and non-conventional processes." *Journal of Hazardous Materials* 187 (1):24–36. doi: https://doi.org/10.1016/j.jhazmat.2011.01.057.

5 Chiral Personal Care Products
Occurrence, Fate, and Toxicity

Edmond Sanganyado

CONTENTS

- 5.1 Introduction 105
  - 5.1.1 Usage of Personal Care Products 106
  - 5.1.2 Scope of Chapter 107
- 5.2 Regulatory Challenges of Personal Care Products 107
- 5.3 Sources and Pathways of Personal Care Products 107
- 5.4 Azole Fungicides 110
  - 5.4.1 Physicochemical Properties 110
  - 5.4.2 Environmental Behavior of Chiral Azole Fungicides 112
  - 5.4.3 Toxicity 113
- 5.5 Synthetic Musks 113
  - 5.5.1 Implications of Chirality on Synthetic Musks 116
  - 5.5.2 Physicochemical Properties 116
  - 5.5.3 Synthetic Musks in the Environment 118
  - 5.5.4 Toxicity 118
- 5.6 Organic UV Filters 119
  - 5.6.1 Physicochemical Properties 120
  - 5.6.2 Organic UV Filters in the Environment 120
  - 5.6.3 Toxicity 124
- 5.7 Concluding Remarks and Future Trends 125
- References 126

5.1 INTRODUCTION

Synthetic chemicals in products are released into the environment during the life cycle of the product. They may cause adverse effects to humans and the environment. There has been a significant increase in knowledge pertaining to the deleterious effects of chemical pollutants in the past 60 years. However, chemical pollution is one of the leading causes of premature deaths and noncommunicable diseases worldwide (Brooks et al., 2020). In 2015 and 2017, exposure to contaminated air, water, soil, and food contributed to the death of around 9 million and 8.3 million people globally, respectively (Landrigan et al., 2018). The adverse effects of chemical pollution are higher in developing countries. Poor sanitation, industrial contamination of drinking water and air, indoor air pollution from the burning of fossil fuels, and a lack of adequate waste management infrastructure are prevalent in developing countries (Briggs, 2003). These factors increase the risk of chemical pollution. In addition, in low- and middle-income countries, pollution-related diseases cause a 2% reduction in gross domestic product per year due to loss of productivity (Das and Horton, 2018). However, the actual disease and death toll due to chemical pollution might be higher because current estimates do not include data from poorly understood health effects of known pollutants and emerging pollutants (Figure 5.1). There are currently more than 350,000 registered chemicals and mixtures worldwide (Wang et al., 2020), of which 5,000 high-production volume chemicals are

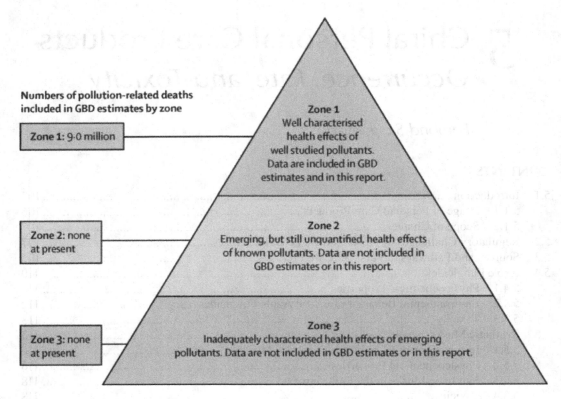

FIGURE 5.1 The death toll of chemical pollutants is estimated from known pollutants with known health effects (Zone 1) and not known (Zone 2) or emerging pollutants (Zone 3) with unknown or unclear health effects. GBD stands for the Global Burden of Disease study. (From Landrigan et al., 2018. Used with permission from Elsevier Ltd.)

ubiquitous in the environment (Landrigan et al., 2018). Less than half of these high-production volume chemicals underwent toxicity testing. Chemical pollution can cause ecological changes that can indirectly affect human health, however (Das and Horton, 2018). For example, chemical pollution can cause loss of biodiversity, which may cause loss of important food sources for humans. Uncharacterized chemicals and ecological changes might be responsible for much of the diseases, deaths, and ecological changes associated with chemical pollution (Landrigan et al., 2018). Unfortunately, the discharge of chemicals into the environment and their corresponding health effects are going to increase since the global production of chemicals is expected to double by 2030 (Brooks et al., 2020).

5.1.1 Usage of Personal Care Products

A wide range of chemical pollutants present at trace levels in the environment emerged in the past two decades due to advances in sample preparation, separation, and detection. Pharmaceuticals and personal care products are an example of one such group. An active pharmaceutical ingredient is a chemical substance that imparts pharmacological activity to a pharmaceutical product or aids in the cure, treatment, diagnosis, mitigation, or prevention of disease through shifting the physiological functions of a person (Sanganyado et al., 2017). Personal care products (PCPs) are a diverse group of bioactive chemicals that are directly used on the human skin and body to improve the quality of life by altering or enhancing human sensory such as taste, smell, or touch (Ying, 2012). They are widely used in health, beauty, and cleaning household products such as disinfectants, fragrances, food additives, insect repellents, hair products, preservatives, organic UV filters, shampoos, and detergents (Hopkins and Blaney, 2016; Montes-Grajales et al., 2017). Most PCPs contain bioactive active ingredients such as alcohols, aromatic hydrocarbons, and surfactants.

It is estimated that an average household spends more than $750 on PCPs per year, and the annual revenue from PCPs exceeds $189 billion in the United States (Hopkins and Blaney, 2016). Previous studies have shown that an average person uses 6–12 PCPs per day (Alani et al., 2013). As standards of living continue to improve globally, the number and quantity of PCPs an individual uses per day are expected to increase significantly. Each year, hundreds of new PCPs are developed to meet the demands of the increasingly sophisticated global market (Zhou et al., 2017). Nevertheless, the effects of most of these new chemicals on human and environmental health are poorly understood. The increase in usage of PCPs will result in increased discharge of PCPs into the environment.

5.1.2 Scope of Chapter

Personal care products are regarded as contaminants of emerging concern because they are bioactive and high production volume compounds. However, some PCPs such as synthetic musks, phthalates, azole fungicides, and organic UV filters are chiral compounds. Chiral compounds exist at least as two enantiomers that have disparate behaviors in chiral systems (Sanganyado et al., 2020). Hence, enantiomers of chiral PCPs may have a different distribution, fate, and toxicity in humans and the environment. This chapter focuses on synthetic musks, organic UV filters, and azole fungicides since more studies have investigated their enantioselective behaviors. Furthermore, this chapter explores the impact of the physicochemical properties on their distribution and fate in the environment. The routes of entry and potential effects of PCPs in the environment are also examined. The chapter concludes by offering recommendations for future studies.

5.2 REGULATORY CHALLENGES OF PERSONAL CARE PRODUCTS

The development, marketing, and safety of PCPs in the EU, the United States, Canada, China, and Japan are regulated under drug and cosmetic regulations (Nohynek et al., 2010). Prior to the 1960s, PCPs were considered safe as it was assumed the human skin was an impervious barrier that protects the human body from environmental pollutants. However, recent studies have shown that topically applied PCPs can penetrate the human skin and result in both contact and systemic toxicological effects. Several toxicity tests were developed to assess the safety of PCPs for human use. In several countries around the globe, PCP manufacturers and distributors are required by the law to ensure that their products are safe and have minimal adverse effects (Kwan et al., 2014; Schneider, 2009). However, regulations of PCPs across the globe vary due to differences in classifications of compounds as well as differences in fundamental definitions. For example, organic UV filters are cosmetics or over the counter drugs according to the EU and U.S. classifications, respectively. Table 5.1 shows the critical differences between cosmetic regulations in Europe, the United States, and Singapore. Briefly, the United States defines cosmetics based on the intended use, while the EU and Singapore are more specific as they incorporate the site of application (Kwan et al., 2014). In addition, the EU minimizes ambiguity by clearly defining the person responsible for ensuring the safety of cosmetics.

5.3 SOURCES AND PATHWAYS OF PERSONAL CARE PRODUCTS

The routes of entry of PCPs into the environment are dependent mainly on their method of application and physicochemical properties. About 90–95% of a topically administered PCP is directly removed from the skin or hair by washing, and it is then discharged into the wastewater system.

TABLE 5.1
Critical Differences between Cosmetic Regulation in the European Union, the United States, and Singapore

Regulation Framework	European Union	United States	Singapore	Reference*
Authority	European Commission (EC)	Food and Drug Administration (FDA)	Health Science Authority (HSA); Cosmetic Control Unit (CCU)	
Legislation	European Union (EU) Cosmetic Directive (76/768/EEC) EU Regulation (1223/2009)	Federal Food, Drug, and Cosmetic Act (FD&C Act) Fair Packaging and Labeling Act (FP&L Act) Title 21, Code of Federal Regulations (21 CFR)	Health Products Act (Amendment of First Schedule) (No. 2) Order 2007 Health Products Regulations (Cosmetic Products – ASEAN Cosmetic Directive) (2007 and 2010)	
Definition of cosmetics	*Any substance or mixture intended to be placed in contact with the external parts of the human body (epidermis, hair system, nails, lips and external genital organs) or with the teeth and the mucous membranes of the oral cavity with a view exclusively or mainly to cleaning them, perfuming them, changing their appearance, protecting them, keeping them in good condition or correcting body odors. A substance or mixture intended to be ingested, inhaled, injected, or implanted into the human body shall not be considered to be a cosmetic product*	*Articles (other than soaps consisting of an alkali salt of a fatty acid and making no claims other than cleansing) intended to be rubbed, poured, sprinkled, or sprayed on, introduced into, or otherwise applied to the human body or any part thereof for cleansing, beautifying, promoting attractiveness, or altering the appearance*	*Any substance or preparation that is intended to be placed in contact with the various external parts of the human body (epidermis, hair system, nails, lips, eyes and external genital organs) or with the teeth and the mucous membranes of the oral cavity with a view exclusively or mainly to cleaning them, perfuming them, changing their appearance, correcting body odors, protecting them or keeping them in good condition*	HSA (2019); Nwaogu and Vernon (2004)
Classification of cosmetics	Based on the formulation and function of the cosmetic products	Based on the function of the cosmetic products only	Based on the formulation and function of the cosmetic products. Adapted from the EU	HSA (2011); Nwaogu and Vernon (2004)
Premarket approval	Not required as "free movement" is being practiced, but notification to the Commission is required prior to placing the product on the market	Not required except for: a. colorants b. products that are violated	Not required, but notification using the online and acknowledgment of the notification from HSA must be received before the product can be marketed	USFDA, 2005; EC, 2009; HSA, 2011

TABLE 5.1 *(Continued)*
Critical Differences between Cosmetic Regulation in the European Union, the United States, and Singapore

Regulation Framework	European Union	United States	Singapore	Reference*
Responsibility of ensuring product safety	A "responsible person" which includes: a. manufacturer established within the Community b. a person within the Community who is designated by the manufacturer within the Community c. a person within the Community who is designated by the manufacturer, which is not within the Community d. an importer who is responsible for imported cosmetics e. a person within the Community who is designated by the importer f. a distributor with the marketed cosmetic product under his name or tradename or with products that have been modified and comply with the applicable requirements of the regulation	Responsibility of product safety assurance lies on: a. manufacturer b. packager c. distributor	The company or person who is placing a cosmetic product in the market: a. manufacturer b. assembler c. importer d. distributor	EC (2009); HSA (2019); USFDA (2012a)

Source: Kwan et al. (2014). Used with Permission from Elsevier Ltd.
* References can be found in Kwan et al. (2014).

However, 5–10% of the PCP is adsorbed into the body where it can undergo distribution, metabolism, and subsequently excretion into the wastewater system. Orally applied PCPs are typically washed off using running water and enter the wastewater treatment system. In wastewater treatment plants, PCPs are not entirely removed, and they are subsequently discharged into rivers (Tanwar et al., 2014). However, some PCPs sorb onto the biosolids in wastewater treatment plants. Biosolids and reclaimed water are often used in agriculture fields to amend soils and for irrigation, respectively (Fu et al., 2016; Meyer et al., 2019). As a result, PCPs enter the agriculture systems where they can be taken up by crops, thus potentially posing a risk to human health. In as much as wastewater treatment plants are the main route of entry, some PCPs enter the environment through atmospheric deposition and urban runoff (Dodgen et al., 2014; Wu et al., 2010, 2015). Volatile PCPs such as synthetic musks that are used through spraying are released to the environment during application (Chiaia-Hernandez et al., 2013). These PCPs can later be deposited into the environment as parent compound or transformation products. Topically administered PCPs can also be washed off during recreational activities such as swimming in natural waters (Molins-Delgado et al., 2014). They can

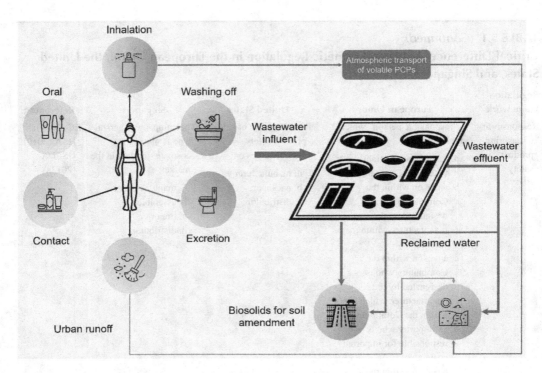

FIGURE 5.2 Sources and pathways of personal care products into the environment

accumulate in the swimming environment if the water is not changed frequently, resulting in other swimmers being exposed to the PCPs. Previous studies have shown that PCPs can elicit allergic reactions, endocrine disruption, and hormonal changes (Li et al., 2020). Figure 5.2 shows the routes of entry of PCPs in the environment.

5.4 AZOLE FUNGICIDES

The PCP classification is not always distinctive. Some personal care products are also active substances in pharmaceutical and agrochemical formulations. A good example is azole fungicides, which are widely used as antifungal agents in personal care products, pharmaceuticals, and agrochemicals (Chen and Ying, 2015). Clotrimazole, climbazole, fluconazole, itraconazole, ketoconazole, miconazole, and posaconazole are used as active pharmaceutical ingredients in tablets, ointments, and nasal sprays to treat fungal infections of the skin or the body. Azole fungicides are used in personal care products such as shower gels, soaps, shampoos, and ointments. For example, ketoconazole and climbazole are used in shampoos and hair products as anti-dandruff agents. It is estimated that up to 1,000 tons and 3,800 tons of climbazole are used per year in the European Union and China, respectively. Therefore, extensive usage of azole fungicides as personal care products can result in increased discharge of the compounds into the environment.

5.4.1 Physicochemical Properties

Azole fungicides contain two or three nitrogen atoms in their five-membered heterocyclic ring(s) (Table 5.2). Most azole fungicide PCPs are classified as imidazoles (e.g., climbazole, econazole, ketoconazole, and miconazole) or triazoles (e.g., itraconazole, posaconazole) (Chen and Ying, 2015). Physicochemical properties of PCPs influence their environmental behavior and choice of analytical methods. Besides climbazole, chiral azole fungicides have low solubility in water (< 1.0 mg L^{-1}),

TABLE 5.2
Physicochemical Properties of Chiral Azole Fungicides

Group	Compound	Structure	Molecular weight, g mol^{-1}	Vapor pressure, mmHg at 25°C	Water solubility, mg L^{-1}	Log K_{OW}	Log K_{OC}	WWTP Removal, %	Log BCF
Azole fungicides	Econazole		381.7	0.0 ± 1.4	0.065	5.61	4.57	89.52	3.62
	Ketoconazole		531.4	6.41 x 10^{-14}	0.29	4.34	3.22	-	2.40
	Miconazole		416.1	0.0 ± 1.4	0.011	6.25	4.79	92.97	4.11
	Climbazole		292.76	0.0 ± 1.1	8.28	3.76	2.75	20.42	2.20
	Posaconazole		700.8	0.0 ± 0.3	0.027	4.77	3.35	-	2.57
	Itraconazole		705.6	0.0 ± 1.7	0.0014	6.16	4.72	-	0.50

probably due to the presence of highly saturated aromatic rings and few polar functional groups. The water solubility of climbazole is estimated at 8.28 mg L^{-1}, suggesting it readily partitions to the aqueous phase when it enters the environment. Chiral azole fungicides do not readily partition to air due to their low volatility; thus, they are less likely to undergo long-range aerial transport. Organic carbon to water partition coefficient (K_{OC}) is a useful tool for measuring the mobility of pollutants in water. Highly hydrophobic substances with log K_{OC} > 4 readily sorb on organic matter and become less bioavailable, rendering them persistent in soil. Ketoconazole, econazole, and posaconazole have log K_{OC} > 4, suggesting that they are probably persistent in soil. Therefore, solid matrices such as soil and sediments can become secondary sources of chiral azole fungicides in terrestrial and aquatic environments, respectively (Kahle et al., 2008). Furthermore, the potential bioaccumulation of chiral azole fungicides is disconcerting since they have high octanol-water partition coefficients (K_{OW}) (Huang et al., 2013; Peng et al., 2012). Compounds with log K_{OW} > are often considered hydrophobic. They are easily removed from the aqueous phase as they tend to readily partition to solid compartments such as soil, sediment, sludge, and plant roots. In contrast, compounds with log K_{OW} < 2 are highly hydrophobic and preferentially partition to the aqueous phase (Chen and Ying, 2015). When hydrophilic substances sorb to a solid, they rapidly desorb into the aqueous phase. The third group of pollutants of interest is dual hydrophilic-hydrophobic compounds. These compounds sorb to solids while readily partitioning to the aqueous phase. Hence, the log K_{OW} of the chiral azole fungicides suggests that climbazole exists in the aqueous and solid phase due to its dual hydrophilic-hydrophobic behavior. In contrast, econazole, ketoconazole, miconazole, posaconazole, and itraconazole are hydrophobic with a high tendency to partition to the solid phase (Chen and Ying, 2015). The mobility of an organic pollutant in terrestrial and aquatic environments is influenced by the sorptive behavior of the pollutant. Pollutants with log K_{OW} < 2 often have a rapid desorption rate compared to compounds with higher log K_{OW}. Hence, it can be expected that the leaching potential in soil or landfills of climbazole is probably higher than that of miconazole or itraconazole, which have log K_{OW} > 6.0 (Chen and Ying, 2015). Pollutants that are lipophilic and nonpolar with log K_{OW} > 4.0 strongly attach to lipids and can readily bioaccumulate in plants and animals and organic matter. A study in wastewater treatment plants found that ketoconazole (log K_{OW} = 4.34) readily underwent biotransformation while econazole (log K_{OW} = 5.61) and miconazole (log K_{OW} = 6.25) partitioned and persisted in the wastewater sludge (Peng et al., 2012). These results show that the hydrophobicity of an organic compound influences its bioavailability. With bioconcentration factors (BCF) above 200, chiral fungicides pose significant risk to predatory wildlife and humans, as they have a great potential to biomagnify. Interestingly, itraconazole has log BCF < 1.0 suggesting low biomagnification potential.

5.4.2 Environmental Behavior of Chiral Azole Fungicides

Several studies have detected various azole fungicides in wastewater treatment plants. Climbazole was detected in a wastewater treatment plant in a German WWTP at maximum concentrations of 1240 ng L^{-1}, 1420 ng L^{-1}, 454 ng L^{-1}, and 600 ng L^{-1}, in sludge, wastewater influent, wastewater effluent, and river water, respectively (Wick et al., 2010). In Switzerland, the concentration of fluconazole, clotrimazole, propiconazole, and tebuconazole in the wastewater treatment plants ranged from 1–110 ng L^{-1} (Kahle et al., 2008). The concentration of the azole fungicides in lakes ranged from not detected to 1.9 ng L^{-1} (Kahle et al., 2008). However, the concentration in wastewater treatment plants in China of azole fungicides was 1–1834 ng L^{-1} (Peng et al., 2012). Climbazole, econazole, ketoconazole, and miconazole were ubiquitous in wastewater treatment plants. Azole fungicides can be removed from the environment via photolysis and biotransformation. Previous studies have shown that azole fungicides are relatively stable in aqueous environments, probably due to the presence of the highly stable aromatic rings. Laboratory studies have shown that climbazole could be removed in aqueous systems by photolysis (Couteau et al., 2000). However, engineering the photolysis technique to wastewater treatment did not yield any significant removal of azole fungicides (Peng et al., 2012). Azole fungicides are recalcitrant due to their antibacterial and antimycotic activities, which tend to inhibit enzyme-mediated biotransformation processes. Azole fungicides often enter soil through the irrigation of agriculture fields using recycled water or following the amendment of soil with biosolids. The half-life of climbazole and miconazole in biosolid amended soil was shown to be 175–192 days and 130–440 days (Chen and Ying, 2015). Climbazole had lower half-life since it has a higher bioavailability compared to miconazole (log K_{OW} > 6.0 suggests a lower rate of desorption from solid particles making it less bioavailable). This is probably the reason higher concentrations of miconazole ranging from 0.53–340 ng g^{-1} have been reported in biosolids compared to climbazole with concentrations of 0.04–48.1 ng g^{-1} (Chen and Ying, 2015). Besides the intrinsic properties of the pollutant, environmental fate of a pollutant is also influenced by local environmental factors such as pH, moisture content, organic carbon content, organic nitrogen content, redox potential, microbial communities, and co-pollutants (Cao et al., 2019; Zhuang et al., 2020, 2019). Previous studies have shown that local environmental factors influenced the dissipation of chiral azole fungicides used for agrochemical purposes (Buser et al., 2002; Celis et al., 2013; Gao et al., 2019; Qin et al., 2014; Zhang et al., 2018a). However, current studies have not explored the effect of local environmental conditions on the environmental fate of azole fungicides.

Climbazole, econazole, ketoconazole, and miconazole are chiral compounds that are commercially available as racemic mixtures. However, most studies on enantioselective behavior focused on azole fungicides used as agrochemicals (Zhang et al., 2018a,b). A previous study showed that climbazole underwent enantioselective biotransformation in water and sludge. The ketone functional group on climbazole was enantiospecifically reduced to an alcohol resulting in the enantiomeric enrichment of one enantiomer. However, the stereoconfiguration of the climbazole enantiomers was not established in the study; therefore, it is difficult to identify the exact enantiomer which was preferentially transformed. A study in the Pearl River Delta in China found that chiral imidazole fungicides (e.g., econazole, ketoconazole, and miconazole) were slightly scalemic in wastewater influent with enantiomeric fraction values of 0.45 to 0.53 (Huang et al., 2013). Following wastewater treatment, the degree of enantiomeric enrichment changed slightly. For example, for miconazole, the enantiomeric fraction in influent changed from 0.470 to 0.505 in the effluent (Huang et al., 2013). Seasonal variation in the enantiomeric composition of both the influent and effluent was observed for econazole and miconazole. These results suggest that biotic processes probably contributed to the enantiomeric enrichment that was observed in the wastewater treatment plant. In surface water and sediment, the chiral imidazole fungicides were predominantly racemic except for miconazole, which had high enantiomeric fractions in surface water during summer. Interestingly, there was a significant difference between the enantiomeric composition of the chiral imidazole fungicides in the dissolved phase and those that were particle attached (Huang et al., 2013). Previous studies on pharmaceuticals and pesticides have shown that enantioselective sorption to particulate matter was possible (Gámiz et al., 2016; Sanganyado et al., 2020, 2016). The differences between the dissolved and particle-attached phases could be due to enantioselective sorption. However, further studies are required to validate this claim.

5.4.3 Toxicity

Although azole fungicides are bioactive compounds that can elicit adverse effects to non-target organisms at low concentrations, data on the environmental toxicity of azole fungicides remain scarce. Table 5.3 shows estimated and experimentally determined toxicity of azole fungicides to fish, invertebrates, plants, and bacteria. Environmentally relevant concentrations of climbazole pose a threat to water lentil since the median effective concentration (EC50) for acute toxicity ranged from 19.0–33.9 $\mu g\ L^{-1}$. In contrast, environmental concentrations of climbazole did not pose a risk of acute toxicity to *Daphnia magna* and zebrafish, considering the EC50 was above 8,000 $\mu g\ L^{-1}$. Similarly, ketoconazole posed a low risk to *D. magna* with EC50 > 8,000 $\mu g\ L^{-1}$. At present, studies exploring the implications of chirality on the environmental toxicity of azole fungicides are scarce. A previous study in rats found that rat liver microsomes did not metabolize ketoconazole enantioselectively (Hamdy and Brocks, 2009). Instead, the observed enantiospecific distribution of ketoconazole was due to the preferential binding of the (+)-enantiomer to plasma protein (Hamdy and Brocks, 2009). Enantiospecificity in the distribution of a toxicant often results in one enantiomer reaching the receptor site, thus contributing to enantioselectivity in toxicity. An investigation on stereoselectivity in the antiangiogenic activity of itraconazole showed that geometric isomerism rather than chirality played a more significant role (Shi et al., 2010). There was a sevenfold difference in potency to human umbilical vein endothelial cells between *cis* and *trans* diastereomers with the cis-isomers more potent (Shi et al., 2010). The most significant enantioselective difference was between (2S,4R,2'S)-itraconazole and (2S,4S,2'R)-itraconazole where a fourfold difference was observed; the latter was more potent. Interestingly, the antifungal activity of itraconazole was highly enantioselective, with differences of up to 32-fold observed (Shi et al., 2010). Enantioselectivity in antifungal activity has also been reported for ketoconazole (Novotna et al., 2014; Rotstein et al., 1992). Therefore, chiral switching to the more potent enantiomer might be beneficial if it will help reduce the loading of azole fungicides into the environment.

5.5 SYNTHETIC MUSKS

For many centuries, the gland extracts from a male musk deer were used as a fragrance (Sommer, 2004). Since natural musks were expensive, scarce, and threatened the population of musk deers, synthetic musks were introduced as a potential replacement. Synthetic musks are a diverse group of compounds widely used as fragrances in deodorants, shampoos, air fresheners, fabric softeners, and detergents (Luckenbach and Epel, 2005). Extraction, isolation, and characterization of natural musks from male musk deer, civet cat, American muskrat, angelica root oil, and ambrette seed oil

TABLE 5.3
A Summary of Environmental Toxicity Data of Selected Chiral Azole Fungicides Frequently Detected in the Environment (µg/L or µg/g)

Compound	Class	Species	Effect	Duration	Endpoint	Value	Reference	Comment
Aquatic toxicity								
Climbazole	Plant	Green algae	–	96 h	EC_{50}	6154	(USEPA, 2014)	Calculated[a]
	Plant	Green algae	–	–	ChV^b	2314	(USEPA, 2014)	Calculated
	Plant	*Lemna minor*	Biomass yield	7 days	EC_{10}	5.6	Richter et al. (2013)	
	Plant	*Lemna minor*	Biomass yield	7 days	EC_{50}	19.0	Richter et al. (2013)	
	Plant	*Lemna minor*	Growth rate	7 days	EC_{10}	11.2	Richter et al. (2013)	
	Plant	*Lemna minor*	Growth rate	7 days	EC_{50}	33.9	Richter et al. (2013)	
	Plant	*Navicula pelliculosa*	Biomass yield	72 h	EC_{10}	64.1	Richter et al. (2013)	
	Plant	*Navicula pelliculosa*	Biomass yield	72 h	EC_{50}	153.6	Richter et al. (2013)	
	Plant	*Navicula pelliculosa*	Growth rate	72 h	EC_{10}	108.3	Richter et al. (2013)	
	Plant	*Navicula pelliculosa*	Growth rate	72 h	EC_{50}	290.6	Richter et al. (2013)	
	Plant	*Pseudokirchneriella subcapitata*	Biomass yield	72 h	EC_{10}	28.7	Richter et al. (2013)	
	Plant	*Pseudokirchneriella subcapitata*	Biomass yield	72 h	EC_{50}	214.4	Richter et al. (2013)	
	Plant	*Pseudokirchneriella subcapitata*	Growth rate	72 h	EC_{10}	314.7	Richter et al. (2013)	
	Plant	*Pseudokirchneriella subcapitata*	Growth rate	72 h	EC_{50}	1191	Richter et al. (2013)	
	Invertebrate	Daphnia	Immobility	48 h	EC_{50}	4219	(USEPA, 2014)	Calculated
	Invertebrate	Daphnia	Immobility	–	ChV	647	(USEPA, 2014)	Calculated
	Invertebrate	*Daphnia magna*	Immobility	48 h	EC_{10}	9660	Richter et al. (2013)	
	Invertebrate	*Daphnia magna*	Immobility	48 h	EC_{50}	15,990	Richter et al. (2013)	
	Fish	Fish	Mortality	96 h	EC_{50}	6316	(USEPA, 2014)	Calculated
	Fish	Fish	Mortality	–	ChV	748	(USEPA, 2014)	Calculated
	Fish	*Danio rerio*	Embryos mortality	48 h	EC_{10}	1070	Richter et al. (2013)	
	Fish	*Danio rerio*	Embryos mortality	48 h	EC_{50}	8200	Richter et al. (2013)	
Ketoconazole	Plant	Green algae	–	96 h	EC_{50}	4362	(USEPA, 2014)	Calculated
	Plant	Green algae	–	–	ChV	1852	(USEPA, 2014)	Calculated

TABLE 5.3 *(Continued)*
A Summary of Environmental Toxicity Data of Selected Chiral Azole Fungicides Frequently Detected in the Environment (μg/L or μg/g)

Compound	Class	Species	Effect	Duration	Endpoint	Value	Reference	Comment
	Invertebrate	Daphnia	Immobility	48 h	EC_{50}	2388	(USEPA, 2014)	Calculated
	Invertebrate	Daphnia	Immobility	–	ChV	426	(USEPA, 2014)	Calculated
	Invertebrate	*Daphnia magna*	Immobility	24 h	EC_{50}	8100	(Haeba et al., 2008)	
	Invertebrate	*Daphnia magna*	Immobility	48 h	EC_{50}	1510	(Haeba et al., 2008)	
	Fish	Fish	Mortality	96 h	EC_{50}	3385	(USEPA, 2014)	Calculated
	Fish	Fish	Mortality	–	ChV	427	(USEPA, 2014)	Calculated
Miconazole	Plant	Green algae	–	96 h	EC_{50}	165	(USEPA, 2014)	Calculated
	Plant	Green algae	–	–	ChV	104	(USEPA, 2014)	Calculated
	Invertebrate	Daphnia	Immobility	48 h	EC_{50}	44	(USEPA, 2014)	Calculated
	Invertebrate	Daphnia	Immobility	–	ChV	13	(USEPA, 2014)	Calculated
	Fish	Fish	Mortality	96 h	EC_{50}	52	(USEPA, 2014)	Calculated
	Fish	Fish	Mortality	–	ChV	8	(USEPA, 2014)	Calculated
Terrestrial toxicity								
Climbazole	Plant	*Avena sativa*	Biomass yield	17 days	EC_{10}	4.9	Richter et al. (2013)	In LUFA
	Plant	*Avena sativa*	Biomass yield	17 days	EC_{50}	18.5	Richter et al. (2013)	In LUFA
	Plant	*Avena sativa*	Biomass yield	17 days	EC_{10}	22.1	Richter et al. (2013)	In artificial soil
	Plant	*Avena sativa*	Biomass yield	17 days	EC_{50}	83.8	Richter et al. (2013)	In artificial soil
	Plant	*Brassica napus*	Biomass yield	17 days	EC_{10}	15.0	Richter et al. (2013)	In LUFA
	Plant	*Brassica napus*	Biomass yield	17 days	EC_{50}	30.7	Richter et al. (2013)	In LUFA
	Plant	*Brassica napus*	Biomass yield	17 days	EC_{10}	10.5	Richter et al. (2013)	In artificial soil
	Plant	*Brassica napus*	Biomass yield	17 days	EC_{50}	45.5	Richter et al. (2013)	In artificial soil
	Microorganism	*Arthrobacter globiformis*	Dehydrogenase activity	15–45 min	EC_{10}	400	Richter et al. (2013)	
	Microorganism	*Arthrobacter globiformis*	Dehydrogenase activity	15–45 min	EC_{50}	456	(Richter et al., 2013)	

Source: Chen and Ying (2015). Used with Permission from Elsevier Ltd.
[a] Acute and chronic values are calculated by ECOSAR v1.11, EPI Suite v4.10 (USEPA, 2014).
[b] ChV, chronic value.

increased research interest on synthesis of musks (Sommer, 2004). Synthetic musks are added to PCPs for their fruity, animal-like, or flowery fragrance. In 1987, 7,000 tons of synthetic musks were produced, of which 61%, 35%, and 4% were polycyclic musks, nitro musks, and macrocyclic musks, respectively (Sommer, 2004). Increased regulatory interest in nitro musks due to reports of photoallergen activities resulted in a sharp decrease in the global production of nitro musks between 1987 and 1996. The presence of nitro functional group on nitro musks imparted several deleterious physicochemical characteristics such as poor stability alkaline media such as that prevalent in soap, shampoo, and detergents (David, 2020). Nitro musks were slightly soluble in water and highly photosensitive, which made them undesirable as active ingredients in cosmetics (David, 2020). By 1998, musk ambrette, musk moskene, and musk tibetene were banned from use in cosmetics by the European Commission (Sommer, 2004). In 1999, the European Commission recommended that the use of other nitro musks such as musk ketone and musk xylene should be limited (Sommer, 2004). Polycyclic musks were first synthesized in the 1950s (David, 2020). They quickly dominated the market owing to their superior photo- and chemical stability, cheaper synthesis pathways, and stronger fabric binding characteristics (Sommer, 2004). Macrocyclic musks are increasingly becoming important since most of them are natural products, unlike nitro and polycyclic musks. There is a global push for the use of natural products due to the misconception among consumers who think natural products are safer than synthetic products. Unlike nitro and polycyclic musks, macrocyclic musks are more biodegradable. However, macrocyclic musks are predominantly used in perfumes since they have stronger musky odor and higher costs of production (Homem et al., 2015). In 2018, it was estimated that consumers spent more than $39 billion on fragrances only (Bomgardner, 2019). As standards of living continue to improve, demand for synthetic musks will continue to grow. Hence, synthetic musks are considered high production volume compounds. Synthetic musks are contaminants of emerging concern since they inadvertently enter the environment through wastewater treatment plants and atmospheric deposition.

5.5.1 Implications of Chirality on Synthetic Musks

Some polycyclic, macrocyclic, and alicyclic musks are chiral compounds. Olfactory receptors are three-dimensional structures that can be enantiospecific (Ahmed et al., 2018; Bentley, 2006). Previous studies found that the odor properties of chiral synthetic musks were enantiospecific. For example, (−)-(3S)-tonalid has a strong musky odor while (+)-(3R)-tonalid has a light but sweet aromatic smell (David, 2020). Other studies found that muscone, galaxolide, and vulcanolide had one enantiomer that had a strong musky smell and a weakly musky antipode (Brenna et al., 2003; David, 2020; Kraft and Fráter, 2001). Discovery of enantioselectivity in odor characteristics promoted the development of enantioselective synthesis pathways. For example, (−)-(3S)-tonalid was produced from enantiopure chrysanthemic acid via Friedel-Crafts, Kochi's decarboxylation, and finally reductive dehalogenation and acetylation (David, 2020). Some synthetic musks have three-dimensional structural configurations that are similar to those of naturally occurring pheromones. For example, galaxolide was shown to be structurally similar to the mammalian steroid pheromone androstenone (Kallenborn et al., 1999). Androstenone plays a significant role in sexual attraction. Response to pheromones is dependent on the enantiomeric composition of the pheromone. Hence, occurrence of galaxolide enantiomers may affect the reproductive activity of some mammals.

5.5.2 Physicochemical Properties

Chiral synthetic musks currently commercially available include polycyclic (e.g., cashmeran, galaxolide, phantolide, tonalide, and traseolide), macrocyclic (e.g., muscone), and alicyclic (e.g., helvetolide) musks (Table 5.4). Besides cashmeran, which has the chemical formula $C_{17}H_{22}O$, chiral polycyclic musks have the formula $C_{18}H_{26}O$ (Liu et al., 2020). The later polycyclic musks comprise a central phenyl group fused to at least one pentyl (galaxolide, phantolide, traseolide) or hexyl ring

Chiral Personal Care Products

TABLE 5.4
Physicochemical Properties of Chiral Synthetic Musks

Compound	Structure	Molecular weight, g mol^{-1}	Vapor pressure, mmHg at 25°C[a]	Water solubility, mg L^{-1}	Log K_{OW}	Log K_{OC}	WWTP Removal, %	Log BCF
Muscone		238.4	0.000469	0.22	5.96	3.78	92.06	3.89
Helvetolide		284.4	0.000327	0.30	5.51	2.92	88.41	3.54
Galaxolide		258.4	0.000252	0.49	5.43	3.80	90.2	3.48
Tonalide		258.4	0.00016	0.24	5.80	3.72	91.16	3.76
Phantolide		244.4	0.000417	0.75	5.30	3.46	85.44	3.38
Traseolide		258.3	0.000195	0.26	5.76	3.77	90.89	3.74
Cashmeran		206.3	0.00403	5.94	4.49	2.77	57.33	2.76

[a] Vapor pressure determined using the modified grain method.

(e.g., tonalide). Chiral synthetic musks are slightly soluble with water solubilities ranging from 0.22 to 0.75 mg L^{-1}, except for cashmeran, which is more soluble at 5.94 mg L^{-1}. Hence, apart from cashmeran, synthetic musks have a low tendency for partitioning to the aqueous phase when they enter the environment. Synthetic musks are semi-volatile and can partition to air from other environmental compartments (Liu et al., 2020). In addition, since they are sometimes applied through spraying, they can attach to indoor dust. Previous studies found synthetic musks on indoor dust at concentrations ranging from 4.42 to 992 ng g^{-1} (Kubwabo et al., 2012; Lu et al., 2011). The median sum of galaxolide, tonalide, cashmeran, celestolide, phantolide, and traseolide was shown to be highest in air samples collected from wastewater treatment plants followed by indoor air, urban areas, and rural areas, in descending order (Wong et al., 2019). The long-range transport potential of synthetic musks is not well established. Synthetic musks rapidly undergo photolysis with half-lives of tens of hours, suggesting low long-range transport potential (Wong et al., 2019). However, a study in Greenland detected galaxolide, cashmeran, and traseolide in polar bears (Vorkamp et al., 2004). A mechanistic study involving measuring synthetic musks present in air and seawater in the Arctic found the air-sea gas exchange of galaxolide and tonalide were dominated by deposition fluxes (Xie et al., 2007). These results suggest that atmospheric deposition influenced the amount of synthetic musks detected in remote locations. Chiral synthetic musks are highly hydrophobic compounds since they have log K_{OW} > 4.0. This indicates that synthetic musks have low mobility in soil and high potential to partition into solids such as sediments, soils, and biota.

5.5.3 Synthetic Musks in the Environment

Synthetic musks have been detected in various environmental matrices such as wastewater, indoor dust, sediment, surface water, soil, and biota. They enter the environment primarily through discharges from wastewater treatment plants. Hence, several studies have found that the concentration of synthetic musks was higher in wastewater influent, followed by wastewater effluent and, lastly, surface water. For example, a study in China found that the galaxolide and tonalide concentrations in wastewater effluent were five- to six-fold higher than in a creek (Zhang et al., 2008). Zhang et al. (2008) estimated that the total galaxolide and tonalide discharged from wastewater treatment plants in Shanghai, China, were 0.38 tons and 1.26 tons, respectively. A study in Portugal found that the concentration of galaxolide and tonalide wastewater effluent was 87.7–282.5 ng L^{-1} and 1600–3855 ng L^{-1}, respectively (Ramos et al., 2019). However, in surface water, the concentrations were 5.2–14.4 ng L^{-1} and not detected to 65.1 ng L^{-1}, respectively (Ramos et al., 2019). Similar concentrations obtained in a river in Germany ranging from 60 to 260 ng L^{-1} were obtained downstream from a wastewater treatment plant. Seasonal variations between the ratio of galaxolide and its transformation product galaxolide-lactone indicated that changes in local environmental factors such as water chemistry affected the distribution of galaxolide by influencing its biotransformation (Lange et al., 2015). Interestingly, Lange et al. (2015) found low concentrations of galaxolide and tonalide upstream of the wastewater treatment plants as well as in a tributary of the river. These results suggested that apart from wastewater discharge, there was another route of entry for synthetic musks into the aquatic environment, possibly atmospheric deposition. A biomonitoring study in Beijing frequently detected galaxolide and tonalide in pine needles (Wang et al., 2019). Higher concentrations were reported in places proximal to chemical plants and industrial plants, suggesting synthetic musks can enter the environment through industrial discharges and dust (Wang et al., 2019). In the United States, galaxolide and tonalide were detected in water, sediment, fish, and mussel in the Hudson River. The galaxolide and tonalide concentrations were higher in sediment (2.8–388 and 113–544 ng g^{-1}) than water (3.95–25.8 and 5.09–22.8 ng L^{-1}), respectively (Reiner and Kannan, 2011). Galaxolide and tonalide have log K_{OW} > 4, suggesting a preference to partitioning to sediments and biota tissue. In fact, the bioaccumulation factor for galaxolide based on lipid weight concentration ranged from 261 to 12,900 (Reiner and Kannan, 2011). The broad range in bioaccumulation factor could be due to interspecies differences in trophic transfer, feeding habits, and metabolism as well as local environmental conditions and the physicochemical properties of the synthetic musk (Lange et al., 2015; Reiner and Kannan, 2011; Yao et al., 2019).

5.5.4 Toxicity

The presence of synthetic musks in the environment is concerning because they can cause adverse ecological effects. Galaxolide and tonalide reduced the growth rate of juvenile freshwater mussels with an EC50 of 153–831 µg L^{-1} and 108–1034 µg L^{-1}, respectively (Gooding et al., 2006). The mechanism of toxicity of synthetic musks involves inhibiting multixenobiotic resistance transporters in aquatic organisms. Multixenobiotic resistance transporters remove xenobiotic compounds from cells. They act as the first line of defense against chemical pollutants. A previous study showed that polycyclic musks inhibited multixenobiotic resistance transporters in mussels (Luckenbach et al., 2004). Synthetic musks (e.g., galaxolide, tonalide, and celestolide) significantly inhibited the development of marine crustacean larvae, *Acartia tonsa*, at low concentrations (Wollenberger et al., 2003). Several studies have shown that environmental concentrations of galaxolide and tonalide may cause oxidative stress and genotoxicity in marine clams *Ruditapes philippinarum* (Ehiguese et al., 2020) and zebra mussels *Dreissena polymorpha* (Parolini et al., 2015).

The toxicity of synthetic musks is often influenced by several factors, such as physicochemical characteristics of the musks, presence of co-pollutants, and biological traits of the target organism. Exposing a rainbow trout cell line to synthetic musks showed that polycyclic musks (e.g., celestolide, galaxolide, and tonalide) were more cytotoxic than nitro musks (e.g., musk xylene and musk ketone) (Schnell et al., 2009). There was a significant relationship between cytotoxicity and the log K_{OW} ($p < 0.05$). With log $K_{OW} > 6.0$, polycyclic musks readily partitioned into the rainbow trout cells compared to nitro musks, which have a log $K_{OW} < 5.0$. Interestingly, previous studies showed that the biotransformation of galaxolide and tonalide in fathead minnows was low, resulting in high bioaccumulation of the musks (Lefebvre et al., 2017). If the elimination and metabolism of pollutants are low, then the pollutant has a high probability of reaching the site of toxic action. A simulated study on urban runoff, exposing goldfish *Carassius auratus* to galaxolide and cadmium resulted in a synergistic increase in oxidative stress after 21 days (Chen et al., 2012). The results demonstrate that metal co-pollutants may alter the toxicity profile of synthetic musks. Local environmental factors can also influence the toxicity of the synthetic musks, particularly in aquatic environments. In the aquatic environment, biota competes for the hydrophobic musks with solids such as sediments and suspended matter. Sorption of highly hydrophobic synthetic musks to these abiotic solids reduces their bioavailability. Studies on chiral toxicity of synthetic musks remain scarce. Current toxicological data suggest that synthetic musks may cause adverse effects on organisms in the environment.

Exposing zebrafish embryo to muscone showed that (*R*)-muscone was more potent than (*S*)-muscone as it caused higher mortality and pericardial edema incidents as well as decreasing hatching rate and heartbeat rate (M. Li et al., 2020). The results show that there is a need for determining the enantiomeric composition of synthetic musks in environmental risk assessments. Future studies should explore the enantioselective toxicity of other synthetic musks, particularly galaxolide and tonalide, which are used more abundantly worldwide.

5.6 ORGANIC UV FILTERS

Although exposure to ultraviolet light is vital for vitamin D biosynthesis, long-term exposure can cause harmful effects to the skin such as skin cancer (Jansen et al., 2013). As the standard of living increased globally, demand for PCPs that offers protection to the skin against UV radiation has surged (Gilbert et al., 2013; Yang et al., 2020). An increasing number of PCPs such as cosmetics, lotions, hair sprays, lipsticks, shampoos, and sunscreens contain UV-filtering active ingredients. The active ingredients protect the skin by absorbing or reflecting UVA, UVB, and UVC (Gilbert et al., 2013). Inorganic UV filters such as TiO_2 and ZnO protect the skin by reflecting UV radiation while organic UV filters such as benzophenone-3, 4-*p*-aminobenzoic acid, ethylexyl methoxycinnamate, 4-methylbenzylidene camphor, diethylhexyl butamido triazone, drometrizole trisiloxane, phenylbenzimidazole sulfonic acid, and butyl methoxydibenzoyl methane by absorbing UVA and UVB. Organic UV filters can be classified according to their chemical structures (Silvia Díaz-Cruz et al., 2008). Organic UV filters are frequently used filters since they protect against a broad UV wavelength and can be mixed with other active ingredients to improve the filtering efficiency. There are currently 25 organic UV filters approved for use in cosmetics by the European Commission (Regulation no. 1223/2009 of the European Commission) (Nohynek et al., 2010). In the United States, the Sunscreen Ingredients Regulated by Over the Counter (OTC) Monograph (21 CFR 352.10) regulates 22 organic UV filters.

5.6.1 Physicochemical Properties

Organic UV filters contain highly conjugated chromophores that absorb UV radiation. Most organic UV filters contain aromatic moieties conjugated with methoxy and carbonyl functional groups (Jansen et al., 2013). Highly conjugated systems are more beneficial as they increase the absorption of longer wavelength light. Often, aromatic compounds with large molecular weight have relatively higher degrees of conjugation. Hence, broad-spectrum and UVA (315–400 nm) filters often have relatively larger molecular weight than UVB (280–315 nm) and UVC (100–280 nm) filters (Jansen et al., 2013). Unfortunately, the high conjugation and the presence of aromatic groups desirable in organic UV filters makes them highly persistent, recalcitrant, and bioaccumulative. Organic UV filters are classified as *p*-aminobenzoic acid derivatives, benzimidazole derivatives, benzophenones, benzotriazoles, benzylmalonate derivatives, camphor derivatives, cinnamates, crylenes, dibenzoylmethane derivatives, and salicylates (Cadena-Aizaga et al., 2020).

The water solubility, vapor pressure, and log K_{OW} provide essential data on the environmental partitioning of the organic UV filters. Table 5.5 shows the physicochemical properties of organic UV filters. Organic UV filters have low solubility and vapor pressure, suggesting they do not readily partition to the aqueous phase. Except for 2-ethylhexyl salicylate and 2-ethylhexyl 4-(dimethylamino)benzoate, the log K_{OC} of most organic UV filters is greater than 4.0. This log K_{OC} suggests organic UV filters readily partition to organic matter, thus may persist in soil and sediments. Generally, organic pollutants with log $K_{OW} > 4.0$ such as ethylhexyl triazone, diethylhexyl butamido triazone, homosalate, and 4-methylbenzylidene camphor are considered lipophilic. This implies the compounds can readily partition to biota when they are discharged into the environment.

5.6.2 Organic UV Filters in the Environment

It is estimated that around 100,000 tons of organic UV filters are produced each year globally (Jurado et al., 2014). When organic UV filters are discharged into the wastewater treatment system, and subsequently enter wastewater treatment plants, different amounts are lost from the aqueous phase due to phase partitioning or transformation (Zhang et al., 2011). It is estimated that between 90–93% of the total organic UV filters that enter wastewater treatment plants are transformed (Table 5). Highly lipophilic organic UV filters such as 4-methylbenzylidene camphor (log K_{OW} = 5.92), bis-ethylhexyloxyphenol methoxyphenyl triazine (log K_{OW} = 9.29), diethylhexyl butamido triazone (log K_{OW} = 14.03), and ethylhexyl triazone (log K_{OW} = 17.05) can readily partition to sludge. Indeed, the concentration of 4-methylbenzylidene camphor, 2-ethylhexyl 2-cyano-3,3-diphenylacrylate (log K_{OW} = 6.88), ethylhexyl triazone, and diethylhexyl butamido triazone in treated sludge ranged 106–3100 ng g^{-1} dry weight (dw), 138–41,610 ng g^{-1} dw, 1700–2700 ng g^{-1} dw, and 54–136 ng g^{-1} dw, respectively (Ramos et al., 2016). Organic UV filters in treated sludge contaminate soil and subsequently groundwater when they are dumped at landfills or used for soil amendment in agricultural fields. In agricultural fields, organic UV filters can be taken up by crops, and pose a human health risk.

Discharge from wastewater treatment plants to rivers and lakes contain the remaining organic UV filters partitioned to the aqueous phase or the suspended organic matter. Since organic UV filters are highly lipophilic with log $K_{OW} > 4.0$ and recalcitrant, they bioaccumulate in organisms. Previous studies have shown that organic UV filters bioaccumulate in aquatic organisms such as fish and mussels. Previous studies found that 2-ethylhexyl-2-cyano-3,3-diphenyl-2-propenoate was

TABLE 5.5
Physicochemical Properties of Chiral Organic Ultraviolet Light (UV) Filters

Family	Compound (Abbreviation) CAS Number	Chemical Structure	UVA	UVB	Boiling Point (°C)	Log K$_{ow}$	Solubility in Water at 25 °C (mg/L)	UV Filters Allowed by: Europe	USA	Japan
p-Aminobenzoic acids derivatives	2-Ethylhexyl 4-(dimethylamino)benzoate (EDP) CAS: 21245-02-3			X	344.5	5.77	2.00×10^{-1}	X	X	X
	Ethoxylated ethyl-p-aminobenzoate (PEG-25 PABA) CAS: 116242-27-4			X	?	?	?	X		
Camphor derivatives	3-(4'-Methylbenzylidene) camphor (4-MBC) CAS: 36861-47-9			X	349.4	5.92	2.00×10^{-1}	X		
	Terephthalylidene dicamphor sulfonic acid (Ecamsule) (TDSA) CAS: 92761-26-7		X		757.5	3.83	1.50×10^{-1}	X		X
	Camphor benzalkonium methosulfate (CBM) CAS: 52793-97-2			X	638.2	3.11	7.34×10^{0}	X		
	Benzylidene camphor sulfonic acid (BCSA) CAS: 56039-58-8			X	472.0	2.22	1.20×10^{2}	X		
	Polyacrylamidomethyl benzylidene camphor (PBC) CAS: 113783-61-2			X	?	?	?	X		

(Continued)

TABLE 5.5 (Continued)
Physicochemical Properties of Chiral Organic Ultraviolet Light (UV) Filters

Family	Compound (Abbreviation) CAS Number	Chemical Structure	UVA	UVB	Boiling Point (°C)	Log K_{ow}	Solubility in Water at 25 °C (mg/L)	UV Filters Allowed by: Europe	USA	Japan
	3-Benzylidene-camphor (3BC) CAS: 15087-24-8			X	337.6	5.37	6.90×10^{-1}	X		
Benzotriazole derivatives	Drometrizole Trisiloxane (DTS) CAS: 155633-54-8		X	X	528.6	10.82	6.40×10^{-7}	X		X
Salicylate derivatives	2-Ethylhexyl salicylate (ES) CAS: 118-60-5			X	344.9	5.97	7.20×10^{-1}	X	X	X
	3,3,5-trimethylciclohexyl salicylate (Homosalate) (HMS) CAS: 118-56-9			X	355.9	6.16	4.20×10^{-1}	X	X	X
	Diethylhexyl butamido triazone (DBT) CAS: 154702-15-5			X	893.5	14.03	1.33×10^{-11}	X		
Triazine derivatives	Ethylhexyl triazone (EHT) CAS: 88122-99-0			X	874.4	17.05	1.45×10^{-14}	X		X

TABLE 5.5 (Continued)
Physicochemical Properties of Chiral Organic Ultraviolet Light (UV) Filters

Family	Compound (Abbreviation) CAS Number	Chemical Structure	UVA	UVB	Boiling Point (°C)	Log K_{ow}	Solubility in Water at 25 °C (mg/L)	UV Filters Allowed by: Europe	UV Filters Allowed by: USA	UV Filters Allowed by: Japan
	Bis-ethylhexyloxyphenol Methoxyphenyl triazine (BEMT) CAS: 187393-00-6		X		786.6	9.29	1.45×10^{-16}	X		X
Cinnamates derivatives	2-Ethylhexyl 4-methoxycinnamate (EMC) CAS: 5466-77-3			X	360.5	5.80	1.50×10^{-1}	X	X	X
Dibenzoyl methane derivative	4-tert-butil-4'-methoxydibenzoylmethane (Avobenzone) (BMDM) CAS: 70356-09-1		X		409.3	4.51	1.52×10^{0}	X	X	X
Crylene derivative	2-Ethylhexyl 2-cyano-3,3-diphenylacrylate (OC) CAS: 6197-30-4			X	472.9	6.88	3.81×10^{-3}	X	X	X
Benzalmalonate derivative	Dimethicodiethylbenzalmalonate (Polysilicone-15) (BMP) CAS: 207574-74-1				?	?	?	X		X

Data on Chiral Organic UV Filters is from Ramos et al. (2016). Used with Permission from Elsevier Ltd.

the most frequently detected organic UV filter in biota even though its concentration was relatively lower than homosalate and octocrylene (Gago-Ferrero et al., 2012). Interestingly, organic UV filters such as benzophenone-3 and ethylhexyl dimethyl PABA were not detected in biota. Since previous studies have detected benzophenone-3 and ethylhexyl dimethyl PABA in various environmental matrices such as surface water, freshwater sediment, marine sediment, and seawater, these results suggest these organic UV filters are readily metabolized in biota (Gago-Ferrero et al., 2012). Interestingly, benzophenone-3, ethylhexyl dimethyl PABA, octocrylene, 4-methylbenzylidene camphor, and ethylhexyl methoxycinnamate were the most frequently detected organic UV filters in marine biota (Cadena-Aizaga et al., 2020). Octocrylene, which has a high log K_{OW} of around 6.88, was the most frequently detected organic UV filter. The difference between freshwater and marine environments is probably due to differences in water chemistry. A previous study on the bioaccumulation of ethylhexyl dimethyl PABA in crucian carp found that bioconcentration decreased significantly when dissolved organic matter was present (Ma et al., 2018). These results suggest that dissolved organic matter reduced the bioavailability of ethylhexyl dimethyl PABA. These results suggest the amount of organic UV filter that bioaccumulates in tissue is not only influenced by the usage of the organic UV filters but by their log K_{OW} values—another critical factor to consider regarding bioaccumulation of organic UV filters in the presence of dissolved organic matter.

Of the 43 organic UV filters approved in Europe, the United States, and Japan, 18 contain at least one chiral center. However, studies on the enantioselective behavior in the environment have been limited to 4-methylbenzylidene camphor and 2-ethylhexyl 4-dimethylaminobenzoate. Commercial sunscreens contain at least 99% (*E*)-4-methylbenzylidene camphor as a racemic mixture (Buser et al., 2005). The (*E*)-isomer can undergo isomerization following exposure to light to yield the (*Z*)-isomer. Hence, environmental samples are expected to comprise four stereoisomers of 4-methylbenzylidene camphor. Buser et al. (2005) found that (*E*)-4-methylbenzylidene camphor was more abundant in wastewater than the (*Z*)-isomer. This indicates that the organic UV filter entered the wastewater system as the parent compound. However, for both the (*E*)- and the (*Z*)-isomer, a scalemic mixture was detected in the wastewater treatment plants (Buser et al., 2005). Considering that biotransformation processes are often enantiospecific, these results suggest that 4-methylbenzylidene camphor underwent an enantioselective transformation. Enantiomeric enrichment of the (*R*)-enantiomers was observed in lakes with enantiomeric excess ranging from 1.70–1.83 (Buser et al., 2005). The bioaccumulation of 4-methylbenzylidene camphor exhibited enantiospecificity in perch with enantiomeric enrichment of the (*R*)-enantiomer. In contrast, bioaccumulation in roach was slightly enantiomeric, and these results show that enantiospecificity in the bioaccumulation of 4-methylbenzylidene camphor in an aquatic organism is species dependent (Buser et al., 2005). Therefore, the results of the enantioselective accumulation of organic UV filters cannot be extrapolated to other species.

5.6.3 Toxicity

Previous studies have shown that exposing aquatic organisms to organic UV filters could cause adverse health effects towards mobility, reproduction, and development (Lozano et al., 2020; Novais et al., 2018; Quintaneiro et al., 2019). In humans, several studies showed that organic UV filters could cause anti-estrogenic, estrogenic, antiandrogenic, and androgenic activity, in addition to skin irritations (Table 5.6). Camphor-related organic UV filters such as 4-methylbenzylidene camphor have been shown to elicit vitellogenin induction, gonad alteration, fertility decrease, and feminization of male fish. For example, a recent study showed that exposing Japanese medaka to 50 and 500 µg L^{-1} of 4-methylbenzylidene camphor caused reproductive and developmental toxicity (Liang et al., 2020). The exposure concentrations were not environmentally relevant, considering there were orders of magnitude above those detected in fish globally. In Japanese clam, 4-methylbenzylidene camphor showed acute toxicity at low concentrations with an LC50

TABLE 5.6
Hormonal Activities of Chiral Organic Ultraviolet Light (UV) Filters

Compound	Estrogenic Activity	Antiestrogenic Activity	Androgenic Activity	Antiandrogenic Activity
4-Methylbenzylidene camphor	—	+++	—	+++
3-Benzylidene camphor	+	+++	—	+++
Isopentyl-4-methoxycinnamate	—	+++	++	+++
Octyl methoxycinnamate	—	+++	++	+++
Octocrylene	—	+++	+	+++
Homosalate	—	+++	+++	+++
Octyl salicylate	—	+++	++	+++
Ethoxylated ethyl 4-amino benzoate	—	+	—	—

Source: Data on Selected Chiral Organic UV Filters is from Díaz-Cruz and Barceló (2009). Used with permission from Elsevier Ltd.

of 7.71 µg L^{-1}. Even though the metabolism of organic UV filters in mammals is enantioselective, there are currently no studies that have explored the effect of chirality on the toxicity of organic UV filters. A previous study showed that the transformation of 2-ethylhexyl 4-dimethylamino-benzoate in rabbit kidneys and liver resulted in an enrichment of the (−)-enantiomer (Liang et al., 2017). There is a need for more studies exploring enantioselectivity in the human and environmental toxicity of organic UV filters.

The +++, ++, and + symbols represent a maximal dose-response curves with ≥80% efficacy, ≥30% efficacy, and <30% efficacy while — represents not detected.

5.7 CONCLUDING REMARKS AND FUTURE TRENDS

Chiral PCPs are pollutants of emerging concern. Enantiomers of chiral pollutants have different properties in biological systems. Most studies on enantioselectivity in the biotransformation of chiral PCPs have been focused on wastewater treatment plants. The concentrations of PCPs are higher in wastewater treatment plants. Since wastewater treatment plants play a crucial role in mitigating pollution, research interest has been focused on them. However, demonstrating enantioselectivity in the removal of PCPs in wastewater treatment plants is not enough to fully understand the implications of chirality on the risk posed by chiral PCPs.

Future research should investigate enantioselective fate of chiral PCPs in surface water, sediments, and marine environments. The natural attenuation of chiral PCPs in surface water determines the exposure of aquatic organisms to the PCPs. If natural attenuation preferentially transforms the less toxic enantiomer, then enantiomeric enrichment of the more potent antipode occurs. In such a case, ignoring chirality results in underestimating the risk posed by the chiral PCP. If the scenario is inverse, then the less toxic antipode is enriched. Ignoring chirality in such a case will result in overly aggressive mitigation strategies. Therefore, for recommendation from an environmental risk assessment to be accurate, the enantioselective behavior and toxicity of the PCPs should be considered.

There is a need for development of enantioselective analysis techniques for other high production volume PCPs such as phthalates. There are currently no studies that have investigated the enantioselective distribution, fate, and toxicity of chiral phthalates such as DEHP. This dearth of studies is probably because of the challenges of separating the enantiomers of DEHP using chromatographic techniques. An *in silico* study showed that (*R,R*)-DEHP had higher endocrine disruption activity then the (*S,S*)-enantiomer or the (*R,S*)-diastereomer. Since several studies demonstrated that phthalates pose significant risk to environmental health, there is a need for incorporating its stereochemistry in risk assessments.

REFERENCES

Ahmed, L., Zhang, Y., Block, E., Buehl, M., Corr, M. J., Cormanich, R. A., Gundala, S., Matsunami, H., O'Hagan, D., Ozbil, M., Pan, Y., Sekharan, S., Ten, N., Wang, M., Yang, M., Zhang, Q., Zhang, R., Batista, V. S., Zhuang, H., 2018. Molecular mechanism of activation of human musk receptors OR5AN1 and OR1A1 by (R)-muscone and diverse other musk-smelling compounds. *Proc. Natl. Acad. Sci. U. S. A.* 115, E3950–E3958. https://doi.org/10.1073/pnas.1713026115

Alani, J. I., Davis, M. D. P., Yiannias, J. A., 2013. Allergy to cosmetics: A literature review. *Dermatitis* 24, 283–290. https://doi.org/10.1097/DER.0b013e3182a5d8bc

Bentley, R., 2006. The nose as a stereochemist. Enantiomers and odor. *Chem. Rev.* 106, 4099–4112. https://doi.org/10.1021/cr050049t

Bomgardner, M. M., 2019. Perfumers seek a natural balance. *C&EN* 97, 30–35.

Brenna, E., Fuganti, C., Serra, S., 2003. Enantioselective perception of chiral odorants. *Tetrahedron Asymmetry* 14, 1–42. https://doi.org/10.1016/S0957-4166(02)00713-9

Briggs, D., 2003. Environmental pollution and the global burden of disease. *Br. Med. Bull.* 68, 1–24. https://doi.org/10.1093/bmb/ldg019

Brooks, B. W., Sabo-Attwood, T., Choi, K., Kim, S., Kostal, J., LaLone, C. A., Langan, L. M., Margiotta-Casaluci, L., You, J., Zhang, X., 2020. Toxicology advances for 21st Century chemical pollution. *One Earth* 2, 312–316. https://doi.org/10.1016/j.oneear.2020.04.007

Buser, H. R., Müller, M. D., Balmer, M. E., Poiger, T., Buerge, I. J., 2005. Stereoisomer composition of the chiral UV filter 4-methylbenzylidene camphor in environmental samples. *Environ. Sci. Technol.* 39, 3013–3019. https://doi.org/10.1021/es048265r

Buser, H. R., Müller, M. D., Poiger, T., Balmer, M. E., 2002. Environmental behavior of the chiral acetamide pesticide metalaxyl: Enantioselective degradation and chiral stability in soil. *Environ. Sci. Technol.* 36, 221–226. https://doi.org/10.1021/es010134s

Cadena-Aizaga, M. I., Montesdeoca-Esponda, S., Torres-Padrón, M. E., Sosa-Ferrera, Z., Santana-Rodríguez, J. J., 2020. Organic UV filters in marine environments: An update of analytical methodologies, occurrence and distribution. *Trends Environ. Anal. Chem.* 25, e00079. https://doi.org/10.1016/j.teac.2019.e00079

Cao, J., Sanganyado, E., Liu, W., Zhang, W., Liu, Y., 2019. Decolorization and detoxification of Direct Blue 2B by indigenous bacterial consortium. *J. Environ. Manage.* https://doi.org/10.1016/j.jenvman.2019.04.067

Celis, R., Gámiz, B., Adelino, M. A., Hermosín, M. C., Cornejo, J., 2013. Environmental behavior of the enantiomers of the chiral fungicide metalaxyl in Mediterranean agricultural soils. *Sci. Total Environ.* 444, 288–297. https://doi.org/10.1016/j.scitotenv.2012.11.105

Chen, F., Gao, J., Zhou, Q., 2012. Toxicity assessment of simulated urban runoff containing polycyclic musks and cadmium in *Carassius auratus* using oxidative stress biomarkers. *Environ. Pollut.* 162, 91–97. https://doi.org/10.1016/j.envpol.2011.10.016

Chen, Z. F., Ying, G. G., 2015. Occurrence, fate and ecological risk of five typical azole fungicides as therapeutic and personal care products in the environment: A review. *Environ. Int.* 84, 142–153. https://doi.org/10.1016/j.envint.2015.07.022

Chiaia-Hernandez, A. C., Krauss, M., Hollender, J., 2013. Screening of lake sediments for emerging contaminants by liquid chromatography atmospheric pressure photoionization and electrospray ionization coupled to high resolution mass spectrometry. *Environ. Sci. Technol.* 47, 976–986. https://doi.org/10.1021/es303888v

Couteau, C., Jadaud, M., Peigne, F., Coiffard, L. J. M., 2000. Influence of pH on the photodegradation kinetics under UV light of climbazole solutions. *Analusis* 28, 557–560. https://doi.org/10.1051/analusis:2000171

Das, P., Horton, R., 2018. Pollution, health, and the planet: Time for decisive action. *Lancet* 391, 407–408. https://doi.org/10.1016/S0140-6736(17)32588-6

David, O. R. P., 2020. A chemical history of polycyclic musks. *Chem. Eur. J. Chem.* https://doi.org/10.1002/chem.202000577

Díaz-Cruz, M. S., Barceló, D., 2009. Chemical analysis and ecotoxicological effects of organic UV-absorbing compounds in aquatic ecosystems. *Trends Anal. Chem.* 28, 708–717. https://doi.org/10.1016/j.trac.2009.03.010

Dodgen, L. K., Li, J., Wu, X., Lu, Z., Gan, J. J., 2014. Transformation and removal pathways of four common PPCP/EDCs in soil. *Environ. Pollut.* 193, 29–36. https://doi.org/10.1016/j.envpol.2014.06.002

Ehiguese, F. O., Alam, M. R., Pintado-Herrera, M. G., Araújo, C. V. M., Martin-Diaz, M. L., 2020. Potential of environmental concentrations of the musks galaxolide and tonalide to induce oxidative stress and genotoxicity in the marine environment. *Mar. Environ. Res.* 105019. https://doi.org/10.1016/j.marenvres.2020.105019

Fu, Q., Sanganyado, E., Ye, Q., Gan, J., 2016. Meta-analysis of biosolid effects on persistence of triclosan and triclocarban in soil. *Environ. Pollut.* 210, 137–144. https://doi.org/10.1016/j.envpol.2015.12.003

Gago-Ferrero, P., Díaz-Cruz, M. S., Barceló, D., 2012. An overview of UV-absorbing compounds (organic UV filters) in aquatic biota. *Anal. Bioanal. Chem.* 404, 2597–2610. https://doi.org/10.1007/s00216-012-6067-7

Gámiz, B., Facenda, G., Celis, R., 2016. Evidence for the effect of sorption enantioselectivity on the availability of chiral pesticide enantiomers in soil. *Environ. Pollut.* 213, 966–973. https://doi.org/10.1016/j.envpol.2016.03.052

Gao, B., Zhang, Z., Li, L., Kaziem, A. E., He, Z., Yang, Q., Qing, P., Zhang, Q., Wang, M., 2019. Stereoselective environmental behavior and biological effect of the chiral organophosphorus insecticide isofenphos-methyl. *Sci. Total Environ.* 648, 703–710. https://doi.org/10.1016/j.scitotenv.2018.08.182

Gilbert, E., Pirot, F., Bertholle, V., Roussel, L., Falson, F., Padois, K., 2013. Commonly used UV filter toxicity on biological functions: Review of last decade studies. *Int. J. Cosmet. Sci.* 35, 208–219. https://doi.org/10.1111/ics.12030

Gooding, M. P., Newton, T. J., Bartsch, M. R., Hornbuckle, K. C., 2006. Toxicity of synthetic musks to early life stages of the freshwater mussel *Lampsilis cardium*. *Arch. Environ. Contam. Toxicol.* 51, 549–558. https://doi.org/10.1007/s00244-005-0223-4

Haeba, M. H., Hilscherová, K., Mazurová, E., Bláha, L., 2008. Selected endocrine disrupting compounds (vinclozolin, flutamide, ketoconazole and dicofol): Effects on survival, occurrence of males, growth, molting and reproduction of *Daphnia magna*. *Environ. Sci. Pollut. Res.* 15, 222–227. https://doi.org/10.1065/espr2007.12.466

Hamdy, D. A., Brocks, D. R., 2009. Nonlinear stereoselective pharmacokinetics of ketoconazole in rat after administration of racemate. *Chirality* 21, 704–712. https://doi.org/10.1002/chir.20669

Health Science Authority (HSA). (2019). Regulatory Guidance: Guidelines on the Control of Cosmetic Products [online]. Available: https://www.hsa.gov.sg/docs/default-source/hprg-cosmetics/guidelines-on-the-control-of-cosmetic-products.pdf (accessed 08 March 2020).

Homem, V., Silva, J. A., Ratola, N., Santos, L., Alves, A., 2015. Long lasting perfume – A review of synthetic musks in WWTPs. *J. Environ. Manage.* 149, 168–192. https://doi.org/10.1016/j.jenvman.2014.10.008

Hopkins, Z. R., Blaney, L., 2016. An aggregate analysis of personal care products in the environment: Identifying the distribution of environmentally-relevant concentrations. *Environ. Int.* 92–93, 301–316. https://doi.org/10.1016/j.envint.2016.04.026

Huang, Q., Wang, Z., Wang, C., Peng, X., 2013. Chiral profiling of azole antifungals in municipal wastewater and recipient rivers of the Pearl River Delta, China. *Environ. Sci. Pollut. Res. Int.* 20, 8890–9.

Jansen, R., Osterwalder, U., Wang, S. Q., Burnett, M., Lim, H. W., 2013. Photoprotection: Part II. sunscreen: Development, efficacy, and controversies. *J. Am. Acad. Dermatol.* 69, 867.e1–867.e14. https://doi.org/10.1016/j.jaad.2013.08.022

Jurado, A., Gago-Ferrero, P., Vàzquez-Suñé, E., Carrera, J., Pujades, E., Díaz-Cruz, M. S., Barceló, D., 2014. Urban groundwater contamination by residues of UV filters. *J. Hazard. Mater.* 271, 141–149. https://doi.org/10.1016/j.jhazmat.2014.01.036

Kahle, M., Buerge, I. J., Hauser, A., Müller, M. D., Poiger, T., 2008. Azole fungicides: Occurrence and fate in wastewater and surface waters. *Environ. Sci. Technol.* 42, 7193–7200. https://doi.org/10.1021/es8009309

Kallenborn, R., Gatermann, R., Rimkus, G. G., 1999. Synthetic musks in environmental samples: Indicator compounds with relevant properties for environmental monitoring. *J. Environ. Monit.* 1. https://doi.org/10.1039/a903408k

Kraft, P., Fráter, G., 2001. Enantioselectivity of the musk odor sensation. *Chirality* 13, 388–394. https://doi.org/10.1002/chir.1050

Kubwabo, C., Fan, X., Rasmussen, P. E., Wu, F., 2012. Determination of synthetic musk compounds in indoor house dust by gas chromatography-ion trap mass spectrometry. *Anal. Bioanal. Chem.* 404, 467–477. https://doi.org/10.1007/s00216-012-6124-2

Kwan, Y. H., Tung, Y. K., Kochhar, J. S., Li, H., Poh, A.-L., Kang, L., 2014. Regulation of Cosmetics, in: Heng, K. Y., Kei, T. Y., Singh, K. J., Hairui, L., Ai-Ling, P., Lifeng, K. (Eds.), *Handbook of Cosmeceutical Excipients and Their Safeties*, Woodhead Publishing Series in Biomedicine. Woodhead Publishing, Waltham, MA, pp. 7–22. https://doi.org/10.1533/9781908818713.7

Landrigan, P.J., Fuller, R., Acosta, N.J.R., Adeyi, O., Arnold, R., Basu, N. (Nil), Baldé, A.B., Bertollini, R., Bose-O'Reilly, S., Boufford, J.I., Breysse, P.N., Chiles, T., Mahidol, C., Coll-Seck, A.M., Cropper, M.L., Fobil, J., Fuster, V., Greenstone, M., Haines, A., Hanrahan, D., Hunter, D., Khare, M., Krupnick, A., Lanphear, B., Lohani, B., Martin, K., Mathiasen, K. V., McTeer, M.A., Murray, C.J.L., Ndahimananjara, J.D., Perera, F., Potočnik, J., Preker, A.S., Ramesh, J., Rockström, J., Salinas, C., Samson, L.D., Sandilya,

K., Sly, P.D., Smith, K.R., Steiner, A., Stewart, R.B., Suk, W.A., van Schayck, O.C.P., Yadama, G.N., Yumkella, K., Zhong, M. 2018. The Lancet Commission on pollution and health. *Lancet* 391, 462–512. https://doi.org/10.1016/S0140-6736(17)32345-0

Lange, C., Kuch, B., Metzger, J. W., 2015. Occurrence and fate of synthetic musk fragrances in a small German river. *J. Hazard. Mater.* 282, 34–40. https://doi.org/10.1016/j.jhazmat.2014.06.027

Lefebvre, C., Kimpe, L. E., Metcalfe, C. D., Trudeau, V. L., Blais, J. M., 2017. Bioconcentration of polycyclic musks in fathead minnows caged in a wastewater effluent plume. *Environ. Pollut.* 231, 1593–1600. https://doi.org/10.1016/j.envpol.2017.09.062

Li, M., Yao, L., Chen, H., Ni, X., Xu, Y., Dong, Wengjing, Fang, M., Chen, D., Aowuliji, Xu, L., Zhao, B., Deng, J., Kwok, K.W., Yang, J., Dong, Wu, 2020. Chiral toxicity of muscone to embryonic zebrafish heart. *Aquat. Toxicol.* 222, 105451. https://doi.org/10.1016/j.aquatox.2020.105451

Li, Y., Chen, L., Li, H., Peng, F., Zhou, X., Yang, Z., 2020. Occurrence, distribution, and health risk assessment of 20 personal care products in indoor and outdoor swimming pools. *Chemosphere* 254, 126872. https://doi.org/10.1016/j.chemosphere.2020.126872

Liang, M., Yan, S., Chen, R., Hong, X., Zha, J., 2020. 3-(4-Methylbenzylidene) camphor induced reproduction toxicity and antiandrogenicity in Japanese medaka (*Oryzias latipes*). *Chemosphere* 249, 126224. https://doi.org/10.1016/j.chemosphere.2020.126224

Liang, Y., Zhan, J., Liu, X., Zhou, Z., Zhu, W., Liu, D., Wang, P., 2017. Stereoselective metabolism of the UV-filter 2-ethylhexyl 4-dimethylaminobenzoate and its metabolites in rabbits in vivo and vitro. *RSC Adv.* 7, 16991–16996. https://doi.org/10.1039/c7ra00431a

Liu, J., Zhang, W., Zhou, Q., Zhou, Qingqin, Z., Zhang, Y., Zhu, L., 2020. Polycyclic musks in the environment: A review of their concentrations and distribution, ecological effects and behavior, current concerns and future prospects. *Crit. Rev. Environ. Sci. Technol.* 0, 1–55. https://doi.org/10.1080/10643389.2020.1724748

Lozano, C., Matallana-Surget, S., Givens, J., Nouet, S., Arbuckle, L., Lambert, Z., Lebaron, P., 2020. Toxicity of UV filters on marine bacteria: Combined effects with damaging solar radiation. *Sci. Total Environ.* 722, 137803. https://doi.org/10.1016/j.scitotenv.2020.137803

Lu, Y., Yuan, T., Yun, S. H., Wang, W., Kannan, K., 2011. Occurrence of synthetic musks in indoor dust from China and implications for human exposure. *Arch. Environ. Contam. Toxicol.* 60, 182–189. https://doi.org/10.1007/s00244-010-9595-1

Luckenbach, T., Corsi, I., Epel, D., 2004. Fatal attraction: Synthetic musk fragrances compromise multixenobiotic defense systems in mussels. *Mar. Environ. Res.* 58, 215–219. https://doi.org/10.1016/j.marenvres.2004.03.017

Luckenbach, T., Epel, D., 2005. Nitromusk and polycyclic musk compounds as long-term inhibitors of cellular xenobiotic defense systems mediated by multidrug transporters. *Environ. Health Perspect.* 113, 17–24. https://doi.org/10.1289/ehp.7301

Ma, B., Lu, G., Yang, H., Liu, J., Yan, Z., Nkoom, M., 2018. The effects of dissolved organic matter and feeding on bioconcentration and oxidative stress of ethylhexyl dimethyl p-aminobenzoate (OD-PABA) to crucian carp (*Carassius auratus*). *Environ. Sci. Pollut. Res.* 25, 6558–6569. https://doi.org/10.1007/s11356-017-1002-2

Meyer, M. F., Powers, S. M., Hampton, S. E., 2019. An evidence synthesis of pharmaceuticals and personal care products (PPCPs) in the environment: Imbalances among compounds, sewage treatment techniques, and ecosystem types. *Environ. Sci. Technol.* 53, 12961–12973. https://doi.org/10.1021/acs.est.9b02966

Molins-Delgado, D., Díaz-Cruz, M. S., Barceló, Damià, 2014. Introduction: Personal care products in the aquatic environment, in: Diaz-Cruz, M. S., Barceló, Damia (Eds.), *Personal Care Products in the Aquatic Environment*. Springer International Publishing, Switzerland, pp. 1–34. https://doi.org/10.1007/698_2014_302

Montes-Grajales, D., Fennix-Agudelo, M., Miranda-Castro, W., 2017. Occurrence of personal care products as emerging chemicals of concern in water resources: A review. *Sci. Total Environ.* 595, 601–614. https://doi.org/10.1016/j.scitotenv.2017.03.286

Nohynek, G. J., Antignac, E., Re, T., Toutain, H., 2010. Safety assessment of personal care products/cosmetics and their ingredients. *Toxicol. Appl. Pharmacol.* 243, 239–259. https://doi.org/10.1016/j.taap.2009.12.001

Novais, C., Campos, J., Freitas, A. R., Barros, M., Silveira, E., Coque, T. M., Antunes, P., Peixe, L., 2018. Water supply and feed as sources of antimicrobial-resistant Enterococcus spp. in aquacultures of rainbow trout (*Oncoryncus mykiss*), Portugal. *Sci. Total Environ.* 625, 1102–1112. https://doi.org/10.1016/j.scitotenv.2017.12.265

Novotna, A., Korhonova, M., Bartonkova, I., Soshilov, A. A., Denison, M. S., Bogdanova, K., Kolar, M., Bednar, P., Dvorak, Z., 2014. Enantiospecific effects of ketoconazole on aryl hydrocarbon receptor. *PLoS One* 9, 1–12. https://doi.org/10.1371/journal.pone.0101832

Nwaogu, T. A., Vernon, J. (2004). Comparative Study on Cosmetics Legislation in the EU and Other Principal Markets with Special Attention to so-called Borderline Products. European Commission, United Kingdom.

Parolini, M., Magni, S., Traversi, I., Villa, S., Finizio, A., Binelli, A., 2015. Environmentally relevant concentrations of galaxolide (HHCB) and tonalide (AHTN) induced oxidative and genetic damage in Dreissena polymorpha. *J. Hazard. Mater.* 285, 1–10. https://doi.org/10.1016/j.jhazmat.2014.11.037

Peng, X., Huang, Q., Zhang, K., Yu, Y., Wang, Z., Wang, C., 2012. Distribution, behavior and fate of azole antifungals during mechanical, biological, and chemical treatments in sewage treatment plants in China. *Sci. Total Environ.* 426, 311–317. https://doi.org/10.1016/j.scitotenv.2012.03.067

Qin, F., Gao, Y. X., Guo, B. Y., Xu, P., Li, J. Z., Wang, H. L., 2014. Environmental behavior of benalaxyl and furalaxyl enantiomers in agricultural soils. *J. Environ. Sci. Heal. Part B* 49, 738–746. https://doi.org/10.1080/03601234.2014.929482

Quintaneiro, C., Teixeira, B., Benedé, J. L., Chisvert, A., Soares, A. M. V. M., Monteiro, M. S., 2019. Toxicity effects of the organic UV-filter 4-Methylbenzylidene camphor in zebrafish embryos. *Chemosphere* 218, 273–281. https://doi.org/10.1016/j.chemosphere.2018.11.096

Ramos, S., Homem, V., Alves, A., Santos, L., 2016. A review of organic UV-filters in wastewater treatment plants. *Environ. Int.* 86, 24–44. https://doi.org/10.1016/j.envint.2015.10.004

Ramos, S., Homem, V., Santos, L., 2019. Simultaneous determination of synthetic musks and UV-filters in water matrices by dispersive liquid-liquid microextraction followed by gas chromatography tandem mass-spectrometry. *J. Chromatogr. A* 1590, 47–57. https://doi.org/10.1016/j.chroma.2019.01.013

Reiner, J. L., Kannan, K., 2011. Polycyclic musks in water, sediment, and fishes from the upper Hudson River, New York, USA. *Water. Air. Soil Pollut.* 214, 335–342. https://doi.org/10.1007/s11270-010-0427-8

Richter, E., Wick, A., Ternes, T. A., Coors, A., 2013. Ecotoxicity of climbazole, a fungicide contained in antidandruff shampoo. *Environ. Toxicol. Chem.* 32, 2816–2825. https://doi.org/10.1002/etc.2367

Rotstein, D. M., Kertesz, D. J., Walker, K. A. M., Swinney, D. C., 1992. Stereoisomers of ketoconazole: Preparation and biological activity. *J. Med. Chem.* 35, 2818–2825. https://doi.org/10.1021/jm00093a015

Sanganyado, E., Fu, Q., Gan, J., 2016. Enantiomeric selectivity in adsorption of chiral β-blockers on sludge. *Environ. Pollut.* 214, 787–794. https://doi.org/10.1016/j.envpol.2016.04.091

Sanganyado, E., Lu, Z., Fu, Q., Schlenk, D., Gan, J., 2017. Chiral pharmaceuticals: A review on their environmental occurrence and fate processes. *Water Res.* 124, 527–542. https://doi.org/10.1016/j.watres.2017.08.003

Sanganyado, E., Lu, Z., Liu, W., 2020. Application of enantiomeric fractions in environmental forensics: Uncertainties and inconsistencies. *Environ. Res.* 184, 109354. https://doi.org/10.1016/j.envres.2020.109354

Schneider, S. B., 2009. Global cosmetic regulations? A long way to go, First Edit. ed, *Global Regulatory Issues for the Cosmetics Industry*. William Andrew Inc., Norwich, NY. https://doi.org/10.1016/B978-0-8155-1569-2.50016-6

Schnell, S., Bols, N. C., Barata, C., Porte, C., 2009. Single and combined toxicity of pharmaceuticals and personal care products (PPCPs) on the rainbow trout liver cell line RTL-W1. *Aquat. Toxicol.* 93, 244–252. https://doi.org/10.1016/j.aquatox.2009.05.007

Shi, W., Nacev, B. A., Bhat, S., Liu, J. O., 2010. Impact of absolute stereochemistry on the antiangiogenic and antifungal activities of itraconazole. *ACS Med. Chem. Lett.* 1, 155–159. https://doi.org/10.1021/ml1000068

Silvia Díaz-Cruz, M., Llorca, M., Barceló, D., 2008. Organic UV filters and their photodegradates, metabolites and disinfection by-products in the aquatic environment. *Trends Anal. Chem.* 27, 873–887. https://doi.org/10.1016/j.trac.2008.08.012

Sommer, C., 2004. The role of musk and musk compounds in the fragrance industry, in: Rimkus, G.G. (Ed.), *Handbook of Environmental Chemistry*. Springer, Berlin, Heidelberg, pp. 1–16. https://doi.org/10.1007/b14130

Tanwar, S., Di Carro, M., Ianni, C., Magi, E., 2014. Occurrence of PCPs in natural waters from Europe, in: Diaz-Cruz, M.S., Barceló, D. (Eds.), *Personal Care Products in the Aquatic Environment*. Springer-Verlag, Berlin Heidelberg, pp. 37–71. https://doi.org/10.1007/698_2014_276

USEPA, 2014. EPI (Estimation Programs Interface) Suite™ for Microsoft® Windows, v 4.11 [WWW Document]. URL https://www.epa.gov/tsca-screening-tools/epi-suitetm-estimation-program-interface (accessed 5.28.20).

Vorkamp, K., Dam, M., Riget, F., Fauser, P., Bossi, R., Hansen, A. B., 2004. Screening of "new" contaminants in the marine environment of Greenland and the Faroe Islands, NERI Technical Report. Ministry of the Environment.

Wang, X. T., Zhou, Y., Hu, B. P., Fu, R., Cheng, H. X., 2019. Biomonitoring of polycyclic aromatic hydrocarbons and synthetic musk compounds with Masson pine (*Pinus massoniana* L.) needles in Shanghai, China. *Environ. Pollut.* 252, 1819–1827. https://doi.org/10.1016/j.envpol.2019.07.002

Wang, Z., Walker, G. W., Muir, D. C. G., Nagatani-Yoshida, K., 2020. Toward a global understanding of chemical pollution: A first comprehensive analysis of national and regional chemical inventories. *Environ. Sci. Technol.* 54, 2575–2584. https://doi.org/10.1021/acs.est.9b06379

Wick, A., Fink, G., Ternes, T. A., 2010. Comparison of electrospray ionization and atmospheric pressure chemical ionization for multi-residue analysis of biocides, UV-filters and benzothiazoles in aqueous matrices and activated sludge by liquid chromatography-tandem mass spectrometry. *J. Chromatogr. A* 1217, 2088–2103. https://doi.org/10.1016/j.chroma.2010.01.079

Wollenberger, L., Breitholtz, M., Ole Kusk, K., Bengtsson, B.-E., 2003. Inhibition of larval development of the marine copepod *Acartia tonsa* by four synthetic musk substances. *Sci. Total Environ.* 305, 53–64. https://doi.org/10.1016/S0048-9697(02)00471-0

Wong, F., Robson, M., Melymuk, L., Shunthirasingham, C., Alexandrou, N., Shoeib, M., Luk, E., Helm, P., Diamond, M. L., Hung, H., 2019. Urban sources of synthetic musk compounds to the environment. *Environ. Sci. Process. Impacts* 21, 74–88. https://doi.org/10.1039/c8em00341f

Wu, C., Spongberg, A. L., Witter, J. D., Fang, M., Czajkowski, K. P., 2010. Uptake of pharmaceutical and personal care products by soybean plants from soils applied with biosolids and irrigated with contaminated water. *Environ. Sci. Technol.* 44, 6157–6161. https://doi.org/10.1021/es1011115

Wu, X., Dodgen, L. K., Conkle, J. L., Gan, J., 2015. Plant uptake of pharmaceutical and personal care products from recycled water and biosolids: A review. *Sci. Total Environ.* 536, 655–666. https://doi.org/10.1016/j.scitotenv.2015.07.129

Xie, Z., Ebinghaus, R., Temme, C., Heemken, O., Ruck, W., 2007. Air-sea exchange fluxes of synthetic polycyclic musks in the North Sea and the Arctic. *Environ. Sci. Technol.* 41, 5654–5659. https://doi.org/10.1021/es0704434

Yang, H., Lu, G., Yan, Z., Liu, J., Dong, H., Bao, X., Zhang, X., Sun, Y., 2020. Residues, bioaccumulation, and trophic transfer of pharmaceuticals and personal care products in highly urbanized rivers affected by water diversion. *J. Hazard. Mater.* 391, 122245. https://doi.org/10.1016/j.jhazmat.2020.122245

Yao, L., Lv, Y. Z., Zhang, L. J., Liu, W. R., Zhao, J. L., Yang, Y. Y., Jia, Y. W., Liu, Y. S., He, L. Y., Ying, G. G., 2019. Bioaccumulation and risks of 24 personal care products in plasma of wild fish from the Yangtze River, China. *Sci. Total Environ.* 665, 810–819. https://doi.org/10.1016/j.scitotenv.2019.02.176

Ying, G., 2012. Personal Care Products, in: Nollet, L. M. L. (Ed.), *Analysis of Endocrine Disrupting Compounds in Food*. Wiley-Blackwell, Oxford, UK, pp. 413–428. https://doi.org/10.1002/9781118346747.ch18

Zhang, Q., Xiong, W., Gao, B., Cryder, Z., Zhang, Z., Tian, M., Sanganyado, E., Shi, H., Wang, M., 2018a. Enantioselectivity in degradation and ecological risk of the chiral pesticide ethiprole. L. Degrad. Dev. https://doi.org/10.1002/ldr.3179

Zhang, Q., Zhang, Z., Tang, B., Gao, B., Tian, M., Sanganyado, E., Shi, H., Wang, M., 2018b. Mechanistic insights into stereospecific bioactivity and dissipation of chiral fungicide triticonazole in agricultural management. *J. Agric. Food Chem.* 66, 7286–7293. https://doi.org/10.1021/acs.jafc.8b01771

Zhang, X., Yao, Y., Zeng, X., Qian, G., Guo, Y., Wu, M., Sheng, G., Fu, J., 2008. Synthetic musks in the aquatic environment and personal care products in Shanghai, China. *Chemosphere* 72, 1553–1558. https://doi.org/10.1016/j.chemosphere.2008.04.039

Zhang, Z., Ren, N., Li, Y. F., Kunisue, T., Gao, D., Kannan, K., 2011. Determination of benzotriazole and benzophenone UV filters in sediment and sewage sludge. *Environ. Sci. Technol.* 45, 3909–3916. https://doi.org/10.1021/es2004057

Zhou, C., Chiang, C., Brehm, E., Flaws, J. A., 2017. Personal care products and cosmetics, in: Gupta, R. C. (Ed.), *Reproductive and Developmental Toxicology*. Academic Press, Cambridge, MA, pp. 857–899. https://doi.org/10.1016/B978-0-12-804239-7.00045-7

Zhuang, M., Sanganyado, E., Li, P., Liu, W., 2019. Distribution of microbial communities in metal-contaminated nearshore sediment from Eastern Guangdong, China. *Environ. Pollut.* 250, 482–492. https://doi.org/10.1016/j.envpol.2019.04.041

Zhuang, M., Sanganyado, E., Zhang, X., Xu, L., Zhu, J., Liu, W., Song, H., 2020. Azo dye degrading bacteria tolerant to extreme conditions inhabit nearshore ecosystems: Optimization and degradation pathways. *J. Environ. Manage.* https://doi.org/10.1016/j.jenvman.2020.110222

6 Chiral Halogenated Organic Contaminants of Emerging Concern

Edmond Sanganyado

CONTENTS

6.1 Introduction 131
    6.1.1 Halogenated Organic Contaminants of Emerging Concern 132
    6.1.2 Implications of Chirality 133
    6.1.3 Scope of Chapter 133
6.2 Chiral Per- and Polyfluoroalkyl Substances 133
    6.2.1 Physicochemical Properties of PFAS 135
    6.2.2 Sources and Transport of PFOS in the Environment 136
    6.2.3 Enantioselective Occurrence and Fate of PFOS 139
        6.2.3.1 Chirality in Perfluorooctane Sulfonates 139
        6.2.3.2 Chiral Analysis 139
        6.2.3.3 Enantioselectivity in the Environment 140
6.3 Chiral Brominated Flame Retardants 140
    6.3.1 Physicochemical Properties of Brominated Flame Retardants 141
        6.3.1.1 Polybrominated Biphenyls 141
        6.3.1.2 Hexabromocyclododecane 141
        6.3.1.3 Other Brominated Cycloalkanes 142
    6.3.2 Enantioselective Analysis and Environmental Behavior 142
        6.3.2.1 Polybrominated Biphenyls 144
        6.3.2.2 Hexabromocyclododecane 144
        6.3.2.3 Other Brominated Cycloalkanes 145
6.4 Conclusion and Future Directions 145
Acknowledgments 145
References 146

6.1 INTRODUCTION

Halogenated organic contaminants are a diverse group of compounds that have at least one halogen attached to their alkyl chain, aromatic ring, or cyclic ring. Halogens are a group of non-metallic elements found in group 17 of the periodic table. Chlorine, fluorine, and iodine are the most common halogens found in halogenated organic contaminants. For example, pesticides such as aldrin, endrin, and dichloro-diphenyl-trichloroethane (DDTs) have chlorine substituting hydrogen atoms on their benzene rings while surfactants such as perfluorooctane sulfonate have fluorine substituting the hydrogen atoms.

    Halogenation of organic compounds is probably one of the most critical processes in the chemical industry. It is estimated that halogenated organic compounds account for 30% and 20% of pharmaceuticals and agrochemicals in current use (Cantillo and Kappe, 2017). The presence of carbon-halogen bonds imparts highly beneficial properties to organic compounds. For example, chlorination improves the efficacy of agrochemicals, bromination may impart thermal stability, and fluorination often imparts therapeutic properties to drug candidates. Overall, halogen atoms may enhance membrane permeability, biological activity, and stability toward chemical transformation, hydrolysis, and photolysis by altering the polarity, lipophilicity, solubility, and ionizability of the organic compound (Cantillo and Kappe, 2017). In addition to industrial halogenation, halogenated organic compounds are also produced as by-products of waste treatment and industrial combustion. Chlorination is a critical process in wastewater disinfection, which often results in the formation of chlorinated transformation products. Furthermore, industrial combustion at high temperatures results in the formation of furans and dioxins.

The presence of halogenated organic compounds in the environment is an issue of major concern due to their high persistence, bioaccumulative potential, and toxicity. Halogenated organic compounds can be classified according to their sources as anthropogenic or natural halogenated organic compounds. It is estimated that there are more than 5,000 natural halogenated organic compounds in the environment (Hauler et al., 2013; Pangallo and Reddy, 2010; Vetter, 2012). Natural halogenated organic compounds are produced biogenically and geogenically via biosynthesis by microorganisms, plants, and marine biota and natural processes such as volcanic eruption, geothermal events, and forest fires, respectively (Häggblom and Bossert, 2005). Anthropogenic halogenated organic compounds are a diverse group of predominantly chlorinated and sometimes brominated and fluorinated synthetic compounds. Their chemical structure ranges from simple halogenated aliphatic hydrocarbons such as chloroethylene to complex aromatic and polycyclic compounds such as polychlorinated biphenyls (PCBs) and DDTs (González-Mariño et al., 2016; Liu et al., 2017a; Luo et al., 2015). Anthropogenic halogenated organic compounds are extensively used in industry, agriculture, households, and pharmaceutical applications. They are frequently used as active ingredients or additives in pesticides, plasticizers, flame retardants, solvents, and personal care products. Due to their widespread usage, anthropogenic halogenated organic compounds are ubiquitous in the environment where they pose a significant risk to humans and wildlife (Rajput et al., 2018; Sanganyado et al., 2018). The present chapter focuses on anthropogenic halogenated organic compounds, here forth represented as HOCs.

6.1.1 Halogenated Organic Contaminants of Emerging Concern

Anthropogenic HOCs can be classified according to the history of their industrial usage and their subsequent detection in the environment. Legacy contaminants are chemicals that have been regulated for more than two decades due to their known persistence, bioaccumulative potential, and toxicity. Often, the use of the legacy contaminants has been banned or restricted, and they are present in the environment as historical contaminants. Examples of legacy HOCs include organochlorine pesticides, PCBs, and halogenated polycyclic aromatic carbons. In contrast, emerging contaminants, which are also called contaminants of emerging concern, are a diverse group of chemicals that pose a potential human and environmental health risk but remain unregulated or have been recently regulated. Examples of HOCs of emerging concern (HOC-ECs) include pharmaceutical and personal care products, halogenated flame retardants, and surfactants.

Emerging contaminants are not necessarily chemicals that were recently released into the environment. They are called emerging contaminants because they were recently detected in environmental matrices due to recent advancements in trace analysis. In the last three decades, there has been considerable advancement in chemical analysis with the development of highly selective, sensitive, and robust equipment such as high-resolution mass spectrometers (Chibwe et al., 2017; Mackintosh et al., 2016; Tian et al., 2020). The advent of equipment that can detect compounds at concentrations below ng L^{-1} in environmental matrices led to the discovery that HOC-ECs were ubiquitous in the environment (Ort et al., 2010). The presence of HOC-ECs in the environment is concerning because subsequent human and environmental toxicity studies at environmentally relevant concentrations have found that they are often chronically toxic (i.e., adverse biological effects occur after a long period following exposure).

6.1.2 Implications of Chirality

Some congeners of HOCs are chiral compounds; that is, they have one or more chiral centers, and their corresponding isomers have identical chemical structures, yet the three-dimensional spatial arrangements of their atoms are different (Ribeiro et al., 2012b; Sanganyado et al., 2017). The isomers of chiral compounds are called enantiomers or diastereomers. Enantiomers are the stereoisomeric pairs of chiral compounds that are non-superimposable mirror images, while diastereomers are stereoisomers that are not mirror images (Qin et al., 2006; Sanganyado et al., 2017; Stalcup, 2010). Enantiomers of a chiral compound have similar physical and chemical properties except for their optical activity in an achiral environment. Under achiral conditions, the solubility, vapor pressure, diffusivity, and lipophilicity of enantiomers are similar. It is then expected that the partitioning of enantiomers between different environmental compartments will be similar (Kasprzyk-Hordern, 2010). Hence, under achiral conditions, no difference in the volatilization of enantiomers of a HOC from river water to the atmosphere or diffusion in the river sediments can be expected. In a chiral environment, the physical and chemical properties of enantiomers often vary significantly (Kasprzyk-Hordern, 2010; Ribeiro et al., 2012a,c). The differences in the three-dimensional configuration of the enantiomers have significant consequences on the fate, transport, and toxicity of the chiral HOCs in the environment because nature is notoriously chiral and achiral environments are not the norm (Sanganyado, 2019; Smith, 2009; Zhang et al., 2018). Most biomolecules, such as sugars, amino acids, carbohydrates, proteins, and DNA, are chiral molecules (Ebeling et al., 2018). As shown in Figure 6.1, the interactions of enantiomers in chiral environments, such as in biological systems, is often enantiospecific due to enantioselectivity in binding, catalysis, and stabilization of the enantiomers (Kasprzyk-Hordern, 2010; Sanganyado et al., 2017, 2016). Hence, enantiomers of HOCs may exhibit significant differences in their enzymatic transformation, metabolism, distribution, and toxicity.

6.1.3 Scope of Chapter

This chapter discusses the physicochemical properties, analysis, occurrence, environmental behavior, and toxicity of chiral HOC-ECs in the environment. In this context, the environment includes both the natural (e.g., rivers, lakes, estuaries, groundwater, atmosphere, and biota) and built (e.g., wastewater treatment plants and constructed wetlands) environments. The chiral HOC-ECs that will be discussed in this chapter are tetrabromoethylcyclohexane, hexabromocyclododecanes, and per- and polyfluoroalkyl substances. Table 6.1 shows the list of chiral emerging contaminants discussed in this chapter and their uses. Most chiral HOC-ECs that have been studied to date are brominated flame retardants. This is probably because of their high persistence, bioaccumulative potential, and potential toxicity compared to other emerging contaminants.

6.2 CHIRAL PER- AND POLYFLUOROALKYL SUBSTANCES

Per- and polyfluoroalkyl substances (PFASs) are fluorinated organic compounds that have wide applications in consumer and industrial products (Ding and Peijnenburg, 2013). Due to their high thermal and chemical stability, PFASs are widely used in food packaging, textile products, firefighting foams, plastic and rubber additives, and insulation films (Liu et al., 2019). They are also extensively used in cooking utensils, cleaning products, and stain repellents since their oil resistance imparts non-stick properties to materials (Ding et al., 2018).

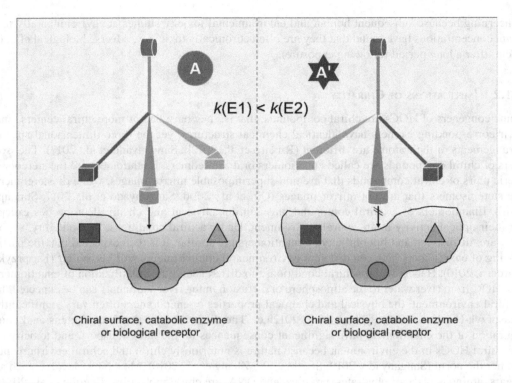

FIGURE 6.1 Enantiomer specific interactions between enantiomers and chiral receptors such as chiral surfaces in sludge or soil, microbial enzymes, and biological receptors. (Adapted from Sanganyado et al., 2020.)

TABLE 6.1
Examples and Applications of Chiral Halogenated Organic Contaminants of Emerging Concern and Personal Care Products (Eljarrat et al., 2008)

Contaminant Class		Examples	Application
Brominated flame retardants	Hexabromocyclododecanes	α-, β- and γ-HBCD; 1,2-dibromo-4-(1,2-dibromoethyl)-cyclohexane	Building materials, polystyrene foam insulation, electronics, and upholstery textiles
	Polybrominated biphenyls	PBB132 and PBB149	Textiles, electronics, and plastics
	Tetrabromoethylcyclohexane	α, β, γ and δ	Plastics, particularly polystyrene and polyurethane
Surfactant, surface treatment	Per- and polyfluoroalkyl substances	1m-, 3m-, 4m-, and 5m-PFOS	Food packaging, textile products, firefighting foams, plastic and rubber additives, and insulation films

6.2.1 Physicochemical Properties of PFAS

The chemical structure of PFASs imparts unique beneficial physical, chemical, and biological characteristics to the compounds. PFASs are alkyl compounds with hydrogen atoms entirely or partially substituted with fluorine atoms and a hydrophilic functional group (e.g., carboxylates, sulfonates, sulfonamides, and alcohols) (Buck et al., 2011). The physicochemical properties of PFASs are influenced by the length, the number of fluorine atoms, and the terminal functional group on the alkyl chain (Ding and Peijnenburg, 2013; Krafft and Riess, 2015; Yeung et al., 2013). Table 6.2 shows the physicochemical properties of PFASs and their associated chemical structure characteristics. The most commonly used PFASs have a sulfonate and carboxylate terminal groups. Examples of such PFAS include perfluorinated sulfonate acids (PFSAs), perfluorooctane sulfonate (PFOS), and perfluorooctanoic acid (PFOA). PFASs have high thermal stability because the C-F bond has a high bond energy of around 105.4 kcal mol^{-1} (Liu et al., 2019). Although the fluorine atom has three pairs of electrons, the electrons cannot receive a proton or form resonance structures like oxygen because of the high electronegativity of fluorine (Ding and Peijnenburg, 2013; Giesy and Kannan, 2002; Liu et al., 2019). Interestingly, these fluorine atoms shield the carbon chain rendering the PFAS stable toward chemical oxidation or reduction, hydrolysis, photolysis, acid-base reactions, and biotransformation.

Furthermore, fluorine atoms significantly reduce the surface potential of PFASs because they are highly ionizable even though they have low polarizability (Ding and Peijnenburg, 2013). Furthermore, PFAS has an aliphatic chain that imparts hydrophobic and lipophilic characteristics and a terminal functional group that imparts hydrophilic characteristics (Buck et al., 2011; Giesy and Kannan, 2002). As a result, PFASs are widely used as surfactants. PFOS are some of the most widely used PFASs in consumer and industrial products (Giesy and Kannan, 2002). Increasing the number of C-F bonds increases the chemical and thermal stability of PFAS; PFASs are often classified as long-chain or short-chain PFAS with long-chain perfluoroalkane sulfonic acids and perfluoroalkyl carboxylic acids and having at least six and seven perfluorinated carbons, respectively

TABLE 6.2
Role of Physicochemical Properties on Industrial Applications of Per- and Polyfluoroalkyl Substances

Physicochemical Properties	Associated Chemical Structure	Example of Application
Thermal stability	• Strong C-F bond • Length of aliphatic chain • The number of fluorine atoms in the chain	Firefighting foams and aviation hydraulic fluids
Chemical stability	• High fluorine electronegativity • Strong C-F bond • Presence of non-bonding electron pairs • The number of fluorine atoms in the chain • Hydrophilic terminal functional group	Firefighting foams, textile products, upholstery, food packaging, insulation films, and aviation hydraulic fluids
Surfactant	• High ionizability of fluorine • Low polarizability of fluorine • Length of the hydrophobic aliphatic chain • Hydrophilic terminal functional group	Water- and stain-repellents for paper, textile, food packaging, and electronic industry as well as emulsifiers

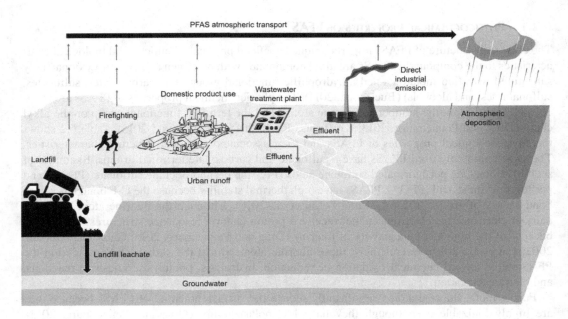

FIGURE 6.2 The sources and transport of perfluorooctane sulfonates and its precursors in the environment.

(Teaf et al., 2019). PFOS are often considered long-chain PFAS since the commercially available homologs often have seven perfluorinated carbons.

6.2.2 Sources and Transport of PFOS in the Environment

PFOS enter the environment through various routes of entry, such as industrial effluent, municipal wastewater discharge, atmospheric deposition, and landfill leachate (Figure 6.2). They are released to various environmental compartments during the life cycle of the PFOS-containing product (Lim et al., 2011; Liu et al., 2019; Liu et al., 2017b). It is estimated that direct industrial emission during the fluoropolymer manufacturing stage accounts for 80% of the total PFOS direct pollution (Liu et al., 2019). However, the global emission of PFOS and related substances has been decreasing due to production bans and restrictions (Buck et al., 2011; Shen et al., 2018; Yamashita et al., 2005). PFOS are highly persistent, bioaccumulative, and potentially toxic. In 2006, several manufacturers in the United States volunteered to reduce the production of PFOS by 95% in less than five years (Lim et al., 2011; Teaf et al., 2019). This industrial commitment was further augmented in 2017 by the addition of PFOS and related salts into the list of persistent organic pollutants of the Stockholm Convention. The European Union subsequently restricted the use of PFOA and related salts (Danish Environmental Protection Agency, 2015; Lim et al., 2011; Wang et al., 2011).

PFOS is a by-product of the electrochemical fluorination process that produces POSF. The POSF is used to manufacture PFOS precursors such as perfluorooctane sulfonamidoethanol (FOSE) and perfluorooctane sulphonamide (FOSA) (Lim et al., 2011). PFOS precursors such as FOSE are sometimes used to manufacture fluoropolymers used in surface treatment. These PFOS precursors often contain PFOS, which is not covalently bound. This residual PFOS can be directly released into the environment during product distribution, use, and disposal. Globally, direct industrial emission is estimated to release of 450–2700 tons of PFOS into the environment (Ding and Peijnenburg, 2013). In 2010, it was estimated that directly industrial emission contributed 70 tons of PFOS into the environment with the metal plating, textile, firefighting, fluoropolymer manufacturing, and the semiconductor industries contributing the largest, in descending order (Liu et al., 2015; Xie et al., 2013). PFOSs are also directly emitted at the point of use sites such as fire-training facilities, airports, and military bases. Direct emissions often result in atmospheric, surface water, groundwater, and soil pollution.

Indirect emissions often occur at disposal sites such as landfills, wastewater treatment plants, and application sites for biosolids (Gallen et al., 2018, 2017; Sinclair and Kannan, 2006). Following a precipitation event, residual PFOS in household and industrial products in landfills is leached into the ground as leachate since PFOS has high water solubility. A study in Norway found the concentration of PFOS in landfill leachate ranged from 15–160 ng L^{-1} (Knutsen et al., 2019). In China, the concentration of PFOS in landfill leachate was considerably higher, ranging from 1150 to 6020 ng L^{-1} (Yan et al., 2015). Other studies in China showed that soil leaching contributed 50% of PFOS found in contaminated groundwater (Liu et al., 2017b). Ambient air in landfills has been shown to contain relatively higher concentrations of PFAS, particularly volatile and semi-volatile PFAS (Hamid et al., 2018). Hence, PFOS can be transported from landfill via groundwater, surface runoff, or ambient air.

Wastewater treatment plants collate indirect indoor emissions of PFOS and its precursors from consumer products. When PFOS and its precursors are released from consumer products indoors as dust, they are collated by various cleaning activities and end up entering the domestic wastewater drain, which ultimately enters wastewater treatment plants (Lim et al., 2011; Xiao et al., 2012). The PFOS precursors are transformed in wastewater treatment plants due to biological and chemical treatment to yield PFOS (Boulanger et al., 2005; Eriksson et al., 2017; Y. Wang et al., 2009). Considering conventional wastewater treatment plants were not designed to remove ionizable organic pollutants efficiently, varying concentrations of PFOS and its precursors are subsequently discharged into rivers as part of the wastewater effluent (Boulanger et al., 2004; Eriksson et al., 2017; Hamid et al., 2018). Recent studies have shown that PFOSs originating from urban environments is mainly transported into the environment via urban runoff (e.g., snowmelt and urban stormwater) (Müller et al., 2020; Xiao et al., 2012). The PFOS in urban runoff comes from atmospheric deposition of contaminated dust, drainage surfaces (runoff of contaminated dust on roads and pavements), and anthropogenic activities (e.g., car washing, firefighting, household waste disposal) (Müller et al., 2020; Xiao et al., 2012). A study in Minnesota, USA, found that the concentration of PFOS on urban surfaces ranged from 2.4 to 47 ng L^{-1} (Simcik and Dorweiler, 2005). It has been shown that the non-atmospheric sources were the major contributor to PFOS found in the urban runoff from industrial and commercial areas (Xiao et al., 2012). The partition coefficient of PFOS between sediments/suspended particles (K_d) has been shown to range from 4.3–97 L kg^{-1} depending on the physicochemical characteristics of the solid particles (Paul et al., 2012). Based on the K_d of PFOS, it is anticipated that partitioning to suspended particles should be <20% (Zhang et al., 2012). However, Xiao et al. (2012) found concentrations of particle-attached PFOSs in urban runoff beyond those expected from the partitioning behavior of PFOS on solids. This suggested that the particles on which PFOSs were attached came from industrial products.

Ultimately, rivers act as a critical pathway for transporting and mobilizing PFOS and its precursors into the ocean where they are archived in marine sediments, taken up in the food chain, or (bio) transformed (Houde et al., 2006; Mcdougall, 2016; Yamashita et al., 2005). A study in Spain found the mass loading of PFOS in the Tagus River per annum ranged from <0.01 to 46 kg (Navarro et al., 2020). However, higher concentrations were found in other European countries. For example, the PFOS mass loading by Rhine River and Danube River into the ocean were found to be 1000 kg and 900 kg, respectively (Lindim et al., 2016). However, estimates based on experimental data showed that PFOS discharge from the Rhine River to the North Sea ranged 420–2200 kg per year (Paul et al., 2012). Lindim et al. (2016) showed that river discharge contributed up to 30 times of PFOS in seas than atmospheric deposition, whereas atmospheric deposition contributed up to 10 times of PFOA than river discharge, and this was because PFOA has higher vapor pressure than PFOS. For that reason, the ratio of the concentrations of PFOA and PFOS is currently used as a molecular marker of atmospheric deposition, particularly long-range transport in the Arctic.

Emission of PFOS and its precursors into the atmosphere during manufacturing, distribution, or use of PFOS-containing products is often accompanied by indirect deposition into freshwater, coastal, and terrestrial environments (Liu et al., 2017b). PFOS deposited onto terrestrial surfaces

TABLE 6.3
Experimental and Predicted Physicochemical Properties of Perfluorooctane Sulfonates (Ding and Peijnenburg, 2013)

Property	Value	Determination Method	Implication on Environmental Behavior
Water solubility, mg L^{-1}	0.21	Predicted using QSPR	PFOS readily dissolves in water. This suggests it has a high chance of partitioning into the water phase.
Vapor pressure, P, Pa	3.2	Predicted using QSPR	PFOS are semi-volatile, which suggests it does not partition readily into the gaseous phase.
Hydrophobicity, $\log K_{OW}$	2.45	Ion-transfer cyclic voltammetry	Compounds with $\log K_{OW} > 5.0$ are often considered to be highly bioaccumulative and have the potential to biomagnify in a food chain.
Octanol/air partition coefficient, $\log K_{oa}$	7.8	Calculated	The $\log K_{oa}$ value suggests that PFOS can readily partition into organic materials in aerosols, soil, and plants. This indicates that aerial deposition can occur following atmospheric pollution.
Organic Carbon/Water Partition Coefficient, $\log K_{oc}$	2.68	Aqueous loss method	This suggests PFOS readily partitions into the organic phase found in soil, sediment, suspended organic particles from the water phase.

QSPR: quantitative structure-activity relationship.

can be further transported into aquatic systems via surface runoff (Müller et al., 2020; Xiao et al., 2012). However, in some cases, following a precipitation event, PFOS can leach into the soil where it can remain sorbed to soil particles, degraded by microbes, taken up by plants, or leach into the groundwater (Lim et al., 2011). A previous study in Europe found that PFOS accounted for 66% of all PFAS detected in groundwater (Loos et al., 2010). PFOS precursors are also a significant indirect source of PFOS in the environment. Several studies have shown that PFOS precursors such as FOSE can undergo chemical, biological, photocatalytic, and hydrolytic transformation in various environmental compartments (e.g., wastewater treatment plants, aerosols, sediments, surface water, and soils) to yield PFOS (Boulanger et al., 2005, 2004; Martin et al., 2006; Sinclair and Kannan, 2006).

Aquatic systems are the primary sink of PFOS and its related substances because of their high water solubility and low vapor pressure (Table 6.3). When PFOS are released into the terrestrial environment, they are readily transported in the water phase as surface runoff, landfill leachate, or in groundwater since they are highly soluble in water. PFOS can partition into organic carbons found in soil, sediments, aerosols, or suspended particles in surface water, as shown by their high octanol/water partition coefficient (Ding and Peijnenburg, 2013). Particle attached PFOS in water phases eventually settles in sediments. Hence, sediments are often the ultimate sink of PFOS. Since PFOS have a high organic carbon/air partition coefficient, they can readily sorb onto organic carbon found in the atmosphere (aerosols), plants (waxy cuticles), and soil (organic materials) (Ding and Peijnenburg, 2013). Previous studies have detected PFOS and related substances in an arctic glacier at concentrations of up to 15 pg L^{-1} (Kwok et al., 2013). These results suggested that long-range transport was a major source of PFOS in the Arctic (Martin et al., 2006). However, point of use sources such as firefighting training and wastewater discharges have been shown to contribute to PFOS pollution in the Arctic (Skaar et al., 2019; Stroski et al., 2020).

6.2.3 Enantioselective Occurrence and Fate of PFOS

6.2.3.1 Chirality in Perfluorooctane Sulfonates

Some PFOSs that are frequently and often abundantly detected in humans and the environment are chiral compounds (Asher et al., 2012; Wang et al., 2011; Xu et al., 2018; Zhao et al., 2020, 2019). Theoretically, PFOS has 89 isomers, of which 66 have at least one chiral center (Rayne et al., 2008). Chiral centers are found predominantly in branched PFOS but not in linear PFOS. Hence, chirality has been observed in PFOS containing isopropylmethylbutyl-, tetramethylbutyl-, methyl-1-propylbutyl-, trimethyl pentyl-, ethylhexyl-, dimethylhexyl-, and monomethyl-branches (Rayne et al., 2008). However, monomethyl-branched PFOS, such as 1m-, 3m-, 4m-, and 5m-PFOSs are commercially available and have subsequently been detected in the environment (Asher et al., 2012). The m in the name of the chiral PFOS isomer refers to the branching position of the trifluoromethyl group along the aliphatic chain (Rayne et al., 2008).

6.2.3.2 Chiral Analysis

Enantioselective analysis of PFOS is influenced by the physicochemical properties of PFOS and the location of the chiral center. Since PFOS have low vapor pressure, enantioselective analysis is often limited to liquid chromatography and supercritical fluid chromatography. Gas chromatography is rarely used in environmental monitoring and characterization of branched PFOS. This is because PFOS has low volatility and requires an additional derivatization step to make it amenable to gas chromatography. The derivatization step is challenging because PFOS is highly ionizable, hydrophobic, and hydrophilic (Langlois et al., 2007). The aliphatic chain is relatively chemically inert since it is often saturated with fluorine atoms (Ding and Peijnenburg, 2013). A previous study separated branched PFOS isomers following derivatization using *iso*-propanol under acidic conditions (Langlois et al., 2007). However, to the best of our knowledge, no studies are employing enantioselective gas chromatography as a result.

Enantioselective analysis of chiral PFOS in environmental matrices is often conducted using liquid chromatography. However, studies on enantiomer specific distribution and fate of PFOS have been limited to 1m-PFOS. Previous studies separated 1m-PFOS and its precursors using an anion exchange chiral column that was based on cellulose tris(3,5-dichlorophenyl)carbamate (Asher et al., 2012; Y. Wang et al., 2009) or a quinine derivate (Chiralpak QN-AX) (Wang et al., 2011; Zhao et al., 2020, 2019). Quinine derivative stationary phases have a three-dimensional cleft that is surrounded by aromatic groups that can interact with the enantiomers via weak hydrophobic or van der Waals interaction. The cleft is also surrounded by hydrogen binding sites. However, within the cleft, there is a tertiary amine capable of forming strong ionic and hydrogen bonds with the enantiomers. Although Chiralpak QN-AX can be used in normal or polar organic mode, current studies have predominantly used it in reverse-phase (Wang et al., 2011). Normal-phase mode is generally avoided in enantioselective analysis of environmental matrices because of the challenges in making the mobile phase compatible with mass spectrometry (Sanganyado et al., 2014). However, Chiralpak QN-AX has been successfully used in separating the enantiomers of 1m-PFOS using supercritical fluid chromatography in human placenta (Zhao et al., 2020). Supercritical fluid chromatography has been shown to have shorter elution time and better enantioresolution compared to liquid chromatography. Another study found enantiomers of 3m, 4m-, 5m-, and 4,5m$_2$-PFOS could not be separated by supercritical fluid chromatography on a Chiralpak QN-AX column (Zhao et al., 2019). Enantioseparation was only achieved for 1m-PFOS. At the 1m branch, the chiral center is close to the sulfonate functional group. The sulfonate group can form strong ionic or hydrogen bonds with the chiral stationary phase, thus enhancing the chiral discrimination of the 1m-PFOS enantiomers. The potential for chiral discrimination decreases when the distance between the chiral center and the terminal functional group increases. This is because chiral centers far from the terminal functional group are predominated by the hydrophobic attraction, which is often weaker than the ionic interactions at the sulfonate group.

6.2.3.3 Enantioselectivity in the Environment

Studies on the enantiomeric distribution of chiral PFOS in the environment are scarce. Asher et al. (2012) found the enantiomeric distribution of 1m-PFOS in Lake Ontario was racemic in water and sediment but nonracemic in zooplankton, *Mysis relicta*, *Diporeia spp.*, and fish. When enantiomers as 1m-PFOS enter the environment, they are taken up by the primary producers, which are then consumed by invertebrate predators. These invertebrate predators are subsequently consumed by fish. Chiral PFOS that accumulates in tissue is transferred across the food chain. However, within each species, chiral PFOS undergoes metabolism, distribution, and elimination. Each one of these processes has been shown to be enantioselective in some species. For example, a study found that the binding affinity of 1m-PFOS enantiomers to human serum albumin was enantioselective, and this suggested the distribution of 1m-PFOS varied significantly between the two enantiomers (Zhao et al., 2020). Hence, enantioselectivity in metabolism, distribution, and elimination may have accounted for the observed nonracemic distribution of 1m-PFOS in aquatic organisms. However, there is a need for additional studies to establish if it is indeed the main reason.

The enantiomeric composition of PFOS in various environmental matrices can be used as a tool for source tracking. Technical PFOS and PFOS generated from abiotic transformations such as chemical oxidation, photolysis, and hydrolysis of PFOS precursors is a racemic mixture. However, PFOS that is generated from PFOS precursors following biological transformation in humans, microorganisms, and wildlife is often a nonracemic mixture. For example, the biotransformation of a model PFOS-precursor isomer $C_6F_{13}C^*F(CF_3)SO_2N(H)CH_2(C_6H_4)OCH_3$ in liver microsomes was shown to be enantioselective (Y. Wang et al., 2009). The metabolic rate of one enantiomer was $k = 6.5 \times 10^{-2}$ min^{-1} and for the other one $k = 5.2 \times 10^{-2}$ min^{-1}. Hence, when the enantiomeric distribution of PFOS in humans and wildlife is nonracemic, it suggests the organism was exposed to PFOS precursors. In fact, a study in Canada found human serum collected from people living in a house that regularly used a water repellent to treat carpet and upholstery had PFOS with enantiomeric fractions that ranged from 0.35 to 0.43 (Wang et al., 2011). This commercial water repellent (ScotchGard) was previously shown to contain various types of PFOS precursors, such as urethane-linked N-methyl perfluorooctanesulfonamidoethanol (Martin et al., 2010). In summary, previous studies on the enantioselective behavior of PFOS and its precursors show that biotic processes in the environment are often enantioselective.

6.3 CHIRAL BROMINATED FLAME RETARDANTS

Flame retardants are a diverse array of compounds that prevent or slow material combustion by interfering with the heating and ignition processes. They are often classified according to their active phase, mode of action, chemical structure, and agent incorporation mechanism. Halogenated flame retardants are the most widely used class of flame retardants. They act in the gaseous phase via a chemical mechanism involving generation of halogen radicals. The halogen radicals react with highly reactive species present during the combustion process such as the hydrogen and hydroxyl radicals to yield molecules that are less reactive or inert. Halogenated flame retardants can be incorporated into materials as additive or reactive agents. Additive flame retardants are added to the materials through physical processes. Since they are not covalently bound to the materials, they can easily leach into the environment during production, distribution, use, and disposal. Reactive flame retardants are often incorporated to the materials through a chemical reaction. Unlike additives, reactive flame retardants can significantly alter the physicochemical properties of the materials.

PCBs were the earliest halogenated flame retardants used in industry. Since PCBs were highly toxic and persistent in the environment, they were replaced with polybrominated biphenyls. The only difference between the compounds was the replacement of chlorine atoms with bromine atoms on the aromatic rings. Production of polybrominated biphenyls was eventually restricted since it was shown to be an immunotoxicant. However, the rapid growth of the plastics industry since the 1970s resulted in an increased demand for halogenated flame retardants. For three decades, the flame retardants market was dominated by polybrominated diphenyl ethers. However, concern regarding their toxicity and persistence arose as early as 1989 when a proposal was submitted to the European Union for their ban. Two decades later, some congeners of polybrominated diphenyl ethers were added to the Stockholm Convention and their production was banned or restricted.

There are currently more than 175 distinct types of halogenated flame retardants, of which at least 80 are brominated. It is estimated that brominated flame retardants comprise 25% of the global annual production of flame retardants. Brominated flame retardants are often aliphatic, aromatic, and alicyclic compounds. Some brominated flame retardants contain at least one chiral center and are sold as racemic mixtures. Examples of chiral brominated flame retardants include polybrominated biphenyls, hexabromocyclododecane, and tetrabrominated cycloalkanes.

6.3.1 Physicochemical Properties of Brominated Flame Retardants

6.3.1.1 Polybrominated Biphenyls

The chemical and physical properties of polybrominated biphenyls significantly vary among the congeners depending on the number of bromine atoms on the aromatic rings. PBBs with high bromine content have higher melting points, chemical stability, and lipophilicity. Most PBBs have low solubility and volatility, but log K_{OW} values > 7. This suggests that PBBs do not readily partition into the aqueous or gaseous phase. Thus, when PBBs enter the environment, they tend to partition to organic matter found in sediment and suspended particles in rivers and lakes, organic particles in soils, and aerosol particles in the atmosphere. Although PBBs have low solubility, they can be transported in aqueous media since their solubility can be enhanced by co-solubilization. For example, the solubility of PBBs in landfill leachate was shown to be more than 200 times higher than in water. This was probably because landfill leachate comprises a cocktail of chemical contaminants such as surfactants which are known to enhance chemical solubilization.

Although none of the 209 congeners of PBBs contain a stereogenic center, congeners with two to four bromine atoms often contain an axis of chirality. The axis of chirality occurs when the phenyl-phenyl bond does not rotate due to steric hindrance and the bromine atoms on the phenyl groups are asymmetrically substituted. This type of chirality which arises due to an impediment in free rotation about a covalent bond is called atropisomerism. Previous studies have shown that PCBs have 19 atropisomers that are environmentally relevant. Since bromine is bulkier than chlorine, it exerts more steric hindrance on the phenyl-phenyl bond thus increasing the chance of more environmentally relevant PBB atropisomers.

6.3.1.2 Hexabromocyclododecane

HBCD is a brominated cycloalkane commonly used in plastics as an additive flame retardant. It is a nonaromatic compound with a molecular weight of 641.7 g mol^{-1} (Gao et al., 2018; Heeb et al., 2015). HBCD is widely used in Europe and the US as a replacement for the restricted polybrominated diphenyl ethers. In 2001, the annual demand of HBCDs was estimated as 16,700 tons (Birnbaum and Staskal, 2004). However, HBCD consumption in the European Union increased significantly between 2001 and 2007 from 9,500 to 11,000 tons (Marvin et al., 2011). The physicochemical properties of HBCDs are like those of other brominated flame retardants. HBCDs have low water solubility (3.4–8 µg L^{-1} at 25°C), low vapor pressure (<1,000 Pa), and high log K_{ow} value (5.81–7.0) (Table 6.4). As a result, HBCD is highly hydrophobic and readily partitions to organic phases in biotic, atmospheric, or aquatic systems (Covaci et al., 2006; Heeb et al., 2005; Marvin et al., 2011; Tomy et al., 2004). For example, when HBCDs are discharged into a river or a lake, they tend to readily partition on suspended particles and sediments (Eljarrat et al., 2004; Harrad et al., 2009; He et al., 2013). The same happens when they are discharged into the atmosphere, they sorb to the organic matter found on aerosols. Despite their low volatility and solubility, HBCDs have a half-life ranging from 0.4–5.2 days in air and 130–2,025 days in water which suggests they have the potential to undergo long range transport (Marvin et al., 2011). Previous studies have detected HBCDs in remote areas such as the Arctic suggesting long range transport is possible.

TABLE 6.4
Physicochemical Characteristics of Brominated Flame Retardants That Have Isomers Containing Chiral Centers

Characteristics	HBCD*	TBECH	TBCO	Comments
Chemical structure	(structure)	(structure)	(structure)	
Molecular weight, g mol^{-1}	641.7	427.8	427.8	-
Boiling point, °C	505.2	371.2	364.2	HBCD had higher boiling point since it was bulkier and had more thermally stable C-Br bonds.
Vapor pressure, 10^{-6} Pa	62.7	105	94.3	TBECH and TBCO have higher chance for aerial transport than HBCD. All the values are less than the regulatory threshold of 1000.
Water solubility, µg L^{-1}	3.4-8	6.92	6.92	All the compounds have a low tendency for partitioning into the aqueous phase; thus, reducing the probability of transport in water systems.
Log K_{OW}	5.4-5.8	5.24	5.24	All compounds have log K_{OW} > 5.0 suggesting a high potential for bioaccumulation in biota.
Log K_{OA}	-	8.01	8.01	TBECH and TBCO readily partition to organic matter on aerosols in the atmosphere.
Log K_{OC}	4.86	3.63	3.61	HBCD has higher potential than TBECH and TBCO to partition to organic matter. Hence, it can readily be transported in water systems attached to particles.
Half-life in water, years	>0.5	86.4	>1000	All the compounds are persistent in water. Their half-lives are above the regulatory threshold of 60 days.
Half-life in air, days	0.4-5.2	2.2	3.2	All the compounds are persistent in air as they have half-lives above the regulatory threshold of 2 days.
Log BCF	3.9-4.3	3.33	3.33	HBCD has higher bioaccumulation factor than TBECH and TBCO. However, TBECH and TBCO are estimated to be slightly within the regulatory threshold of 3.7
Removal in wastewater, %	70	83.98	83.98	All the compounds persist in wastewater treatment plants and can be discharged to the environment.

* Data was obtained from Marvin et al. (2011).

HBCDs are industrially produced by brominating cyclododecatriene isomers. The bromination process results in the formation of six stereogenic centers, and this theoretically results in 16 stereo-isomers. The technical mixture contains six pairs of enantiomers. Technical grade HBCD mainly consists of at least 75%, 10%, and 1% γ-, α-, and β-HBCD (Heeb et al., 2005). Trace levels of δ- and ε-HBCD have been detected also in some technical grade HBCDs. HBCDs are thermally labile (Marvin et al., 2011). At temperatures above 160°C, which can be found during production and processing, they can undergo isomerization resulting in a mixture dominated by α-HBCD (78%) followed by β-HBCD (13%) and γ-HBCD (9%) (Covaci et al., 2006). X-ray diffraction analysis showed that the absolute stereoconfiguration of (−) α-HBCD, (+) β-HBCD, and (−) γ-HBCD were (1R,2R,5S,6R,9R,10S), (1S,2S,5S,6R,9S,10R), and (1S,2S,5S,6R,9R,10S), respectively (Heeb et al., 2007). Unlike their respective enantiomers which have similar physicochemical properties, α-, β-, and γ-HBCD have widely different water solubilities of 48.8, 14.7, and 2.1 μg L^{-1}, respectively due to their different polarities. As a result, α-, β-, and γ-HBCD often have different half-lives in air, water, and soil (Covaci et al., 2006).

6.3.1.3 Other Brominated Cycloalkanes

Tetrabromoethylcyclohexane (TBECH) and tetrabromocyclooctane (TBCO) are brominated cycloalkanes that are used as additive flame retardants in plastics. TBCEH has a molecular weight of 427.8 g mol^{-1}, low water solubility (6.92 μg L^{-1} at 25°C), low vapor pressure (0.11 Pa), and a high log K_{OW} value of 5.24 (Table 6.4). TBCO has similar water solubility and log K_{OW} with TBCO except for the slightly lower vapor pressure (0.07 Pa). This is probably because both TBCEH and TBCO have comparable molecular weight and number of bromine atoms attached to an aliphatic chain or ring. TBCEH and TBCO have similar environmental behavior to the other brominated flame retardants, that is, it readily partitions to organic matter in either atmospheric or aqueous phase. Technical-grade TBECH comprises of equimolar amounts of α- and β-TBECH diastereoisomers. The α- and β-TBECH are thermally labile and can isomerize to γ- and δ-TBECH at elevated temperatures. TBCO has four chiral centers and exists as two diastereoisomers with four enantiomers.

6.3.2 Enantioselective Analysis and Environmental Behavior

Enantiomers of chiral brominated flame retardants are often separated using liquid chromatography. Permethylated β-cyclodextrin-based chiral stationary phases such as Nucleodex β-PM column are commonly used for enantioselective analysis of HBCDs in environmental matrices (Guerra et al., 2008; Heeb et al., 2007; Köppen et al., 2010; Li et al., 2017). Enantioseparation of six PBB atropisomers was previously achieved using liquid chromatography (Berger et al., 2002). Low molecular weight flame retardants like TBECH and TBCO are often analyzed using gas chromatography since they have lower boiling points (Ruan et al., 2018a, 2019). Previous studies separated the enantiomers of TBECH and TBCO using β-cyclodextrin-based capillary columns containing 2,6-di-O-pentyl-3-trifluoroacetyl or 2,3-dimethyl-6-tert-butyldimethylsilyl derivates. To date, only PBB 149 has been successfully separated using enantioselective gas chromatography (Berger et al., 2002; Götsch et al., 2005). Previous studies showed that electron capture negative ion tandem mass spectrometry (GC/ECNI-MS/MS) was highly selective and sensitive for the detection of PBB 149 enantiomers (von der Recke et al., 2005). In contrast, another study observed enantioselectivity in detection of PBB 149 in environmental samples using electron ionization tandem mass spectrometry (GC/EI-MS/MS) but not when using GC/ECNI-MS (Götsch et al., 2005). The lack of enantioselectivity observed using GC/ECNI-MS was probably due to coelution of unknown organobromines. Enantioseparation is difficult to achieve using gas chromatography because PBB congeners have high molecular weight (> 540 g mol^{-1}) and boiling points (>200°C) (Berger et al., 2002). As a result, vaporization of the congeners in gas chromatography would require high elution temperatures. However, PBB atropisomers are thermally labile and can undergo enantiomerization at elevated temperatures, and this may introduce uncertainties in the analysis (Sanganyado et al., 2020).

6.3.2.1 Polybrominated Biphenyls

Previous studies detected PBBs in various environmental matrices such as soil, sediment, air, and water. In South Africa, PBB 153 concentration ranged from 0.25 to 1.56 ng L^{-1} along a river (Daso et al., 2013a). In sediments, Daso et al. (2013b) found concentrations of up to 1.21 µg kg^{-1}. The higher concentrations in sediments were probably because of the high log K_{OC} value of 5.50, concentrations in sediments are expected to be higher than in the water column. Another study in South Africa found the total concentration of 16 PBB congeners in office dust was up to 196 µg kg^{-1} dry weight (dw) (Kefeni and Okonkwo, 2012). In Thailand, the concentration of PBB 209 was an order of magnitude higher than the sum PBB concentration in South Africa with concentrations ranging from 20–2,300 µg kg^{-1} (Muenhor et al., 2010). PBBs readily partition to the organic matter on aerosols considering their high log K_{OA} values which are often greater than 8.50 (Hongxia et al., 2009). These results suggest the presence of PBB in aquatic environment remains of great concern despite the global restrictions on their production. This is probably because of continued use of PBBs despite the ban or release of the chemicals from historically contaminated sites. However, a study in China found the sum concentration of PBBs at e-waste disposal sites ranged from 108.78 µg kg^{-1} to 1,943.86 µg kg^{-1} suggesting continued use of PBBs (H. M. Wang et al., 2009). A previous study found PBB concentrations of up to 2,116 µg kg^{-1} in fish from the Baltic and North Seas (Gieroń et al., 2010). PBBs have log K_{OW} values > 5.0 which suggest they can readily bioaccumulate in biota. Estimates indicate PBB 153 has a log BCF = 5.53 (where BCF is bioaccumulation factor), which is greater than the regulatory threshold of 3.7.

There are few studies on the enantioselective distribution and behavior of PBBs in the environment. This is probably because of the lack of sensitive and selective methods for environmental analysis. Previous studies found that the sensitivity of GC/ECNI-MS/MS decreased with degree of bromination in congeners (Gonzalez-Gago et al., 2015). Recent studies have shown that two-dimensional gas chromatography coupled to triple quadrupole mass spectrometry (GC × GC-MS/MS) is a highly robust, selective, and sensitive technique for enantioselective analysis of PCBs in a wide range of environmental matrices (Bucheli and Brändli, 2006). Considering PCBs are structurally analogous to PBBs, GC×GC-MS/MS is potentially a powerful technique for determining the enantiomeric composition of PBBs in the environment. However, additional studies are required to validate the technique.

6.3.2.2 Hexabromocyclododecane

Previous studies have detected HBCDs in various environmental matrices such as air, water, sediments, and soil. A study on the diastereomeric and enantiomeric distribution of HBCDs in urban air showed that around 69–97% of HBCDs were particle attached (Yu et al., 2008). HBCDs have a high log K_{OA} value of around 5.4; thus, HBCDs have a high potential for long range aerial transport. Studies in China found racemic concentrations of β- and γ-HBCD in urban air but slight enrichment of (−)-α-HBCD enantiomer (Li et al., 2012; Yu et al., 2008). A study in surface soils impacted by e-waste found the concentrations of α-, β, and γ-HBCDs were nonracemic (Gao et al., 2011). This suggested there was no enantioselective transformation of the three isomers in surface soils. Interestingly, the enantiomeric fractions of γ-HBCD in river sediments (EF = 0.431–0.479) differed significantly from the technical grade HBCD compared to those from e-waste disposal sites (Feng et al., 2012). Similarly, enantiomeric enrichment of (+)-α-, and (+)-γ-HBCDs was observed in sediments from the Cinca River in Spain (Guerra et al., 2008). These results suggest enantioselective transformation probably occurred in sediments.

A previous study found the bioaccumulation and elimination of HBCDs in zebrafish (*Danio rerio*) was enantioselective (Du et al., 2012). Du et al. (2012) found there was enantiomeric enrichment of (+)-α-HBCD and (+)-γ-HBCD relative to their antipodes throughout the 60-d study period. Similar enantiomeric enrichment of (−)-α-HBCD was observed in herring gull eggs from the German coastal regions (Esslinger et al., 2011a). Enantiomeric enrichment of (−)-α -, (−)-β-, and

(+)-γ-HBCD was observed in fish collected from Etnefjorden, Norway (Köppen et al., 2010). In contrast, a previous study found (−)-α-HBCD enantiomer was preferentially degraded in rat and trout liver with enantiomeric fractions of 0.321 and 0.419, respectively (Abdallah et al., 2014). Exposing HBCD enantiomers to earthworms resulted in enantiomeric enrichment of (−)-α-, (−)-β- and (−)-γ-HBCDs. Another study in rats found preferential degradation of (+)-α- and (−)-γ-HBCD (Esslinger et al., 2011b). These results show that the metabolism and elimination of HBCDs in biota is often enantioselective even though the enantioselective behavior cannot be extrapolated to other species. Since the bioaccumulation and elimination of HBCDs is often enantioselective, it is possible for the trophic magnification to be enantioselective as well. An investigation on trophic magnification in a marine food web found preferential trophic magnification of (−)-α- and (+)-α-HBCD (Ruan et al., 2018b). The trophic magnification factors (TMFs) of the enantiomers were around 12 and 9, respectively (Ruan et al., 2018b).

6.3.2.3 Other Brominated Cycloalkanes

There are few studies on the environmental behavior of TBCEH and TBCO. Effluent from wastewater treatment plants in Hong Kong was found to enrich the second eluting α- and β-TBECH (Ruan et al., 2019). These results suggest there was preferential transformation of the first eluting α- and β-TBECH enantiomers probably due microbial degradation. In contrast, the second eluting α- and β-TBECH were preferentially transformed in the first 90 days of incubation (Wong et al., 2012). Interestingly, the second eluting β-TBCO enantiomer was preferentially degraded for the first 90 days (EF = 0.54 on day 90) while the first eluting enantiomers was preferred from day 90 to 400 (EF = 0.465 on day 400) (Wong et al., 2012). These results show that the enantioselective behavior of tetrabrominated alkanes varies between different environmental matrices. Although TBCEH and TBCO have high potential to bioaccumulate, they are few studies investigating their occurrence in biota. A study on the trophic magnification of TBECH found that there was enantiomeric enrichment the first eluting δ-TBECH in higher trophic level organisms in a marine food web (Ruan et al., 2018b).

6.4 CONCLUSION AND FUTURE DIRECTIONS

Chiral BFRs and PFOS are important contaminants of emerging concern frequently detected in the environment. They are widely used for various applications particularly in plastic and textile industries. It is important to understand the behavior of the enantiomers of BFRs and PFOS in the environment because enantiomers often have different biological properties. Several studies have shown that enantiomers of chiral contaminants such as pharmaceuticals and pesticides exhibit different environmental toxicity, fate, and transport. Most studies on chiral BFRs have been limited to HBCDs. There is need for more studies on the enantioselective distribution and behavior of other BFRs such TBECH, TBCO, and novel BFRs such as isobutoxypentabromocyclododecanes (iBPBCDs). A previous study separated and detected 16 iBPBCD stereoisomers in various environmental samples and biota using liquid chromatography (Heeb et al., 2010). Furthermore, future studies should explore the impact of chirality on bioaccumulation and biomagnification of the chiral halogenated organic contaminants of emerging concern in different organisms. More comprehensive studies on the environmental toxicity of BFRs and PFOS are required, particularly on differences in toxicity between enantiomers. This is essential to better understand the impact of these contaminants on ecosystems.

ACKNOWLEDGMENTS

This study was supported by Shantou University Research Start-Up Program and has been approved (Grant Number NTF20002).

REFERENCES

Abdallah, M. A.-E., Uchea, C., Chipman, J. K., Harrad, S., 2014. Enantioselective biotransformation of hexabromocyclododecane by in vitro rat and trout hepatic sub-cellular fractions. *Environ. Sci. Technol.* 48, 2732–2740. https://doi.org/10.1021/es404644s

Asher, B. J., Wang, Y., De Silva, A. O., Backus, S., Muir, D. C. G., Wong, C. S., Martin, J. W., 2012. Enantiospecific perfluorooctane sulfonate (PFOS) analysis reveals evidence for the source contribution of PFOS-precursors to the Lake Ontario foodweb. *Environ. Sci. Technol.* 46, 7653–7660. https://doi.org/10.1021/es301160r

Berger, U., Vetter, W., Götsch, A., Kallenborn, R., 2002. Chromatographic enrichment and enantiomer separation of axially chiral polybrominated biphenyls in a technical mixture. *J. Chromatogr. A* 973, 123–133. https://doi.org/10.1016/S0021-9673(02)01112-3

Birnbaum, L. S., Staskal, D. F., 2004. Brominated flame retardants: Cause for concern? *Environ. Health Perspect.* 112, 9–17. https://doi.org/10.1289/ehp.6559

Boulanger, B., Vargo, J., Schnoor, J. L., Hornbuckle, K. C., 2004. Detection of perfluorooctane surfactants in Great Lakes water. *Environ. Sci. Technol.* 38, 4064–4070. https://doi.org/10.1021/es0496975

Boulanger, B., Vargo, J. D., Schnoor, J. L., Hornbuckle, K. C., 2005. Evaluation of perfluorooctane surfactants in a wastewater treatment system and in a commercial surface protection product. *Environ. Sci. Technol.* 39, 5524–5530. https://doi.org/10.1021/es050213u

Bucheli, T. D., Brändli, R. C., 2006. Two-dimensional gas chromatography coupled to triple quadrupole mass spectrometry for the unambiguous determination of atropisomeric polychlorinated biphenyls in environmental samples. *J. Chromatogr. A* 1110, 156–164. https://doi.org/10.1016/j.chroma.2006.01.069

Buck, R. C., Franklin, J., Berger, U., Conder, J. M., Cousins, I. T., Voogt, P. De, Jensen, A. A., Kannan, K., Mabury, S. A., van Leeuwen, S. P. J., 2011. Perfluoroalkyl and polyfluoroalkyl substances in the environment: Terminology, classification, and origins. *Integr. Environ. Assess. Manag.* 7, 513–541. https://doi.org/10.1002/ieam.258

Cantillo, D., Kappe, C. O., 2017. Halogenation of organic compounds using continuous flow and microreactor technology. *React. Chem. Eng.* 2, 7–19. https://doi.org/10.1039/c6re00186f

Chibwe, L., Titaley, I. A., Hoh, E., Simonich, S. L. M., 2017. Integrated framework for identifying toxic transformation products in complex environmental mixtures. *Environ. Sci. Technol. Lett.* 4, 32–43. https://doi.org/10.1021/acs.estlett.6b00455

Covaci, A., Gerecke, A. C., Law, R. J., Voorspoels, S., Kohler, M., Heeb, N. V., Leslie, H., Allchin, C. R., de Boer, J., 2006. Hexabromocyclododecanes (HBCDs) in the environment and humans: A review. *Environ. Sci. Technol.* 40, 3679–3688. https://doi.org/10.1021/es0602492

Danish Environmental Protection Agency, 2015. Perfluoroalkylated substances: PFOA, PFOS and PFOSA Evaluation of health hazards and proposal of a health based quality criterion for drinking water, soil and ground water. Danish Environmental Protection Agency, Copenhagen, Denmark.

Daso, A. P., Fatoki, O. S., Odendaal, J. P., 2013a. Occurrence of polybrominated diphenyl ethers (PBDEs) and 2,2',4,4',5,5'-hexabromobiphenyl (BB-153) in water samples from the Diep River, Cape Town, South Africa. *Environ. Sci. Pollut. Res.* 20, 5168–5176. https://doi.org/10.1007/s11356-013-1503-6

Daso, A. P., Fatoki, O. S., Odendaal, J. P., 2013b. Polybrominated diphenyl ethers (PBDEs) and hexabromobiphenyl in sediments of the Diep and Kuils rivers in South Africa. *Int. J. Sediment Res.* 31, 61–70. https://doi.org/10.1016/j.ijsrc.2013.10.001

Ding, G., Peijnenburg, W. J. G. M., 2013. Physicochemical properties and aquatic toxicity of poly- and perfluorinated compounds. *Crit. Rev. Environ. Sci. Technol.* 43, 598–678. https://doi.org/10.1080/10643389.2011.627016

Ding, G., Xue, H., Zhang, J., Cui, F., He, X., 2018. Occurrence and distribution of perfluoroalkyl substances (PFASs) in sediments of the Dalian Bay, China. *Mar. Pollut. Bull.* 127, 285–288. https://doi.org/10.1016/j.marpolbul.2017.12.020

Du, M., Lin, L., Yan, C., Zhang, X., 2012. Diastereoisomer- and enantiomer-specific accumulation, depuration, and bioisomerization of hexabromocyclododecanes in zebrafish (*Danio rerio*). *Environ. Sci. Technol.* 46, 11040–11046. https://doi.org/10.1021/es302166p

Ebeling, D., Šekutor, M., Stiefermann, M., Tschakert, J., Dahl, J. E. P., Carlson, R. M. K., Schirmeisen, A., Schreiner, P. R., 2018. Assigning the absolute configuration of single aliphatic molecules by visual inspection. *Nat. Commun.* 9, 1–8. https://doi.org/10.1038/s41467-018-04843-z

Eljarrat, E., de la Cal, A., Raldua, D., Duran, C., Barcelo, D., 2004. Occurrence and bioavailability of polybrominated diphenyl ethers and hexabromocyclododecane in sediment and fish from the Cinca River, a tributary of the Ebro River (Spain). *Environ. Sci. Technol.* 38, 2603–2608. https://doi.org/10.1021/es0301424

Eljarrat, E., Guerra, P., Barceló, D., 2008. Enantiomeric determination of chiral persistent organic pollutants and their metabolites. *TrAC Trends Anal. Chem.* 27, 847–861. https://doi.org/10.1016/j.trac.2008.08.010

Eriksson, U., Haglund, P., Kärrman, A., 2017. Contribution of precursor compounds to the release of per- and polyfluoroalkyl substances (PFASs) from waste water treatment plants (WWTPs). *J. Environ. Sci.* 61, 80–90. https://doi.org/10.1016/j.jes.2017.05.004

Esslinger, S., Becker, R., Jung, C., Schröter-Kermani, C., Bremser, W., Nehls, I., 2011a. Temporal trend (1988–2008) of hexabromocyclododecane enantiomers in herring gull eggs from the German coastal region. *Chemosphere* 83, 161–167. https://doi.org/10.1016/j.chemosphere.2010.12.047

Esslinger, S., Becker, R., Maul, R., Nehls, I., 2011b. Hexabromocyclododecane enantiomers: Microsomal degradation and patterns of hydroxylated metabolites. *Environ. Sci. Technol.* 45, 3938–3944. https://doi.org/10.1021/es1039584

Feng, A. H., Chen, S. J., Chen, M. Y., He, M. J., Luo, X. J., Mai, B. X., 2012. Hexabromocyclododecane (HBCD) and tetrabromobisphenol A (TBBPA) in riverine and estuarine sediments of the Pearl River delta in southern China, with emphasis on spatial variability in diastereoisomer- and enantiomer-specific distribution of HBCD. *Mar. Pollut. Bull.* 64, 919–925. https://doi.org/10.1016/j.marpolbul.2012.03.008

Gallen, C., Drage, D., Eaglesham, G., Grant, S., Bowman, M., Mueller, J. F., 2017. Australia-wide assessment of perfluoroalkyl substances (PFASs) in landfill leachates. *J. Hazard. Mater.* 331, 132–141. https://doi.org/10.1016/j.jhazmat.2017.02.006

Gallen, C., Eaglesham, G., Drage, D., Nguyen, T. H., Mueller, J. F., 2018. A mass estimate of perfluoroalkyl substance (PFAS) release from Australian wastewater treatment plants. *Chemosphere* 208, 975–983. https://doi.org/10.1016/j.chemosphere.2018.06.024

Gao, H., Zhang, L., Lu, Z., He, C., Li, Q., Na, G., 2018. Complex migration of antibiotic resistance in natural aquatic environments. *Environ. Pollut.* 232, 1–9. https://doi.org/10.1016/j.envpol.2017.08.078

Gao, S., Wang, J., Yu, Z., Guo, Q., Sheng, G., Fu, J., 2011. Hexabromocyclododecanes in surface soils from e-waste recycling areas and industrial areas in south China: Concentrations, diastereoisomer- and enantiomer-specific profiles, and inventory. *Environ. Sci. Technol.* 45, 2093–2099. https://doi.org/10.1021/es1033712

Gieroń, J., Grochowalski, A., Chrzaszcz, R., 2010. PBB levels in fish from the Baltic and North seas and in selected food products from Poland. *Chemosphere* 78, 1272–1278. https://doi.org/10.1016/j.chemosphere.2009.12.031

Giesy, J. P., Kannan, K., 2002. Perfluorochemical surfactants in the environment. *Environ. Sci. Technol.* 36, 146A-152A. https://doi.org/10.1007/s13398-014-0173-7.2

Gonzalez-Gago, A., Pröfrock, D., Prange, A., 2015. Comparison of GC-NCI MS, GC-ICP-MS, and GC-EI MS-MS for the determination of PBDEs in water samples according to the Water Framework Directive. *Anal. Bioanal. Chem.* 407, 8009–8018. https://doi.org/10.1007/s00216-015-8973-y

González-Mariño, I., Carpinteiro, I., Rodil, R., Rodríguez, I., Quintana, J. B., 2016. High-resolution mass spectrometry identification of micropollutants transformation products produced during water disinfection with chlorine and related chemicals. *Compr. Anal. Chem.* 71, 283–334. https://doi.org/10.1016/bs.coac.2016.01.007

Götsch, A., Mariussen, E., von der Recke, R., Herzke, D., Berger, U., Vetter, W., 2005. Analytical strategies for successful enantioselective separation of atropisomeric polybrominated biphenyls 132 and 149 in environmental samples. *J. Chromatogr. A* 1063, 193–199. https://doi.org/10.1016/j.chroma.2004.11.059

Guerra, P., Eljarrat, E., Barceló, D., 2008. Enantiomeric specific determination of hexabromocyclododecane by liquid chromatography-quadrupole linear ion trap mass spectrometry in sediment samples. *J. Chromatogr. A* 1203, 81–87. https://doi.org/10.1016/j.chroma.2008.07.027

Häggblom, M. M., Bossert, I. D., 2005. Halogenated organic compounds - A global perspective, in: *Dehalogenation*. Kluwer Academic Publishers, Boston, pp. 3–29. https://doi.org/10.1007/0-306-48011-5_1

Hamid, H., Li, L. Y., Grace, J. R., 2018. Review of the fate and transformation of per- and polyfluoroalkyl substances (PFASs) in landfills. *Environ. Pollut.* 235, 74–84. https://doi.org/10.1016/j.envpol.2017.12.030

Harrad, S., Abdallah, M. A. E., Rose, N. L., Turner, S. D., Davidson, T. A., 2009. Current-use brominated flame retardants in water, sediment, and fish from English lakes. *Environ. Sci. Technol.* 43, 9077–9083. https://doi.org/10.1021/es902185u

Hauler, C., Martin, R., Knölker, H. J., Gaus, C., Mueller, J. F., Vetter, W., 2013. Discovery and widespread occurrence of polyhalogenated 1,1'-dimethyl-2,2'- bipyrroles (PDBPs) in marine biota. *Environ. Pollut.* 178, 329–335. https://doi.org/10.1016/j.envpol.2013.03.025

He, M. J., Luo, X. J., Yu, L. H., Wu, J. P., Chen, S. J., Mai, B. X., 2013. Diastereoisomer and enantiomer-specific profiles of hexabromocyclododecane and tetrabromobisphenol A in an aquatic environment in a highly industrialized area, South China: Vertical profile, phase partition, and bioaccumulation. *Environ. Pollut.* 179, 105–110. https://doi.org/10.1016/j.envpol.2013.04.016

Heeb, N. V., Graf, H., Bernd Schweizer, W., Lienemann, P., 2010. Isobutoxypentabromocyclododecanes (iBPBCDs): A new class of polybrominated compounds. *Chemosphere.* 78, 950–957. https://doi.org/10.1016/j.chemosphere.2009.12.045

Heeb, N. V., Schweizer, W. B., Kohler, M., Gerecke, A. C., 2005. Structure elucidation of hexabromocyclododecanes - A class of compounds with a complex stereochemistry. *Chemosphere.* 61, 65–73. https://doi.org/10.1016/j.chemosphere.2005.03.015

Heeb, N. V., Schweizer, W. B., Mattrel, P., Haag, R., Gerecke, A. C., Kohler, M., Schmid, P., Zennegg, M., Wolfensberger, M., 2007. Solid-state conformations and absolute configurations of (+) and (-) α-, β-, and γ-hexabromocyclododecanes (HBCDs). *Chemosphere.* 68, 940–950. https://doi.org/10.1016/j.chemosphere.2007.01.032

Heeb, N. V., Wyss, S. A., Geueke, B., Fleischmann, T., Kohler, H. P. E., Bernd Schweizer, W., Moor, H., Lienemann, P., 2015. Stereochemistry of enzymatic transformations of (+)β- and (-)β-HBCD with LinA2 - A HCH-degrading bacterial enzyme of *Sphingobium indicum* B90A. *Chemosphere.* 122, 70–78. https://doi.org/10.1016/j.chemosphere.2014.11.008

Hongxia, Z., Jingwen, C., Xie, Q., Baocheng, Q., Xinmiao, L., 2009. Octanol-air partition coefficients of polybrominated biphenyls. *Chemosphere.* 74, 1490–1494. https://doi.org/10.1016/j.chemosphere.2008.11.041

Houde, M., Bujas, T. A. D., Small, J., Wells, R. S., Fair, P. A., Bossart, G. D., Solomon, K. R., Muir, D. C. G., 2006. Biomagnification of perfluoroalkyl compounds in the bottlenose dolphin (Tursiops truncatus) food web. *Environ. Sci. Technol.* 40, 4138–4144. https://doi.org/10.1021/es060233b

Kasprzyk-Hordern, B., 2010. Pharmacologically active compounds in the environment and their chirality. *Chem. Soc. Rev.* 39, 4466. https://doi.org/10.1039/c000408c

Kefeni, K. K., Okonkwo, J. O., 2012. Analysis of major congeners of polybromobiphenyls and polybromodiphenyl ethers in office dust using high resolution gas chromatography-mass spectrometry. *Chemosphere.* 87, 1070–1075. https://doi.org/10.1016/j.chemosphere.2012.02.014

Knutsen, H., Mæhlum, T., Haarstad, K., Slinde, G. A., Arp, H. P. H., 2019. Leachate emissions of short- and long-chain per- and polyfluoralkyl substances (PFASs) from various Norwegian landfills. *Environ. Sci. Process. Impacts* 21, 1970–1979. https://doi.org/10.1039/c9em00170k

Köppen, R., Becker, R., Esslinger, S., Nehls, I., 2010. Enantiomer-specific analysis of hexabromocyclododecane in fish from Etnefjorden (Norway). *Chemosphere.* 80, 1241–1245. https://doi.org/10.1016/j.chemosphere.2010.06.019

Krafft, M. P., Riess, J. G., 2015. Per- and polyfluorinated substances (PFASs): Environmental challenges. *Curr. Opin. Colloid Interface Sci.* 20, 192–212. https://doi.org/10.1016/j.cocis.2015.07.004

Kwok, K. Y., Yamazaki, E., Yamashita, N., Taniyasu, S., Murphy, M. B., Horii, Y., Petrick, G., Kallerborn, R., Kannan, K., Murano, K., Lam, P. K. S., 2013. Transport of perfluoroalkyl substances (PFAS) from an arctic glacier to downstream locations: Implications for sources. *Sci. Total Environ.* 447, 46–55. https://doi.org/10.1016/j.scitotenv.2012.10.091

Langlois, I., Berger, U., Zencak, Z., Oehme, M., 2007. Mass spectral studies of perfluorooctane sulfonate derivatives separated by high-resolution gas chromatography. *Rapid Commun. Mass Spectrom.* 21, 3547–3553. https://doi.org/10.1002/rcm.3241

Li, B., Chen, H., Sun, H., Lan, Z., 2017. Distribution, isomerization and enantiomer selectivity of hexabromocyclododecane (HBCD) diastereoisomers in different tissue and subcellular fractions of earthworms. *Ecotoxicol. Environ. Saf.* 139, 326–334. https://doi.org/10.1016/j.ecoenv.2017.01.004

Li, H., Mo, L., Yu, Z., Sheng, G., Fu, J., 2012. Levels, isomer profiles and chiral signatures of particle-bound hexabromocyclododecanes in ambient air around Shanghai, China. *Environ. Pollut.* 165, 140–146. https://doi.org/10.1016/j.envpol.2012.02.015

Lim, T. C., Wang, B., Huang, J., Deng, S., Yu, G., 2011. Emission inventory for PFOS in China: Review of past methodologies and suggestions. *Sci. World J.* 11, 1963–1980. https://doi.org/10.1100/2011/868156

Lindim, C., van Gils, J., Cousins, I. T., 2016. Europe-wide estuarine export and surface water concentrations of PFOS and PFOA. *Water Res.* 103, 124–132. https://doi.org/10.1016/j.watres.2016.07.024

Liu, S., Lu, Y., Xie, S., Wang, T., Jones, K. C., Sweetman, A. J., 2015. Exploring the fate, transport and risk of perfluorooctane sulfonate (PFOS) in a coastal region of China using a multimedia model. *Environ. Int.* 85, 15–26. https://doi.org/10.1016/j.envint.2015.08.007

Liu, W., Wu, J., He, W., Xu, F., 2019. A review on perfluoroalkyl acids studies: Environmental behaviors, toxic effects, and ecological and health risks. *Ecosyst. Heal. Sustain.* 5, 1–19. https://doi.org/10.1080/20964129.2018.1558031

Liu, X., Wu, W., Zhang, Y., Wang, T., Zhao, J., Chen, Z., 2017a. Occurrence, profiles, and ecological risks of polybrominated diphenyl ethers in mangrove sediments of Shantou, China. *Environ. Sci. Pollut. Res.* 24, 3608–3617. https://doi.org/10.1016/j.chemosphere.2015.02.064

Liu, Z., Lu, Y., Wang, P., Wang, T., Liu, S., Johnson, A. C., Sweetman, A. J., Baninla, Y., 2017b. Pollution pathways and release estimation of perfluorooctane sulfonate (PFOS) and perfluorooctanoic acid (PFOA) in central and eastern China. *Sci. Total Environ.* 580, 1247–1256. https://doi.org/10.1016/j.scitotenv.2016.12.085

Loos, R., Locoro, G., Comero, S., Contini, S., Schwesig, D., Werres, F., Balsaa, P., Gans, O., Weiss, S., Blaha, L., Bolchi, M., Gawlik, B. M., 2010. Pan-European survey on the occurrence of selected polar organic persistent pollutants in ground water. *Water Res.* 44, 4115–4126. https://doi.org/10.1016/j.watres.2010.05.032

Luo, J., Hu, J., Wei, X., Fu, L., Li, L., 2015. Dehalogenation of persistent halogenated organic compounds: A review of computational studies and quantitative structure-property relationships. *Chemosphere* 131, 17–33. https://doi.org/10.1016/j.chemosphere.2015.02.013

Mackintosh, S. A., Dodder, N. G., Shaul, N. J., Aluwihare, L. I., Maruya, K. A., Chivers, S. J., Danil, K., Weller, D. W., Hoh, E., 2016. Newly identified DDT-related compounds accumulating in Southern California bottlenose dolphins. *Environ. Sci. Technol.* 50, 12129–12137. https://doi.org/10.1021/acs.est.6b03150

Martin, J. W., Asher, B. J., Beesoon, S., Benskin, J. P., Ross, M. S., 2010. PFOS or PreFOS? Are perfluorooctane sulfonate precursors (PreFOS) important determinants of human and environmental perfluorooctane sulfonate (PFOS) exposure? *J. Environ. Monit.* 12, 1979–2004. https://doi.org/10.1039/c0em00295j

Martin, J. W., Ellis, D. A., Mabury, S. A., Hurley, M. D., Wallington, T. J., 2006. Atmospheric chemistry of perfluoroalkanesulfonamides: Kinetic and product studies of the OH radical and Cl atom initiated oxidation of N-ethyl perfluorobutanesulfonamide. *Environ. Sci. Technol.* 40, 864–872. https://doi.org/10.1021/es051362f

Marvin, C. H., Tomy, G. T., Armitage, J. M., Arnot, J. A., McCarty, L., Covaci, A., Palace, V., 2011. Hexabromocyclododecane: Current understanding of chemistry, environmental fate and toxicology and implications for global management. *Environ. Sci. Technol.* 45, 8613–8623. https://doi.org/10.1021/es201548c

Mcdougall, M. R. R., 2016. *Developing a Trophic Bioaccumulation Model for PFOA and PFOS in a Marine Food Web*. Simon Fraser University.

Muenhor, D., Harrad, S., Ali, N., Covaci, A., 2010. Brominated flame retardants (BFRs) in air and dust from electronic waste storage facilities in Thailand. *Environ. Int.* 36, 690–698. https://doi.org/10.1016/j.envint.2010.05.002

Müller, A., Österlund, H., Marsalek, J., Viklander, M., 2020. The pollution conveyed by urban runoff: A review of sources. *Sci. Total Environ.* 709, 136125. https://doi.org/10.1016/j.scitotenv.2019.136125

Navarro, I., de la Torre, A., Sanz, P., Martínez, M. de los Á., 2020. Perfluoroalkyl acids (PFAAs): Distribution, trends and aquatic ecological risk assessment in surface water from Tagus River basin (Spain). *Environ. Pollut.* 256. https://doi.org/10.1016/j.envpol.2019.113511

Ort, C., Lawrence, M. G., Rieckermann, J., Joss, A., 2010. Sampling for pharmaceuticals and personal care products (PPCPs) and illicit drugs in wastewater systems : Are your conclusions valid ? A critical review. *Environ. Sci. Technol.* 44, 6024–6035.

Pangallo, K. C., Reddy, C. M., 2010. Marine natural products, the halogenated 1'-methyl-1,2'-bipyrrols, biomagnify in a northwestern Atlantic food web. *Environ. Sci. Technol.* 44, 5741–5747.

Paul, A. G., Scheringer, M., Hungerbühler, K., Loos, R., Jones, K. C., Sweetman, A. J., 2012. Estimating the aquatic emissions and fate of perfluorooctane sulfonate (PFOS) into the river Rhine. *J. Environ. Monit.* 14, 524–530. https://doi.org/10.1039/c1em10432b

Qin, S., Budd, R., Bondarenko, S., Liu, W., Gan, J., 2006. Enantioselective degradation and chiral stability of pyrethroids in soil and sediment. *J. Agric. Food Chem.* 54, 5040–5045. https://doi.org/10.1021/jf060329p

Rajput, I. R., Xiao, Z., Yajing, S., Yaqoob, S., Sanganyado, E., Ying, H., Fei, Y., Liu, W., 2018. Establishment of pantropic spotted dolphin (Stenella attenuata) fibroblast cell line and potential influence of polybrominated diphenyl ethers (PBDEs) on cytokines response. *Aquat. Toxicol.* 203, 1–9. https://doi.org/10.1016/j.aquatox.2018.07.017

Rayne, S., Forest, K., Friesen, K. J., 2008. Congener-specific numbering systems for the environmentally relevant C 4 through C8 perfluorinated homologue groups of alkyl sulfonates, carboxylates, telomer alcohols, olefins, and acids, and their derivatives. *J. Environ. Sci. Heal. - Part A Toxic/Hazardous Subst. Environ. Eng.* 43, 1391–1401. https://doi.org/10.1080/10934520802232030

Ribeiro, A. R., Afonso, C. M., Castro, P. M. L., Tiritan, M. E., 2012a. Enantioselective HPLC analysis and biodegradation of atenolol, metoprolol and fluoxetine. *Environ. Chem. Lett.* 11, 83–90. https://doi.org/10.1007/s10311-012-0383-1

Ribeiro, A. R., Castro, P. M. L., Tiritan, M. E., 2012b. Environmental fate of chiral pharmaceuticals: Determination, degradation and toxicity, in: Lichtfouse, E., Schwarzbauer, J., Robert, D. (Eds.), *Environmental Chemistry for a Sustainable World*. Springer Netherlands, Dordrecht, pp. 5–45.

Ribeiro, A. R., Castro, P. M. L., Tiritan, M. E., 2012c. Chiral pharmaceuticals in the environment. *Environ. Chem. Lett.* 10, 239–253.

Ruan, Y., Lam, J. C. W., Zhang, X., Lam, P. K. S., 2018a. Temporal changes and stereoisomeric compositions of 1,2,5,6,9,10-Hexabromocyclododecane and 1,2-Dibromo-4-(1,2-dibromoethyl)cyclohexane in marine mammals from the South China Sea. *Environ. Sci. Technol.* 52, 2517–2526. https://doi.org/10.1021/acs.est.7b05387

Ruan, Y., Zhang, K., Lam, J. C. W., Wu, R., Lam, P. K. S., 2019. Stereoisomer-specific occurrence, distribution, and fate of chiral brominated flame retardants in different wastewater treatment systems in Hong Kong. *J. Hazard. Mater.* 374, 211–218. https://doi.org/10.1016/j.jhazmat.2019.04.041

Ruan, Y., Zhang, X., Qiu, J. W., Leung, K. M. Y., Lam, J. C. W., Lam, P. K. S., 2018b. Stereoisomer-specific trophodynamics of the chiral brominated flame retardants HBCD and TBECH in a marine food web, with implications for human exposure. *Environ. Sci. Technol.* 52, 8183–8193. https://doi.org/10.1021/acs.est.8b02206

Sanganyado, E., 2019. Comments on "Chiral pharmaceuticals: Environment sources, potential human health impacts, remediation technologies and future perspective." *Environ. Int.* 122, 412–415. https://doi.org/10.1016/j.envint.2018.11.032

Sanganyado, E., Fu, Q., Gan, J., 2016. Enantiomeric selectivity in adsorption of chiral β-blockers on sludge. *Environ. Pollut.* 214, 787–794. https://doi.org/10.1016/j.envpol.2016.04.091

Sanganyado, E., Lu, Z., Fu, Q., Schlenk, D., Gan, J., 2017. Chiral pharmaceuticals: A review on their environmental occurrence and fate processes. *Water Res.* 124, 527–542. https://doi.org/10.1016/j.watres.2017.08.003

Sanganyado, E., Lu, Z., Gan, J., 2014. Mechanistic insights on chaotropic interactions of liophilic ions with basic pharmaceuticals in polar ionic mode liquid chromatography. *J. Chromatogr. A* 1368, 82–88. https://doi.org/10.1016/j.chroma.2014.09.054

Sanganyado, E., Lu, Z., Liu, W., 2020. Application of enantiomeric fractions in environmental forensics: Uncertainties and inconsistencies. *Environ. Res.* 184, 109354. https://doi.org/10.1016/j.envres.2020.109354

Sanganyado, E., Rajput, I. R., Liu, W., 2018. Bioaccumulation of organic pollutants in Indo-Pacific humpback dolphin: A review on current knowledge and future prospects. *Environ. Pollut.* 237, 111–125. https://doi.org/10.1016/j.envpol.2018.01.055

Shen, A., Lee, S., Ra, K., Suk, D., Moon, H. B., 2018. Historical trends of perfluoroalkyl substances (PFASs) in dated sediments from semi-enclosed bays of Korea. *Mar. Pollut. Bull.* 128, 287–294. https://doi.org/10.1016/j.marpolbul.2018.01.039

Simcik, M. F., Dorweiler, K. J., 2005. Ratio of perfluorochemical concentrations as a tracer of atmospheric deposition to surface waters. *Environ. Sci. Technol.* 39, 8678–8683. https://doi.org/10.1021/es0511218

Sinclair, E., Kannan, K., 2006. Mass loading and fate of perfluoroalkyl surfactants in wastewater treatment plants. *Environ. Sci. Technol.* 40, 1408–1414. https://doi.org/10.1021/es051798v

Skaar, J. S., Ræder, E. M., Lyche, J. L., Ahrens, L., Kallenborn, R., 2019. Elucidation of contamination sources for poly- and perfluoroalkyl substances (PFASs) on Svalbard (Norwegian Arctic). *Environ. Sci. Pollut. Res.* 26, 7356–7363. https://doi.org/10.1007/s11356-018-2162-4

Smith, S. W., 2009. Chiral toxicology: It's the same thing…only different. *Toxicol. Sci.* 110, 4–30. https://doi.org/10.1093/toxsci/kfp097

Stalcup, A. M., 2010. Chiral separations. *Annu. Rev. Anal. Chem.* 3, 341–63. https://doi.org/10.1146/annurev.anchem.111808.073635

Stroski, K. M., Luong, K. H., Challis, J. K., Chaves-Barquero, L. G., Hanson, M. L., Wong, C. S., 2020. Wastewater sources of per- and polyfluorinated alkyl substances (PFAS) and pharmaceuticals in four Canadian Arctic communities. *Sci. Total Environ.* 708, 134494. https://doi.org/10.1016/j.scitotenv.2019.134494

Teaf, C. M., Garber, M. M., Covert, D. J., Tuovila, B. J., 2019. Perfluorooctanoic acid (PFOA): environmental sources, chemistry, toxicology, and potential risks. *Soil Sediment Contam.* 28, 258–273. https://doi.org/10.1080/15320383.2018.1562420

Tian, Z., Peter, K. T., Gipe, A. D., Zhao, H., Hou, F., Wark, D. A., Khangaonkar, T., Kolodziej, E. P., James, C. A., 2020. Suspect and nontarget screening for contaminants of emerging concern in an urban estuary. *Environ. Sci. Technol.* 54, 889–901. https://doi.org/10.1021/acs.est.9b06126

Tomy, G. T., Budakowski, W., Halldorson, T., Whittle, D. M., Keir, M. J., Marvin, C., MacInnis, G., Alaee, M., 2004. Biomagnification of alpha- and gamma-hexabromocyclododecane isomers in a Lake Ontario food web. *Environ. Sci. Technol.* 38, 2298–2303. https://doi.org/10.1021/es034968h

Vetter, W., 2012. *Polyhalogenated Alkaloids in Environmental and Food Samples, The Alkaloids: Chemistry and Biology*. Elsevier. https://doi.org/10.1016/B978-0-12-398282-7.00003-5

von der Recke, R., Mariussen, E., Berger, U., Götsch, A., Herzke, D., Vetter, W., 2005. Determination of the enantiomer fraction of PBB 149 by gas chromatography/electron capture negative ionization tandem mass spectrometry in the selected reaction monitoring mode. *Rapid Commun. Mass Spectrom.* 19, 3719–3723. https://doi.org/10.1002/rcm.2258

Wang, H. M., Yu, Y. J., Han, M., Yang, S. W., Li, Q., Yang, Y., 2009. Estimated PBDE and PBB congeners in soil from an electronics waste disposal site. *Bull. Environ. Contam. Toxicol.* 83, 789–793. https://doi.org/10.1007/s00128-009-9858-6

Wang, Y., Arsenault, G., Riddell, N., McCrindle, R., McAlees, A., Martin, J. W., 2009. Perfluorooctane sulfonate (PFOS) precursors can be metabolized enantioselectively: Principle for a new PFOS source tracking tool. *Environ. Sci. Technol.* 43, 8283–8289. https://doi.org/10.1021/es902041s

Wang, Y., Beesoon, S., Benskin, J. P., De Silva, A. O., Genuis, S. J., Martin, J. W., 2011. Enantiomer fractions of chiral perfluorooctanesulfonate (PFOS) in human sera. *Environ. Sci. Technol.* 45, 8907–8914. https://doi.org/10.1021/es2023434

Wong, F., Kurt-Karakus, P., Bidleman, T. F., 2012. Fate of brominated flame retardants and organochlorine pesticides in urban soil: Volatility and degradation. *Environ. Sci. Technol.* 46, 2668–2674. https://doi.org/10.1021/es203287x

Xiao, F., Simcik, M. F., Gulliver, J. S., 2012. Perfluoroalkyl acids in urban stormwater runoff: Influence of land use. *Water Res.* 46, 6601–6608. https://doi.org/10.1016/j.watres.2011.11.029

Xie, S., Wang, T., Liu, S., Jones, K. C., Sweetman, A. J., Lu, Y., 2013. Industrial source identification and emission estimation of perfluorooctane sulfonate in China. *Environ. Int.* 52, 1–8. https://doi.org/10.1016/j.envint.2012.11.004

Xu, C., Lin, X., Yin, S., Zhao, L., Liu, Y., Liu, K., Li, F., Yang, F., Liu, W., 2018. Enantioselectivity in biotransformation and bioaccumulation processes of typical chiral contaminants. *Environ. Pollut.* 243, 1274–1286. https://doi.org/10.1016/j.envpol.2018.09.095

Yamashita, N., Kannan, K., Taniyasu, S., Horii, Y., Petrick, G., Gamo, T., 2005. A global survey of perfluorinated acids in oceans. *Mar. Pollut. Bull.* 51, 658–668. https://doi.org/10.1016/j.marpolbul.2005.04.026

Yan, H., Cousins, I. T., Zhang, C., Zhou, Q., 2015. Perfluoroalkyl acids in municipal landfill leachates from China: Occurrence, fate during leachate treatment and potential impact on groundwater. *Sci. Total Environ.* 524–525, 23–31. https://doi.org/10.1016/j.scitotenv.2015.03.111

Yeung, L. W. Y., De Silva, A. O., Loi, E. I. H., Marvin, C. H., Taniyasu, S., Yamashita, N., Mabury, S. A., Muir, D. C. G., Lam, P. K. S., 2013. Perfluoroalkyl substances and extractable organic fluorine in surface sediments and cores from Lake Ontario. *Environ. Int.* 59, 389–397. https://doi.org/10.1016/j.envint.2013.06.026

Yu, Z., Chen, L., Mai, B., Wu, M., Sheng, G., Fu, J., Peng, P., 2008. Diastereoisomer- and enantiomer-specific profiles of hexabromocyclododecane in the atmosphere of an urban city in South China. *Environ. Sci. Technol.* 42, 3996–4001. https://doi.org/10.1021/es7027857

Zhang, Q., Zhang, Z., Tang, B., Gao, B., Tian, M., Sanganyado, E., Shi, H., Wang, M., 2018. Mechanistic insights into stereospecific bioactivity and dissipation of chiral fungicide triticonazole in agricultural management. *J. Agric. Food Chem.* 66, 7286–7293. https://doi.org/10.1021/acs.jafc.8b01771

Zhang, Y., Meng, W., Guo, C., Xu, J., Yu, T., Fan, W., Li, L., 2012. Determination and partitioning behavior of perfluoroalkyl carboxylic acids and perfluorooctanesulfonate in water and sediment from Dianchi Lake, China. *Chemosphere.* 88, 1292–1299. https://doi.org/10.1016/j.chemosphere.2012.03.103

Zhao, L., Chen, F., Guo, F., Liu, W., Liu, K., 2019. Enantioseparation of chiral perfluorooctane sulfonate (PFOS) by supercritical fluid chromatography (SFC): Effects of the chromatographic conditions and separation mechanism. *Chirality.* 31, 870–878. https://doi.org/10.1002/chir.23120

Zhao, L., Chen, F., Yin, S., Xie, J., Aamir, M., Liu, S., Liu, W., 2020. Enantioselectivity in transplacental transfer of perfluoro-1-methylheptanesulfonate (1m-PFOS): Human biomonitoring and in silico study. *Environ. Pollut.* 261, 114136. https://doi.org/10.1016/j.envpol.2020.114136

7 Persistent Organic Pollutants

Leo M.L. Nollet

CONTENTS

7.1 The Stockholm Convention 153
    Annex A Elimination 154
    Annex B Restriction 154
    Annex C Unintentional Production 154
7.2 Chiral POPs 155
7.3 Detection Methods 156
7.4 Chiral Separation 157
7.5 1,2,5,6,9,10-Hexabromocyclododecane 158
7.6 Decabromodiphenyl Ether 160
7.7 Pentachlorobenzene (PeCB) 168
7.8 Hexachlorobutadiene 169
7.9 Short-Chain Chlorinated Paraffins 171
7.10 Polychlorinated Naphtalenes 172
7.11 PFASs (Perfluoroalkyl Substances) 177
7.12 Polybrominated Biphenyls 180
References 181

7.1 THE STOCKHOLM CONVENTION

The Stockholm Convention (1) is a global treaty with the objective of protecting human health and the environment from persistent organic pollutants (POPs). Following paragraphs may be consulted at their website.

Persistent Organic Pollutants (POPs) are organic chemical substances, that is, they are carbon-based. They possess a particular combination of physical and chemical properties such that, once released into the environment, they:

- remain intact for exceptionally long periods of time (many years);
- become widely distributed throughout the environment as a result of natural processes involving soil, water and, most notably, air;
- accumulate in the fatty tissue of living organisms including humans, and are found at higher concentrations at higher levels in the food chain; and
- are toxic to both humans and wildlife.

As a result of releases to the environment over the past several decades due especially to human activities, POPs are now widely distributed over large regions (including those where POPs have never been used) and, in some cases, they are found around the globe. This extensive contamination of environmental media and living organisms includes many foodstuffs and has resulted in the sustained exposure of many species, including humans, for periods of time that span generations, resulting in both acute and chronic toxic effects (1).

In addition, POPs concentrate in living organisms through another process called bioaccumulation. Though not soluble in water, POPs are readily absorbed in fatty tissue, where concentrations can become magnified by up to 70,000 times the background levels. Fish, predatory birds, mammals, and humans are high up the food chain and so absorb the greatest concentrations. When they travel, the POPs travel with them. As a result of these two processes, POPs can be found in people and animals living in regions such as the Arctic, thousands of kilometers from any major POP source.

Specific effects of POPs can include cancer, allergies and hypersensitivity, damage to the central and peripheral nervous systems, reproductive disorders, and disruption of the immune system. Some POPs are also considered to be endocrine disrupters, which, by altering the hormonal system, can damage the reproductive and immune systems of exposed individuals as well as their offspring; they can also have developmental and carcinogenic effects.

The chemicals targeted by the Stockholm Convention are listed in the 3 annexes of the convention text.

Annex A Elimination
Parties must take measures to eliminate the production and use of the chemicals listed under Annex A.

Aldrin, Chlordane, Chlordecone, Decabromodiphenyl ether (commercial mixture, c-decaBDE), Dicofol, Dieldrin, Endrin, Heptachlor, Hexabromobiphenyl, Hexabromocyclododecane (HBCDD), **Hexabromodiphenyl ether and heptabromodiphenyl ether, Hexachlorobenzene (HCB),** Hexachlorobutadiene, **Alpha hexachlorocyclohexane, Beta hexachlorocyclohexane, Lindane, Mirex,** Pentachlorobenzene, Pentachlorophenol and its salts and esters, **Polychlorinated biphenyls (PCB),** Polychlorinated naphthalenes, Perfluorooctanoic acid (PFOA), its salts and PFOA-related compounds, Short-chain chlorinated paraffins (SCCPs), Technical endosulfan and its related isomers, Tetrabromodiphenyl ether and pentabromodiphenyl ether, Toxaphene.

Annex B Restriction
Parties must take measures to restrict the production and use of the chemicals listed under Annex B in light of any applicable acceptable purposes and/or specific exemptions listed in the Annex.

DDT, Perfluorooctane sulfonic acid, its salts and perfluorooctane sulfonyl fluoride.

Annex C Unintentional Production
Parties must take measures to reduce the unintentional releases of chemicals listed under Annex C with the goal of continuing minimization and, where feasible, ultimate elimination.

Hexachlorobenzene (HCB), Hexachlorobutadiene (HCBD), Pentachlorobenzene, Polychlorinated biphenyls (PCB), Polychlorinated dibenzo-*p*-dioxins (PCDD) Polychlorinated dibenzofurans (PCDF), Polychlorinated naphthalenes.

Measurable levels of POPs are present in all living organisms and evidence of POPs contamination in, e.g., human blood and breast milk has been still documented worldwide (2–4). Despite their hazards, these chemicals continue to be produced, used, and stored in many countries, and restrictions and national bans cannot protect citizens from exposure to POPs that have migrated from other regions where these chemicals are still in use.

7.2 CHIRAL POPs

Enantiomers of chiral compounds commonly undergo enantioselective transformation in most biologically mediated processes. As chiral POPs are extensively distributed in the environment, differences between enantiomers in biotransformation should be carefully considered to obtain exact enrichment and specific health risks. Zhang et al. (5) provide an overview of in vivo biotransformation of chiral POPs currently indicated in the Stockholm Convention and their chiral metabolites.

Peer-reviewed journal articles focused on the research question were thoroughly searched. A set of inclusion and exclusion criteria were developed to identify relevant studies. The authors mainly compared the results from different animal models under controlled laboratory conditions to show the difference between enantiomers in terms of distinct transformation potential. Interactions with enzymes involved in enantioselective biotransformation, especially cytochrome P450 (CYP), were discussed.

Limited evidence for a few POPs has been found. Enantioselective biotransformation of α-hexachlorocyclohexane (α-HCH), chlordane, dichlorodiphenyltrichloroethane (DDT), heptachlor, hexabromocyclododecane (HBCD), polychlorinated biphenyls (PCBs), and toxaphene, has been investigated using laboratory mammal, fish, bird, and worm models. Tissue and excreta distributions, as well as bioaccumulation and elimination kinetics after administration of racemate and pure enantiomers, have been analyzed in these studies. Changes in enantiomeric fractions have been considered as an indicator of enantioselective biotransformation of chiral POPs in most studies. Results of different laboratory animal models revealed that chiral POP biotransformation is seriously affected by chirality. Pronounced results of species-, tissue-, gender-, and individual-dependent differences are observed in in vivo biotransformation of chiral POPs. Enantioselective biotransformation of chiral POPs is dependent on enzyme amounts and activities. However, the role of cytochrome P450 in enantioselective biotransformation has not yet been confirmed.

Understanding the enantioselectivity is critical for the assessment of environmental behavior and toxicological effect, as well as the ecological risk. As chirality is an important consideration in green chemistry, recent studies demonstrated that enantioselectivity of typical chiral organic pollutants, especially for chiral pesticides, exists in most processes such as aquatic toxicity and degradation. However, to further clarify the metabolism mechanism of enantioselectivity, especially when catalyzed by cytochrome P450 enzymes, development of computational methods is of great importance in this field.

Many outstanding achievements around the hot field and prosperous direction for the development of chiral compounds in computational analysis were reviewed (6). A promising combination of theoretical methods, named hybrid quantum mechanical–molecular mechanics (QM/MM), will potentially make a significant contribution to metabolism mechanism of enantioselectivity by P450s.

First was clarified the enantioselectivity of chiral pollutants and environmental safety with twenty-two studies, and the remaining (twenty-one) were about cytochromes P450-mediated metabolism of typical chiral compounds. Furthermore, a series of theoretical method for cytochromes P450-mediated metabolism with sixteen essays were enumerated. This review illustrates the importance of the enantioselective in environmental safety, and then discussed the enantioselectivity of interactions mechanism with cytochromes P450 metabolic. Finally theoretical calculations became available and enabled various hypotheses to be tested and new solutions to be offered, especially for QM/MM methods.

Elimination of POPs under national and international controls reduces "primary" emissions, but "secondary" emissions continue from residues deposited in soil, water, ice, and vegetation during former years of high usage (7). Secondary sources are expected to dominate in the future, when POPs transport and accumulation will be controlled by air–surface exchange and the biogeochemical cycle of organic carbon. Climate change is likely to affect mobilization of POPs through, e.g., increased temperature, loss of ice cover in polar regions, melting glaciers, and changes in soil and water microbiology which affect degradation and transformation. Chiral compounds offer advantages for following transport and fate pathways because of their ability to distinguish racemic (newly released or protected from microbial attack) and nonracemic (microbially altered) sources. The rationale for this approach is explained and applications where chiral POPs could aid investigation of climate–mediated exchange and degradation processes are suggested. Examples include distinguishing agricultural vs. non–agricultural and recently used vs. residual pesticides, degradation and sequestration processes in soil, historical vs. recent atmospheric deposition, sources in arctic air, and influence of ice cover on volatilization.

More and more chiral POPs have emerged as industry has developed. Studying only racemic POPs will lead to overestimations and underestimations of their ecological safety and potential to cause health risks (8). This is because, although the enantiomers of a molecule have the same chemical properties, they can interact with enzymes in different ways. The enantiomeric ratio (ER) and the enantiomeric fraction (EF) are often used to identify the sources of, and the effects of microbial degradation on, chiral chemicals in the environment. These parameters can also be used to trace the migration of chiral substances through the environment and the transformation processes that have acted upon them. EF values for chiral POPs in environmental matrixes commonly decrease in the order soil > water > air. The EF values of chiral POPs are also often altered in organisms, and the biggest EF deviations (from the values found in the wider environment) in animals have been found in brain tissue, followed by liver tissue, with greater deviations in animals at high trophic levels than at low trophic levels. This may be related to enantiomer-specific biological processes, such as uptake and degradation, and the enzymes involved in the transformation of POPs in organisms. Chiral POPs show enantiomer-specific toxicity (both acute and chronic, and including endocrine disruption effects), which may also be caused by the different ways enantiomers interact with enzymes. Regulating authorities may not, therefore, get a true picture of the risks and effects of POPs if they do not consider changes in the ERs or EFs in the environment and biota and the individual toxicities of the enantiomers.

Chiral organochlorine compounds (OCs) were measured in various environmental matrices (air, soil, and vegetation) from western Antarctica using high-resolution gas chromatography coupled with high-resolution mass spectrometry (HRGC/HRMS) (9). They were generally detected at a global background level compared with the previous studies. α-HCH and PCB-183 was observed in all the matrices except PCB-183 in two soil samples, while PCB-95, -136, -149, -174, -176, and o,p'-DDT were detected in most air but only a few solid matrices. Enantiomeric fractions (EFs) indicated that nonracemic residues of chiral OCs occurred in all the matrices and a wide variation of the EF values was observed in the vegetation. There was significant discrepancy between the EF values of PCB-183 and the racemic values, indicating that stereoselective depletion of PCB-183 was probably associated with the water-air exchange. The EFs values of α-HCH were generally lower than the racemic values but no statistical difference was obtained in all the matrices except lichen, supporting the assumption that water-air exchange may make influence on long-range transport of α-HCH.

7.3 DETECTION METHODS

Mass spectrometry (MS) coupled with chromatography is the most widely applied technique for POP quantification in food and environmental matrices. Since food-based matrices are complex, selectivity is the primary concern (10 and references in 10).

Gas chromatography (GC) is the most commonly used technique for separation. The GC separation is dependent on the boiling points of the compounds and their interactions with the stationary phase of the column. Most POPs are semi-volatile, and their polarities are between moderate and non-polar. These physicochemical properties make most POPs well suited to being measured by GC–MS, except for PFAS-related chemicals, which are always measured using LC–MS/MS method. However, there is no single column that can separate all congeners of PCBs and dioxins/

furans. To overcome this difficulty, comprehensive two-dimensional GC was introduced. When passing two columns, there are two degrees of separation based on different physiochemical properties. Compared to single columns, the two-dimensional GC can significantly improve selectivity (peak capacity) and sensitivity. The majority of the analytical methods for the detection of POPs is GC coupled with conventional detectors such as an electron capture detector and MS operated in different ionization modes. The electron capture detector is a low-cost detector and is mostly used in the analysis of PCBs and OCPs in different foodstuff. However, recently, MS has become the most commonly used detector for POP analysis. Several techniques are used in MS to generate ions. One widely used technique for the detection of POPs in food is GC coupled with MS in the selected ion mode. The selected ion mode could improve selectivity by focusing on a selected number of relevant masses corresponding to analytes. Electron capture negative chemical ionization is an alternative softer ionization and is very useful for the detection of POPs. PCB and PBDEs can be analyzed using GC–MS in either the electron capture negative ionization mode or the electron ionization mode. For liquid chromatography, atmospheric ionization such as electrospray and atmospheric pressure chemical ionization are two widely used techniques for the detection of POPs in food. Atmospheric pressure chemical ionization was also coupled with GC to analyze dioxins and PBDEs. GC coupled with 13C-labeled isotope dilution high-resolution mass spectrometry (HRMS) is considered a standard method for the detection of specific POPs such as dioxins and furans. The direct 13C-labeled isotope dilution provides reliable quantification. However, due to the high cost of the equipment and the need for skilled technicians for 13C-labeled isotope dilution HRMS, other MS instruments such as time-of-flight mass spectrometry (TOF-MS) have been utilized. TOF–MS is promising when coupled with suitable GC methods such as comprehensive two-dimensional gas chromatography. GC×GC TOF-MS has been successfully applied to detect dioxins and PCBs in food. The specificity of GC×GCTOF–MS could be improved by either operating the instrument in tandem model (MS/MS) or improving the chromatographic separation. Recent studies have shown that GC coupled with triple-quadrupole tandem MS had a high-performance similar to GC-HRMS for the detection of POPs in food and feedstuff samples such as vegetable oil and fish. Atmospheric pressure gas chromatography (APGC) triple quadrupole has also proved to have sufficient sensitivity and selectivity in the analysis of dioxins and PCBs in food and feed samples.

POPs can be grouped into subclasses based on their original intended uses or chemical properties. In reference (11), five such groups of compounds are considered: polycyclic aromatic hydrocarbons, chlorinated aromatic compounds, pesticides, brominated flame retardants, and perfluoroalkyl and polyfluoroalkyl substances. This reference provides an overview of some of the measurement issues and liquid chromatography methods used in the determination of persistent organic contaminants in environmentally relevant samples. Emphasis has been placed on research carried out since 2000. Purely gas chromatography-based methods have been excluded from this discussion, as have reports that emphasize sample cleanup and processing research.

7.4 CHIRAL SEPARATION

In general, there are two types of methods for chiral separation: indirect and direct.

In the indirect approach, chiral derivatizing reagents form stable diastereomers, which could subsequently be separated using non-chiral stationary phases. In this procedure the two compounds react with a chiral selector (CS) forming strong diastereomers where strong bonds are formed. This time-consuming process has some advantages, such as an increase in sensitivity when using UV or mass spectrometry (MS) detectors due to the possibility of introducing additional special groups.

On the contrary, in the direct separation, the CS is included in the separation system (chiral stationary phases, CSPs) forming continuous transient diastereomeric complexes. Direct methods are widely employed nowadays. This method is based on the use of a CSP where two analytes interact continuously forming diastereoisomeric complexes with bonds.

The number of CSs is large and varied but no universal one exists. Peptides, chiral amino acids, cyclodextrins, polysaccharide derivatives, quinine-based, glycopeptides antibiotics, and chiral crown-ethers are examples of chiral selectors.

Examples of HPLC columns for chiral separations are Acquity UPC² Trefoil AMY1 2.5 μm, Acquity UPC² Trefoil CEL1 2.5 μm, and Acquity UPC² Trefoil CEL2 2.5 μm from Waters and Lux® 3 μm Cellulose-2 or Lux® 3 μm Amylose-1 from Phenomenex.

Polysaccharide-derived chiral stationary phases (CSPs) are well known for their versatility when used for high-performance liquid chromatography and supercritical fluid chromatography separations. Among the various polysaccharides available, amylose and cellulose derivatives are most useful as CSPs.

Another CS widely applied are cyclodextrins (CDs) and their derivatives.

7.5 1,2,5,6,9,10-HEXABROMOCYCLODODECANE

Hexabromocyclododecane (HBCD or HBCDD) is a brominated flame retardant. Its primary application is in extruded (XPS) and expanded (EPS) polystyrene foam that is used as thermal insulation in the building industry.

None of the studies on HBCD have addressed effects at the level of individual HBCD stereoisomers or enantiomers (120).

The review of Covaci et al. (12) summarizes HBCD concentrations in several environmental compartments and analyzes these data in terms of point sources versus diffuse sources, biomagnification potential, stereoisomer profiles, time trends, and global distribution. Generally, higher concentrations were measured in samples (air, sediment, and fish) collected near point sources (plants producing or processing HBCDs), while lower concentrations were recorded in samples from locations with no obvious sources of HBCDs. High concentrations were measured in top predators, such as marine mammals and birds of prey (up to 9600 and 19 200 ng/g lipid weight, respectively), suggesting a biomagnification potential for HBCDs. Relatively low HBCD concentrations were reported in the few human studies conducted to date (median values varied between 0.35 and 1.1 ng/g lipid weight). HBCD levels in biota are increasing slowly and seem to reflect the local market demand. One important observation is the shift from the high percentage of the γ-HBCD stereoisomer in the technical products to a dominance of the α-HBCD stereoisomer in biological samples. A combination of factors such as variations in solubility, partitioning behavior, uptake, and, possibly, selective metabolism of individual isomers may explain the observed changes in stereoisomer patterns.

LC/MS/MS is the preferred analytical technique, because it allows both diastereo- and enantioselective determination of HBCD in environmental samples. Gas chromatography or high-performance LC MS are commonly used for the quantitative determination of HBCD (13). However, both of these well-established methods have limitations. As with many other halogenated persistent organic pollutants, HBCD has been analyzed via GC/MS, with detection in both the positive and negative ion modes. Because of the higher sensitivity, the mass spectrometer is usually operated in the electron-capture negative-ion (ECNI) mode, with only the bromide ion typically monitored during chromatographic runs. However, larger fragment ions, which are necessary for structural confirmation of HBCD, do not form in ECNI mode. GC separation has its own serious limitations. HBCD diastereomers interconvert at temperatures >160°C. Because of this thermal isomerization and the fact that HBCD elutes from the GC column at temperatures >160°C, a broad, unresolved peak is observed. Therefore, the total amount of HBCD can be determined by GC but not the quantities of each of the isomers. Moreover, pure HBCD undergoes decomposition by elimination of hydrogen bromide at temperatures >240°C. Not surprisingly, partial breakdown and even complete loss of HBCD have been reported in GC systems. The thermal exposure of HBCD must therefore be minimized during analysis. Cold on-column injection, short GC columns, thin-film stationary phases, and high flow rates are several measures that can minimize the risk of thermal degradation and reduce the elution temperature.

Stereoisomers of 1,2,5,6,9,10-hexabromocyclododecane and 1,2-dibromo-4-(1,2-dibromoethyl)-cyclohexane (TBECH) were determined in sediments and 30 marine species in a marine food web to investigate their trophic transfer (14). Lipid content was found to affect the bioaccumulation of ΣHBCD and ΣTBECH in these species. Elevated biomagnification of each diastereomer from prey species to marine mammals was observed. For HBCD, biota samples showed a shift from γ- to α-HBCD when compared with sediments and technical mixtures; trophic magnification potential of (−)-α- and (+)-α-HBCD were observed in the food web, with trophic magnification factors (TMFs) of 11.8 and 8.7, respectively. For TBECH, the relative abundance of γ- and δ-TBECH exhibited an increasing trend from abiotic matrices to biota samples; trophic magnification was observed for each diastereomer, with TMFs ranging from 1.9 to 3.5. The enantioselective bioaccumulation of the first eluting enantiomer of δ-TBECH in organisms at higher TLs was consistently observed across samples.

α-, β-, and γ-Hexabromocyclododecanes were subjected to in vitro biotransformation experiments with rat and trout liver S9 fractions for different incubation times (10, 30, and 60 min) at 2 concentration levels (1 and 10 μM) (15). The metabolic degradation of target HBCDs followed first order kinetics. Whereas β-HBCD undergoes rapid biotransformation ($t_{0.5}$ = 6.4 and 38.1 min in rat and trout, respectively), α-HBCD appears the most resistant to metabolic degradation ($t_{0.5}$ = 17.1 and 134.9 min). The biotransformation rate in trout was slower than in rat. Investigation of HBCD degradation profiles revealed the presence of at least 3 pentabromocyclododecene (PBCD) and 2 tetrabromocyclododecadiene (TBCD) isomers indicating reductive debromination as a metabolic pathway for HBCDs. Both mono- and di-hydroxyl metabolites were identified for parent HBCDs, while only mono hydroxyl metabolites were detected for PBCDs and TBCDs. Interestingly, δ-HBCD was detected only in trout S9 fraction assays indicating metabolic interconversion of test HBCD diastereomers during biotransformation in trout. Finally, enantioselective analysis showed significant enrichment of the (−)-α-HBCD.

Hexabromocyclododecane has frequently been detected in wildlife (16). However, there is limited research on its bioaccumulation and biomagnification in insect-dominated aquatic and terrestrial food webs. The occurrence of HBCDD in insects and their predators collected from a former e-waste contaminated pond and its surrounding region was investigated. The concentrations of ΣHBCDD (sum concentrations of α-, β-, and γ-HBCDDs) ranged from nd to 179 ng g^{-1} lipid weight. α-HBCDD was the predominant diastereoisomer in all biotic samples, and the contribution of α-HBCDD was higher in predators than in prey insects. A significantly positive linear relationship was found between ΣHBCDD concentrations (lipid weight) and trophic levels based on $δ^{15}$N in aquatic organisms ($p < 0.05$), while trophic dilution was observed in the terrestrial food web. This result indicates an opposite trophic transfer tendency of HBCDD in terrestrial and aquatic ecosystems. The biomagnification factor (BMF) for α-HBCDD was higher in terrestrial birds (2.03) than in frogs (0.29), toads (0.85), and lizards (0.63). This may be due to differences between poikilotherms and homeotherms in terrestrial ecosystems.

In the study of Sun et al. (17), α-, β-, and γ-HBCDs were measured in several fish species from rivers and an electronic waste (e-waste) recycling site in Pearl River Delta, South China. The concentrations of HBCDs were 12.8 to 640, 5.90 to 115, and 34.3 to 518 ng/g lipid weight (lw) in mud carp (*Cirrhinus molitorella*), tilapia (*Tilapia nilotica*), and plecostomus (*Hypostomus plecostomus*), respectively. Plecostomus showed the highest HBCD concentrations among three fish species. The contributions of α-HBCD to total HBCDs were 78% to 97%, 93% to 99%, and 87% to 98% in carp, tilapia, and plecostomus, respectively. Fish samples from a harbor and the e-waste site exhibited the highest HBCD concentrations among all samples.

α-, β-, and γ-HBCDs were analyzed on a system of an Agilent 1200 series liquid chromatography and an Agilent 6410 triple quadruple mass spectrometer with an electrospray interface working in negative ionization mode. HBCD diastereoisomers were separated by an XDB-C18 column.

7.6 DECABROMODIPHENYL ETHER

Decabromodiphenyl ether (decaBDE, deca-BDE, DBDE, deca, decabromodiphenyl oxide, DBDPO, or bis(pentabromophenyl) ether) is a brominated flame retardant which belongs to the group of polybrominated diphenyl ethers (PBDEs). Commercial decaBDE is a technical mixture of different PBDE congeners, with PBDE congener number 209 (decabromodiphenyl ether) and nonabromodiphenyl ether being the most common.

Since the beginning of this century, an increasing number of laboratories have become involved in the analysis of brominated flame retardants (BFRs) in the environment, humans, and food. To assure as well as improve the quality of these analyses, a series of international interlaboratory exercises has been organized (18). It appeared that the differences between the BFRs create more difficulties for their analysis than, for example, that of polychlorinated biphenyls. Problems that arise with the determination of decabromodiphenyl ether and hexabromocyclododecane, due to instability at higher temperatures, high background values, and other factors, make their analysis much more demanding than, for example, that of 2,4,2′,4′-tetrabromodiphenylether. A number of these pitfalls can be identified from the evaluation of the interlaboratory study results.

Decabromodiphenyl ether (BDE-209) is present at relatively high levels in sediments and macroinvertebrates of the River Po (19). Since it was demonstrated that BDE-209 can be biotransformed to smaller and more toxic PBDEs, the main objective of this study was to assess whether the large quantities of BDE-209 present in the River Po are bioavailable to the higher levels of the food web and are biotransformed in feral fishes. To this aim, 23 cyprinids, mainly common carp, were analyzed for the hepatic contents of PBDEs. Contrary to sediments and invertebrates of the same area, no fish sample contained detectable levels of BDE-209. All fishes contained typical PBDE representatives, e.g., BDE-47, BDE-99, BDE-100, BDE-153, and BDE-154, but more importantly they contained three congeners, i.e., BDE-179, BDE-188, and BDE-202, which are not present in any technical formulations and are known products of BDE-209 debromination in fish. The age of carps had no effects on the bioaccumulation of PBDEs. Conversely, the contents of PCBs, which also were determined in the same fish samples, showed a positive correlation with age. Both groups of chemicals displayed a tendency to a higher contamination in male fish.

Analysis of BDE-209 was performed using a gas chromatograph Trace GC 2000 equipped with a PTV injector coupled with a PolarisQ Ion-Trap mass spectrometer. Analysis was carried out using an Rxi-5MS capillary column.

Polybrominated diphenyl ethers are hydrophobic chemicals and can biomagnify in food chains. Plankton, *Diporeia*, lake whitefish, lake trout, and Chinook salmon were collected from Lake Michigan in 2006 between April and August (20). Fish liver and muscle and whole invertebrates were analyzed for six PBDEs (BDE-47, 99, 100, 153, 154, and 209). Carbon and nitrogen stable isotope ratios ($\delta^{13}C$ and $\delta^{15}N$) were also quantified in order to establish the trophic structure of the food web. Geometric means of $\Sigma PBDE \Sigma PBDE$ concentrations in fish ranged from 0.562 to 1.61 µg/g-lipid. BDE-209 concentrations ranged from 0.184 to 1.23 µg/g-lipid in all three fish species. $\Sigma BDE-47 \Sigma BDE-47$, 99, and 209 comprised 80–94% of $\Sigma PBDE \Sigma PBDE$ molar concentration. Within each fish species, there were no significant differences in PBDE concentrations between liver and muscle. The highest concentration of BDE-209 (144 µg/g-lipid) was detected in *Diporeia*. Based on analysis of $\delta^{15}N$ and PBDE concentrations, BDE-47 and 100 were found to biomagnify, whereas BDE-209 did not. A significant negative correlation between BDE-209 and trophic level was found in this food web. Biomagnification factors were also calculated and again BDE-47 and 100 biomagnified between food web members whereas BDE-209 did not. *Diporeia* could be one of the main dietary sources of BDE-209 for fish in Lake Michigan; BDE-47 and 100 biomagnified within this food chain; the concentration of BDE-209 decreased at higher trophic levels, suggesting partial uptake and/or biotransformation of BDE-209 in the Lake Michigan food web.

Samples were analyzed in a Varian 3800 gas chromatography equipped with ECD, a split/splitless injector, and a VF-5ht capillary column.

Polybrominated diphenyl ethers are contaminants that are widely distributed in the environment and in food samples. A highly sensitive biotin–streptavidin system-based real-time immuno-polymerase chain reaction assay (BA-IPCR) was developed to detect 2,2′,4,4′-tetrabromodiphenyl ether (BDE-47) (21). The BDE-47 immunogen and coating antigen were prepared, and a specific polyclonal anti-BDE-47 antibody was obtained. Several physiochemical factors in the immunoassay were optimized. The established BA-IPCR method was used to detect BDE-47 with a linearity ranging from 10 pg L−1 to 100 ng L−1. The limit of detection was 2.96 pg L−1. The recovery of the spiked samples was 92.97–107.19% and the coefficient of variation was 6.93–11.06%. The proposed method was applied to successfully detect BDE-47 in marine fish. The BA-IPCR data for the detection of BDE47 were consistent with gas chromatography–mass spectrometry findings. The BA-IPCR method is accurate, sensitive, and suitable for detecting trace amount of BDE-47 in food samples.

Polybrominated diphenyl ethers (PBDEs) are toxic chemicals widely distributed in the environment, but few studies are available on their potential toxicity to rice at metabolic level (22). Therefore ten rice (*Oryza sativa*) varieties were exposed to 2,20-, 4,40-tetrabromodiphenyl ether (BDE-47), a predominant congener of PBDEs, in hydroponic solutions with different concentrations. Two varieties that showed different biological effects to BDE-47, YY-9 and LJ-7, were screened as sensitive and tolerant varieties according to changes of morphological and physiological indicators. Metabolic research was then conducted using gas chromatography-mass spectrometry combined with diverse analysis. Results showed that LJ-7 was more active in metabolite profiles and adopted more effective antioxidant defense machinery to protect itself against oxidative damages induced by BDE-47 than YY-9. For LJ-7, the contents of 13 amino acids and 24 organic acids, especially L-glutamic acid, beta-alanine, glycolic acid, and glyceric acid were upregulated significantly which contributed to scavenging reactive oxygen species. In the treatment of 500 mg/L BDE-47, the contents of these four metabolites increased by 33.6-, 19.3-, 10.6-, and 10.2-fold, respectively. The levels of most saccharides (such as D-glucose, lactulose, maltose, sucrose, and D-cellobiose) also increased by 1.7–12.4 fold which promoted saccharide-related biosynthesis metabolism. Elevation of tricarboxylic acid cycle and glyoxylate and dicarboxylate metabolism enhanced energy producing processes. Besides, the contents of secondary metabolites, chiefly polyols and glycosides, increased significantly to act on defending oxidative stress induced by BDE-47. In contrast, the levels of most metabolites decreased significantly for YY-9, especially those of 13 amino acids (by 0.9%–67.1%) and 19 organic acids (by 7.8%–70.0%). The positive metabolic responses implied LJ-7 was tolerant to BDE-47, while the down-regulation of most metabolites indicated the susceptible nature of YY-9.

The freeze-dried solid samples (roots and leaves of the two screened varieties) were extracted with 20 mL of hexane/DCM (1:1, v/v). After ultrasonic extraction, adding acid silica gel into the collected extracts and nitrogen-blowing, the samples were prepared for GC-MS analysis.

In Table 7.1, a number of quantification methods of PBDEs in foods are presented (23).

Compound-specific and enantiomer-specific carbon isotope composition was investigated in terms of biotransformation of polychlorinated biphenyls (PCBs) and polybrominated diphenyl ethers (PBDEs) as well as atropisomers of chiral PCB congeners in fish by exposing common carp (*Cyprinus carpio*) to certain PCB and PBDE congeners (72). The calculated carbon isotope enrichment factors (εC) for PCB 8, 18, and 45 were −1.99, −1.84, and −1.70‰, respectively, providing evidence of the metabolism of these congeners in fish. The stable carbon isotopic compositions of PBDE congeners clearly reflect the debromination of PBDEs in carp. Significant isotopic fractionation was also observed during the debromination process of BDE 153 (εC = −0.86‰). Stereoselective elimination of chiral PCB congeners 45, 91, and 95 was observed, indicating a stereoselective biotransformation process. The similar εC values for E1-PCB 45 (−1.63‰) and E2-PCB 45 (−1.74‰) indicated that both atropisomers were metabolized by the same reaction mechanisms and stereoselection did not occur at carbon bond cleavage. However, the εC values of (+)-PCB 91 (−1.5‰) and (−)-PCB 95 (−0.77‰) were significantly different from those of (−)-PCB 91 and (+)-PCB 95, respectively. In the latter, no significant isotopic fractionations were observed, indicating

TABLE 7.1
Selected Papers Concerning PBDEs Determination in Food of Animal Origin

Year	Main/First Author	Matrix	Sample Pretreatment	Sample Weight	Extraction Technique	Extraction Solvent	Clean Up	Detection Method	Number of Citations*	Reference
2001	Alaee, M.	Lake trout, Herring, Sockeye salmon	—	10 g w.w.	Column extraction with Na_2SO_4	DCM	GPC deactivated silica gel	HRMS	62	24
2002	Huwe, J.K.	Chicken fat	—	1 g	—	—	EPA method 1613 modification, Hx: H_2SO_4 digested ACS	LRMS	55	25
2002	Jacobs, M.N.	Salmon, fish oil	—	10 g w.w.	Na_2SO_4 Soxhlet 2 h	75 ml Hx:DCM:Acetone (3:1:1 v + v + v)	2 × SPE acid/neutral silica gel: deactivated alumina	LRMS	156	26
2003	Pirard C.	Fortified beef fat	—	1 g	—	—	ACS	IT	49	27
2003	De Boer J.	Flounder, bream, marine mussels, freshwater mussels	—	100 g w.w.	Na_2SO_4 Soxhlet 12 h	Hx/acetone (3:1 v + v) 70 °C	GPC H_2SO_4 digestion Silica gel column	LRMS	28	28
2003	Voorspoels S.	Benthic invertebrates, benthic fish, benthic flatfish, gadoid fish	—	1–10 g w.w.	Na_2SO_4 Soxhlet 2,5 h	Hx/acetone (3:1 v + v)	Jacobs, M. N., 2002	LRMS	135	29
2004	Fernandes A.	Food	Freeze-drying	Equivalent of up to 10 l.w.	Solvent extraction	Hx	Multilayered silica connected with carbon column; Silica and alumina mini columns	HRMS	46	30

TABLE 7.1 (Continued)
Selected Papers Concerning PBDEs Determination in Food of Animal Origin

Year	Main/First Author	Matrix	Sample Pretreatment	Sample Weight	Extraction Technique	Extraction Solvent	Clean Up	Detection Method	Number of Citations*	Reference
2004	Papke O.	Fish, meat	—	10–100 g w.w.	Na_2SO_4 Solvent extraction	Cyclohexane: DCM	Drying on sodium sulphate, acid digestion, activated silica gel, alumina	HRMS		31
2005	Johnson-Restrepo B.	Nine species of marine fish, dolphins	—	20 g or 3 g l.w.	Na_2SO_4 Soxhlet 16 h	DCM:Hx (3:1 v + v)	Multilayer silica	HRMS/ECD		32
2006	Labandeira A.	Carp	Freeze-drying	1 g d.w.	Ground with alumina 2:1 W + w ASE	DCM:Hx	During extraction	LRMS		33
2006	Liu H.	Fish tissue	Freeze-drying	—	Acid silica Soxhlet 24 h	DCM:Hx	Silica gel with $Ag\,NO_3$	HRMS	33	34
2006	Gómara B.	Diary, eggs, sea fish, shellfish, meat	Freeze-drying	6 g d.w.	MSPD with silica gel	Acetone: Hx	Acid silica and basic alumina columns	IT/ECD	54	35
2007	Akutsu K.	125 types of food	—	5 g w.w.	Digested with 1 M KOH by 2 h LLE	Hx	Acid silica gel	LRMS	9	36
2007	Luo Q.	Carp, grass carp, silver carp, mud carp, carp	Freeze-drying	1–5 g d.w.	Na_2SO_4 Soxhlet 12 h	DCM: acetone	H_2SO_4 digestion, acid silica and alumina columns	IT/LRMS	28	37
2007	Pulkrabová J.	Fish: chub, barbell, bream, perch, and trout	—	30 g w.w.	Na_2SO_4 Soxhlet 8 h	Hx:DCM	GPC: H_2SO_4 digestion	LRMS	11	38
2008	Shaw S.D.	Harbor seals blubber	—	5 g	+ Na_2SO_4 + silica gel	Hx:DCM	ACS	HRMS/LRMS	25	39

(Continued)

TABLE 7.1 (Continued)
Selected Papers Concerning PBDEs Determination in Food of Animal Origin

Year	Main/First Author	Matrix	Sample Pretreatment	Sample Weight	Extraction Technique	Extraction Solvent	Clean Up	Detection Method	Number of Citations[*]	Reference
2008	Tapie N.	Trout, eel, mysids	Freeze-drying	0.5 g d.w.	MAE 3O W10 min purification with sulfuric acid	DCM	Filtration 5 x shaking with H_2SO_4; silica gel	ECD	21	40
2008	Wang D.	Seals blubber	Freeze-drying	2 g d.w.	ASE + Na_2SO_4	Hx:DCM	Multilayered silica column	IT/ECD	11	41
2009	Losada S.	Trout, salmon and horse mackerel	Freeze-drying	0.5 g d.w.	+ 10 g Na_2SO_4 + 25 g Florisil ASE	Hx:DCM (9:1 v+v)	During extraction	IT	19	42
2010	Li Y.F.	Pig muscle, fat, liver and small intestine	Freeze-drying	0.5–8 g w.w.	Soxhlet 24 h	Hx:DCM	Multilayered silica column $AgNO_3$ silica column	LRMS	6	43
2010	Losada S.	Eel, conger eel, spotted flounder, sardine, gilthead seabream, clam, mussel	Freeze-drying	2.5 g d.w.	Soxhlet 24 h	DCM:Hx	Multilayered silica and alumina columns	HRMS		44
2010	Lacorte S.	Fish and mil	—	5–10 g w.w.	PLE	DCM:Hx	GPC, Florisil columns	HRMS	6	45
2010	Blanco S.L.	Fish oil	—	1 g	—	—	ACS Acid silica and alumina columns	It	7	46
2010	Chen M.Y.Y.	Fish, seafood, meat, poultry, eggs, dairy products	—	1-150 g w.w.	+ Na_2SO_4 solvent extraction	Hx:DCM	Silica gel column ACS Acid silica and alumina columns	HRMS	8	47

Persistent Organic Pollutants

TABLE 7.1 (Continued)
Selected Papers Concerning PBDEs Determination in Food of Animal Origin

Year	Main/First Author	Matrix	Sample Pretreatment	Sample Weight	Extraction Technique	Extraction Solvent	Clean Up	Detection Method	Number of Citations*	Reference
2010	Labadie P.	Eel	Freeze-drying	3 g d.w.	+ acid silica gel MSPD + USAL 2 x 20 min (130 W; 42 Hz)	Hx:DCM	Centrifugation H_2SO_4 digestion Centrifugation Multilayered silica column	MS-MS	14	48
2010	Wang J.	Honey	—	20 g w.w.	+ Na_2SO_4 And wrapped with a filter paper ASE	Acetone:DCM	Dried with Na_2SO_4 Aluminum/silica column	IT/ECD	6	49
2011	Zhang Z.	Sheep liver	Freeze-drying	0.8 g d.w.	+ 20 g of acid silica gel ASE	Isoheksan:DCM (9:1 v + v)	During extraction	LRMS	13	50
2011	Kalachova K.	Trout, salmon, shrimps	—	10 g w.w.	QuECherS	Hx	GPC Multilayered silica column with $AgNO_3$	TOF	11	51
2011	Fontana A.R.	Fish tissue, chicken breast muscle, eggs	—	1 g w.w.	USAL	Hx:DCM (8:2 v + v)	DSPE (C18) SPE (silica gel)	IT	7	52
2012	Sprague M.	Atlantic Bluefin tuna: muscle, liver, gonad, adipose tissue	Free-drying	25–50 g w.w.	ASE	Isohexane	ACS	LRMS	0	53
2012	Roszko M.	Milk fat	—	5 g l.w.	Lipid gravity separation + Na_2SO_4 SPM (semi-permeable membrane)	Hx	GPC Multilayered silica column with $AgNO_3$	IT		54

(Continued)

TABLE 7.1 (Continued)
Selected Papers Concerning PBDEs Determination in Food of Animal Origin

Year	Main/First Author	Matrix	Sample Pretreatment	Sample Weight	Extraction Technique	Extraction Solvent	Clean Up	Detection Method	Number of Citations*	Reference
2012	Waszak I.	Flounder	Freeze-drying	10 g d.w.	ASE	DCM:Hx	Multilayered silica column	ECD		55
2012	Mackintosh S.A.	Lake trout, yellow perch, round gobies	—	4 g w.w.	+ Na$_2$SO$_4$ ASE	DCM:Hx	GPC Acid silica column	MS-MS		56
2013	Jürgens M.D.	Fish: Roach, Bleak, Eel	—	5 g w.w.	+ Na$_2$SO$_4$ Soxhlet overnight	DCM	GPC Acid silica column	LRMS		57
2013	Kalachova K.	Fish	—	10 g w.w.	QuEChersS	Ethyl acetate	Mini silica column	MS-MS		58
2013	Zacs D.	Wild salmon	Freeze-drying	-	Soxhlet 16 h	DCM:Hx	GPC, acid silica and florisil columns Carbon:cellite (1:1) Digested by 37N H$_2$SO$_4$ Layers separation by centrifugation	HRMS		59
2013	Sühring R.	Eel	Mixed with Na$_2$SO$_4$	1 g f.w.	+ Na$_2$SO$_4$ ASE	DCM	GPC, 10% deactivated silica gel	LRMS		60
2014	Baron E.	Fish, dolphin blubber, bird eggs	Freeze-drying	1 g d.w.	Alumina PLE	DCM:Hx	H$_2$SO$_4$ digestion SPE alumina (Al-N, 5 g)	MS-MS		61
2014	Poma G.	Zebra mussels, fish	Freeze-drying	0.1–1 g d.w.	Soxhlet	Hx:Acetone (3:1 v + v)	Multilayered silica/Florisil column	IT		62
2015	Sapozhnikova Y.	Croaker, Salmon	—	4 g w.w.	QuEChersS	MeCN	During extraction, filtered	MS-MS		63
2015	Couderc M.	Eel	Freeze-drying	1 g d.w.	ASE	Toluene:acetone (7:3, v + v)	Acid silica, Florisil, cellite/carbon columns	HRMS		64

TABLE 7.1 (Continued)
Selected Papers Concerning PBDEs Determination in Food of Animal Origin

Year	Main/First Author	Matrix	Sample Pretreatment	Sample Weight	Extraction Technique	Extraction Solvent	Clean Up	Detection Method	Number of Citations*	Reference
2015	Portolés T.	Fish, prawn, squid	—	6–200 g w.w.	MSPD + silica gel/ Na_2SO_4	Acetone/hx	Multilayered silica Carbon	MS-MS		65
2016	Lin Y.	Human Milk	Freeze-drying	40 g w.w.	ASE	DCM:Hx	D-SPE (acid silica) Tandem SPE (acid silica and alumina)	HRMS		66
2016	Dosis I.	Mussel	—	10 g w.w.	LLE	Isopropanol-diethylether (2.5: 1 v + v) and Hx-diethylether (9:1)	H_2SO_4 (98 %) digestion Acid silica column	LRMS		67
2016	Martinelli T.	Meat, eggs, milk, cheese, fish, fish oil, mussels	—	—	ASE	Hx:acetone (8:2, v + v)	ACS	LRMS		68
2016	Zeng Y.-H.	Chicken and goose eggs	Freeze-drying	1 g d.w.	Soxhlet	Hx:acetone (1:1, v + v)	Florisil/acid silica gel columns	LRMS		69
2016	Poma G.	Fish, meat, oil, eggs, milk	Freeze-drying	0.2, 1, or 2 g d.w. (<20%, 2–20, or <2 fat)	USAL	MeCN:toluene (9:1, v + v)	Florisil/acid silica gel columns	LRMS		70
2016	Bichon E.	Crustacean and mollusks, milk, fish, eggs, ovine liver, muscle	Freeze-drying	—	PLE	Toluene/acetone (70:30, v + v)	Acid silica gel/Florisil	MS-MS		71

ACS—automatic cleanup system; *—according to Web of Knowledge: +—indicates addition to sample or extraction cell.

that the stereoselective elimination of PCB 91 and 95 could be caused by a different reaction mechanism in the two atropisomers.

Investigations were performed with GC-C-isotope ratio mass spectrometry (IRMS).

7.7 PENTACHLOROBENZENE (PeCB)

PeCB was once used industrially for a variety of uses, as an intermediate in the manufacture of pesticides and a component of a mixture of chlorobenzenes added to products containing polychlorinated biphenyls in order to reduce viscosity.

A magnetic solid-phase extraction method for the pre-concentration of three organochlorine pesticides from water samples has been proposed, based on magnetic phosphatidylcholine (MPC) as adsorbents. The proposed method was employed for analysis of pentachlorobenzene, α-hexachlorocyclohexane, and β-endosulfan in the surface water from two rivers in northeast China (73). The extraction procedure was carried out in a single step by stirring the mixture of MPC and water samples. Subsequently, the MPC was collected by an external magnetic field without additional centrifugation or filtration. The analytes were desorbed from the MPC and finally analyzed by gas chromatography–tandem mass spectrometry. The influence of various parameters on OCPs recoveries was studied. Results show that phosphatidylcholine amount and extraction time were critical in enhancing extraction performance, and the presence of humic acid was shown to significantly reduce the extraction efficiency. The limits of detection obtained were in the range of 0.1–0.15 ng L^{-1}. Recoveries of spiked water samples ranged from 76.2% to 101.5% with relative standard deviations varying from 3.8% to 7.7%.

Data on the occurrence of POPs in food is scarce. Validated analytical methodology was developed to investigate the occurrence of hexachlorobutadiene (HCBD), pentachlorobenzene (PCBz), hexachlorobenzene (HCB), pentachlorophenol (PCP), and polychlorinated naphthalenes (PCNs) in 120 retail foods and 19 total diet study samples (74). The foods covered the range of commonly consumed dietary items including dairy products, eggs (hen and other species), poultry, meat, fish, vegetables, etc. HCBD showed a low frequency of detection, whereas PCBz, HCB, and PCNs occurred in most samples (ranges: <0.01 to 0.19 µg/kg; <0.01 to 3.16 µg/kg, and 0.1 to 166 ng ΣPCNs/kg, respectively). PCP (<0.01 to 1.9 µg/kg) was detected more frequently in meat products, offal, and eggs. Fish, shellfish, eggs from all species, animal fats, meat, offal, and meat products showed higher contamination levels, which is normal when investigating lipophilic POPs. These levels of occurrence are similar to more recently reported literature levels but perhaps lower, relative to historic data. This is not unexpected, given the restrictions/limitations on these chemicals within the United Kingdom and Western Europe. The estimated human exposure to population groups through dietary intake is correspondingly low, and based on current toxicological knowledge, the levels in the examined samples do not suggest a cause for health concern. The data also provide a current baseline for HCBD, PCBz, and PCP, and update existing data for PCN and HCB occurrence in foods.

HRGC-HRMS measurements were performed on a Waters Autospec Ultima High mass spectrometer (EI mode) coupled to a Hewlett-Packard 6890 N gas chromatograph fitted with a 60 m x 0.25 mm i.d. J&W DB-5 MS fused silica capillary column (2.25 µm thickness).

Chang (75) examined residual concentrations in water, sediments, and fishes as well as the association between the health risks of OCPs and fish consumption in the Taiwanese population. Various water and sediment samples from Taiwanese aquaculture and fish samples from different sources were collected and analyzed through gas chromatography tandem mass spectrometry (Agilent JW Scientific VF-5MS capillary column – 0.25 33 × 30 cm × 0.25 µm) to determine the concentrations of 20 OCPs, among which were hexachlorobenzene; alpha-hexachlorocyclohexane; beta-hexachlorocyclohexane, and pentachlorobenzene. None of the analyzed samples was positive for OCP contamination, suggesting no new input pollution from the land through washing into Taiwanese aquaculture environments. However, OCP residues were detected in fishes caught along the coast, namely, skipjack tuna and bigeye barracuda, and in imported fishes, such as codfish and salmon. The risk was assessed in terms of the estimated daily intake (EDI) for potential adverse indices; the EDI of OCP residues was lower than 1% of the acceptable daily intake established by the Food and Agriculture Organization of the United Nations and World Health Organization. The assessed risk was negligible and considered to be at a safe level, suggesting no association between fish consumption and risks to human health in Taiwan.

A rapid and simple analytical method for the determination of ten chlorinated priority substances (hexachloro-1,3-butadiene, pentachlorobenzene, hexachlorobenzene, hexachlorocyclohexane isomers, heptachlor, and heptachlor epoxides) in fish samples using QuEChERS extraction, dual dispersive solid-phase extraction (dSPE) clean up, and GC analysis was developed (76). For the extraction, two published extraction/partitioning procedures were evaluated, and the recoveries obtained for the analytes (in range 54–98% with RSDs ≤15%) were in favor of the conventional QuEChERS method. The use of the dual dSPE clean up yields cleaner extracts than in the case of single dSPE, which enables the use of ECD for the detection of the analytes and simplifies the maintenance of the GC system. The method was optimized using homogenates of chub fish that is frequently sampled for monitoring purposes. The linearity of the method was evaluated using matrix-matched calibration curves (in the range 2–50 µg kg^{-1}), and correlation coefficients (r^2) in the range 0.9927–0.9992 and RSDs of the relative response factors (RRF) below the value of 20% were achieved. LODs ranged from 0.5 to 1.1 µg kg^{-1}, while LOQs ranged from 1.5 to 3.5 µg kg^{-1}. The accuracy of the method was verified by the analysis of the NIST standard reference material SRM 1946 (Lake Superior Fish Tissue), and most of the analytes of interest presented good agreement with the certified values. An Agilent Technologies 7890A GC system with a split/splitless injector and a microelectron capture detector (µECD) was used for the analysis of fish homogenate extracts. The chromatograph was equipped with an HP-5 capillary column (30 m × 0.32 mm I.D. × 0.25 µm film thickness).

7.8 HEXACHLOROBUTADIENE

One of the primary applications of hexachlorobutadiene is as a solvent for chlorine.

Hexachlorobutadiene (HCBD) is a potential persistent organic pollutant that has been found in abiotic environments and organisms. However, information on HCBD in soils and its accumulation in terrestrial food chains is scarce. The accumulation of HCBD in soils, plants, and terrestrial fauna in a typical agricultural area in Eastern China was investigated, and comparisons with organochlorine pesticides were drawn (77). The HCBD concentrations in soils were <0.02–3.1 ng/g dry weight, which were similar to α-endosulfan concentrations but much lower than the concentrations of some other OCPs. The HCBD soil-plant accumulation factors, 8.5–38.1, were similar to those of o,p'-DDT and higher than those of HCHs and p,p'-DDT, indicating that HCBD is strongly bioaccumulated by rice and vegetables. HCBD concentrations of 1.3–8.2 ng/g lipid weight were found in herbivorous insects, earthworms, and Chinese toads. The biomagnification factor, the ratio between the lipid-normalized concentrations in the predator and the prey, was found to be 0.16–0.64 for different food chains of Chinese toads, so HCBD was found not to biomagnify, which is in contrast with OCPs.

The quantitative analysis of HCBD and OCPs was carried out using an Agilent gas chromatograph 7890 with a HP column and a ^{63}Ni-ECD detector.

The contamination of organochlorine pesticides and hexachlorobutadiene in soils, plants, and terrestrial fauna from a former pesticide-producing area in Southwest China was investigated (78). High levels of OCP residues (ΣHCHs and ΣDDTs of 3.89–13,386 and 23.3–11,186 ng g^{-1} dw, respectively) were found in the soils within the producing factory, indicating that the former pesticide factory brownfield site poses a high environment risk and that effective soil remediation is needed. The OCP concentrations in soils surrounding the pesticide factory were 1–2 orders of magnitude higher than in the other agricultural areas from China. In the study area, the concentrations of HCBD were

<0.02–5.59, 0.03–24.6 ng g^{-1} dw, and 1.65–3.80 ng g^{-1} lw in soils, plants, and animals, respectively, which were relatively low and not sufficient to cause observable adverse biological effects.

Organochlorine pesticides and HCBD were identified using an Agilent 7890 gas chromatograph equipped with a Ni63 electron capture detector and an Agilent HP-5 column (30 m 9 0.32 mm i.d., film thickness: 0.25 lm).

Since 2007, about 200 to 300 fish per year—generally roach (*Rutilus rutilus*), also a few bleak (*Alburnus alburnus*), and eels (*Anguilla anguilla*)—have been collected from a number of English river sites and stored at −80°C to build up a Fish Tissue Archive as a resource for the monitoring of pollutants (79). Some of the fish from the Fish Tissue Archive from the years 2007–2011 were analyzed for substances in current and proposed European legislation regarding environmental quality standards (EQS) in biota. The legacy fungicide hexachlorobenzene (HCB) was below the EQS of 10 μg/kg in all fish analyzed, with a maximum of 6 μg/kg in some eels. The legacy solvent hexachlorobutadiene (HCBD) was below the EQS of 55 μg/kg, being < 0.2 μg/kg in all samples where it was measured. The sums of six polybrominated diphenyl ethers (PBDEs) were several orders of magnitude higher than the new proposed 0.0085 μg/kg biota EQS. This study showed that the regular collection and analysis of whole body homogenate samples of relatively small native pelagic fish is suitable for the monitoring of contaminants capable of bioaccumulation.

The extracts were analyzed by gas chromatography–mass spectrometry (Thermo Trace, electron impact mode, single ion monitoring, source temperature 250°C, splitless injection. Columns: 50 m CPSil8, 0.25mm ID, 0.12 μm film (Varian) for HCB and HCBD).

The paper of Majoris (80) summarizes the validation strategy and the results obtained for the simultaneous determination of hexachlorobenzene and hexachlorobutadiene in fish tissue with a maximum of about 10% m/m fat content using a GC–IDMS technique. The method is applicable for the determination of HCB and HCBD at trace levels in different kinds of fish tissue samples in accordance with the requirements of the EU Directive 2008/105/EC establishing Environmental Quality Standard (EQS) levels for biota in aquatic ecosystems (10 ng/g for HCB and 55 ng/g for HCBD).

The method validation aimed to assess performance parameters such as linearity, limit of detection/limit of quantification (LOD/LOQ), trueness, selectivity, intermediate precision, repeatability, stability of the extracts, and robustness. The validation experiments have been performed by using uncontaminated fish tissue. Trueness was evaluated by using a certified reference material (NIST SRM 1947) (where applicable) and by the standard addition method. Very good linear signal-concentration curves were obtained for both analytes over the whole range of calibration. The repeatability and the intermediate precision of the method, expressed as relative standard deviation (RSD) and calculated at the EQS level, were estimated to be below 3% both for HCB and HCBD. The limits of quantification were 3.7 ng/g for HCB and 15.7 ng/g for HCBD in the fish.

HCBD's toxicity and propensity for long-range transport means there is special concern for its potential impacts on Arctic ecosystems. The review of Balmer et al. (81) comprehensively summarizes all available information of the occurrence of HCBD in the Arctic environment, including its atmospheric, terrestrial, freshwater, and marine ecosystems and biota. Overall, reports of HCBD in Arctic environmental media are scarce. HCBD has been measured in Arctic air collected from monitoring stations in Finland and Canada, yet there is a dearth of data for other abiotic matrices (i.e., soils, sediments, glacier ice, freshwaters, and seawater). Low HCBD concentrations have been measured in Arctic terrestrial and marine biota, which is consistent with laboratory studies that indicate that HCBD has the potential to bioaccumulate, but not to biomagnify. Available data for Arctic biota suggest that terrestrial birds and mammals and seabirds have comparatively higher HCBD concentrations than fish and marine mammals, warranting additional research. Although spatial and temporal trends in HCBD concentrations in the Arctic are currently limited, future monitoring of HCBD in the Arctic will be important for assessing the impact of global regulations newly imposed by the Stockholm Convention on POPs.

The paper of Surma-Zadora (82) focuses upon the use of semipermeable membranes (SPM) as a clean up method for the determination of 4,4′-DDD, 4,4′-DDE, 4,4′-DDT, Aldrin, Dieldrin, Isodrin, Lindane (γ -HCH), 1,2,4-trichlorobenzene (1,2,4-TCB), 1,2,3-trichlorobenzene (1,2,3-TCB), 1,2,3,4-tetrachlorobenzene (1,2,3,4-TCB), 1,2,4,5-tetrachlorobenzene (1,2,4,5-TCB), Pentachlorobenzene (PeCBz), Hexachlorobenzene, and Hexachlorobutadiene in high fat food samples. Pork fat, beef fat, butter, egg yolks, and chocolate were all used as high fat food samples. The procedure consists of three steps: the first is dialysis in an SPM tube, using n-hexane as an external solvent. The second step is a clean up procedure using a silica gel column, and the third step is GC/ECD analysis. This experiment shows that recovery values obtained for individual compounds were in the range of 55–100%. The conclusion drawn is that the SPM technique is an efficient method of preparation of high fat food samples for the determination of POPs by GC/ECD methods.

7.9 SHORT-CHAIN CHLORINATED PARAFFINS

A novel solid-phase microextraction (SPME) method coupled to gas chromatography with electron capture detection (GC–ECD) was developed as an alternative to liquid–liquid and solid-phase extraction for the analysis of short-chain chlorinated paraffins (SCCPs) in water samples (83). The extraction efficiency of five different commercially available fibers was evaluated and the 100-μm polydimethylsiloxane coating was the most suitable for the absorption of the SCCPs. Optimization of several SPME parameters, such as extraction time and temperature, ionic strength, and desorption time, was performed. Quality parameters were established using Milli-Q, tap water, and river water. Linearity ranged between 0.06 and 6 μg l^{-1} for spiked Milli-Q water and between 0.6 and 6 μg l^{-1} for natural waters. The precision of the SPME–GC–ECD method for the three aqueous matrices was similar and gave relative standard deviations (RSD) between 12 and 14%. The limit of detection (LOD) was 0.02 μg l^{-1} for Milli-Q water and 0.3 μg l^{-1} for both tap water and river water.

Polychlorinated naphthalenes (PCNs), short-chain chlorinated paraffins (SCCPs), and polychlorinated biphenyls (PCBs) were analyzed in marine sediment samples collected from the coastal area of Barcelona (Spain) and near a submarine emissary coming from a waste water treatment plant located at the mouth of the Besòs River (Barcelona) (84). An integrated sample treatment based on Soxhlet extraction followed by a simple clean up with Florisil and graphitized carbon cartridge was employed. Gas chromatography coupled to ion-trap tandem mass spectrometry (GC–MS/MS) and gas chromatography–mass spectrometry in electron capture negative ionization mode were used for PCN and SCCP determinations, respectively, while for PCB analysis gas chromatography with electron capture detection (GC–ECD) was used. The method developed provided low limits of detection (0.001–0.003 ng g^{-1} dry weight (dw) for PCNs, 1.8 ng g^{-1} for SCCPs, and 0.006–0.014 ng g^{-1} dw for PCBs) and good run-to-run precisions (lower than RSD 8%) for the analysis of sediment samples.

Chlorinated paraffins (CPs) are industrially produced in large quantities in the Liaohe River Basin. Short-chain CPs (SCCPs) were analyzed in sediments, paddy soils, and upland soils from the Liaohe River Basin, with concentrations ranging from 39.8 to 480.3 ng/g dry weight (85). A decreasing trend in SCCP concentrations was found with increasing distance from the cities, suggesting that local industrial activity was the major source of SCCP contamination. The congener group profiles showed that the relative abundances of shorter chain and lower chlorinated CP congeners (C_{10}–CPs with 5 or 6 chlorine atoms) in soils in rural areas were higher than in sites near cities, which demonstrated that long-range atmospheric transportation could be the major transport pathway. Environmental degradation of SCCPs might occur, where higher chlorinated congeners could dechlorinate to form the lower chlorinated congeners.

Nine bivalve and two gastropod species were collected in 2009 to evaluate the spatial distributions and potential factors influencing the bioaccumulation of SCCPs in mollusks in the Chinese Bohai Sea (86). The concentrations of Σ SCCPs in the mollusks were in the range 64.9–5510 ng/g (dry weight) with an average chlorine content of 61.1%. C_{10} and C_{11} were the predominant homologue groups of SCCPs, which accounted for about 29.7% and 34.9% of Σ SCCPs, respectively. Six and seven chlorinated substituents were the main congener groups. *Mya arenaria* (Mya), *Mactra veneriformis* (Mac), and *Crassostrea talienwhanensis* (Oyster, Ost) had higher average concentrations of SCCPs than other species, implying that these bivalves could be used as sentinels to indicate SCCPs contamination in this coastal region. A significant positive linear relationship was found between SCCP concentrations and lipid content of the mollusks, whereas the lipid-normalized SCCP concentrations were negatively linear-related to the trophic levels (TL), which implied that SCCPs did not show biomagnification in mollusks in this region.

The analytical determination of SCCPs is very challenging. Although there is at present no fully validated measurement procedure that might be applied in routine monitoring, the European Union Water Framework Directive (WFD) has required regular monitoring of this class of compounds at river-basin scale since 2007.

To assess the *status quo* of the analysis of SCCPs in relation to the requirements of the WFD, an interlaboratory comparison on the quantification of SCCPs in an extract of an industrial soil was organized (87). Six laboratories participated in the exercise using three different techniques (i.e., gas chromatography coupled to mass spectrometry in electron-capture negative ionization mode, GC with atomic emission detection, and carbon-skeleton GC-MS). The results reported were in the range 8.5–3200 mg/L. This confirms that reliable quantification of SCCPs is still very difficult to achieve and that the comparability of SCCP data reported to the European Commission is at least questionable.

7.10 POLYCHLORINATED NAPHTALENES

Polychlorinated naphtalenes (PCNs) were used in insulating coatings for electrical wires, as well as other applications.

Recent advances in measurement techniques have allowed a greater characterization of PCN occurrence, yielding more specific data including individual PCN congener concentrations (88). Emerging data on food shows widespread occurrence in most commonly consumed foods from different parts of the world. Concurrently, toxicological studies have also allowed a greater insight into the potencies of some congeners, a number of which are known to elicit potent, aryl hydrocarbon receptor (AhR) mediated responses, often referred to as dioxin-like toxicity. The dietary pathway is widely recognized as the most likely route to nonoccupational human exposure. In Table 7.2, some of the more recent findings on PCN occurrence in food, biota, and human tissues are shown.

Concentrations and patterns of 75 PCN congeners in feed raw materials of animal and plant origin were investigated (106). Six types of feed raw materials of animal origin and three types of feed raw materials of plant origin from China were collected in 2016. The total concentrations of PCNs in the collected materials ranged from 147 to 1009 ng kg-1, with the highest occurring in fish meal. The mean PCNs concentration in feed raw materials of animal origin (551 ng kg^{-1}) was higher than in those of plant origin (294 ng kg^{-1}). Additionally, lower chlorinated PCNs were the main homologues in raw feed materials, while Di-CNs were the predominant homologues in all samples (mean: 53%), followed by tri-CNs (mean: 28%). The most abundant congeners were CN5/7 and 24/14. Additionally, the toxicity equivalencies (TEQs) of PCNs in the feed raw materials ranged from 0.010 to 0.046 ng TEQ kg-1, with the highest TEQ concentrations of PCNs detected in gluten meal. Together, CN5/7, 66/67, 65/70, and 73 contributed approximately 64% of the total PCN TEQs in feed raw materials. Concentrations of polychlorinated dibenzo-p-dioxins (PCDDs), polychlorinated dibenzofurans (PCDFs), and polychlorinated biphenyls (PCBs) in the feed raw materials were detected to compare the TEQ distribution of those dioxin-like compounds. The mass concentrations of PCNs were 1–3 orders of magnitude higher than those of PCDD, PCDFs, and PCBs, while the TEQ concentrations of PCNs contributed 2.0%–6.5% of the total TEQs of PCNs, PCDDs, PCDFs and PCBs in the feed raw materials.

TABLE 7.2
A Summary of Recent PCN Data on Concentrations in Food

Food Type	Number of Samples	Sampling Period	Geographical Origin	Report/ Quantitation Type	PCN Concentration Sum—ng/kg Whole (wet) Weight	Notes	Reference
Fish	22	1996–97	Great Lakes and inland Michigan N. America	Homologue totals (tri-octa)/ individual congeners	19–31, 400	Nine species, carp, trout, pike, etc.	89, 90
	16	2000	Catalonia, Spain	Homologue totals tetra-octa	39 (average)	Retail (including shellfish)	91
	52	2001–03	Baltic sea	Sum of 16, tetra-octa congeners	20–450	Baltic herring and salmon	92
	45	2001–03	Remote lakes Finland	Sum of 16, tetra-octa congeners	2–66	Nine freshwater species	92
	53	2001–03	Baltic sea	Sum of 16, tetra-octa congeners	1–190	Baltic fish species (various)	92
	8	2003–04	East and South China Sea	Homologue totals tri-octa	137–545 (lipid wt., average)	Four species, retail	93
	27	2005	Catalonia, Spain	Homologue totals tetra-octa	13–227	Retail, composites	94
	9	2006	Catalonia, Spain	Homologue totals tetra-octa	12.8–226.9	Nine retail species	95
	9 (max)	2006	China	Homologue totals/ congeners	225–640 (lipid wt., average)	Pooles samples coastal water	96
	12	2007	United Kingdom	Sum of 12 penta-octa congeners*	0.73–37.3	Wild and farmed, retail	97
	12	2007–08	Ireland	Sum of 12 penta-octa congeners	2.08–59.3	Wild and farmed species	98
	16	2008	Scotland	Sum of 12 penta-octa congeners	2.5–103	Five freshwater species	99
	32	2008	Scotland	Sum of 12 penta-octa congeners	0.3–63	Eighteen deep-sea species	99
	46	2008–09	UK fresh waters	Sum of 12 penta-octa congeners	1.1–1198	Fifteen freshwater species	100
	31	2013–14	UK marine waters	Sum of 12 penta- octa congeners	0.7–265	Seven common marine species	101

(Continued)

TABLE 7.2 (Continued)
A Summary of Recent PCN Data on Concentrations in Food

Food Type	Number of Samples	Sampling Period	Geographical Origin	Report/ Quantitation Type	PCN Concentration Sum—ng/kg Whole (wet) Weight	Notes	Reference
Shellfish	15	2003–04	East and South China Sea	Homologue totals tri-octa	94–1300 (lipid wt., average)	Seven shellfish, two cephalopod species	93
	15	2005	Catalonia, Spain	Homologue totals tetra-octa	3–22	Retail, composites	94
	5	2006	Catalonia, Spain	Homologue totals tetra-octa	2.7–21.8	Retail, composites	95
	5	2007–08	Ireland	Sum of 12 penta-octa congeners	0.18–2.34	Oysters	98
	5	2008	Scotland	Sum of 12 penta-octa congeners	0.8–6.5	Mussels	99
Eggs and poultry	4	2000	Catalonia, Spain	Homologue totals tetra-octa	23 (average)	Retail eggs	91
	1	2006	Catalonia, Spain	Homologue totals tetra-octa	1.7–4.3	Chicken, retail eggs, composites	95
	9 (max)	2006	China	Homologue totals tetra-octa	43.8 (lip dwt., average)	Pooled samples coastal waters	96
	9	2007	United Kingdom	Sum of 12 penta-octa congeners	0.48–8.29	Retail	97
	15	2007–08	Ireland	Sum of 12 penta-octa congeners	0.23–2.22	Composite eggs, including barn, free range etc.	98
Milk	4	2000	Catalonia, Spain	Homologue totals tetra-octa	0.4 (average)	Retail, composites	91
	2	2006	Catalonia, Spain	Homologue totals tetra-octa	0.5–1.2	Retail, composites	95
	2	2007	United Kingdom	Sum of 12 penta-octa congeners	0.19–0.22	Retail	97
	15	2007–08	Ireland	Sum of 12 penta-octa congeners	0.09–0.38	Composite from farm	98
Dairy products	4	2000	Catalonia, Spain	Homologue totals tetra-octa	36 (average)	Retail, composites	91

TABLE 7.2 (Continued)
A Summary of Recent PCN Data on Concentrations in Food

Food Type	Number of Samples	Sampling Period	Geographical Origin	Report/Quantitation Type	PCN Concentration Sum—ng/kg Whole (wet) Weight	Notes	Reference
	2	2006	Catalonia, Spain	Homologue totals tetra-octa	0.8–22.7	Retail, composites	95
	7	2007	United Kingdom	Sum of 12 penta-octa congeners	0.52–6.09	Retail	97
	6	2007–08	Ireland	Sum of 12 penta-octa congeners	0.68–3.13	Retail	98
	7 + 7	-	China		5.6–103 (cheese) 5.0–199 (butter)	-	102
Meat	30	2000	Catalonia, Spain	Homologue totals tetra-octa	18 (average)	Retail, composites, including meat products	90
	9	2006	Catalonia, Spain	Homologue totals tetra-octa	1.8–5.8	Retail, composites, including meat products	95
	4	2007	United Kingdom	Sum of 12 penta-octa congeners	0.19–5.69	Retail meat	97
	1	2007	Tibet-Qinghai Plateau, China	Tri- to octa individual, congeners	13.6	Muscles meat of yak	103
Animal fat	8	2001	Japan	Tri- to octa individual, congeners	125–308	Fats—pork and chicken	109
	21	2007–08	Ireland	Sum of 12 penta-octa congeners	1.6–13.4	Fats—beef, pork, lamb, avian	98
	1	2007	Tibet-Qinghai Plateau, China	Tri- to octa individual, congeners	49.1	Marbling—intramuscular fat of yak	71
Meat products and offal	5	2007	United Kingdom	Sum of 12 penta-octa congeners	0.34–4.88	Retail	97
	12	2007–08	Ireland	Sum of 12 penta-octa congeners	0.18–4.26	Retail liver, 5 species	98
	2	2007–08	Ireland	Sum of 12 penta-octa congeners	0.13–0.54	Processed meat	98
Fats and oils	6	2000	Catalonia, Spain	Homologue totals tetra-octa	447 (average)	Retail, composites	90
	4	2006	Catalonia, Spain	Homologue totals tetra-octa	20.4–22.3	Retail, composites	95

(Continued)

TABLE 7.2 (Continued)
A Summary of Recent PCN Data on Concentrations in Food

Food Type	Number of Samples	Sampling Period	Geographical Origin	Report/ Quantitation Type	PCN Concentration Sum—ng/kg Whole (wet) Weight	Notes	Reference
Cereals and bread	12	2000	Catalonia, Spain	Homologue totals tetra-octa	3–71 (average)	Retail pulses, cereals, composites	95
	6	2006	Catalonia, Spain	Homologue totals tetra-octa	5.3–15.1	Retail, composites, including pulses	95
	1	2007	United Kingdom	Sum of 12 penta-octa congeners	0.28	Retail	97
	4	2007–08	Ireland	Sum of 12 penta-octa congeners	0.25–0.77	Retail	98
	28 + 28	2013	Pakistan	Sum of 39 congeners	20–210 (wheat); 20–1210 (rice)	Agricultural land along river tributaries	105
Fruit, vegetables	32	2000	Catalonia, Spain	Homologue totals tetra-octa	0.7–4 (average)	Retail, composites	91
	9	2006	Catalonia, Spain	Homologue totals tetra-octa	0.9–3.9	Retail, composites	95
	2	2007	United Kingdom	Sum of 12 penta-octa congeners	0.15–0.16	Retail	97
	8	2007–08	Ireland	Sum of 12 penta-octa congeners	0.16–2.84	Retail	98

* PCNs 52/60, 53, 66/67, 64/68, 69, 71/72, 74 and 75.

The concentrations and distribution of polychlorinated naphthalenes in the whole blood of eight typical terrestrial meat animals (chicken, duck, rabbit, pig, cattle, sheep, horse, and donkey) consumed daily in our life were investigated (107). The total concentrations (on a liquid volume basis) of PCNs were in a range from 305 to 987 pg/L. Donkey blood contained the highest PCN concentrations. Mono-CNs were the dominant homolog group, accounting for 38%–71% PCNs. Apart from the mono-CNs and tri-CNs homolog groups, two hepta-CNs (mean: 9.5%) contributed most, followed by tetra-CNs (mean: 6.5%). The congeners CN1, 5/7, 24/14, 27/30, 52/60, 66/67, and 73 were the most abundant congeners or congener groups. The highest toxicity equivalencies (TEQs) were observed in cattle blood (117.4 fg TEQ/L) then chicken blood (117.1 fg TEQ/L). CN73 contributed 65% to total TEQs, followed by CN70 (20%) and CN66/67 (14%). The dietary intakes of PCNs were also estimated. Chicken meat, which forms the second largest component of meat product consumption in China, contributed most to the total TEQs (61%), followed by beef (27%) and pork (5.9%). The consumption of chicken might pose the highest risk from exposure to PCNs than other types of meat to populations who prefer to eat chicken meat.

Persistent Organic Pollutants

The PCNs were analyzed by a gas chromatograph (GC) coupled to a DFS mass spectrometer (Thermo Fisher Scientific, Hudson, NH, USA) using an electron impact (EI) source.

Relatively limited knowledge exists on the abundance of PCNs in the edible portion of a variety of Great Lakes fish to aid in understanding their potential risk to human consumers (108). Levels, patterns, trends, and significance of PCNs in a total 470 fillet samples of 18 fish species collected from the Canadian waters of the Great Lakes between 2006 and 2013 were studied. A limited comparison of fillet and whole-body concentrations in carp and bullhead was also conducted. The $\sum$ PCN ranged from 0.006–6.7 ng/g wet weight (ww) and 0.15–190 ng/g lipid weight (lw) with the dominant congeners being PCN-52/60 (34%), PCN-42 (21%), and PCN-66/67 (15%). The concentrations spatially varied in the order of the Detroit River > Lakes Erie > Ontario > Huron > Superior. PCN-66/67 was the dominating congener contributing on average 76–80% of toxic equivalent concentration (TEQ_{PCN}). Contribution of TEQ_{PCN} to TEQ_{Total} ($TEQ_{Dioxins + Furans + dioxin-likePCBs + PCNs}$) was mostly < 15%, especially at higher TEQ_{Total}, and PCB-126 remains the major congener contributing to TEQ_{Total}. The congener pattern suggests that impurities in PCB formulations and thereby historical PCB contamination, instead of unintentional releases from industrial thermal processes, could be an important source of PCNs in Great Lakes fish. A limited temporal change analysis indicated declines in the levels of PCN-66/67 between 2006 and 2012, complemented by previously reported decrease in PCNs in Lake Ontario Lake Trout between 1979 and 2004. The whole body concentrations were 1.4–3.2 fold higher than the corresponding fillets of carp and bullhead. Overall, the study results suggest that only targeted monitoring of PCNs in Great Lakes fish, especially at the Detroit River, Lake Erie, and Lake Ontario, is necessary to assess continued future improvements of this group of contaminants of concern.

HRGC-HRMS measurements were performed on a Waters Autospec Ultima High mass spectrometer(EI mode) coupled to a Hewlett-Packard 6890 N gas chromatograph fitted with an RTX-5 column.

Polychlorinated naphthalenes were measured in 33 seafood species including fish, mollusks, and crustaceans purchased from local markets in five Korean cities between 2012 and 2013 (109). Five samples were collected from each species for the measurements. Thirty-seven PCN congeners from tetra-CN to octa-CN were measured. Octa-CN (octachloronaphthalene) was not detected in any of the samples. Tetra-CN and penta-CN were the predominant homologues of PCNs in seafood samples with PCN 51 and PCN 52/60 being the most abundant congeners in the samples. Total PCNs concentrations and their corresponding dioxin-like toxic equivalent (TEQ) values ranged from non-detection (ND) to 110 pg/g on a wet weight (ww) basis and from ND to 0.14 pg-TEQ/g ww, respectively. The estimated daily intake of total PCNs based on an absolute content and TEQ potency were estimated for the Koreans to be 570 pg/day and 0.44 pg-TEQ/day, respectively. However, the estimated TEQ value of PCNs intake from seafood represented only a small fraction (3.0%) of the total TEQ intake from consumption of seafood in Korean population. This is the first report to exhibit the presence of PCNs in seafood samples collected from local markets in Korea and their intake by general population.

The analysis method was HRMS (DB5-MS capillary column (60 M × 0.25 mm i.d. × 0.25 μm film thickness).

7.11 PFASs (PERFLUOROALKYL SUBSTANCES)

A lot of papers on PFASs have been published in recent years.

PFOA (perfluorooctanoate) has widespread applications. PFOA is used as a water and oil in different products. The compound is used in insulators for electric wires, firefighting foam, and outdoor clothing.

As a salt, its dominant use is as an emulsifier for the emulsion polymerization of fluoropolymers such as PTFE, polyvinylidene fluoride, and fluoroelastomers.

Perfluoroalkylated substances (PFAS) is the collective name for a vast group of fluorinated compounds, including oligomers and polymers, which consist of neutral and anionic surface active compounds with high thermal, chemical, and biological inertness (110). Perfluorinated compounds are generally hydrophobic but also lipophobic and will therefore not accumulate in fatty tissues as is usually the case with other persistent halogenated compounds. An important subset is the (per) fluorinated organic surfactants, to which perfluorooctane sulfonate (PFOS) and perfluorooctanoic acid (PFOA) belong. The analytical detection method of choice for PFOS and PFOA is currently liquid chromatography-mass spectrometry/mass spectrometry (LC-MS/MS), whereas both LC-MS/MS and gas chromatography-mass spectrometry (GC-MS) can be used for the determination of precursors of PFOS and PFOA.

Concentrations of perfluorinated compounds in food and the dietary intake of perfluorooctane sulfonate (PFOS) and perfluorooctanoate (PFOA) in The Netherlands were checked (111). The concentrations of perfluorinated compounds in food were analyzed in pooled samples of foodstuffs randomly purchased in several Dutch retail store chains with nation-wide coverage. The concentrations analyzed for PFOS and PFOA were used to assess the exposure to these compounds in The Netherlands. As concentrations in drinking water in The Netherlands were missing for these compounds, conservative default concentrations of 7 pg/g for PFOS and 9 pg/g for PFOA, as reported by European Food Safety Authority, were used in the exposure assessment. In food, 6 out of 14 analyzed perfluorinated compounds could be quantified in the majority of the food categories (perfluoroheptanoic acid (PFHpA), PFOA, perfluorononanoic acid (PFNA), perfluorodecanoic acid (PFDA), perfluoro-1-hexanesulfonate (PFHxS), and PFOS). The highest concentration of the sum of these six compounds was found in crustaceans (825 pg/g product, PFOS: 582 pg/g product) and in lean fish (481 pg/g product, PFOS: 308 pg/g product). Lower concentrations were found in beef, fatty fish, flour, butter, eggs, and cheese (concentrations between 20 and 100 pg/g product; PFOS, 29–82 pg/g product) and milk, pork, bakery products, chicken, vegetable, and industrial oils (concentration lower than 10 pg/g product; PFOS not detected). The median long-term intake for PFOS was 0.3 ng/kg bw/day and for PFOA 0.2 ng/kg bw/day. The corresponding high level intakes (99th percentile) were 0.6 and 0.5 ng/kg bw/day, respectively. These intakes were well below the tolerable daily intake values of both compounds (PFOS, 150 ng/kg bw/day; PFOA, 1500 ng/kg bw/day). The intake calculations quantified the contribution of drinking water to the PFOS and PFOA intake in The Netherlands. Important contributors of PFOA intake were vegetables/fruit and flour. Milk, beef, and lean fish were important contributors of PFOS intake.

The levels of 21 perfluoroalkyl substances (PFASs) in 283 food items (38 from Brazil, 35 from Saudi Arabia, 174 from Spain, and 36 from Serbia) among the most widely consumed foodstuffs in these geographical areas were assessed (112). These countries were chosen as representatives of the diet in South America, Western Asia, Mediterranean countries, and South-Eastern Europe. The analysis of foodstuffs was carried out by turbulent flow chromatography (TFC) combined with liquid chromatography with triple quadrupole mass spectrometry (LC-QqQ-MS) using electrospray ionization (ESI) in negative mode. The analytical method was validated for the analysis of different foodstuff classes (cereals, fish, fruit, milk, ready-to-eat foods, oil, and meat). The analytical parameters of the method fulfill the requirements specified in the Commission Recommendation 2010/161/EU. Recovery rates were in the range between 70% and 120%. For all the selected matrices, the method limits of detection (MLOD) and the method limits of quantification (MLOQ) were in the range of 5 to 650 pg/g and 17 to 2000 pg/g, respectively. In general trends, the concentrations of PFASs were in the pg/g or pg/mL levels. The more frequently detected compounds were perfluorooctane sulfonic acid (PFOS), perfluorooctanoic acid (PFOA), and perfluorobutanoic acid (PFBA). The prevalence of the eight-carbon chain compounds in biota indicates the high stability and bioaccumulation potential of these compounds. But, at the same time, the high frequency of the shorter chain compounds is also an indication of the use of replacement compounds in the new fluorinated materials. When comparing the compounds profile and their relative abundances in the samples from diverse origin, differences were identified. However, in absolute amounts of total PFASs no large differences were found between the studied countries. Fish and seafood were identified as the major PFASs contributors to the diet in all the countries. The total sum of PFASs in fresh fish and seafood was in the range from the MLOQ to 28 ng/g ww. According to the FAO-WHO diets composition, the daily intake (DI) of PFASs was calculated for various age and gender groups in the different diets. The total PFASs food intake was estimated to be between 2300 and 3800 ng/person per day for the different diets. Finally, the risk intake (RI) was calculated for selected relevant compounds. The results have indicated that by far in no case the tolerable daily intake (TDI) (150, 1500, 50,000, 1,000,000, 150, 1500 ng/kg body weight, for perfluorohexanesulfonate (PFHxS), fluorotelomer alcohol (FTOH), perfluorobutanesulfonic acid (PFBS), perfluorobutanoic acid (PFBA), PFOS, and PFOA, respectively) was exceeded.

To comprise the future requirements to detect low levels of perfluoroalkyl acids (PFAAs) including branched and linear perfluorooctane sulfonic acid, perfluorooctanoic acid, and perfluorohexane sulfonic acid (PFHxS) in food items, analytical methods for their determination in six different food matrices (cow milk, butter, chicken egg, chicken meat, beef, and fish) were developed and validated (113). After optimization, the method applied on foods of animal origin includes alkaline digestion, extraction, and clean up with solid phase extraction and adsorption on granular carbon where necessary. The method was shown effective to eliminate taurodeoxycholic acid (TDC), a bile acid that is an endogenous interference compound in egg samples causing ionization suppression and false positive result for PFOS when 499 > 80 transition was used for quantification. The validation was performed and resulted in recoveries >70% for all three PFAAs, the limits of quantification (LOQs) in all matrices were 3.1 pg g^{-1}, 3.4 pg g^{-1}, and 4.9 pg g^{-1} for PFHxS, PFOA, and L-PFOS, respectively. The optimized method was successfully applied to 53 food samples from the Swedish market and from developing countries. PFOS and PFOA were detected in all samples, and PFHxS was detected in 80% of the samples. With this method, concentrations in the low pg g^{-1} range in food samples of animal origin were quantified including the branched PFOS isomers. This method can be applied to enforce potential future limit values for PFOS and PFOA as a consequence of the recent European Food Safety Authority (EFSA) recommendation where the tolerable intakes have been drastically lowered.

The determination of perfluorooctanoic acid and perfluorooctanesulfonate in food and beverages sold in Turkey was carried out using liquid chromatography–tandem mass spectrometry (LC-MS/MS) (114). A total of 123 samples of selected food and beverages such as fish, meat, offal, egg, cracker, chips, cake, chocolate, vegetable, milk, and juice were examined. The highest PFOA concentrations were determined in cow meat (5.15 ng g^{-1}), cow kidney (5.65 ng g^{-1}), cow spleen (5.06 ng g^{-1}), and chicken liver (5.02 ng g^{-1}). The highest PFOS levels were found in horse mackerel (52.43 ng g^{-1}), pike-perch (45.87 ng g^{-1}), sardine (42.83 ng g^{-1}), and black cod (41.33 ng g^{-1}). Fish was found to be major source of the PFOS intake, while meat and offal were found to be major sources of the PFOA intake.

Aquatic ecosystems represent the final reservoir for PFASs due to their high affinity for sedimentary and living organic matter. Recommendation 161/2010/EU stated that the Member States should carry out analysis of perfluoroalkylated substances in order to detect the presence of PFOS and PFOA by making use of an analytical method that has been proven to generate reliable results. A LC-MS/MS method has been developed for detecting PFOS and PFOA in fish and cereals (115). Validation of this method resulted in limits of quantification (LOQs) varying from 0.50 µg/kg for PFOA to 0.70 µg/kg for PFOS, fulfilling recoveries (95–109%), and showing a good linearity (R2 = 0.99). Precision varied from 4.43% to 8.97% of repeatability for PFOA and from 6.30% to 12.33% of repeatability for PFOS.

Fruit and vegetables play a major role in human nutrition due to richness of nutrients, dietary fiber, and phytochemicals. As dietary intake is identified as one of the dominant exposure pathways to perfluoroalkyl substances (PFASs), a cross-sectional study involving determination of their levels in food of plant origin has been conducted (116). Locally grown and imported fruit and vegetable samples collected in 2016 were inspected for 10 perfluoroalkyl acids using QuEChERS as sample pre-treatment procedure followed by micro-HPLC-MS/MS. Three of 10 target analytes, perfluorobutanoic acid (PFBA), perfluorooctanoic acid, and perfluorooctane sulfonate, were quantitatively determined. The detection frequency for PFASs across the 55 samples analyzed was less than 10%. The major contributor of the total PFASs concentration in the investigated group was PFBA for which the concentration, reported only for banana, apple, and orange samples, was 50.740 ng g^{-1} ww. The most often detected compound was PFOA. The origin and growing region are possible factors with the potential to influence PFASs distribution profile and their levels in food.

7.12 POLYBROMINATED BIPHENYLS

Polybrominated biphenyls (PBBs), also called brominated biphenyls or polybromobiphenyls, are a group of manufactured chemicals that consist of polyhalogenated derivatives of a biphenyl core. Their chlorine analogs are the PCBs.

PBBs are used as flame retardants of the brominated flame retardant group.

Three types of commercial PBB mixtures were: hexabromobiphenyl (hexaBB), octabromobiphenyl (octaBB), and decabromobiphenyl (decaBB).

Two gas chromatography/mass spectrometry (GC/MS) methods for the determination of polybrominated biphenyls by isotope dilution analysis (IDA) using $^{13}C_{12}$-PBB 153 in the presence of polybrominated diphenyl ethers (PBDEs) were compared (117). Recovery of $^{13}C_{12}$-PBB 153 which was added to the extracted lipids before sample purification was commenced ranged from 88–117% (mean value 98.2±8.9%). Nevertheless, IDA analysis of PBBs using $^{13}C_{12}$-labelled congeners is limited by the potential co-elution of PBBs with polybrominated diphenyl ethers (PBDEs). The pair PBB 153 and BDE 154 was inspected since M$^+$ and [M−2Br]$^+$ ions of $^{13}C_{12}$-PBB 153 and BDE 154 were only separated by 4 u. Gas chromatography/electron ionization high-resolution mass spectrometry with selected ion monitoring (GC/EI-HRMS-SIM) was suitable when m/z 475.7449 and m/z 477.7429 were used for $^{13}C_{12}$-PBB 153 because they are below the monoisotopic peak of the [M−2Br]$^+$ fragment ion of hexaBDEs at m/z 479.7. Gas chromatography/electron capture negative ion tandem mass spectrometry selected reaction monitoring (GC/ECNI-MS/MS-SRM) measurements could be applied because $^{13}C_{12}$-PBB 153 and BDE 154 were separated by GC on a 25-m Factor Four CP-Sil 8MS column.

Comparative measurements with GC/EI-HRMS-SIM and GC/ECNI-MSMS-SRM were carried out with samples of Tasmanian devils from Tasmania (Australia), an endangered species due to a virus epidemic which has already proved fatal for half of the population. Both techniques verified concentrations of PBB 153 in the range 0.3–11 ng/g lipids with excellent agreement of the levels in all but two samples. The PBB residue pattern demonstrated that PBB pollution originated from the previous discharge with technical hexabromobiphenyl which is dominated by PBB 153. Other congeners such as PBB 132 and PBB 138 were detected in the Tasmanian devils but the proportions relative to PBB 153 were lower than in the technical product. Samples of healthy and affected Tasmanian devils showed no significant difference in the PBB pollution level. The PBB concentrations in the Tasmanian devils were significantly below those causing toxic effects. On the other hand, PBB concentrations were one level or even higher than PBDEs.

A fast and simple pressurized liquid extraction (PLE) method combined with gas chromatography coupled to ion trap tandem mass spectrometry (GC–ITMS-MS) for the determination of polybrominated biphenyls in fish samples was described (118). The method is based on a simultaneous extraction/clean up step to reduce analysis time and solvent consumption. The effect of several PLE operating conditions, such as solvent type, extraction temperature and time, number of cycles, and lipid retainer, was optimized to obtain maximum recovery of the analytes with the minimum presence of matrix-interfering compounds. The best conditions were obtained at 100°C with n-hexane using 15 g of silica modified with sulfuric acid (44%, w/w) as sorbent for lipid removal. Quality parameters of the GC–ITMS-MS method were established, achieving good linearity ($r > 0.998$), between 1 and 500 ng ml^{-1}, and low instrumental limits of detection (0.14–0.76 pg injected). For the whole method, limits of detection ranging from 0.03 to 0.16 ng g^{-1} wet weight and good precision (RSD < 16%) were obtained.

Individual eggs of six species of birds from Norway representing different food chains were analyzed for residues of polybrominated biphenyls (118). In all species, the residue pattern was dominated by hexaBBs. The dominating congeners were PBB 153, PBB 154, and PBB 155. Whereas PBB 153 is present in technical hexabromobiphenyl, PBB 154 and PBB 155 are formed by the reductive debromination of decabromobiphenyl. This was evidenced by the detection of several heptaBBs and octaBBs all of which are typical degradation intermediates of PBB 209. Hepta- and octaBBs were more than one order of magnitude less abundant than the hexaBBs. The second most prevailing homologue group was pentaBBs. The most relevant pentabrominated isomers were PBB 99 and PBB 101. Concentrations of the three hexaBBs—PBB 153, PBB 154, and PBB 155—amounted to 1.3–13 ng/g wet weight or 3–23% of the contamination with polybrominated diphenyl ethers.

The applied method was GC with ECNI and MS/MS with SRM.

As part of a study to investigate the spatial distribution of these contaminants, data on polybrominated diphenlyethers (PBDEs) and polybrominated biphenyls were collected and analyzed by introducing a web-based resource which enables efficient spatial, species, and concentration level representations (119). Furthermore, hierarchical cluster analysis permits correlations within the data to be predicted. The data provide current information on levels of PBDE and PBB occurrence, allowing identification of locations that show higher contaminant levels. One hundred thirty-five fish samples of various species were analyzed from UK marine waters, encompassing the waters around Norway in the North and to the Algarve in the South. PBDEs were observed in all samples with the majority of measured congeners being detected. The concentrations ranged from 0.087 µg/kg to 8.907 µg/kg whole weight (ww) for the sum of all measured PBDE congeners. PBBs occurred less frequently showing a corresponding range of <0.02 µg/kg to 0.97 µg/kg ww for the sum of seven PBB congeners. Concentrations vary depending on species and locations where landed, e.g., PBBs occurred more frequently and at higher levels in grey mullet from French waters. The high frequency of PBDE occurrence makes it prudent to continue the monitoring of these commonly consumed marine fish species. The web-based resource provides a flexible and efficient tool for assessors and policy-makers to monitor and evaluate levels within caught fish species improving evidenced-based decision processes.

Analytical measurement was carried out using high-resolution gas chromatography–high-resolution mass spectrometry (HRGC-HRMS).

Several PBB congeners, containing 2-4 bromine orthosubstituents cannot rotate about the intraannular phenyl-phenyl bond due to steric hindrance (120). Some of the PBBs found in the environment are axially chiral (PBB 132 en PBB 149). However, no studies investigating the fate of atropisomeric PBBs in the environment have been published.

REFERENCES

1. http://chm.pops.int/TheConvention/ThePOPs/ListingofPOPs/tabid/2509/Default.aspx
2. González, N., Marquès, M., Nadal, M., Domingo, J.L. Occurrence of Environmental Pollutants in Foodstuffs: A Review of Organic vs. Conventional Food. *Food and Chemical Toxicology* 2019, 125, 370–375.
3. Li, X., Zhang, Q., Wang, P., Fu, J., Jiang, G. Post Dioxin Period for Feed: Cocktail Effects of Emerging POPs and Analogues. *Environmental Science & Technology* 2020, 54, 1, 6–8.
4. Ssebugere, P., Sillanpää, M., Menatavu, H., Mubiru, E. Human and Environmental Exposure to PCDD/Fs and Dioxin-Like PCBs in Africa: A review. *Chemosphere* 2019, 223, 486–493.
5. Zang, Y., Ye, J., Liu, M. Enantioselective Biotransformation of Chiral Persistent Organic Pollutants. *Current Protein and Peptide Science* 2017, 18 (1), 48–56.
6. Guo, F., Zhang, J., Wang, C. Enantioselectivity in Environmental Safety and Metabolism of Typical Chiral Organic Pollutants. *Current Protein and Peptide Science* 2017, 18, 1, 4–9.
7. Bidleman, T.F., Jantunen, L.M., Kurt-Karakus, P.B., Wong, F. Chiral Persistent Organic Pollutants as Tracers of Atmospheric Sources and Fate: Review and Prospects for Investigating Climate Change Influences. *Atmospheric Pollution Research* 2012, 3, 4, 371–382.

8. https://www.chemeurope.com/en/news/143076/environmental-significance-of-chiral-persistent-organic-pollutants.html
9. Wang, P, Zhang, Q., Li, Y., Zhu, C., Chen, Z., Zheng, S., Sun, H., Liang, Y., Jiang, G. Occurrence of Chiral Organochlorine Compounds in the Environmental Matrices from King George Island and Ardley Island, west Antarctica. *Scientific Reports* 2015, 5, 13913, DOI: 10.1038/srep13913.
10. Guo, W., Pan, B., Sakkiah, S., Yavas, G., Ge, W., Zou, W., Tong, W., Hong, H. Persistent Organic Pollutants in Food: Contamination Sources, Health Effects and Detection Methods. *International Journal of Environmental Research and Public Health* 2019, 16, 22, 4364390.
11. Sander, L.C., Schantz, M.M., Wise, S.A. Chapter 14 - Environmental Analysis: Persistent Organic Pollutants. *Liquid Chromatography* (2nd ed.), 2017, 401–449.
12. Covaci, A., Gerecke, A.C., Law, R.J., Voorspoels, S., Kohler, M., Heeb, N.V., Leslie, H., Allchin, C.R., de Boer, J. Hexabromocyclododecanes (HBCDs) in the Environment and Humans: A Review. *Environmental Science & Technology* 2006, 40, 12, 3679–3688.
13. Law, R.J., Kohler, M., Heeb, N.V., Gerecke, A.C., Schmid, P., Voorspoels, S., Covaci, A., Becher, G., Janak, K., Thomsen, C. Hexabromocyclododecane Challenges Scientists and Regulators. *Environmental Science & Technology* 2005, 281–287A
14. Ruan, Y., Zhang, X., Qiu, J.-W., Leung, K.M.Y., Lam, J.C.W., Lam, P.K.S. Stereoisomer-Specific Trophodynamics of the Chiral Brominated Flame Retardants HBCD and TBECH in a Marine Food Web, with Implications for Human Exposure. *Environmental Science & Technology* 2018, 52, 15, 8183–8193.
15. Abdallah, M. A.-E., Uchea, C., Chipman, J.K., Harra, S. Enantioselective Biotransformation of Hexabromocyclododecane by in Vitro Rat and Trout Hepatic Sub-Cellular Fractions. *Environmental Science & Technology* 2014, 48, 5, 2732–2740.
16. Liu, Y., Luo, X., Zeng, Y., Deng, M, Tu, W., Wu, Y., Mai, B. Bioaccumulation and Biomagnification of Hexabromocyclododecane (HBCDD) in Insect-dominated Food Webs from a Former e-waste Recycling Site in South China. *Chemosphere* 2020, 240, 124813.
17. Sun, R., Luo, X., Zheng, X., Cao, Z., Peng, P., Li, Q.X., Mai, B. Hexabromocyclododecanes (HBCDs) in fish: Evidence of Recent HBCD input into the Coastal Environment. *Marine Pollution Bulletin* 2018, 126, 357–362.
18. de Boer, J., Wells, D.F. Pitfalls in the Analysis of Brominated Flame Retardants in Environmental, Human and Food Samples – Including Results of Three International Interlaboratory Studies. *TrAC: Trends in Analytical Chemistry* 2006, 25, 4, 364–372.
19. Viganò, L., Roscioli, C., Guzzella, L. Decabromodiphenyl Ether (BDE-209) Enters the Food Web of the River Po and is Metabolically Debrominated in Resident Cyprinid Fishes. *Science of the Total Environment* 2011, 409, 23, 4966–4972.
20. Kuo, Y.-M., Sepulveda, M.S., Hua, I., Ochoa-Acuña, H., Setton, T.M. Bioaccumulation and Biomagnification of Polybrominated Diphenyl Ethers in a Food Web of Lake Michigan. *Ecotoxicology* 2010, 19, 623–634.
21. Ma, Z., Zhuang, H. Biotin–Streptavidin System-Based Real-Time Immunopolymerase Chain Reaction for Sensitive Detection of 2,2´,4,4´- Tetrabromodiphenyl Ether in Marine Fish. *Food and Agricultural Immunology* 2018, 29, 1, 1012–1027.
22. Chen, J., Li, K., Le, X. C., Zhu, L. Metabolomic Analysis of Two Rice (Oryza sativa) Varieties Exposed to 2, 20, 4, 40 -Tetrabromodiphenyl Ether. *Environmental Pollution* 2018, 237, 308–317.
23. Pietroń, W.J., Małagocki, P. Quantification of Polybrominated Diphenyl Ethers (PBDEs) in food. A review. *Talanta* 2017, 167, 411–427.
24. Alaee, M., Sergeant, D.B., Ikonomou, M.G., Luross, J.M. A Gas Chromatography/Highresolution Mass Spectrometry (GC/HRMS) Method for Determination of Polybrominated Diphenyl Ethers in Fish. *Chemosphere* 2001, 44, 1489–1495.
25. Huwe, J.K., Lorentzsen, M., Thuresson, K., Bergman, Å. Analysis of Mono- to Decabrominated Diphenyl Ethers in Chickens at the Part per Billion Level, *Chemosphere* 2002, 46, 635–640.
26. Jacobs, M.N., Covaci, A., Schepens, P. Investigation of Selected Persistent Organic Pollutants in Farmed Atlantic Salmon (Salmo salar), Salmon Aquaculture Feed, and Fish Oil Components of the Feed. *Environmental Science & Technology* 2002, 36, 2797–2805.
27. Pirard, E., De Pauw, E, Focant, J.-F. New Strategy for Comprehensive Analysis of Polybrominated Diphenyl Ethers, Polychlorinated Dibenzo-p-dioxins, Polychlorinated Dibenzofurans and Polychlorinated Biphenyls by Gas Chromatography Coupled with Mass Spectrometry, *Journal of Chromatography A* 2003, 998, 169–181.

28. de Boer, J., Wester, P.G., van der Horst, A., Leonards, P.E. Polybrominated Diphenyl Ethers in Influents, Suspended Particulate Matter, Sediments, Sewage Treatment Plant and Effluents and Biota from the Netherlands. *Environmental Pollution* 2003, 122, 63–74.
29. Voorspoels, S., Covaci, A., Schepens, P. Olybrominated Diphenyl Ethers in Marine Species from the Belgian North Sea and the Western Scheldt Estuary: Levels, Profiles, and Distribution Polybrominated Diphenyl Ethers in Marine Species from the Belgian North Sea and the Western Scheldt Estuary. *Environmental Science & Technology* 2003, 37, 4348–4357.
30. Fernandes, A., White, S., D'Silva, K., Rose, M. Simultaneous Determination of PCDDs, PCDFs, PCBs and PBDEs in food, *Talanta* 2004, 63, 1147–1155.
31. Päpke, O., Fürst, P., Herrmann, T. Determination of Polybrominated Diphenylethers (PBDEs) in Biological Tissues with Special Emphasis on QC/QA Measures, *Talanta* 2004, 63, 1203–1211.
32. Johnson-Restrepo, B., Kannan, K., Addink, R., Adams, D.H. Polybrominated Diphenyl Ethers and Polychlorinated Biphenyls in a Marine Foodweb of Coastal Florida. *Environmental Science & Technology* 2005, 39, 8243–8250.
33. Labandeira, A., Eljarrat, E., Barceló, D. Congener Distribution of Polybrominated Diphenyl Ethers in Feral Carp (Cyprinus carpio) from the Llobregat River, Spain. *Environmental Pollution* 2007, 146, 188–195.
34. Liu, H., Zhang, Q., Cai, Z., Li, A., Wang, Y., Jiang, G. Separation of Polybrominated Diphenyl Ethers, Polychlorinated Biphenyls, Polychlorinated Dibenzo-p-dioxins and Dibenzo-furans in Environmental Samples Using Silica gel and Florisil Fractionation Chromatography. *Analytica Chimica Acta* 2006, 557, 314–320.
35. Gómara, B., Herrero, L., González, M.J., Survey of Polybrominated Diphenyl Ether Levels in Spanish Commercial Foodstuffs. *Environmental Science & Technology* 2006, 40, 7541–7547.
36. Akutsu, K., Takatori, S., Nakazawa, H., Hayakawa, K., Izumi, S., Makino, T. Dietary Intake Estimations of Polybrominated Diphenyl Ethers (PBDEs) Based on a Total Diet Study in Osaka, Japan. *Food Additives & Contaminants: Part B* 2008, 1, 58–68.
37. Luo, Q., Wong, M., Cai, Z. Determination of Polybrominated Diphenyl Ethers in Freshwater Fishes from a River Polluted by e-wastes. *Talanta* 2007, 72, 1644–1649.
38. Pulkrabová, J., Hajšlová, J., Poustka, J., Kazda, R. Fish as Biomonitors of Polybrominated Diphenyl Ethers and Hexabromocyclododecane in Czech Aquatic Ecosystems: Pollution of the Elbe River Basin. *Environ. Health Perspect.* 2007, 115, 28–34.
39. Shaw, S.D., Brenner, D., Berger, M.L., Fang, F., Hong, C., Addink, R., Hilker, D. Bioaccumulation of Polybrominated Diphenyl Ethers in Harbor Seals from the Northwest Atlantic, *Chemosphere* 2008, 73, 1773–1780.
40. Tapie, N., Budzinski, H., Le Menach, K., Fast and Efficient Extraction Methods for the Analysis of Polychlorinated Biphenyls and Polybrominated Diphenyl Ethers in Biological Matrices. *Anal. Bioanal. Chem.* 2008, 391, 2169–2177.
41. Wang, D., Atkinson, S., Hoover-Miller, A., Shelver, W.L., Li, Q.X. Simultaneous use of Gas Chromatography/ion Trap Mass Spectrometry – Electron Capture Detection to Improve the Analysis of Bromodiphenyl Ethers in Biological and Environmental Samples. *Rapid Commun. Mass Spectrom.* 2008, 22, 647–656.
42. Losada, S., Santos, F.J., Galceran, M.T. Selective Pressurized Liquid Extraction of Polybrominated Diphenyl Ethers in Fish. *Talanta* 2009, 80, 839–845.
43. Li, Y.F., Yang, Z.Z., Wang, C.H., Yang, Z.J., Qin, Z.F., Fu, S. Tissue Distribution of Polybrominated Diphenyl Ethers (PBDEs) in Captive Domestic Pigs, Sus Scrofa, from a Village Near an Electronic Waste Recycling site in South China. *Bull. Environ. Contam. Toxicol.* 2010, 84, 208–211.
44. Losada, S., Parera, J., Abalos, M., Abad, E., Santos, F.J., Galceran, M.T. Suitability of Selective Pressurized Liquid Extraction Combined with Gas Chromatography-ion-trap Tandem Mass Spectrometry for the Analysis of Polybrominated Diphenyl Ethers. *Analytica Chimica Acta* 2010, 678, 73–81.
45. Lacorte, I.M.G.F.M. S, Lacorte, S., Ikonomou, M.G., Fischer, M. A Comprehensive Gas Chromatography Coupled to High Resolution Mass Spectrometry Based Method for the Determination of Polybrominated Diphenyl Ethers and their Hydroxylated and Methoxylated Metabolites in Environmental Samples. *J. Chromatogr. A* 2010, 1217, 337–347.
46. Blanco, S.L., Vieites, J.M. Single-run Determination of Polybrominated Diphenyl Ethers (PBDEs) di- to deca-brominated in Fish Meal, Fish and Fish Feed by Isotope Dilution: Application of Automated Sample Purification and Gas Chromatography/ion trap Tandem Mass Spectrometry (GC/ITMS). *Analytica Chimica Acta* 2010, 672, 137–146.

47. Chen, M.Y.Y., Tang, A.S.P., Ho, Y.Y., Xiao, Y. Dietary Exposure of Secondary School Students in Hong Kong to Polybrominated Diphenyl Ethers from Foods of Animal Origin, *Food Addit. Contam. Part A Chem. Anal. Control Expo. Risk Assess.* 2010, 27, 521–529.
48. Labadie, P., Alliot, F., Bourges, C., Desportes, A., Chevreuil, M. Determination of Polybrominated Diphenyl Ethers in Fish Tissues by Matrix Solid-phase Dispersion and Gas Chromatography Coupled to Triple Quadrupole Mass Spectrometry: Case Study on European eel (Anguilla anguilla) from Mediterranean Coastal Lagoons. *Analytica Chimica Acta* 2010, 675, 97–105.
49. Wang, J., Kliks, M.M., Jun, S., Li, Q.X. Residues of Polybrominated Diphenyl Ethers in Honeys from Different Geographic Regions. *Journal of Agricultural and Food Chemistry* 2010, 58, 3495–3501.
50. Zhang, Z., Ohiozebau, E., Rhind, S.M. Simultaneous Extraction and Clean-up of Polybrominated Diphenyl Ethers and Polychlorinated Biphenyls from Sheep Liver Tissue by Selective Pressurized Liquid Extraction and Analysis by Gas Chromatography-mass Spectrometry. *Journal of Chromatography A* 2011, 1218, 1203–1209.
51. Kalachova, K., Pulkrabova, J., Drabova, L., Cajka, T., Kocourek, V., Hajslova, J. Simplified and Rapid Determination of Polychlorinated Biphenyls, Polybrominated Diphenyl Ethers, and Polycyclic Aromatic Hydrocarbons in Fish and Shrimps Integrated into a Single Method. *Analytica Chimica Acta* 2011, 707, 84–91.
52. Fontana, A.R., Camargo, A., Martinez, L.D., Altamirano, J.C. Dispersive Solid-phase Extraction as a Simplified Clean-up Technique for Biological Sample Extracts. Determination of Polybrominated Diphenyl Ethers by Gas Chromatography–Tandem Mass Spectrometry. *Journal of Chromatography A* 2011, 1218, 18, 2490–2496.
53. Sprague, M., Dick, J.R.R., Medina, A., Tocher, D.R.R., Bell, J.G.G., Mourente, G. Lipid and Fatty Acid Composition, and Persistent Organic Pollutant Levels in Tissues of Migrating Atlantic Bluefin tuna (Thunnus thynnus, L.) Broodstock, *Environmental Pollution* 2012, 171, 61–71.
54. Roszko, M., Rzepkowska, M., Szterk, A., Szymczyk, K., Jedrzejczak, R., Bryła, M. Application of Semi-permeable Membrane Dialysis/ion Trap Mass Spectrometry Technique to Determine Polybrominated Diphenyl Ethers and Polychlorinated Biphenyls in Milk Fat. *Analytica Chimica Acta* 2012, 748, 9–19.
55. Waszak, I., Dabrowska, H., Góra, A. Bioaccumulation of Polybrominated Diphenyl Ethers (PBDEs) in Flounder (Platichthys flesus) in the Southern Baltic Sea, *Mar. Environ. Res.* 2012, 79, 132–141.
56. Mackintosh, A., Perez-Fuenteaja, A., Zimmerman, L.R., Pacepavicius, G., Clapsadl, M., Alaee, M., Aga, D.S. Analytical Performance of a Triple Quadrupole Mass Spectrometer Compared to a High Resolution Mass Spectrometer for the Analysis of Polybrominated Diphenyl Ethers in Fish. *Analytica Chimica Acta* 2012, 747, 67–75.
57. Jurgens, M.D., Johnson, A.C., Jones, K.C., Hughes, D., Lawlor, A.J. The Presence of EU Priority Substances Mercury, Hexachlorobenzene, Hexachlorobutadiene and PBDEs in Wild Fish from Four English Rivers. *Science of the Total Environment* 2013, 461–462, 441–452.
58. Kalachova, K., Pulkrabova, J., Cajka, T., Drabova, L., Stupak, M., Hajslova, J. Gas Chromatography-triple Quadrupole Tandem Mass Spectrometry: A Powerful Tool for the (Ultra)trace Analysis of Multiclass Environmental Contaminants in Fish and Fish Feed Rapid Detection in Food and Feed, *Analytical and Bioanalytical Chemistry* 2013, 405, 7803–7815.
59. Zacs, D., Rjabova, J., Bartkevics, V. Occurrence of Brominated Persistent Organic Pollutants (PBDD/DFs, PXDD/DFs, and PBDEs) in Baltic Wild Salmon (Salmo Salar) and Correlation with PCDD/DFs and PCBs, *Environmental Science & Technology* 2013, 47, 9478–9486.
60. Sühring, R., Möller, A., Freese, M., Pohlmann, J.D., Wolschke, H., Sturm, R., Xie, Z., Hanel, R., Ebinghaus, R. Brominated Flame Retardants and Dechloranes in Eels from German Rivers. *Chemosphere* 2013, 90, 118–124.
61. Baron, E., Eljarrat, E., Barcelo, D. Gas Chromatography/Tandem Mass Spectrometry Method for the Simultaneous Analysis of 19 Brominated Compounds in Environmental and Biological Samples. *Analytical and Bioanalytical Chemistry* 2014, 406, 7667–7676.
62. Poma, G., Volta, P., Roscioli, C., Bettinetti, R., Guzzella, L. Concentrations and Trophic Interactions of Novel Brominated Flame Retardants, HBCD, and PBDEs in Zooplankton and Fish from Lake Maggiore (Northern Italy). *Science of the Total Environment* 2014, 481, 401–408.
63. Sapozhnikova, Y., Simons, T., Lehotay, S.J. Evaluation of a Fast and Simple Sample Preparation Method for Polybrominated Diphenyl ether (PBDE) Flame Retardants and Dichlorodiphenyltrichloroethane (DDT) Pesticides in Fish for Analysis by ELISA Compared with GC-MS/MS. *Journal of Agricultural and Food Chemistry* 2015, 63, 4429–4434.
64. Couderc, M., Poirier, L., Zalouk-Vergnoux, A., Kamari, A., Blanchet-Letrouvé, I., Marchand, P., Vénisseau, A., Veyrand, B., Mouneyrac, C., Le Bizec, B. Occurrence of POPs and Other Persistent Organic Contaminants in the European eel (Anguilla anguilla) From the Loire Estuary, France. *Science of the Total Environment* 2015, 505, 199–215.

65. Portolés, T., Sales, C., Gómara, B., Sancho, J.V., Beltrán, J., Herrero, L., González, M.J., Hernández, F. Novel Analytical Approach for Brominated Flame Retardants Based on the Use of Gas Chromatography-Atmospheric Pressure Chemical IonizationTandem Mass Spectrometry with Emphasis in Highly Brominated Congeners. *Analytical Chemistry* 2015, 87, 9892–9899.
66. Lin, Y., Feng, C., Xu, Q., Lu, D., Qiu, X., Jin, Y., Wang, G., Wang, D., She, J., Zhou, Z. A Validated Method for Rapid Determination of Dibenzo-p-dioxins/furans (PCDD/Fs), Polybrominated Diphenyl Ethers (PBDEs) and Polychlorinated Biphenyls (PCBs) in Human Milk: Focus on Utility of Tandem Solid Phase Extraction (SPE) Cleanup. *Analytical and Bioanalytical Chemistry* 2016, 408, 4897–4906.
67. Dosis, I., Athanassiadis, I., Karamanlis, X. Polybrominated Diphenyl Ethers (PBDEs) in Mussels From Cultures and Natural Population. *Marine Pollution Bulletin* 2016, 107, 92–101.
68. Martellini, T., Diletti, G., Scortichini, G., Lolini, M., Lanciotti, E., Katsoyiannis, A., Cincinelli, A. Occurrence of Polybrominated Diphenyl Ethers (PBDEs) in Foodstuffs in Italy and Implications for Human Exposure. *Food Chem. Toxicol.* 2016, 89, 32–38.
69. Zeng, Y.-H., Luo, X.-J., Tang, B., Mai, B.-X. Habitat- and Species-dependent Accumulation of Organohalogen Pollutants in Home-produced Eggs from an Electronic Waste Recycling site in South China: Levels, Profiles, and Human Dietary Exposure. *Environmental Pollution* 2016, 216, 64–70.
70. Poma, G., Malarvannan, G., Voorspoels, S., Symons, N., Malysheva, S. V., Van Loco, J., Covaci, A. Determination of Halogenated Flame Retardants in Food: Optimization and Validation of a Method Based on A Two-Step Clean-Up and Gas Chromatography-Mass Spectrometry. *Food Control* 2016, 65, 168–176.
71. Bichon, E., Guiffard, I., Vénisseau, A., Lesquin, E., Vaccher, V., Brosseaud, A., Marchand, P., Le Bizec, B. Simultaneous Determination of 16 Brominated Flame Retardants in Food and Feed of Animal Origin by fast gas Chromatography Coupled to Tandem Mass Spectrometry Using Atmospheric Pressure Chemical Ionisation. *Journal of Chromatography A* 2016, 1459, 120–128.
72. Tang, B., Luo, X.-J., Luo, X.-J., Zeng, Y.-H., Mai, B.-X. Tracing the Biotransformation of PCBs and PBDEs in Common Carp (Cyprinus carpio) Using Compound-Specific and Enantiomer-Specific Stable Carbon Isotope Analysis. *Environmental Science & Technology* 2017, 51, 5, 2705–2713.
73. Wang, H., Qu, B., Liu, H., Ding, J., Ren, N. Fast Determination of β-endosulfan, α-Hexachlorocyclohexane and Pentachlorobenzene in the River Water from Northeast of China. *International Journal of Environmental Analytical Chemistry* 2018, 98, 5, 413–428.
74. Recently listed Stockholm convention POPs: Analytical methodology, occurrence in food and dietary exposure
75. Chang, G.R. Persistent Organochlorine Pesticides in Aquatic Environments and Fishes in Taiwan and their Risk Assessment. *Environmental Science and Pollution Research* 2018, 25, 7699–7708.
76. Tölgyessy, P., Miháliková, Z., Matulová, M. Determination of Selected Chlorinated Priority Substances in Fish using QuEChERS Method with Dual dSPE Clean-up and Gas Chromatography. *Chromatographia* 2016, 79, 1561–1568.
77. Tang, Z., Huang, Q., Cheng, J., Qu, D., Yang, Y., Guoo, W. Distribution and Accumulation of Hexachlorobutadiene in Soils and Terrestrial Organisms from an Agricultural Area, East China. *Ecotoxicology and Environmental Safety* 2014, 108, 329–334.
78. Tang, Z., Huang, O., Nie, Z., Yang, Y., Yang, J., Qu, D., Cheng, J. Levels and Distribution of Organochlorine Pesticides and Hexachlorobutadiene in Soils and Terrestrial Organisms from a Former Pesticide-Producing Area in Southwest China. *Stochastic Environmental Research and Risk Assessment* 2016, 30, 1249–1262.
79. Jürgens, M.D., Johnson, A.C., Jones, K.C., Hughes, D., Lawlor, A.J. The Presence of EU Priority Substances Mercury, Hexachlorobenzene, Hexachlorobutadiene and PBDEs in Wild Fish from four English Rivers. *Science of the Total Environment* 2013, 461-462, 441–452.
80. Majoros, I., Lava, R., Ricci, M., Binici, B., Sandor, F., Held, A., Emons, H. Full Method Validation for the Determination of Hexachlorobenzene and Hexachlorobutadiene in Fish Tissue by GC–IDMS. *Talanta* 2013, 116, 251–258.
81. Balmer, J.E., Hung, H., Vorkamp, K., Letcher, R.J., Muir, D.C.G. Hexachlorobutadiene (HCBD) Contamination in the Arctic Environment: A Review. *Emerging Contaminants* 2019, 5, 116–122.
82. Surma-Zadora, M., Grochowalski, A. Using a Membrane Technique (SPM) for High fat food Sample Preparation in the Determination of Chlorinated Persistent Organic Pollutants by a GC/ECD Method. *Food Chemistry* 2008, 111, 1, 230–235.
83. Castells, P., Santos, F.J., Galceran, M.T. Solid-Phase Microextraction for the Analysis of Short-Chain Chlorinated Paraffins in Water Samples. *Journal of Chromatography A* 2003, 984, 1, 1–8.
84. Castells, P., Parera, J., Santos, F.J., Galceran, M.T. Occurrence of Polychlorinated Naphthalenes, Polychlorinated Biphenyls and Short-chain Chlorinated Paraffins in Marine Sediments from Barcelona (Spain). *Chemosphere* 2008, 70, 9, 1552–1562.

85. Gao, Y., Zhang, H., Su, F., Tian, Y., Chen, J. Environmental Occurrence and Distribution of Short Chain Chlorinated Paraffins in Sediments and Soils from the Liaohe River Basin, P. R. China. *Environmental Science & Technology* 2012, 46, 7, 3771–3778.
86. Yuan, B., Wang, T., Zhu, N., Zhang, K., Zeng, L., Fu, J., Wang, Y. Jiang, G. Short Chain Chlorinated Paraffins in Mollusks from Coastal Waters in the Chinese Bohai Sea. *Environmental Science & Technology* 2012, 46, 12, 6489–6496.
87. Pellizzato, F., Ricci, M., Held, A., Emons, H., Böhmer, W., Geiss, S., Iozza, S., Mais, S., Petersen, M., Lepom, P. Laboratory Intercomparison Study on the Analysis of Short-chain Chlorinated Paraffins in an Extract of Industrial Soil. *TrAC: Trends in Analytical Chemistry* 2009, 28, 8, 1029–1035.
88. Fernandes, A., Rose, M., Falandysz, J. Polychlorinated Naphthalenes (PCNs) in Food and Humans. *Environmental International* 2017, 104, 1–13.
89. Kannan, K. Blankenship, A.L., Giesy, J.P., Imagaw, T. Polychlorinated Naphthalenes in Soil, Sediment, and Biota Collected near a Former Chloralkali Plant in Coastal Georgia, USA. *Central European Journal of Public Health* 2000, 8 Suppl, 10–12.
90. Kannan, K., Yamashita, N., Imagawa, T., Decoen, W., Khim, J.S., Day, R.M., Summer, C.L. Giesy, J.P. Polychlorinated Naphthalenes and Polychlorinated Biphenyls in Fishes from Michigan Waters Including the Great Lakes. *Environmental Science & Technology* 2000a, 34, 4, 566–572.
91. Domingo, J.L., Falcó, G., Llobet, J.M., Casas, C. Teixido, A., Müller, L. Polychlorinated Naphthalenes in Foods: Estimated Dietary Intake by the Population of Catalonia, Spain. *Environmental Science & Technology* 2003, 37, 11, 2332–2335.
92. Isosaari, P., Hallikainen, A., Kiviranta, H., Pekka, J.V., Parmanne, R., Koistinen, J., Vartiainen, T. Polychlorinated Dibenzo-p-dioxins, Dibenzofurans, Biphenyls, Naphthalenes and Polybrominated Diphenyl Ethers in the Edible Fish Caught from the Baltic Sea and Lakes in Finland. *Environmental Pollution* 2006, 141, 2, 213–225.
93. Jiang, Q., Hanari, N., Miyake, Y., Okazawa, T., Lau, R.K.F., Chen, K., Wyrzykowska, B., So, M.K., Yamashita, N., Lam, P.K.S. Health Risk Assessment for Polychlorinated Biphenyls, Polychlorinated Dibenzo-p-dioxins and Dibenzofurans, and Polychlorinated Naphthalenes in Seafood from Guangzhou and Zhoushan, China. *Environmental Pollution* 2007, 148, 1, 31–39.
94. Llobel, J.M., Domingo, J.L., Bocio, A., Casas, C., Teixidó, A., Mülle, L. Human Exposure to Dioxins Through the Diet in Catalonia, Spain: Carcinogenic and Non-carcinogenic Risk. *Chemosphere* 2003, 50, 9, 1193–1200.
95. Marti-Cid, R., Llobet, J.M., Castell, V., Domingo, J.L. Human Dietary Exposure to Hexachlorobenzene in Catalonia, Spain. *J Food Prot* 2008, 71, 10, 2148–2152.
96. Linag, Y.Y., Jing, P., Hua, Z.X., Duan, L.X., Hui, L.G., Qi, L., De, L.X. Studies on Coplanar-PCBs and PCNs in Edible Fish and Duck in Qingdao and Chongming Island. *Research of Environmental Sciences* 2009, 22, 2, 187–193.
97. Fernandes, A., Mortimer, D., Gem, M., Smith, F., Rose, M., Panton, S., Carr, M. Polychlorinated Naphthalenes (PCNs): Congener Specific Analysis, Occurrence in Food, and Dietary Exposure in the UK. *Environmental Science & Technology* 2010, 44, 9, 3533–3538.
98. Fernandes, A.R., Tlustos, C., Rose, M., Smith, F., Carr, M., Panton, S. Polychlorinated Naphthalenes (PCNs) in Irish Foods: Occurrence and Human Dietary Exposure. *Chemosphere* 2011, 85, 3, 322–328.
99. Fernandes, A., Smith, F., Petch, R., Brereton, N., Bradley, E., Panton, S., Carr, M., Rose, M. Investigation into the Levels of Environmental Contaminants in Scottish Marine and Freshwater Fin Fish and Shellfish Food and Environment Research Agency Sand Hutton YORK YO41 1LZ
100. Rose, M., Fernandes, A., Mortimer, D., Baskaran, C. Contamination of Fish in UK Fresh Water Systems: Risk Assessment for Human Consumption. *Chemopshere* 2015, 122, 183–189.
101. Fernandes, A., Rose, M., Smith, F., Panton, S. Geographical Investigation for Chemical Contaminants in Fish Collected from UK and Proximate Marine Waters
102. Li, L., Sun, S., Dong, S., Jiang, X., Liu, G., Zheng, M. Polychlorinated Naphthalene Concentrations and Profiles in Cheese and Butter, and Comparisons with Polychlorinated Dibenzo-p-dioxin, Polychlorinated Dibenzofuran and Polychlorinated Biphenyl Concentrations, 2015, *International Journal of Environmental Analytical Chemistry* 2015, 95, 3, 203–216.
103. Pan, J., Yang, Y., Taniyasu, S., Yeung, L.W. Y., Falandysz, J., Yamashita, N. Comparison of Historical Record of PCDD/Fs, Dioxin-Like PCBs, and PCNs in Sediment Cores from Jiaozhou Bay and Coastal Yellow Sea: Implication of Different Sources. *Bulletin of Environmental Contamination and Toxicology* 2012, 89, 1240–1246.
104. Guruge, K.S., Seike, N., Yamanaka, N., Miyazaki, S. Accumulation of Polychlorinated Naphthalenes in Domestic Animal Related Samples. *J. Environ. Monit.* 2004, 6, 753–757.

105. Mahmood, A., Malik, R.N., Li, J., Zhang, G. Jones, K.C. PCNs (polychlorinated napthalenes): Dietary Exposure via Cereal Crops, Distribution and Screening-Level Risk Assessment in Wheat, Rice, Soil and Air Along two Tributaries of the River Chenab, Pakistan. *Science of the Total Environment* 2014, 481, 409–417.
106. Dong, S., Li, X., Wang, P., Su, X. Polychlorinated Naphthalene Concentrations and Distribution in Feed Raw Materials. *Chemosphere* 2018, 211, 912–917.
107. Han, Y., Liu, W., Li, H., Lei, R., Liu, G., Gao, L., Su, G. Distribution of Polychlorinated Naphthalenes (PCNs) in the Whole Blood of Typical Meat Animals. *Journal of Environmental Sciences* 2018, 72, 208–212.
108. Gewurtz, S.B., Gandhi, N., Drouillard, K.G., Kolic, T., MacPherson, K., Reiner, E.J., Bhaysar S.P. Levels, Patterns, Trends and Significance of Polychlorinated Naphthalenes (PCNs) in Great Lakes Fish. *Science of the Total Environment* 2018, 624, 499–508.
109. Kim, J., Shin, E.-S., Choi, S.-D., Zhu, J., Chang, Y.-S. Polychlorinated Naphthalenes (PCNs) in Seafood: Estimation of Dietary Intake in Korean Population. *Science of the Total Environment* 2018, 624, 40–47.
110. European Food Safety Authority (EFSA). Perfluorooctane Sulfonate (PFOS), Perfluorooctanoic Acid (PFOA) and their Salts Scientific Opinion of the Panel on Contaminants in the Food Chain. *The EFSA Journal* 2008, 653, 1–131.
111. Noorlander, C.W., van Leeuwen, S.P.J., te Biesebeek, J.D., Mengelers, M.J.B., Zeilmaker, M.J. Levels of Perfluorinated Compounds in Food and Dietary Intake of PFOS and PFOA in The Netherlands. *Journal of Agricultural and Food Chemistry* 2011, 59, 13, 7496–7505.
112. Pérez, F., Llorca, M., Köck-Schulmeyer, M., Škrbić, B., Silva Oliveira, L., da Boit Martinello, K., Al-Dhabi N.A., Antić, I., Farré, M., Barceló, D. Assessment of Perfluoroalkyl Substances in Food Items at Global Scale. *Environmental Research*, 2014, 135, 181–189.
113. Sadia, M., Yeung, L.W.Y., Fiedler, H. Trace Level Analyses of Selected Perfluoroalkyl Acids in Food: Method Development and Data Generation. *Environmental Pollution* 2020, 263, A, 113721.
114. Sungur, S., Köroğlu, M., Turgut, F. Determination of Perfluorooctanoic Acid (PFOA) and Perfluorooctane Sulfonic Acid (PFOS) in Food and Beverages. *International Journal of Environmental Analytical Chemistry* 2018, 98, 4, 360–368.
115. Ciccotelli, V., Abete, M.C., Squadrone, S. PFOS and PFOA in Cereals and Fish: Development and Validation of a High-performance Liquid Chromatography-Tandem Mass Spectrometry Method. *Food Control* 2016, 59, 46–52.
116. Sznajder-Katarzyńska, K., Surma, M., Cieślik, E., Wiczkowski, W. The Perfluoroalkyl Substances (PFASs) Contamination of Fruits and Vegetables Part A Chemistry, Analysis, Control, Exposure & Risk Assessment. *Food additives & contaminants* 2018, 35, 9, 1776–1786.
117. Vetter, W., von der Recke, R., Symons, R., Pyecroft, S. Determination of Polybrominated Biphenyls in Tasmanian Devils (*Sarcophilus harrisii*) by Gas Chromatography Coupled to Electron Capture Negative ion Tandem Mass Spectrometry or Electron Ionization High-resolution Mass Spectrometry. *Rapid Communications in Mass Spectrometry* 2008, 22, 24, 4165–4170.
118. Vetter, W., von der Recke, R., Herzke, D., Nygård, T. Detailed Analysis of Polybrominated Biphenyl Congeners in Bird Eggs from Norway. *Environmental Pollution* 2008, 156, 3, 1204–1210.
119. Zhihua, L., Panton, S., Marshall, L., Ferndes, A., Rose, M., Smith, F., Holmes, M. Spatial Analysis of Polybrominated Diphenylethers (PBDEs) and Polybrominated Biphenyls (PBBs) in Fish Collected from UK and Proximate Marine Waters. *Chemosphere* 2018, 195, 727–734.
120. Eljarrat, E., Guerra, P., Barceló, D. Enantiomeric Determination of Chiral Persistent Organic Pollutants and their Metabolites. *Trends in Analytical Chemistry* 2008, 27, 10, 847–861.
121. Malavia, J., Santos, F.J., Galceran, M.T. Simultaneous Pressurized Liquid Extraction and Clean-up for the Analysis of Polybrominated Biphenyls by Gas Chromatography–Tandem Mass Spectrometry. *Talanta*, 2011, 84 (4), 1155–1162.

Section II

Analysis, Fate, and Toxicity of Chiral Pollutants in Food

Section 1

Analysis, Fate, and Toxicity of Chiral Pollutants in Food

8 Current Trends in Enantioselective Food Analysis

Basil K. Munjanja

CONTENTS

8.1 Introduction 191
    8.1.1 Chirality in Food Safety 193
    8.1.2 Scope of Chapter.. 193
8.2 Mechanism of Enantioseparation 194
8.3 Enantioselective Analysis Techniques Used in Food Science... 196
    8.3.1 Comparison of Enantioselective Techniques.. 196
    8.3.2 Chiral Stationary Phases 196
8.4 Food Safety Assessment 196
    8.4.1 Enantioselective Liquid Chromatography in Food Analysis.............................. 196
    8.4.2 Enantioselective Supercritical Fluid Chromatography in Food Safety 205
    8.4.3 Enantioselective Gas Chromatography in Food Safety 205
    8.4.4 Enantioselective Capillary Electrophoresis in Food Safety 205
8.5 Food Quality Control and Traceability... 214
    8.5.1 Enantioselective Liquid Chromatography in Food Quality................................ 214
    8.5.2 Enantioselective Gas Chromatography in Food Quality 214
        8.5.2.1 Solid Phase Microextraction 215
        8.5.2.2 Multidimensional Gas Chromatography.. 215
    8.5.3 Enantioselective Electrokinetic Chromatography in Food Quality.................... 216
    8.5.4 Enantioselective Capillary Electrophoresis in Food Quality 217
8.6 Conclusion 217
References... 217

8.1 INTRODUCTION

Chirality is a common phenomenon in many biologically active compounds such as polysaccharides and proteins. It arises when a molecule has either an asymmetric carbon or perpendicular lines that are not symmetrical and cannot freely rotate against each other (Sanganyado et al., 2020). Many synthetic compounds available on the market have at least one asymmetric carbon atom, also known as a chiral center, and exist as enantiomers. These enantiomers are not superimposable, and they rotate the plane of polarized light either to the right (dextrorotatory) or to the left (levorotatory). Enantiomers interact differently with different biological receptors, transport systems, and enzymes; hence, their enantiospecific biological activities (Alvarez-Rivera et al., 2020).

Food products owe their taste, texture, and aroma to a diverse group of molecules some of which are chiral. The odor and taste of a foodstuff may depend on the enantiomeric composition of the chiral food molecules (Figure 8.1) (Alvarez-Rivera et al., 2020). For example, spearmint and caraway fruits have different aroma despite arising from the same chemical carvone. The observed differences in aroma are due to the stereochemistry of carvone. (R)-carvone has a distinct spearmint

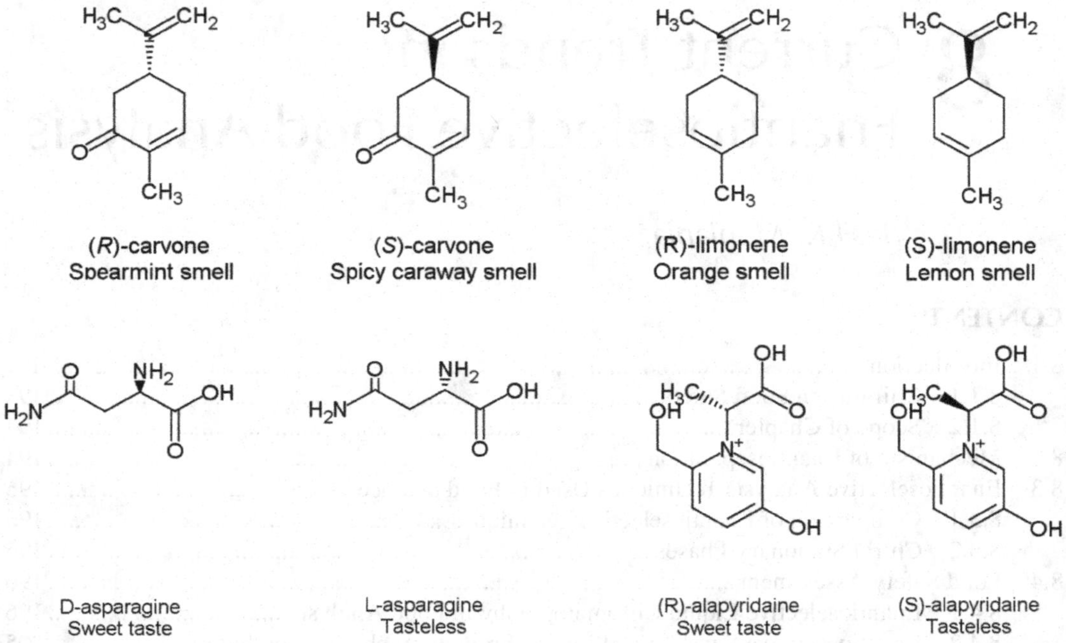

(R)-carvone Spearmint smell	(S)-carvone Spicy caraway smell	(R)-limonene Orange smell	(S)-limonene Lemon smell
D-asparagine Sweet taste	L-asparagine Tasteless	(R)-alapyridaine Sweet taste	(S)-alapyridaine Tasteless

FIGURE 8.1 Examples of stereospecificity in taste and flavor of natural products.

aroma while (S)-carvone has a spicy scent (Rocco et al., 2013). Citrus peel contains a chiral cyclic monoterpene aliphatic hydrocarbon called limonene. (R)-limonene has a distinct orange smell while (S)-limonene a distinct lemon smell. Enantiomeric differences have been observed in the taste and flavor of food chemicals such as asparagine and alapyridaine. The D-asparagine is sweet while the L-enantiomer is tasteless. Alapyridaine (N-(1-carboxyethyl)-6-hydroxymethyl-pyridinium-3-ol inner salt) is produced through the Maillard reaction during food processing when glucose and L-alanine are heated. Despite being tasteless, (S)-alapyridaine can enhance various tastes such as the salty, sweet, and umami tastes of NaCl, glucose, and monosodium L-glutamate, respectively (Soldo et al., 2003). Unlike its antipode, (R)-(-)-alapyridaine did not enhance the sweet, salty, and umami taste of beef bouillon (Hofmann et al., 2005). Human sensory receptors are three-dimensional biomolecules that can stereorecognize enantiomers.

Enantiomeric composition can also influence the nutritional value of foodstuff. Humans oftentimes metabolize one enantiomer while the antipode is eliminated unchanged without contributing any nutritional value. Previous studies showed that the D-form of Vitamin C had lower nutritional value compared to the L-form (Dabbagh and Azami, 2014). The human body can metabolize L-vitamin C but not the D-form.

The enantiomeric composition of food molecules can change during food processing and storage. Previous studies demonstrated that fermentation process shifts the enantiomeric composition of food molecules in addition to adding other chiral molecules (Armstrong et al., 1990). Wine owes its taste from the molecules that come from the grapes, yeast fermentation, and oak wood. Oak wood barrels releases lyoniresinol, a compound with two chiral centers, during wine aging. Previous studies showed that (+)-lyoniresinol had a strong bitter taste, (-)-lyoniresinol was tasteless while its diastereomer *epi*-lyoniresinol exhibited slight sweetness (Cretin et al., 2015). These results showed that enantiomeric composition of lyoniresinol in wine can affect wine taste and consumer perception. Natural flavors and fragrances are chiral compounds. Biosynthetic processes in plants are highly stereospecific and result in the formation of almost enantiopure flavors and fragrances (Sanganyado et al., 2020). In contrast, industrial synthesis of flavors and fragrances yield racemic mixtures. For

Current Trends in Chiral Food Analysis

example, natural apricot flavor is due to (R)-(+)-γ-decalatone (94–100%), however commercial products were found to contain racemic mixtures of γ-decalatone (Ravid et al., 2010). Amino acids in fruit juices are mostly found in the L-form (D'Orazio et al., 2017). Hence, enantioselective analysis can play a key role in food authentication.

8.1.1 Chirality in Food Safety

Organic pollutants discharged from agricultural, industrial, and domestic activities enter the environment and subsequently the food supply. For example, pesticide use in agriculture results in contamination of edible crops. The use of biosolids for soil amendment and recycled wastewater can result in uptake of pharmaceutical and personal care products by the crops (Fu et al., 2016). Antibiotics are extensively used in animal husbandry; hence, several studies have detected antibiotics in fish, beef, pork, and milk (Hernández et al., 2007; S. Li et al., 2018; McEachran et al., 2015). Besides environmental transfer, organic pollutants can be introduced into food during preparation, packaging, storage, and distribution. For example, polycyclic aromatic hydrocarbons can be released when the food is prepared at high temperatures while plasticizers and flame retardants can be released by the food packaging into the food (Shaw et al., 2014). Exposure to environmental organic pollutants through food consumption may cause adverse health effects in humans such as cardiovascular diseases, endocrine disruption, cancer, diabetes, congenital disorders, and dysfunctional immune systems (Guo et al., 2019). However, some organic pollutants such as pesticides, pharmaceuticals, personal care products, and flame retardants are chiral compounds with enantiospecific toxicities to humans (Liu et al., 2019; Sanganyado, 2019; Sanganyado et al., 2017; Stanley et al., 2007, 2006; Stanley and Brooks, 2009). Hence, understanding the enantiomeric composition of organic food contaminants is critical for accurate human risk assessment.

8.1.2 Scope of Chapter

There is a need for precise and accurate analytical methods that can determine the enantiomeric ratios of the compounds present in the food as a measure of quality assurance, food safety, and security (Zor et al., 2019). This chapter discusses the developments in various aspects of the analytical process in the determination of chiral food components and contaminants. The mechanism of enantioseparation will be discussed first followed by the application of different separation techniques in food safety and quality assessment. Table 8.1 shows the current applications of enantioselective analysis in food.

TABLE 8.1
An Overview of Current Applications of Enantioselective Analysis in Food

Application of Enantioselective Analysis in Food and Beverage Studies

Determination of aroma components and development of flavors closely approximating natural flavors
Identification of markers to assess authenticity and adulteration of foods and beverage
Evaluation of processing and storage time effects
Age dating
Investigation of health-promoting compounds
Control and monitoring of fermentation processes, and microbiological activity in general
Analysis of chiral metabolites
Biotransformation of persistent pollutants

Source: Reprinted from *TrAC Trends in Analytical Chemistry*, Vol. 52, Anna Rocco, Zeineb Aturki, Salvatore Fanali, Chiral Separations in Food Analysis, pp. 206–225, 2013. With permission from Elsevier.

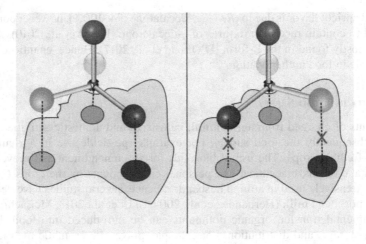

FIGURE 8.2 Chiral molecular interaction showing that the *left* enantiomer has a stronger affinity for the active site on the chiral selector while the *right* enantiomer binds to the active site weakly. (Reprinted from *Water Research*, Vol. 124, Edmond Sanganyado, Zhijiang Lu, Qiuguo Fu, Daniel Schlenk, Jay Gan, Chiral pharmaceuticals: A review on their environmental occurrence and fate processes, pp. 527–542, 2017. With permission from Elsevier.)

8.2 MECHANISM OF ENANTIOSEPARATION

Enantiomers of chiral compounds have similar physicochemical properties in an achiral environment but differ in a chiral environment. Gas chromatography, liquid chromatography, supercritical fluid chromatography, and capillary electrophoresis are frequently used to analyze chiral residues in foods (Alvarez-Rivera et al., 2020; Rocco et al., 2013). Chiral selectors are often used to create a chiral environment in the columns (Lämmerhofer, 2010; Ribeiro et al., 2012a; Scriba, 2016). During enantioselective analysis, enantiorecognition occurs due to the differences in stability of the diastereomeric complexes formed by the enantiomers and the chiral selector (Scriba, 2016). An enantiomer that binds strongly to the chiral selector is retained by the chiral selector longer than the weakly bound antipode. Enantiorecognition is often explained by the three-point interaction model as shown in Figure 8.2.

The molecular interactions between the enantiomer and the chiral selector at the three points are often electric, hydrogen bond, steric hindrance, π-π, ion-dipole, dipole-dipole, dipole-induced-dipole, and van der Waals interactions, in order of decreasing strength (Table 8.2) (Berthod, 2009, 2006; Berthod et al., 2010). Polar ionogenic compounds such as pharmaceuticals often interact with chiral selectors via hydrogen bonding and ionic interactions (Sanganyado et al., 2014). However, steric hindrance and π-π interactions are more important in separation of bulky chiral analytes such as aromatic biaryls atropisomers.

In chromatography and electrophoresis, enantiorecognition is often achieved using a chiral stationary phase in the separation column following mobile phase optimization (Evans and Kasprzyk-Hordern, 2014; Sanganyado et al., 2014; Tang, 1996). A chiral stationary phase contains a chiral selector bound or coated to stationary silica or polymeric materials. In capillary electrophoresis, the chiral selector is added to the background electrolyte or bound to the capillary wall (D'Orazio et al., 2017). The type and degree of enantioresolution achieved by the chiral stationary phase is influenced by the composition and spatial arrangement of the stationary phase and the chiral selector together with the physicochemical properties of the enantiomer (Alvarez-Rivera et al., 2020). In the indirect approach, the enantiomers react with the chiral selector forming diastereoisomers that can be separated using conventional stationary phases (D'Orazio et al., 2017). Table 8.3 shows different

TABLE 8.2
An Overview of the Characteristics on Different Types of Molecular Interaction in the Three-Point Interaction Model

Type of Interaction	Strength	Direction	Range (d)
Coulomb or electric	Very strong	Attractive medium or repulsive	Medium ($1/d^2$)
Hydrogen bond	Very strong	Attractive	Long
Steric hindrance	Very strong	Attractive (donor or acceptor) or repulsive	Very short
π-π	Strong	Attractive	Medium
Ion–dipole	Strong	Attractive	Short
Dipole–dipole	Intermediate	Attractive	Short ($1/d^3$)
Dipole-induced-dipole	Weak	Attractive	Very short ($1/d^6$)
London dispersion or van der Waals	Very weak	Attractive	Very short ($1/d^6$)

Source: Reprinted with permission from Berthod, A. 2006 *Anal. Chem.* 78(7), 2093–2099. Copyright (2006) American Chemical Society.

TABLE 8.3
Mechanisms of Enantiorecognition of Different Types of Chiral Selectors and the Applications

Chiral Selector	Mechanisms and Interactions	Separation Technique
Cyclodextrins	*Inclusion complexation* Hydrogen bond π-π Dipole-dipole	GC, CE, CEC
Polysaccharide derivatives	*Insertion into helical structures* Hydrogen bond π-π	HPLC, SFC
Macrocyclic antibiotics	*Multiple binding sites* Hydrogen bond π-π Dipole-dipole Steric Electrostatic	HPLC, CE, CEC
Ion-exchangers	*Ion pairing* Electrostatic π-π Hydrogen bond	HPLC, SFC

Source: Reprinted from *TrAC Trends in Analytical Chemistry*, Vol. 123, Gerardo Alvarez-Rivera, Mónica Bueno, Diego Ballesteros-Vivas, Alejandro Cifuentes, Chiral Separations in Food Analysis, pp. 115–761, 2020. With permission from Elsevier.

types of chiral sectors used in enantioselective chromatography and capillary electrophoresis and their mechanisms of enantiorecognition.

8.3 ENANTIOSELECTIVE ANALYSIS TECHNIQUES USED IN FOOD SCIENCE

8.3.1 Comparison of Enantioselective Techniques

Gas chromatography (GC), liquid chromatography (LC), supercritical fluid chromatography (SFC), and capillary electrophoresis are the main techniques used in analysis of chiral compounds in foods (Miao et al., 2017; Raimbault et al., 2019; L. Zhao et al., 2019). Recently, the use of sensors has also been reported in chiral food analysis (Zor et al., 2019). The choice of the analytical technique depends on the physicochemical properties of the samples and the analyte, equipment and solvent availability, and cost of analysis. Thermolabile compounds are often analyzed using liquid chromatography, capillary electrophoresis, and supercritical fluid chromatography while gas chromatography is used for thermally stable enantiomers. Liquid chromatography is widely used in food contamination studies while gas chromatography is used in food authentication. Natural flavor and aroma compounds are often thermally stable volatile compounds; hence, the use of gas chromatography in their analysis. Use of capillary electrophoresis in enantioselective analysis has increased in recent years owing to its numerous advantages. Table 8.4 shows the advantages and disadvantages of different enantioselective analysis techniques in current use.

8.3.2 Chiral Stationary Phases

Most chiral separations on liquid chromatography have been carried out using polysaccharide, amylose derivatives, and cyclodextrin chiral stationary phases (Ribeiro et al., 2012b; Sanganyado et al., 2017; Ward and Ward, 2012). Other chiral stationary phases such as protein, ligand, macrocyclic antibiotics, and ion exchange have also been used (Lomenova and Hroboňová, 2018). The same chiral stationary phases used in liquid chromatography apply to supercritical fluid chromatography. On the other hand, capillary electrophoresis separations mainly used cyclodextrins, and micellar separations (Herrero et al., 2010). A few articles have utilized ligand exchange (Knob et al., 2013; Kodama et al., 2013). Additionally, different capillary electrophoresis modes have been applied in food analysis, and these include CD-modified capillary zone electrophoresis, CD-electrokinetic chromatography, and capillary electrochromatography (Domínguez-Vega et al., 2009; Herrero et al., 2010). Another new trend that has developed is the use of miniaturized liquid chromatography systems based on capillary electrophoresis, which offer higher sensitivity than classical liquid chromatography (D'Orazio et al., 2008). In gas chromatography, the most common chiral stationary phases are amino acid derivatives, and diamides, chiral metallic complexes, and cyclodextrin derivatives (Perez-Fernandez et al., 2011a).

8.4 FOOD SAFETY ASSESSMENT

8.4.1 Enantioselective Liquid Chromatography in Food Analysis

Food safety assessment involves determining the presence of contaminants such as pesticides, pharmaceuticals, and personal care products in foods. Some of these pesticides are chiral and require analytical techniques to resolve and accurately determine the enantiomeric ratios in foodstuffs. Most of the studies on pesticide residues have been carried out using liquid chromatography (HPLC), which can be coupled to selective detectors such as UV-Vis (He et al., 2018), diode array detector (DAD) (Dong et al., 2013), or mass spectrometers such as time-of-flight (TOF) (Qi et al., 2016), triple quadrupole (QQQ), and Orbitrap (Zhang et al., 2019). The classes of pesticides that have been studied include fungicides such as triazoles (Perez-Fernandez et al., 2011b; Q. Zhang et al., 2018a,

TABLE 8.4
Comparison of Enantioseparation Techniques Currently Used in Food Analysis

Separation Technique	Advantages	Limitations
GC	– Best choice for volatile compounds analysis – Instrumentation commercially available – Possibility of two-dimensional chromatography	– The variety of CSPs for GC chiral analysis is limited – Mainly used for analytical purposes
HPLC	– Applicability to preparative and analytical purposes – Easy operation mode – Wide range of CSPs available – Possibility of two-dimensional chromatography	– Use of organic solvents in the separation process – Chiral columns are more expensive than the ones for GC
SFC	– Applicability to preparative and analytical purposes – Short equilibration time – Short analysis time – Green features – The CSPs developed for HPLC can be used in SFC – The use of neat CO_2 provides the recovery of solvent-free products	– High pressure equipment – Maintenance costs – The solubility of polar compounds should be enhanced with the use of co-solvents, sometimes not as green as CO_2
CE	– Short equilibration time – Very short analysis time – Low analysis costs – Low consumption of sample, organic solvents and CSs – High flexibility in the selection and combination of CSs	– Used only for analytical purposes – Low sensitivity
CEC	– Low analysis costs – Low consumption of sample, organic solvents – Easier coupling with MS – The instrumentation employed is the same as in CE	– Used only for analytical purposes – Low sensitivity

Source: Reprinted from *TrAC Trends in Analytical Chemistry*, Vol. 123, Gerardo Alvarez-Rivera, Mónica Bueno, Diego Ballesteros-Vivas, Alejandro Cifuentes, Chiral Separations in Food Analysis, pp. 115–761, 2020. With permission from Elsevier.

GC: gas chromatography; HPLC: high-performance liquid chromatography; SFC: supercritical fluid chromatograph; CE: capillary electrophoresis; CEC: capillary electrochromatography; CSPs: chiral stationary phases; CSs: chiral selectors; MS: mass spectrometry.

2018b), synthetic pyrethroids (Liu et al., 2005; Qin et al., 2006), organophosphates (Sun et al., 2012), acylanilides, imidazolinones, and phenoxy propanoic-acid herbicides (Liu et al., 2009; Xie et al., 2016; Ye et al., 2009).

Liquid chromatography coupled to selective detectors such as UV and diode array has been used to determine the enantiomers of pesticides and their metabolites in various food matrices (Table 8.5). In most studies carried out, polysaccharide based stationary phases have been used more than any other

TABLE 8.5
Selected Applications of HPLC to Food Chiral Analysis

Analytes	Matrix	Column / Chiral Selector	Experimental Conditions	Detector	Reference
Nutraceuticals (flavanone, 4'-methoxyflavanone, 7-methoxyflavanone, 2'-hydroxyflavanone, 4'hydroxyflavanone, 6-hydroxyflavanone)	–	No commercial column (250 mm × 4.6 mm I.D., 5 μm) C.S. (immobilized): amylose 3,5-dimethylphenylcarbamate (ADMPC) or cellulose 3,5-dichlorophenylcarbamate (CDPC)	MeOH or EtOH or 2-propanol or AcN or hexane/2-propanol or EtOH 1 mL/min	UV (λ = 220 nm)	Fanali et al. (2016)
Flavanone derivatives	–	No commercial column (100 mm × 4.6 mm I.D.) C.S. (Cu – homochiral metal–organic frameworks (MOFs)): (R)-2,2'-dihydroxy-1,1'-binaphthalene-6,6'-dicarboxylic acid	Hexane/CH$_2$Cl$_2$ (1 mL/min)	UV (λ = 254 nm)	Tanaka et al. (2016)
Flavanone, naringenin and hesperetin	Liquorice, sweet, hot, and chilli pepper	Chiralpak AD-3R (150 mm × 2.1 mm I.D.; 3 μm) C.S.: Amylose tris(3,5-dimethylphenylcarbamate)	MeOH (0.5 mL/min)	MS/MS	Baranowska et al. (2016)
Oxyneolignan	Bamboo	CHIRALCEL-IC (250 mm × 4.6 mm ID) C.S. (immobilized): Cellulose tris(3,5-dichlorophenylcarbamate)	1. Oxyneolignans A: n-hexane:EtOH:TFA (70:30:0.1, v/v/v) (0.8 mL/min) 2. Oxyneolignans B: n-hexane:EtOH:TFA (80:20:0.1, v/v/v) (0.8 mL/min) 3. Oxyneolignans C and D: n-hexane:EtOH:TFA (80:20/0.1, v/v/v) (1 mL/min)	UV λ = 220 nm	Sun et al. (2015)
6-Prenylnaringenin (6PN)	natural health products and dietary supplements	Chiralpak® AD-RH C.S.: Amylose tris(3,5-dimethylphenylcarbamate)	ACN and 10 mM ammonium formate (pH 8.5) (39:61, v/v) (1.25 mL/min)	MS	Martinez and Davies (2015)
Poly-γ-glutamate	Fermented soybean food natto	2D-LC 1D: monolithic octadecylsilane (1000 mm × 0.53 mm I.D.) 2D dimension: Sumichiral OA-2500S (250 mm × 1.5 mm I.D.) C.S.: 3,5-dinitrobenzoyl group	1D: 6% (v/v) ACN and 0.06% (v/v) TFA in water (0.25 mL/min) 2D: mixed solutions of MeOH and ACN containing citric acid or formic acid	Fluorescence $\lambda_{emission}$ = 530 nm $\lambda_{excitation}$ = 470 nm	Ashiuchi et al. (2015)

TABLE 8.5 (Continued)
Selected Applications of HPLC to Food Chiral Analysis

Analytes	Matrix	Column Chiral Selector	Experimental Conditions	Detector	Reference
Flavanone, 6-hydroxyflavanone	—	Chiralpak ID (250 mm × 4.6 mm I.D., 5 µm) or Chiralpak ID 3 (150 mm × 0.30 mm I.D, 3 µm) C.S.: amylose tris-(3-chlorophenylcarbamate)	n-hexane/2-propanol 90:10 v/v	UV $\lambda = 254$ nm	Ahmed et al. (2014)
Flavanone, 7-methoxyflavanone, 6-methoxyflavanone, 7-hydroxyflavanone, 4-hydroxyflavanone, hesperetin, naringenin	—	No commercial column (250 mm × 4.6 mm I.D.) C.S. (immobilizer): perphenylcarbamate cyclodextrin	MeOH/H_2O = 50/50 (v/v), 1.0 mL/min or MeOH/H_2O, 80:20 (v/v), 1.0 mL/min	UV $\lambda = 250$ nm	Pang et al. (2014)
D-amino acids	Cheese	GraceSmart RP 18 (250 mm × 4.6 mm I.D., 5 µm) C.S.: chiral ligand-exchange chromatography (CLEC) stationary phase based on S-trityl-L-cysteine (L-STC) units	0.25 mM Cu(II) nitrate aqueous solution (0.1, 0.3, 0.5 or 1 mL/min)	UV $\lambda = 210$ and 254 nm	Sardella et al. (2013)
D-amino acids	Japanese traditional black vinegars	2D-LC ¹D: monolithic ODS (1000 mm × 0.53 mm I.D.) ²D: Sumichiral OA-2500S (250 mm × 1.5 mm I.D.) and Chiralpak QN-AX (250 mm × 1.5 mm I.D.) C.S.: 3,5-dinitrobenzoyl group or (Immobilized) O-9-(tert-butylcarbamoyl) quinine	MeOH/ACN containing citric acid or formic acid	Fluorescence $\lambda_{emission} = 530$ nm $\lambda_{excitation} = 470$ nm	Miyoshi et al. (2014)
D- and L-lactic acids	Yogurts and fermented milk drinks	L-column 2 ODS (150 mm × 2.1 mm I.D., 5 µm) C.S.: 4-(4,6-dimethoxy-1,3,5-triazin-2-yl)-4-methylmorpholinium chloride (DMT-MM) as an enantioseparation enhancer	ACN containing 0.1% formic acid (0.2 mL/min)	Fluorescence $\lambda_{emission} = 580$ nm $\lambda_{excitation} = 470$ nm	Todoroki et al. (2014)

(Continued)

TABLE 8.5 (Continued)
Selected Applications of HPLC to Food Chiral Analysis

Analytes	Matrix	Column Chiral Selector	Experimental Conditions	Detector	Reference
Spirocyclicterpenoids	—	All columns: 250 mm × 4.6 mm I.D. CHIRALPAK AD, AY, AS, CHIRALCEL OD, OJ, OZ, C.S. (coated): polysaccharide phases CHIRALPAK IA, CHIRALPAK IB, CHIRALPAK IC, CHIRALPAK IF, CHIRALPAK IE, CHIRALPAK ID C.S. (immobilized): polysaccharide phases, Lux Amylose-2, Lux Amylose-4 C.S.: Amylose derivative CSPs Pirkle- or "brush"-type phases (R,R and S,S) Whelk-O1)	MeOH, EtOH, iPrOH, ACN (HPLC, 1 mL/min; SFC, 2.5 mL/min)	DAD	Schaffrath et al. (2014)
Diketopiperazine indole alkaloids	Hemp seed	Chiralpak AS-H (250 mm × 4.6 mm I.D., 5 μm) C.S.: Amylose tris-[(S)-α-methylbenzylcarbamate]	Hexane/isopropanol/diethylamine (4:1:0.05, v/v/v)	MS	Yan et al. (2016)
Methyl jasmonate		Semi-preparative Nucleodex β-PM (30 m × 20.0 mm I.D., 5 μm) C.S.: permethylated β-cyclodextrin	MeOH/water (55:45, v/v) (different flow rates)	DAD λ = 210 nm	Flores et al. (2013)
Lutein and the three stereoisomers of zeaxanthin	Fish flesh	System 1: PVA-Sil semipreparative column (100 mm × 10 mm I.D., 5 μm) System 2: Daicel Chiralpak AD-3 (250 × 4.6 mm I.D., 3 μm) C.S.: Amylose derivative coated on silica-gel	System 1: hexane:isopropanol (90:10, v/v) (2 mL/min) System 2: solvent A: hexane:isopropanol (95:5, v/v), solvent B: hexane:isopropanol (90:10, v/v) (0.5 mL/min)	DAD	Prado-Cabrero et al. (2016)
Flavanol enantiomers: (+)- and (−)-epicatechin and (+)- and (−)-catechin	Cocoa-based matrices	Astec cyclobond I-2000 RSP (250 mm × 4.6 mm I.D., 5 μm) C.S. (immobilized): (R,S)-Hydroxypropyl modified β-cyclodextrin	20 mM ammonium acetate buffer, pH 4.0–MeOH (70:30, v/v) (1 mL/min)	Fluorescence $\lambda_{emission}$ = 316 nm $\lambda_{excitation}$ = 276 nm	Machonis et al. (2012)

TABLE 8.5 (Continued)
Selected Applications of HPLC to Food Chiral Analysis

Analytes	Matrix	Column Chiral Selector	Experimental Conditions	Detector	Reference
Flavanol stereoisomers, caffeine and theobromine	Commercial chocolates	Astec Cyclobond I-2000 RSP (250 mm × 4.6 mm I.D., 5 μm) C.S. (immobilized): (R,S)-Hydroxypropyl modified β-cyclodextrin	A: 50 mM NaH_2PO_4 pH 3.0, B: 80% ACN in 30 mM NaH_2PO_4 pH 3.0 (1 mL/min)	Fluorescence $\lambda_{emission}$ = 310 nm $\lambda_{excitation}$ = 280 nm	Alañón et al. (2016)
Clenbuterol	Swine, beef, lamb meat	Chirobiotic T (150 mm, 2.1 mm I.D., 5 μm) C.S.: teicoplanin	10 mM ammonium formate in MeOH (0.4 mL/min)	MS	Z. L. Wang et al. (2016)
Isocarbophos	Orange pulp, peel, kumquat	– Lux Cellulose-1 (150 mm × 2.0 mm I.D., 3 μm) C.S.: cellulose tris (3,5-dimethylphenyl carbamate) – Lux Cellulose-2 (150 mm × 2.0 mm I.D., 3 μm) C.S.: cellulose tris (3-chloro-4-methyl phenylcarbamate) – Lux Cellulose-3 (150 mm × 2.0 mm I.D., 3 μm) C.S.: cellulose tris (4-methylbenzoate) – Lux Cellulose-4 (150 mm × 2.0 mm I.D., 3 μm) C.S.: cellulose tris (4-chloro-3-methylphenylcarbamate) – Lux Amylose-2 (150 mm × 2.0 mm I.D., 3 μm) C.S.: amylose tris (5-chloro-2-methyl phenylcarbamate) – Chiralpak AD-3R (150 mm × 2.1 mm I.D., 3 μm) C.S.: amylose tris (3,5-dimethylphenyl-carbamate)	MeOH or ACN/water containing 2 mM ammonium formate or 0.1% formic acid (0.25 mL/min)	MS	Qi et al. (2015)
Penconazole	Cucumber, tomato, head cabbage, pakchoi	Lux Cellulose-2 (150 mm × 2.0 mm I.D., 3 μm) C.S.: cellulose tris (3-chloro-4-methylphenylcarbamate)	MeOH/2 mM ammonium acetate buffer solution containing 0.1% formic acid (70:30, v/v) (0.2 mL/min)	MS	Wang et al. (2014)
Metalaxyl	Cucumber, cabbage, spinach, pakchoi	Lux Cellulose-2 (250 mm × 4.6 mm I.D.) C.S.: Cellulose tris(3-chloro-4-methylphenylcarbamate)	ACN-0.1% formic acid solution (40:60, v/v) (1 mL/min)	UV λ = 220 nm	Q. Zhang et al. (2014)

(Continued)

TABLE 8.5 (Continued)
Selected Applications of HPLC to Food Chiral Analysis

Analytes	Matrix	Column Chiral Selector	Experimental Conditions	Detector	Reference
Sulfoxaflor	Brown rice, cucumber, apple	– Chiralcel OD-H (250 mm × 4.6 mm I.D., 5 μm) C.S.: cellulose tris-3,5-dimethylphenylcarbamate – Chiralcel OJ-H (250 mm × 4.6 mm I.D., 5 μm) C.S.: cellulose tris-4-methylbenzoate – Chiralpak AS-H (250 mm × 4.6 mm I.D., 5 μm) C.S: amylose tris-[(S)-a-methylbenzylcarbamate] – Chiralpak AD-H (250 mm × 4.6 mm ID,5 μm) C.S.: amylose tris-3,5-dimethylphenylcarbamate – ChromegaChiral CCA (250 mm × 4.6 mm I.D., 5 μm)	n-hexane/EtOH/MeOH (90:2:8, v/v/v) (1 mL/min)	DAD $\lambda = 230$ nm	Chen et al., (2014)
Cyflumetofen	Cucumber, tomato, and apple	Chiralpak AD-H (250 mm × 4.6 mm I.D., 5 μm) C.S.: amylose tris-3,5-dimethylphenylcarbamate	n-hexane and 2-propanol (95:5, v/v) (1 mL/min)	UV $\lambda = 234$ nm	Chen et al., (2013)
Mandipropamid	Potato, pepper, grape, watermelon	– Lux Cellulose-1 (150 mm × 2.0 mm I.D., 5 μm) C.S.: cellulose tris(3,5-dimethylphenylcarbamate) – Lux Cellulose-2 (150 mm × 2.0 mm I.D., 3 μm) C.S.: cellulose tris(3-chloro-4-methylphenylcarbamate) – Lux Cellulose-3 (150 mm × 2.0 mm I.D., 3 μm) C.S.: cellulose tris(4-methylbenzoate) – Lux Amylose-2 (150 mm × 2.0 mm I.D., 3 μm) C.S.: amylose tris(2-chloro-5-methylphenylcarbamate) – Chiralpak AD-3R (150 mm × 2.1 mm I.D., 3 μm) C.S.: amylose tris (3,5-dimethylphenyl-carbamate)	ACN with 0.1% formic acid or MeOH with 0.1% formic acid combined with 0.1% aqueous formic acid (0.3 mL/min)	MS	H. Zhang et al., (2014)
Dinotefuran	Cucumber and soil	ChromegaChiral CCA (250 mm × 4.6 mm I.D., 5 μm) C.S.: Amylose tris (3,5-dimethylphenylcarbamate)	n-hexane–EtOH–MeOH (85:5:10, v/v/v) (1 mL/min)	DAD $\lambda = 270$ nm	Chen et al. (2015)
Flufiprole	Vegetables, fruits, and soil	Lux Cellulose-2 (250 mm × 4.6 mm I.D., 5 μm) C.S.: cellulose tris (3-chloro-4-methylphenylcarbamate)	ACN/H$_2$O or MeOH H$_2$O (0.7 mL/min)	UV $\lambda = 230$ nm	Tian et al. (2015)

TABLE 8.5 (Continued)
Selected Applications of HPLC to Food Chiral Analysis

Analytes	Matrix	Column Chiral Selector	Experimental Conditions	Detector	Reference
Flutriafol	Cucumber, tomato, grape, pear, wheat, soil, and water	Lux Cellulose-2 (250 mm × 4.6 mm I.D.) C.S.: cellulose tris(3-chloro-4-methyl phenyl carbamate)	ACN/water (40:60, v/v) (0.8 mL/min)	UV $\lambda = 210$ nm	Q. Zhang et al. (2014)
Benalaxyl	Grape and wine	CDMPC chiral stationary phase column (250 mm × 4.6 mm I.D.) C.S.: cellulose tris-(3, 5-dimethylphenyl-carbamate)	n-hexane/i-PrOH (90:10, v/v) (1 mL/min)	UV $\lambda = 230$ nm	Lu et al. (2016)
Triticonazole	Soil, cucumber, tomato, pear, cabbage, and apple	Lux Cellulose-2 (250 mm × 4.6 mm I.D., 3 μm) C.S.: cellulose tris(3-chloro-4-methyl phenyl carbamate)	ACN/water (0.7 mL/min)	UV $\lambda = 260$ nm	Zhang et al. (2016)
Isofenphos-methyl	Vegetables, fruits, soil	Lux Cellulose-3 (250 mm × 4.6 mm I.D., 5 μm) C.S.: Cellulose-tris-(4-methylbenzoate)	ACN/water/MeOH (31:57:12, v/v/v) 0.8 mL/min	UV $\lambda = 228$ nm	Gao et al. (2016)

Source: Reprinted from *TrAC Trends in Analytical Chemistry*, Vol. 96, Giovanni D'Orazio, Chiara Fanali, María Asensio-Ramos, Salvatore Fanali, Chiral Separations in Food Analysis, pp. 151–171, 2017. With permission from Elsevier.

type in the normal or reverse phase (Li et al., 2012). Enantioselective determination of metconazole was carried out in soil and flour using HPLC equipped with a UV detector (He et al., 2018). Of the four-polysaccharide based chiral stationary phases that were evaluated, Enantiopak OD had the best separation efficiency. In a different study, enantioseparation and determination of triticonazole enantiomers in fruits and vegetables was done on HPLC coupled to a UV detector (Zhang et al., 2016). Complete enantioseparation was achieved using a cellulose tris (3-chloro-4-methyl phenyl carbamate column with a relatively good resolution (Rs=14.04). Additionally, the stereochemical structure of the triticonazole enantiomers was determined using a combination of experiments and predicted using electronic circular dichroism spectra (ECD). The same research group used the same technique and a combination of experimental and circular dichroism spectra to determine enantioseparation of flutriafol in fruits and vegetables (Q. Zhang et al., 2014). More recently, two-dimensional liquid chromatography coupled to a diode array detector was applied to determine eight triazole fungicides, and their enantiomers in fruits and vegetables (Zeng et al., 2019). The stationary phases used were C18 and cellulose-tris (3,3-dimethylphenylcarbamate), and the detection limits were between 0.03113 and 0.3525 µg/ml. Liquid chromatography coupled to selective detectors has also been used to investigate the enantiospecific degradation of triazole fungicides (Z. Zhang et al., 2018). In another study, liquid chromatography coupled to a UV detector has been successfully applied to investigate the stereoselective behavior of benalaxyl during grape growth and the wine making process (Lu et al., 2016). It was reported that the degradation of the (*R*) enantiomer degraded faster than the (*S*) enantiomer in grapes, and they were hardly degraded in the fermentation process.

Enantioselective liquid chromatography is widely used in assessing the fate of chiral pollutants in food. Enantiomeric degradation of hexaconazole was investigated in tomato and cucumber using chiral liquid chromatography with tandem mass spectrometry (LC-MS/MS) (Li et al., 2013). Separation was done on a reverse-phase Chiralcel OD-RH column, and the (+) enantiomer showed a faster degradation than the (-) one. Liquid chromatography-tandem mass spectrometry has also been used to determine fenbuconazole and myclobutanil residues, as well as their degradation kinetics in strawberry (Zhang et al., 2011). Resolution of the enantiomers of these compounds was performed on a cellulose tris (3,5-dimethylphenylcarbamate) column. Degradation of fenbuconazole was not enantioselective, while that of myclobutanil showed enantiospecificity with the (+) enantiomer degrading faster than the (-) one. Using the same technique, a Lux Cellulose-2 chiral column was used to separate penconazole enantiomers in vegetables (Wang et al., 2014). Preferential degradation was observed in (-) penconazole in cabbage and pakchoi, but this enantioselective behavior was not observed in cucumber and tomato. Enantiomeric separation of famoxadone enantiomers in tomato, grape, and apple was successfully studied using chiral liquid chromatography tandem mass spectrometry using a Lux Amylose-1 chiral stationary phase (Xu et al., 2018). A distinct advantage of the column was that it improved separation efficiency while decreasing analysis time. Similarly, chiral separation and enantioselective degradation of zoxamide enantiomers was successfully investigated using ultra-high-performance liquid chromatography tandem mass spectrometry on a Lux Amylose-2 column (Pan et al., 2017). Lastly, the degradation of tetraconazole enantiomers during strawberry wine processing was studied using ultraperformance convergence chromatography tandem triple quadrupole mass spectrometry (Liu et al., 2018). It was reported that (-) tetraconazole degraded faster than (+) tetraconazole.

A recent trend has been the analysis of multiple pesticides using liquid chromatography tandem mass spectrometry. Multiresidue analysis, although challenging for compounds that have diverse physicochemical properties, offers quick analysis time and cost effectiveness. A study was carried out to simultaneously determine six chiral pesticides in functional foods using chiral liquid chromatography coupled with tandem mass spectrometry, fitted with a Chiralpak IG column for chiral recognition (Zhao et al., 2019a). In a different study, the same group used the same method to analyze for 22 chiral pesticides in fruits and vegetables within 47 min (Zhao et al., 2019b). It is worth noting that multiresidue analyses of food samples require highly selective and sensitive analytical techniques to detect the analytes which are in low concentrations (Carrão et al., 2020).

8.4.2 ENANTIOSELECTIVE SUPERCRITICAL FLUID CHROMATOGRAPHY IN FOOD SAFETY

Supercritical fluid chromatography is widely used in food safety assessment. Separation in supercritical fluid is achieved by optimizing various factors such as chiral column, composition of mobile phase, flow rate of mobile phase, column temperature, and pressure (R. Li et al., 2018; Tao et al., 2018). The chiral stationary phases that have been used include cellulose (tris-3,5-dimethylphenylcarbamate) (Pan et al., 2016), amylose (tris-3,5-dimethylphenylcarbamate) (Cheng et al., 2016), and cellulose tris(4-methylbenzoate) (Tao et al., 2018). Cellulose based chiral stationary phases were found to give more efficient separation results than amylose-based ones. For instance, the potential of supercritical fluid chromatography coupled to mass spectrometry (SFC-ESI-MS/MS) to efficiently separate the enantiomers of cyhalothrin and metalaxyl was investigated to polysaccharide based and amylose based stationary phases in tomato, orange, leek, and cayenne (Cutillas et al., 2020). The study reported that the polysaccharide based chiral stationary phase gave better separation results than the amylose based chiral stationary phase for both pesticides. Supercritical fluid chromatography coupled to quadrupole time-of-flight mass spectrometry (SFC-Q-TOF/MS) was used to separate the enantiomers of diniconazole in tea, grape, and apple samples (Xinzhong et al., 2018). Optimal separation was obtained using a Chromega chiral CCA column using methanol as a mobile phase modifier. However, incomplete separation was obtained using the same chiral stationary phase and acetonitrile based mobile phase modifier. This has been attributed to the fact that methanol is a hydrogen bond donor and acceptor, whereas acetonitrile is a weak hydrogen bond acceptor (Tao et al., 2014). Chiral supercritical fluid mass spectrometry has also been used to determine other pesticides such as prothioconazole (Jiang et al., 2018), thiacloprid (R. Li et al., 2018), dinotefuran (Chen et al., 2015), and several other neonicotinoid pesticides and their metabolites (Pérez-Mayán et al., 2020).

8.4.3 ENANTIOSELECTIVE GAS CHROMATOGRAPHY IN FOOD SAFETY

Chiral gas chromatography has also been used to determine pesticides in food products, but not as liquid chromatography (Table 8.6). In most of the studies carried out so far, chiral stationary phases based on cyclodextrin were used (Kuang et al., 2010). Enantioselective determination of simeconazole was carried out on a gas chromatography coupled to an ion trap mass spectrometer fitted with a cyclodextrin based chiral stationary phase (BGB-172) (Li et al., 2011). The method successfully resolved the enantiomers of simeconazole in cucumber, tomato, pear, wheat, and rice, and the limits of detection of each enantiomer were between 0.4 and 0.9 µg/kg. Another study was also carried out to resolve the enantiomers of acephate, and its metabolite methamidophos in vegetables using gas chromatography tandem mass spectrometry fitted with a cyclodextrin based chiral stationary phase (CP-Chirasil-Dex CB column) (Wang et al., 2013). The method was successful in resolving the enantiomers and reported their enantioselective metabolism. Lastly, 17 pyrethroids and chlorpyrifos were analyzed in the fatty content of animal products using gas chromatography tandem mass spectrometry (Dallegrave et al., 2016). The diastereoisomers of cyfluthrin and cypermethrin were resolved using a chiral column based on β-CD derivative. The authors reported that chiral analysis of recovery data showed predominance of cis isomers over trans while limits of detection ranged from 0.002 to 6.43 ng/g lipid weight.

8.4.4 ENANTIOSELECTIVE CAPILLARY ELECTROPHORESIS IN FOOD SAFETY

Capillary electrophoresis and capillary electrokinetic chromatography have also been used in food safety assessment because of their low cost, short analysis time, and selectivity, but not as much as the chromatographic techniques (Herrero et al., 2010; Simó et al., 2003). The most widely used chiral selectors for capillary electrophoresis and capillary electrokinetic chromatography have been the cyclodextrins (Table 8.7). In addition, chiral analysis using capillary electrophoresis has

TABLE 8.6
Enantioselective Gas Chromatography in Food Analysis

Analytes	Matrix	Chiral Selector	Experimental Conditions	Mode/Detection	Sample Preparation	Reference
γ- and δ-lactones, α-ionone, linalool, nerolidol, ethyl 2-methylbutyrate, 2-methylbutyric acid, 2-methylbutanol	Peach, coconut, apricot, raspberry, strawberry, melon	Columns coated with $6^{I\text{-}VII}$-O-TBDMS-$2^{I\text{-}VII}$-$3^{I\text{-}VII}$-O-acetyl-β-CD (AcAc-CD) and $6^{I\text{-}VII}$-O-TBDMS-$2^{I\text{-}VII}$-ethyl-$3^{I\text{-}VII}$-O-methyl-b-CD (EtMe-CD), both diluted at 30% in PS086. For each chiral selector three different column dimensions used: 25 m × 0.25 mm I.D. × 0.25 μm film thickness, for AcAc-CD and 25 m × 0.25 mm I.D., 0.15 μm film thickness, for EtMe-CD, and two 0.10 mm I.D. × 0.10 μm film thickness narrow bore columns 11.7 m for AcAc-CD, and 11.3 m for EtMe-CD, and 5 m long, respectively.	He carrier gas T program 50–220°C	MS LODs: 1–3 ppb	Rapid TAS by on line combining headspace-SPME with enantioselective GC-MS	Cagliero et al. (2012)
Monoterpenes (133 compounds which comprise mono- and sesquiterpenes, their oxygenated derivatives; chiral: α-pinene, sabinene, camphene, β-pinene, limonene, β-phellandrene, neomenthol)	Needles and berries of the *Juniperus communis* and *J. oxycedrus*	Cyclodex-B capillary column 30 m × 0.256 mm I.D., 0.25 μm film thickness	He gas carrier T program 40–180°C	MS	Hydrodistillation and headspace-SPME (three beds 50/30 μm divinylbenzene-carboxenpolydimethylsiloxane)	Foudil-Cherif and Yassaa (2012)
Linalool, wine lactone, β-citronellol, *cis*-rose oxide	Sour citrus fruits (Yuzu, Sudachi and Kabosu)	GC × GC: BC-WAX column 30 m × 0.25 mm I.D., 0.25 μm film thickness × β-DEX 225 column 30 m × 0.25 mm I.D., film thickness of 0.25 μm	He carrier gas Different T program	FID	Peel: extraction with the solvent mixture (pentane–diethyl ether, 2:1 by volume). Volatile components isolated by solvent assisted flavor evaporation technique. Juice: squeezed, saturated with brine to inhibit enzymatic reactions and to give the salting-out effect, extracted with the solvent mixture. Volatile components prepared under the same conditions used for peel extracts.	Tomiyama et al. (2012)

TABLE 8.6 (Continued)
Enantioselective Gas Chromatography in Food Analysis

Analytes	Matrix	Chiral Selector	Experimental Conditions	Mode/Detection	Sample Preparation	Reference
Simeconazole	Cucumber, tomato, apple, pear, wheat, rice	BGB-172 column 30 m × 0.25 mm I.D. × 0.25 μm film thickness (20% tert-butyldimethylsilyl-β-CD in 15% diphenyl- and 85% dimethyl-polysiloxane)	He carrier gas T program 150–240°C	EI-MS LODs: 0.4–0.9 μg/kg	Preparative HPLC for single enantiomer. Sample smashed, homogenized with acetonitrile; addition of anhydrous magnesium sulfate (MgSO4) and sodium chloride (NaCl); vortexed; centrifuged; SPE.	Li et al. (2011)
Ethyl 2-methylbutanoate, linalool and 4-hydroxy-2,5-dimethyl-3(2H)-furanone (i.e., furaneol), methyl jasmonate	Strawberry	through oven transfer adsorption–desorption (TOTAD) interface for on-line coupling RPLC–GC (RPLC-TOTAD–GC) RPLC: 150 mm × 4.6 mm I.D., 10 um-C6 column; GC: fused-silica column coated with permethylated -CD (Chirasil–Dex) 25 m × 0.25 mm I.D. a 0.25 μm film thickness	RPLC: linear gradient methanol/water, from 25:75, v/v up to 100% methanol. GC: He carrier gas T program 40–180°C	FID LODs: 0.02–0.07 mg/L	Steam distillation-solvent extraction of volatile compounds of methyl jasmonate treated samples.	de la Peña Moreno et al. (2010)
Cypermethrin and cis-bifenthrin	Tea	BGB-172 chiral column 30 m × 0.25 mm I.D. × 0.25 μm film thickness (20% tert-butyldimethylsilyl-β-CD in 15% diphenyl- and 85% dimethylpolysiloxane, v/v)	He carrier gas T program 160–230°C	ECD	Ground tea powder soaked with hot water; ultrasonic extraction with acetone; partition with hexane; upper solution collected for evaporation; petroleum ether used to reconstitute the residues. Clean up with sandwich glass cartridge.	Kuang et al. (2010)
Chiral: -thujene, camphene, -pinene, -pinene, sabinene, -phellandrene, -phellandrene, limonene, citronellal, linalool, terpinen-4-ol and aterpineol	Mandarin essential oils	For chiral analysis: Megadex DETTBS-(diethyl-tert-butyl-silyl -CD) 25 m × 0.25 mm I.D., 0.25 μm film thickness	H_2 carrier gas T program 50–200°C	isotope ratio MS FID (for chiral)	Dilution in hexane	Schipilliti et al. (2010)

(Continued)

TABLE 8.6 (Continued)
Enantioselective Gas Chromatography in Food Analysis

Analytes	Matrix	Chiral Selector	Experimental Conditions	Mode/Detection	Sample Preparation	Reference
Sotolon	Dry white wine	Fused silica column coated with a 20% β-CD solution in 35% phenylmethylpolysiloxane 30 m × 0.25 mm I.D., 0.25 μm film thickness	He gas carrier T program 50–210°C	MS LOD: 1 mg/L	LLE with CH_2Cl_2 and concentrated with nitrogen	Pons et al. (2010)
Medium-to-high volatility racemates in the flavor and fragrance field (Hydrocarbons, Heterocycles, esters, lactones, alcohols, ketones, aldehydes, acids)	Different essential oils (bergamot, lemon, orange, bitter orange, lavender, peppermint, rosemary and sage)	6^{I-VII}-O-TBDMS-3^{I-VII}O-ethyl-2^{I-VII}-O-methyl--CD (MeEt-CD, 4) and 6^{I-VII}-O-TBDMS-2^{I-VII}O-ethyl-3^{I-VII}-O-methyl--CD (EtMe-CD, 3) Compared with: 6^{I-VII}-O-TBDMS-2^{I-VII}, 3^{I-VII}-O-methyl--CD (MeMe-CD) and 6^{I-VII}-O-TBDMS-2^{I-VII}, 3^{I-VII}O-ethyl--CD (EtEt-CD) (25 m × 0.25 mm I.D.) columns coated with 0.15 um film of CD diluted at 30% in PS086 (polymethylphenylpolysiloxane, 15% phenyl)	He carrier gas T program 50–220°C	MS	Hydrodistillation and dilution	Bicchi et al. (2010)
Ala, Val, Thr, Gly[a] Ile, Pro, Leu, Ser, GABA[a], Asx, Met, Phe, Glx, Tyr, Orn, Lys [a]no chiral AA	white, red, Ice and sparkling wines	Chirasil-$_L$-Val fused silica capillary column (N-propionyl-$_L$-valine tert butyl amide polysiloxane), 25 m × 0.25 mm I.D., 0.12 μm film thickness	He carrier gas T program 70–190°C	MS LOD: 0.05–0.1% D-enantiomer (mg/L)	Sample adjusted to pH 2.3 with 0.1 M HCl, ion exchanger SPE, conversion into N(O)-pentafluoropropionyl 2-propyl esters	Ali et al. (2010)
Ala, Val, Leu, ILe, Pro, α-aminobutyric acid, β-aminoisobutyric acid, pipecolic acid, norleucine, norvaline, 2,3-diaminopropionic acid, N-methylaspartic acid	Kidney and adzuki bean	Indirect method Achiral columns: DB-5 (SE-54 bonded) and DB-17 (OV-17 bonded) fused-silica capillary columns, 30 m × 0.25 mm I.D., 0.25 μm film thickness Ultra-2 (5% phenyl–95% methylpolysiloxane bonded phase cross-linked capillary column, 25 m × 0.20 mm I.D., 0.11 μm film thickness	He gas carrier T program 80–290°C T ramp 150–300°C	FID MS LODs: 2.9–79.4 ng/mL	Extraction with acidified (pH ≤ 2) water, diethyl ether and ethyl acetate in sequence. Dilution with distilled water and subjected to the N-ethoxycarbonylation with subsequent diastereomeric amidation	Paik et al. (2008)

Current Trends in Chiral Food Analysis

TABLE 8.6 (Continued)
Enantioselective Gas Chromatography in Food Analysis

Analytes	Matrix	Chiral Selector	Experimental Conditions	Mode/Detection	Sample Preparation	Reference
Twenty-nine volatile compounds (chiral: ketone, a-ionone, a-pinene, linalool, terpinen-4-ol, δ-octalactone, δ-decalactone, and 6-methyl-5-hepten-2-ol)	'Chilliwack', 'Tulameen', 'Willamette', 'Yellow Meeker', and 'Meeker' raspberries	Cyclosil B column 30 m × 0.25 mm I.D., 0.25 μm film thickness.	He carrier gas T program 40–220°C	MS LOD: 1 μg/kg.	SBSE	Malowicki et al. (2008a)
2-hydroxydecanoic acid (2-OH-10:0), 2-hydroxytetradecanoic acid (2-OH-14:0), γ-decalactone, γ-dodecalactone, δ-decalactone, δ-dodecalactone, and δ-tetradecalacton; 2-hydroxydodecanoic acid (2-OH-12:0), 2-hydroxyhexadecanoic acid (2-OH-16:0), 2-hydroxyoctadecanoic acid (2-OH-18:0), 2-hydroxyeicosanoic acid (2-OH-20:0), 3-hydroxydecanoic acid (3-OH-10:0), 3-hydroxydodecanoic acid (3-OH-12:0), 3-hydroxytetradecanoic acid (3-OH-14:0), 3-hydroxyhexadecanoic acid (3-OH-16:0), 3-hydroxyoctadecanoic acid (3-OH-18:0)	Bovine, human, mare and goat milk, goat cheese, goat cream cheese, Emmental, hand and feta cheese, cow feta cheese, cow mozzarella, buffalo mozzarella, St. John's Wort oil, cedarnut oil, walnut oil, avocado oil, brain of pork, bovine, goat, and sheep.	Indirect method Factor Four VF-5 ms column 30 m, 0.25 mm I.D., 0.25 μm film thickness	He carrier gas T program 40–300°C	ECNI-MS LODs: 0.02–0.42 mg/100 g	Extracted saponifiable lipids converted into methyl esters, and the resulting FAMEs separated into OH-FAMEs (minor fraction) and derivatized with (R)-$(-)$-R-methoxy-R-trifluoromethylphenylacetyl chloride (Mosher's reagent)	Jenske and Vetter (2008)

(Continued)

TABLE 8.6 (Continued)
Enantioselective Gas Chromatography in Food Analysis

Analytes	Matrix	Chiral Selector	Experimental Conditions	Mode/Detection	Sample Preparation	Reference
Chiral: α-ionone, α-pinene, linalool, terpinen-4-ol, δ-octalactone, and δ-decalactone	RBDV-resistant transgenic and wild type 'Meeker' type raspberries	Cyclosil B column 30 m × 0.25 mm I.D., 0.25 μm film thickness	He carrier gas T program 40–220°C	MS	SBSE	Malowicki et al. (2008b)
Sotolon	White wines	Precolumn: fused silica column coated with polar phase BP20; 2 m × 0.25 mm I.D., 0.25 μm film thickness; fused silica column coated with a solution of 20% β-CD in (35% phenyl)-methylpolysiloxane, 30 m × 0.25 mm I.D., 0.25 μm film thickness	He carrier gas T program 50–210°C	MS	Samples with addition of anhydrous sodium sulfate extracted three times with CH_2Cl_2; dried over anhydrous sodium sulfate and concentrated under a nitrogen stream. Preparative-HPLC for sotolon enantiomers with Chiralpak AS-H model 250 × 20 mm, 5 μm.	Pons et al. (2008)
Optically active compounds in the flavor and fragrance	Balm lemon, bergamot, boronia, cornmint, lavender, lemon, peppermint, rosemary essential oils; apple flavor and apricot, peach and coconut extracts	Four columns coated with four CD derivatives as chiral selectors diluted at 30% in PS-086: I) 2,6-di-O-methyl-3-O-pentyl-β-CD (2,6DM3PEN-β-CD) II) 2,3-di-O-methyl-6-O-*tert*-butyldimethylsilyl-β-CD (2,3DM6TBDMS-β-CD) III) 2,3-di-O-ethyl-6-O-*tert*-butyldimethylsilyl-β-CD (2,3DE6TBDMS-β-CD) IV) 2,3-di-O-acetyl-6-O-*tert*-butyldimethylsilyl-β-CD (2,3DA6TBDMS-β-CD), 25 m × 0.25 mm I.D., 0.25 μm film thickness	He carrier gas T program 50–220°C	MS	Headspace-SPME	Liberto et al. (2008)

**TABLE 8.6 (Continued)
Enantioselective Gas Chromatography in Food Analysis**

Analytes	Matrix	Chiral Selector	Experimental Conditions	Mode/Detection	Sample Preparation	Reference
Linalool, linalyl propionate, γ-lactones(C6, C7, C8, C11, C12, C14, C15), δ-octalactone, α-hexachlorohexane, trichlorfon, trans-chlordane, heptachlor, limonene, 2-octanol, camphor, isobornyl acetate, linalyl acetate, 2-methyl-(3Z)-hexenyl butyrate, menthol, hydroxycitronellal, γ-decalactone and δ-decalactone	Lavender and bergamot essential oils	Columns coated with 30% of 2,3-di-O-ethyl-6-O-*tert*-butyldimethylsilyl-β-CD in PS-086 25 m × 0.25 mm I.D., 0.15 μm and 0.25 μm film thickness; 5 m × 0.25 mm I.D., 0.15 μm film thickness; 1, 2, 5 and 10 m × 0.10 mm I.D., 0.10 μm film thickness	He carrier gas Different T program	MS	Hydrodistillation and dilution	Bicchi et al. (2008)

Source: Reprinted from *TrAC Trends in Analytical Chemistry*, Vol. 52, Anna Rocco, Zeineb Aturki, Salvatore Fanali, Chiral Separations in Food Analysis, pp. 206–225, 2013. With permission from Elsevier.

TABLE 8.7
Chiral Separations in Food Analysis Using Electrophoresis, Electrokinetic Chromatography, and Miniaturized Liquid Chromatography

Analytes	Matrix	Column Chiral Selector	Experimental Conditions	Mode/Detection	Reference
D,L-amino acids	Pomegranate juice	Fused silica capillary (75 μm I.D. × L_{tot} = 55 cm; L_{eff} = 45 cm) C.S.: β-CD	BGE: 50 mM borate (pH 9.65), 5 mM SDBS, 10 mM β-CD V: 25 kV Inj: 50 mbar × 6 s	MEKC-LIF	Tezcan et al. (2013)
D,L-amino acids	Beer	Fused silica capillary (50 μm I.D. × L_{tot} = 60 cm; L_{eff} = 40 cm) C.S.: β-CD	Capillary filling: 1.5 M Tris-borate (pH10)/12.5% (v/v) isopropanol/35 mM STDC/35 mM β-CD Running vials: 150 mM Tris-borate (pH 8.5)/12.5% (v/v) isopropanol, 0.5% (w/v) PEO, 35 mM MSTDC/35 mM β-CD V: 18 kV Inj: 20 cm height × 10 s	MEKC-UV λ = 260 nm	Lin et al. (2013)
(+)-Catechin, (−)-epicatechin, (−)-epigallocatechin, (−)-epicatechingallate, (−)-epigallocatechin gallate, (−)-gallocatechin, (−)-gallocatechingallate, (+)-gallocatechin	Green tea	Fused silica capillary (50 μm I.D. × L_{tot} = 30 cm; L_{eff} = 8.5 cm) C.S.: (2-hydroxypropyl)-β-CD (HP-β-CD)	BGE: 12.5 mM borate–phosphate buffer (pH 2.5), 90 mM SDS, 25 m MHP-β CD V: 15 kV Inj: 25 mbar × 5 s	CE-DAD λ = 200 nm	Mirasoli et al. (2014)
(−)-Epigallocatechin gallate, (−)-epigallocatechin, (−)-epicatechingallate, (−)-epicatechin, (+)-catechin, (−)-catechin	Green tea	Fused silica capillary (50 μm I.D. × L_{tot} = 30 cm; L_{eff} = 8.5 cm) C.S.: (2-hydroxypropyl)-β-CD (HP-β-CD)	BGE: 25 mM borate–phosphate buffer (pH 2.5), 90 mM SDS, 25 mM HP-β-CD V: 15 kV Inj: 25 mbar × 5 s	MEKC-DAD λ = 200 nm	Pasquini et al. (2016)
L,L and D,D-neotame	Mango juice, cola soft drink, orange soft drink	Fused silica capillary (75 μm I.D. × L_{tot} = 64.5 cm; L_{eff} = 56.0 cm) C.S.: Heptakis 2,3,6 tri-o-methy-β-CD	BGE: 50 mM phosphate buffer (pH 5.5), 30 mM heptakis 2,3,6 tri-o-methy- β-CD V: +20 kV Inj: 100 mbar × 5 s	CE-DAD λ = 200 nm	Bathinapatla et al. (2014)
D,L-citrulline	Food supplements	Fused silica capillary (50 μm I.D. × L_{tot} = 48.5 cm; L_{eff} = 40 cm) C.S.: sulfated γ-CD	BGE: 100 mM formate buffer (pH 3.0), 10 mM sulfated γ-CD V = −20 kV Inj: cathodic end, 50 mbar × 15 s	CE-DAD λ = 210 nm	Pérez-Míguez et al. (2016)

TABLE 8.7 (Continued)
Chiral Separations in Food Analysis Using Electrophoresis, Electrokinetic Chromatography, and Miniaturized Liquid Chromatography

Analytes	Matrix	Column Chiral Selector	Experimental Conditions	Mode/Detection	Reference
1,3-dimethylamylamine	Food supplements	Fused silica capillary (50 μm I.D. × L_{tot} = 48.5 cm; L_{eff} = 40.0 cm) C.S.: sulfated-α-CD, Sulfated-β-CD	BGE: 5 mM phosphate/Tris (pH 3.0), 10 mM of BTEAC, 1.1% (w/v) sulfated-α-CD, 0.2% (w/v) sulfated-β-CD V: 15 kV Inj:50 mbar × 5s	CE-DAD λ = 214 nm	Přibylka et al. (2015)
L-malic acid, D-malic acid, D-tartaric acid, L-tartaric acid, L-isocitric acid, D-isocitric acid	Apple, grape, orange, pineapple grapefruit and mixed fruit juices	Fused silica capillary (50 μm I.D. × L_{tot} = 64.5 cm; L_{eff} = 56.0 cm) C.S.: D-quinic acid and Cu(II)/Sn(II) ion	BGE: 20 mM acetic acid, 100 mM D-quinic acid, 10 mM CuSO$_4$, 1.8 mM ScCl$_3$ (pH 5.0) V: +15 kV Inj: 50 mbar × 3 s	LE-CE-DAD λ = 250 nm	Kodama et al. (2013)
D, L-tartaric acid, D, L-malic acid	Wine	Fused silica capillary (75 μm I.D. × L_{tot} = 75.0 cm; L_{eff} = 60 cm) C.S.: D-quinic acid and Cu(II)/Al(III)	BGE: 20 mM acetate buffer, 100 mM D-quinic acid, 10 mM Cu(II), 0.5 mM Al(III), 0.5 mM CTA-OH. V: −16 kV Inj:40 mbar × 15s	LE-CE-UV λ = 254 nm	Kamencev et al. (2016)
D,L-tartaric acid	Wine, wine grapes	Fused silica capillary (75 μm I.D. ×L_{tot} = 60.0 cm; L_{eff} = 45.0 cm) C.S.: CuCl$_2$ and trans-4-hydroxy-L-proline,	BGE: 7 mM CuCl$_2$,14 mM trans-4-hydroxy-L-proline, 100 mM ε-aminocaproic acid (pH 5) V: −20 kV Inj: 50 mbar × 5 s	LE-CE–CCD	Knob et al. (2013)
L- and D-malic acid	Apple juice	Open-tubular column (50 μm I.D. × 37 cm) C.S.: (in-situ grafting) L-His-modified HPMA-Cl	Isocratic MP: 60:40 (v/v) ACN/5 mM CuSO$_4$, 20 mM (NH$_4$)$_2$SO$_4$ (pH 3.0) V: −20 kV Inj:−12 kV, 0.05 min	OT-CEC-DAD λ = 214 nm	Aydoğan (2018)
FMOC-D,L-carnosine	Dietary supplements	Monolithic column (100 μm I.D. × L_{packed} = 21.0 cm) C.S.: poly(MQD-co-HEMA-co-EDMA)	MP: 35:65 (v/v) ACN/MeOH, 0.1 M acetic acid, 4 mM TEA (apparent pH 6.0) Flow rate: 10 μL/min Inj: 20 nL	Capillary-LC-UV λ = 254 nm	Q. Wang et al. (2016)

Source: Reprinted from *TrAC Trends in Analytical Chemistry*, Vol. 96, Giovanni D'Orazio, Chiara Fanali, María Asensio-Ramos, Salvatore Fanali, Chiral Separations in Food Analysis, pp. 151–171, 2017. With permission from Elsevier.

been coupled to either selective detectors or mass spectrometry (Sánchez-Hernández et al., 2010). Chiral analysis of the fungicide vinclozolin in wines was carried out using cyclodextrin-modified micellar electrokinetic chromatography using γ-cyclodextrin together with sodium dodecyl sulphate (Kodama et al., 2002). The resolution of racemic vinclozolin was approximately 2.1, and the degradation rates were different between (+)- and (-)-vinclozolin. The same research group applied capillary electrophoresis using 2-hydroxypropyl-β-cyclodextrin as a chiral selector for chiral separation of imazalil enantiomers. In four out of the eight oranges analyzed, degradation of (-)-imazalil was faster than that of (+)-imazalil (Kodama et al., 2003). Separation of herbicides (aryloxypropionic acid, aryloxyphenoxypropionic acid, and aminopropionic acid) was carried out using capillary electrokinetic chromatography equipped with an L-RNA aptamer chiral stationary phase (André et al., 2006). The important feature of the stationary phase was its high efficiency and retention during a prolonged period.

8.5 FOOD QUALITY CONTROL AND TRACEABILITY

8.5.1 ENANTIOSELECTIVE LIQUID CHROMATOGRAPHY IN FOOD QUALITY

Enantioselective liquid chromatography is sometimes used in determining the quality of foods. Phenylalanine is an amino acid, which is necessary for human growth and can be obtainable from dietary supplements. Studies have been carried out to separate its enantiomers in various foods using liquid chromatography coupled to selective detectors (UV, DAD, and variable wavelength detector) in order to maintain reasonable dosages (Guillén-Casla et al., 2010; Wang et al., 2017). For instance, liquid chromatography coupled to a UV has been used to separate its enantiomers in dietary supplements, and energy drinks (Lomenova and Hroboňová, 2018). Enantioseparation (resolution values of 1.59 and 2.75) was obtained using macrocyclic antibiotic teicoplanin and ristocetin based stationary phases in reverse phase mode with mobile phase of acetonitrile and water at ratios of 75:25 and 60:40 v/v respectively. A recent trend in the chiral analysis of amino acids is the use of liquid chromatography coupled to mass-spectrometry. Recently, a liquid chromatography tandem mass spectrometry method was developed for the chiral analysis of 18 amino acids in vinegar (Nakano et al., 2017). Remarkable sensitivity was obtained with the liquid chromatography tandem mass spectrometry method as compared to liquid chromatography time-of-flight mass spectrometry (Konya et al., 2017). In another study, the same research group extended the scope of liquid chromatography tandem mass spectrometry to metabolic profiling of chiral amino acids and their metabolites in cheese (Nakano et al., 2019). The separation was successfully achieved using binaphthyl based crown ether and cinchona alkaloid derived zwitterionic stationary phase.

Determining the enantiomeric composition of natural flavor and taste products can be used to assess food quality and authenticity. The enantiomers of the polyfunctional thiols 3-sulfanylhexan-1-ol and its O-acetate, 3-sulfanylhexyl acetate contribute differently to the aroma of wines (Tominaga et al., 2006). A chiral liquid chromatography tandem mass spectrometry method was developed using stable isotope dilution analysis (Chen et al., 2018). Validation of the method using a Lux Amylose-1 column gave limits of detection less than 0.7 ng/l in various wines. Liquid chromatography coupled to a UV detector was used to separate the enantiomers of three volatile compounds in palm wines (Lasekan, 2018). Noticeable differences in the enantiomeric ratios and concentrations of these enantiomers during storage occurred. In all the wines, concentration of the (S)-form decreased during storage, whereas those of the (R)-form increased.

8.5.2 ENANTIOSELECTIVE GAS CHROMATOGRAPHY IN FOOD QUALITY

Enantioselective gas chromatography mass spectrometry (γ-cyclodextrin) has also been used to assay the enantiomers of 2-methylbutyl acetate in red and white wines from various vintages and origins (Cameleyre et al., 2017). The findings of the study were that the presence of (S)-2-methylbutyl acetate in the fruity aromatic reconstitution led to the enhancement of the perception of black, fresh, and jammy fruit. Moreover, the levels of (S)-2methylbutyl acetate were shown to increase during aging thus contributing to the wine aroma during the aging process. Multidimensional gas chromatography mass spectrometry has also been successfully applied in the evaluation of the enantiomeric composition of lactones that are associated with the "overripe orange" nuances in Bordeaux dessert wines (Stamatopoulos et al., 2016). The authors reported that the (R)-form of 2-nonen-4-olide was dominant in wines that had aged, while the (S)-form was dominant in young wines. Chiral chromatography mass spectrometry was also used to demonstrate that (R)-piperitone was the main compound responsible for the flavor of red wines (Pons et al., 2016). However, the levels of piperitone were not linked to its age or its premature evolution. Head space microextraction followed by heart-cut multidimensional gas chromatography tandem mass spectrometry fitted with a octakis (2,3-di-O-pentyl-6-O.methyl)-γ-cyclodextrin as a chiral selector was successfully applied for enantioselective determination of norisopreonoids such as α-ionone, β-ionone, and damascenone in wines (Langen et al., 2016b). All the wines studied showed higher levels of the α-ionone. Elevated α- and β-ionone levels served as a useful indicator of adulteration. The same technique was also applied to evaluate the merits of 1,2-propanediol as a potential marker for wine adulteration (Langen et al., 2016a). In all the authentic wines studied, 1,2-propanediol showed a higher enantiomeric ratio in favor of the (R)-enantiomer, proving its potential as a marker for the adulteration with flavor extracts based on industrial 1,2-propanediol as a solvent. Several other studies have been carried out using multidimensional gas chromatography mass spectrometry and these using the enantiomeric ratios of volatile organic compounds to distinguish between juniper-flavored spirits from different countries (Pažitná and Špánik, 2014).

8.5.2.1 Solid Phase Microextraction

Solid phase microextraction coupled to gas chromatography mass spectrometry has been used to study enantiomeric composition of chiral terpenes in fruit beverages (Ruiz del Castillo et al., 2003). The study revealed that the chiral terpenes (limonene, linalool, and α-terpineol) contribute to the aroma of fruit beverages, and the (+) enantiomer is the most dominant in most cases. Using solid phase microextraction, and multidimensional gas chromatography mass spectrometry, the enantiomeric ratios of chiral terpenes were also used to determine the botanical origin (Vyviurska et al., 2017). An excess of (R)-linalool and (S)-α-terpineol was found only for pear brandy, and (R)-limonene and the second eluted enantiomer of nerolidol for *Sorbus domestica* and strawberry were dominant, respectively.

Enantiomeric analysis of volatile terpenoids can be used to determine the botanical origin of diverse types of tea. Head space solid phase microextraction (HS-SPME)-chiral gas chromatography mass spectrometry was successfully used to analyze 12 pairs of terpenoid enantiomers in various teas (Zhu et al., 2017). It was reported that diverse enantiomeric ratios and concentrations existed between Chinese and Indonesian tea varieties. Moreover, using partial least squares discriminant analysis, concentration differences existed between large and small leaf origins of black tea, and significant differences were also observed between the concentrations of linalool oxides A, B, and C between green, white, and dark teas. In another study, head space solid phase microextraction and enantioselective gas chromatography mass spectrometry was used to quantify volatile components in 22 tea cultivars from different locations in China (Mu et al., 2018). The key findings of the study were that tea processing influenced the formation of volatile enantiomers, and that enantiomeric distribution of volatile constituents closely relates with the geographical origins, leaf types, and manufacturing suitability of the examined tea cultivars.

8.5.2.2 Multidimensional Gas Chromatography

Solid phase microextraction coupled to two-dimensional gas chromatography mass spectrometry has also been successfully used to determine botanical origin of honey based on enantiomeric distribution of volatile organic compounds (Špánik et al., 2014). The enantiomeric ratios of chiral volatile organic compounds varied up to 10% between honey of different botanical origins. In a related study, a gas chromatography mass spectrometry method was validated for the fingerprinting of headspace volatile compounds (Aliferis et al., 2010). Combined mass spectra of honey samples originating from different plants and geographic origins of Greece were subjected to orthogonal partial least squares discriminant analysis, soft independent modelling of class analogy, and hierarchical cluster analysis. Excellent separation between honey samples according to their botanical origin with misclassification as low as 1.3% were obtained.

Enantiomeric characterization of essential oils is of great importance in quality control to reveal adulteration or the presence of natural components that can change the biological and olfactory properties (Bonaccorsi et al., 2011; Davies et al., 2016). Chiral comprehensive two-dimensional gas chromatography coupled to quadrupole-accurate mass time-of-flight mass spectrometry was applied to the determination of several monoterpenes in cardamom oil (Chin et al., 2014). The enantiomeric resolution of the chiral monoterpenes was achieved by use of 2,3-di-O-methyl-6-t-butylsilyl β-cyclodextrin phase, and high polarity ionic liquid stationary phase as first dimension and second dimension columns, respectively. Analysis of volatile flavor compounds in citrus oils was used to determine the enantiomeric ratios for their discrimination and authentication by multidimensional gas chromatography mass spectrometry (Hong et al., 2017). Lastly, a study was carried out to evaluate the possible variations of the characteristic terpenes in lemon oil with growth stage using multidimensional gas chromatography mass spectrometry (Guadayol et al., 2018). The enantiomeric ratios of α-pinene, β-pinene, and limonene were consistent, but those of linalool varied with growth stage in a spontaneous and natural way. The flavor of pasta is influenced by the presence of aroma active compounds in flour. Heart-cutting multidimensional gas chromatography mass spectrometry was used to determine flavor constituents in flour and derived pasta (Costa et al., 2017). Fingerprints with the definition of markers of quality and traceability were obtained by using a first dimension equipped with a nonpolar column and a flame ionization detector, and a second dimension equipped with a modified β-cyclodextrin stationary phase, and a mass spectrometric detector.

8.5.3 Enantioselective Electrokinetic Chromatography in Food Quality

A chiral micellar electrokinetic chromatography laser induced fluorescence method (MEKC-LIF) for the determination of chiral amino acids in pomegranate juices was developed (Tezcan et al., 2013). The amino acid profiles of the pomegranate juices were compared to apple amino acids, and L-Asn was proposed as a marker for the adulteration of pomegranate juices with apple juices. Chiral nano liquid chromatography has also been used in the analysis of amino acid enantiomers in juices with a derivatization step to overcome the drawbacks such as reduced interaction with the chiral selector (D'Orazio et al., 2008). For instance, excellent separation was obtained in the chiral analysis of amino acid enantiomers in fruit juice using open-tubular nano-liquid chromatography (Aydoğan, 2018). The stationary phase was prepared by in situ polymerization of 3-chloro-2-hydroxypropylmethacrylate and ethylene dimethacrylate, and the reactive chloro groups of the porous stationary phase were reacted with β-cyclodextrin. Lastly, chiral electrophoresis mass spectrometry was used to carry out enantioseparation of amino acids in orange juices (Simó et al., 2005). Good separation was achieved by use of a non-volatile chiral selector as β-cyclodextrin in the background electrolyte together with a physically coated capillary aimed at preventing contamination of the electrospray. Two different derivatization methods using dansyl chloride and fluorescein isothiocyanate gave comparable sensitivity and resolution. To separate nine pairs of amino acid enantiomers in beers, poly (ethylene oxide)-based stacking, β-cyclodextrin-mediated micellar electrokinetic chromatography, and 9-fluoroenylmethylchloroformate (FMOC) derivatization were used (Lin et al., 2013). An important feature of the study was that the nine pairs of FMOC derivatized amino acid enantiomers were baseline separated using a discontinuous system, and they were effectively stacked without losing chiral resolution. Another important trend is the enantioseparation of tartaric and malic acids in wines by ligand exchange capillary electrophoresis using uncoated fused capillary (Kamencev et al., 2016). Enantioselective cyclodextrin-modified micellar electrokinetic chromatography was successfully applied for discrimination of origin of green tea samples from Japan and China (Pasquini et al., 2016). By determining catechins and methylxanthines, Japanese green tea samples were found to be similar to each other, as they were from a relatively limited geographical area, while Chinese ones showed different patterns. Generally, Japanese samples contained lower catechins levels than Chinese ones.

8.5.4 ENANTIOSELECTIVE CAPILLARY ELECTROPHORESIS IN FOOD QUALITY

Quality control of dietary supplements has also been studied using capillary electrophoresis. Capillary electrophoresis using trimethyl-β-cyclodextrin as a chiral selector was used to determine enantioseparation of lipoic acid in dietary supplements (Kodama et al., 2012). The potential of capillary electrophoresis-tandem mass spectrometry in determining the enantiomers of carnitine in dietary food supplements was also explored (Sánchez-Hernández et al., 2010). Improved precision and sensitivity were obtained without using a partial filling technique, and with a low concentration of a chiral selector. Another in-capillary derivatization method by capillary electrophoresis with 6-aminoquinolyl-N-hydroxysuccinimidyl carbamate was developed for the determination of arginine, lysine, and ornithine in dietary supplements and wine (Martínez-Girón et al., 2009). Electrokinetic chromatography coupled to a UV detector was successfully used to determine ornithine enantiomers in food supplements (Domínguez-Vega et al., 2009). An important aspect of the study was the use of an ultrasound probe to accelerate derivatization.

8.6 CONCLUSION

Chirality influences the physicochemical properties of molecules in chiral environments such as biological systems. The techniques that have been mainly used to carry out enantioseparation are liquid chromatography, supercritical fluid chromatography, gas chromatography, and capillary electrophoresis. Of these techniques, most of the studies have been carried out with liquid chromatography more than any of the other techniques. Most of the studies that have been carried out in terms of chiral analysis of food molecules are on food safety assessment and quality control and traceability studies. However, more work still needs to be done in using chiral sensors which may reduce the analytical cost. Additionally, further work still needs to be done to try out newer chiral stationary phases, and increased use of advanced analytical techniques such as ultra-high-performance liquid chromatography mass spectrometry. There is also a need to improve on the analysis of multiple compounds in a single run.

REFERENCES

Ahmed, M., Gwairgi, M., Ghanem, A., 2014. Conventional chiralpak ID vs. capillary chiralpak ID-3 amylose tris-(3-Chlorophenylcarbamate)-based chiral stationary phase columns for the enantioselective HPLC separation of pharmaceutical racemates. *Chirality* 26, 677–682. https://doi.org/10.1002/chir.22390

Alañón, M. E., Castle, S. M., Siswanto, P. J., Cifuentes-Gómez, T., Spencer, J. P. E., 2016. Assessment of flavanol stereoisomers and caffeine and theobromine content in commercial chocolates. *Food Chem.* 208, 177–184. https://doi.org/10.1016/j.foodchem.2016.03.116

Ali, H. S. M., Pätzold, R., Brückner, H., 2010. Gas chromatographic determination of amino acid enantiomers in bottled and aged wines. *Amino Acids* 38, 951–958. https://doi.org/10.1007/s00726-009-0304-1

Aliferis, K. A., Tarantilis, P. A., Harizanis, P. C., Alissandrakis, E., 2010. Botanical discrimination and classification of honey samples applying gas chromatography/mass spectrometry fingerprinting of headspace volatile compounds. *Food Chem.* 121, 856–862. https://doi.org/10.1016/j.foodchem.2009.12.098

Alvarez-Rivera, G., Bueno, M., Ballesteros-Vivas, D., Cifuentes, A., 2020. Chiral analysis in food science. *TrAC Trends Anal. Chem.* 123, 115761. https://doi.org/10.1016/j.trac.2019.115761

André, C., Berthelot, A., Thomassin, M., Guillaume, Y.-C., 2006. Enantioselective aptameric molecular recognition material: Design of a novel chiral stationary phase for enantioseparation of a series of chiral herbicides by capillary electrochromatography. *Electrophoresis* 27, 3254–3262. https://doi.org/10.1002/elps.200500070

Armstrong, D. W., Chang, C. D., Li, W. Y., 1990. Relevance of enantiomeric separations in food and beverage analyses. *J. Agric. Food Chem.* 38, 1674–1677. https://doi.org/10.1021/jf00098a010

Ashiuchi, M., Oike, S., Hakuba, H., Shibatani, S., Oka, N., Wakamatsu, T., 2015. Rapid purification and plasticization of d-glutamate-containing poly-γ-glutamate from Japanese fermented soybean food natto. *J. Pharm. Biomed. Anal* 116, 90–93. https://doi.org/10.1016/j.jpba.2015.01.031

Aydoğan, C., 2018. Chiral separation and determination of amino acid enantiomers in fruit juice by open-tubular nano liquid chromatography. *Chirality* 30, 1144–1149. https://doi.org/10.1002/chir.23006

Baranowska, I., Hejniak, J., Magiera, S., 2016. Development and validation of a RP-UHPLC-ESI-MS/MS method for the chiral separation and determination of flavanone, naringenin and hesperetin enantiomers. *Talanta* 159, 181–188. https://doi.org/10.1016/j.talanta.2016.06.020

Bathinapatla, A., Kanchi, S., Singh, P., Sabela, M. I., Bisetty, K., 2014. Determination of neotame by high-performance capillary electrophoresis using ß-cyclodextrin as a chiral selector. *Anal. Lett.* 47, 2795–2812. https://doi.org/10.1080/00032719.2014.924008

Berthod, A., 2009. Chiral recognition mechanisms with macrocyclic glycopeptide selectors. *Chirality* 21, 167–175.

Berthod, A., 2006. Chiral recognition mechanisms. *Anal. Chem.* 78, 2093–2099.

Berthod, A., Qiu, H. X., Staroverov, S. M., Kuznestov, M. A., Armstrong, D. W., 2010. *Chiral Recognition in Separation Methods*, in: Berthod, A. (Ed.). Springer Berlin Heidelberg, Berlin, Heidelberg, pp. 203–222.

Bicchi, C., Cagliero, C., Liberto, E., Sgorbini, B., Martina, K., Cravotto, G., Rubiolo, P., 2010. New asymmetrical per-substituted cyclodextrins (2-O-methyl-3-O-ethyl- and 2-O-ethyl-3-O-methyl-6-O-t-butyldimethylsilyl-β-derivatives) as chiral selectors for enantioselective gas chromatography in the flavour and fragrance field. *J. Chromatogr. A* 1217, 1106–1113. https://doi.org/10.1016/j.chroma.2009.09.079

Bicchi, C., Liberto, E., Cagliero, C., Cordero, C., Sgorbini, B., Rubiolo, P., 2008. Conventional and narrow bore short capillary columns with cyclodextrin derivatives as chiral selectors to speed-up enantioselective gas chromatography and enantioselective gas chromatography-mass spectrometry analyses. *J. Chromatogr. A* 1212, 114–123. https://doi.org/10.1016/j.chroma.2008.10.013

Bonaccorsi, I., Sciarrone, D., Cotroneo, A., Mondello, L., Dugo, P., Dugo, G., 2011. Enantiomeric distribution of key volatile components in citrus essential oils. *Brazilian J. Pharmacogn.* 21, 841–849.

Cagliero, C., Bicchi, C., Cordero, C., Rubiolo, P., Sgorbini, B., Liberto, E., 2012. Fast headspace-enantioselective GC-mass spectrometric-multivariate statistical method for routine authentication of flavoured fruit foods. *Food Chem.* 132, 1071–1079. https://doi.org/10.1016/j.foodchem.2011.10.106

Cameleyre, M., Lytra, G., Tempere, S., Barbe, J., 2017. 2-Methylbutyl acetate in wines: Enantiomeric distribution and sensory impact on red wine fruity aroma. *Food Chem.* 237, 364–371. https://doi.org/https://doi.org/10.1016/j.foodchem.2017.05.093

Carrão, D. B., Perovani, I. S., de Albuquerque, N. C. P., de Oliveira, A. R. M., 2020. Enantioseparation of pesticides: A critical review. *Trends Anal. Chem.* 122, 115719. https://doi.org/10.1016/j.trac.2019.115719

Chen, L., Capone, D. L., Jeffery, D. W., 2018. Chiral analysis of 3-sulfanylhexan-1-ol and 3-sulfanylhexyl acetate in wine by high-performance liquid chromatography–tandem mass spectrometry. *Anal. Chim. Acta* 998, 83–92. https://doi.org/https://doi.org/10.1016/j.aca.2017.10.031

Chen, X., Dong, F., Liu, X., Xu, J., Li, J., Li, Y., Li, M., Wang, Y., Zheng, Y., 2013. Chiral determination of a novel acaricide cyflumetofen enantiomers in cucumber, tomato, and apple by HPLC. *J. Sep. Sci.* 36, 225–231. https://doi.org/10.1002/jssc.201200636

Chen, X., Dong, F., Xu, J., Liu, X., Wang, Y., Zheng, Y., 2015. Enantioselective degradation of chiral insecticide dinotefuran in greenhouse cucumber and soil. *Chirality* 27, 137–141. https://doi.org/10.1002/chir.22402

Chen, Z., Dong, F., Xu, J., Liu, X., Cheng, Y., Liu, N., Tao, Y., Zheng, Y., 2014. Stereoselective determination of a novel chiral insecticide, sulfoxaflor, in brown rice, cucumber and apple by normal-phase high-performance liquid chromatography. *Chirality* 26, 114–120. https://doi.org/10.1002/chir.22278

Cheng, Y., Zheng, Y., Dong, F., Li, J., Zhang, Y., Sun, S., Li, N., Cui, X., Wang, Y., Pan, X., Zhang, W., 2016. Stereoselective analysis and dissipation of propiconazole in wheat, grapes, and soil by supercritical fluid chromatography tandem mass spectrometry. *J. Agric. Food Chem.* 234–243. https://doi.org/10.1021/acs.jafc.6b04623

Chin, S., Nolvachai, Y., Marriott, P. J., 2014. Enantiomeric separation in comprehensive two-dimensional gas chromatography with accurate mass analysis. *Chirality* 26, 747–753. https://doi.org/10.1002/chir.22280

Costa, R., Albergamo, A., Bua, G. D., Saija, E., Dugo, G., 2017. Determination of flavour constituents in particular types of flour and derived pasta by heart-cutting multidimensional gas chromatography coupled with mass spectrometry and multiple headspace solid-phase microextraction. *Food Sci. Technol.* 86, 99–107.

Cretin, B. N., Sallembien, Q., Sindt, L., Daugey, N., Buffeteau, T., Waffo-Teguo, P., Dubourdieu, D., Marchal, A., 2015. How stereochemistry influences the taste of wine: Isolation, characterization and sensory evaluation of lyoniresinol stereoisomers. *Anal. Chim. Acta* 888, 191–198. https://doi.org/10.1016/j.aca.2015.06.061

Cutillas, V., García-Valverde, M., Gómez-Ramos, M., Díaz-Galiano, F. J., Ferrer, C., Fernández-Alba, A. R., 2020. Supercritical fluid chromatography separation of chiral pesticides: Unique capabilities to study cyhalothrin and metalaxyl as examples. *J. Chromatogr. A* 1620, 461007. https://doi.org/https://doi.org/10.1016/j.chroma.2020.461007

D'Orazio, G., Cifuentes, A., Fanali, S., 2008. Chiral nano-liquid chromatography–mass spectrometry applied to amino acids analysis for orange juice profiling. *Food Chem.* 108, 1114–1121. https://doi.org/https://doi.org/10.1016/j.foodchem.2007.11.062

D'Orazio, G., Fanali, C., Asensio-Ramos, M., Fanali, S., 2017. Chiral separations in food analysis. *Trends Anal. Chem.* 96, 151–171. https://doi.org/https://doi.org/10.1016/j.trac.2017.05.013

Dabbagh, H. A., Azami, F., 2014. Experimental and theoretical study of racemization, stability and tautomerism of vitamin C stereoisomers. *Food Chem.* 164, 355–362. https://doi.org/10.1016/j.foodchem.2014.04.121

Dallegrave, A., Pizzolato, T. M., Barreto, F., Eljarrat, E., Barceló, D., 2016. Methodology for trace analysis of 17 pyrethroids and chlorpyrifos in foodstuff by gas chromatography–tandem mass spectrometry. *Anal. Bioanal. Chem.* 408, 7689–7697. https://doi.org/10.1007/s00216-016-9865-5

Davies, N. W., Larkman, T., Marriott, P. J., Khan, I. A., 2016. Determination of enantiomeric distribution of terpenes for quality assessment of Australian tea tree oil. *J. Agric. Food Chem.* 64, 4817–4819. https://doi.org/10.1021/acs.jafc.6b01803

de la Peña Moreno, F., Blanch, G. P., Flores, G., Ruiz del Castillo, M. L., 2010. Development of a method based on on-line reversed phase liquid chromatography and gas chromatography coupled by means of an adsorption-desorption interface for the analysis of selected chiral volatile compounds in methyl jasmonate treated strawberries. *J. Chromatogr. A* 1217, 1083–1088. https://doi.org/10.1016/j.chroma.2009.10.037

Domínguez-Vega, E., Martínez-Girón, A. B., García-Ruiz, C., Crego, A. L., Marina, M. L., 2009. Fast derivatization of the non-protein amino acid ornithine with FITC using an ultrasound probe prior to enantiomeric determination in food supplements by EKC. *Electrophoresis* 30, 1037–1045. https://doi.org/10.1002/elps.200800358

Dong, F., Li, J., Chankvetadze, B., Cheng, Y., Xu, J., Liu, X., Li, Y., Chen, X., Bertucci, C., Tedesco, D., Zanasi, R., Zheng, Y., 2013. Chiral triazole fungicide difenoconazole: Absolute stereochemistry, stereoselective bioactivity, aquatic toxicity, and environmental behavior in vegetables and soil. *Environ. Sci. Technol.* 47, 3386–3394. https://doi.org/10.1021/es304982m

Evans, S. E., Kasprzyk-Hordern, B., 2014. Applications of chiral chromatography coupled with mass spectrometry in the analysis of chiral pharmaceuticals in the environment. *Trends Environ. Anal. Chem.* 1, e34–e51. https://doi.org/10.1016/j.teac.2013.11.005

Fanali, C., Fanali, S., Chankvetadze, B., 2016. HPLC separation of enantiomers of some flavanone derivatives using polysaccharide-based chiral selectors covalently immobilized on silica. *Chromatographia* 79, 119–124. https://doi.org/10.1007/s10337-015-3014-8

Flores, G., Blanch, G. P., Castillo, M. L. R. Del, 2013. Isolation of the four methyl jasmonate stereoisomers and their effects on selected chiral volatile compounds in red raspberries. *Food Chem.* 141, 2982–2987. https://doi.org/10.1016/j.foodchem.2013.05.117

Foudil-Cherif, Y., Yassaa, N., 2012. Enantiomeric and non-enantiomeric monoterpenes of Juniperus communis L. and Juniperus oxycedrus needles and berries determined by HS-SPME and enantioselective GC/MS. *Food Chem.* 135, 1796–1800. https://doi.org/10.1016/j.foodchem.2012.06.073

Fu, Q., Sanganyado, E., Ye, Q., Gan, J., 2016. Meta-analysis of biosolid effects on persistence of triclosan and triclocarban in soil. *Environ. Pollut.* 210, 137–144. https://doi.org/10.1016/j.envpol.2015.12.003

Gao, B., Zhang, Q., Tian, M., Zhang, Z., Wang, M., 2016. Enantioselective determination of the chiral pesticide isofenphos-methyl in vegetables, fruits, and soil and its enantioselective degradation in pak choi using HPLC with UV detection. *Anal. Bioanal. Chem.* 408, 6719–6727. https://doi.org/10.1007/s00216-016-9790-7

Guadayol, M., Guadayol, J. M., Vendrell, E., Collgrós, F., Caixach, J., 2018. Relationship between the terpene enantiomeric distribution and the growth cycle of lemon fruit and comparison of two extraction methods. *J. Essent. Oil Res.* 30, 244–252. https://doi.org/10.1080/10412905.2018.1435427

Guillén-Casla, V., León-González, M. E., Pérez-Arribas, L. V., Polo-Díez, L. M., 2010. Direct chiral determination of free amino acid enantiomers by two-dimensional liquid chromatography: application to control transformations in E-beam irradiated foodstuffs. *Anal. Bioanal. Chem.* 397, 63–75. https://doi.org/10.1007/s00216-009-3376-6

Guo, W., Pan, B., Sakkiah, S., Yavas, G., Ge, W., Zou, W., Tong, W., Hong, H., 2019. Persistent organic pollutants in food: Contamination sources, health effects and detection methods. *Int. J. Environ. Res. Public Health* 16, 10–12. https://doi.org/10.3390/ijerph16224361

He, R., Fan, J., Tan, Q., Lai, Y., Chen, X., Wang, T., Jiang, Y., Zhang, Y., Zhang, W., 2018. Enantioselective determination of metconazole in multi matrices by high-performance liquid chromatography. *Talanta* 178, 980–986. https://doi.org/10.1016/j.talanta.2017.09.045

Hernández, F., Sancho, J. V., Ibáñez, M., Guerrero, C., 2007. Antibiotic residue determination in environmental waters by LC-MS. *TrAC Trends Anal. Chem.* 26, 466–485. https://doi.org/10.1016/j.trac.2007.01.012

Herrero, M., Simó, C., García-Cañas, V., Fanali, S., Cifuentes, A., 2010. Chiral capillary electrophoresis in food analysis. *Electrophoresis* 31, 2106–2114. https://doi.org/10.1002/elps.200900770

Hofmann, T., Soldo, T., Ottinger, H., Frank, O., Robert, F., Blank, I., 2005. Structural and functional characterization of a multimodal taste enhancer in beef bouillon, in: Frey, C., Rouseff, R. L. (Eds.), *Natural Flavors and Fragrances: Chemistry, Analysis and Production, ACS Symposium Series*. American Chemical Society, Washington, DC, USA, pp. 173–188. https://doi.org/10.1021/bk-2005-0908.ch012

Hong, J. H., Khan, N., Jamila, N., Hong, Y. S., Nho, E. Y., Choi, J. Y., Lee, C. M., Kim, K. S., 2017. Determination of volatile flavour profiles of *Citrus* spp. fruits by SDE-GC–MS and enantiomeric composition of chiral compounds by MDGC–MS. *Phytochem. Anal.* 28, 392–403. https://doi.org/10.1002/pca.2686

Jenske, R., Vetter, W., 2008. Enantioselective analysis of 2- and 3-hydroxy fatty acids in food samples. *J. Agric. Food Chem.* 56, 11578–11583. https://doi.org/10.1021/jf802772a

Jiang, Y., Fan, J., He, R., Guo, D., Wang, T., Zhang, H., Zhang, W., 2018. High-fast enantioselective determination of prothioconazole in different matrices by supercritical fluid chromatography and vibrational circular dichroism spectroscopic study. *Talanta* 187, 40–46. https://doi.org/https://doi.org/10.1016/j.talanta.2018.04.097

Kamencev, M., Komarova, N., Morozova, O., 2016. Enantioseparation of tartaric and malic acids in wines by ligand exchange capillary electrophoresis using uncoated fused silica capillary. *Chromatographia* 79, 927–931. https://doi.org/10.1007/s10337-016-3099-8

Knob, R., Petr, J., Ševčík, J., Maier, V., 2013. Enantioseparation of tartaric acid by ligand-exchange capillary electrophoresis using contactless conductivity detection. *J. Sep. Sci.* 36, 3426–3431. https://doi.org/10.1002/jssc.201300507

Kodama, S., Aizawa, S., Taga, A., Yamamoto, A., Honda, Y., Suzuki, K., Kemmei, T., Hayakawa, K., 2013. Determination of α-hydroxy acids and their enantiomers in fruit juices by ligand exchange CE with a dual central metal ion system. *Electrophoresis* 34, 1327–1333. https://doi.org/10.1002/elps.201200645

Kodama, S., Taga, A., Aizawa, S., Kemmei, T., Honda, Y., Suzuki, K., Yamamoto, A., 2012. Direct enantioseparation of lipoic acid in dietary supplements by capillary electrophoresis using trimethyl-β-cyclodextrin as a chiral selector. *Electrophoresis* 33, 2441–2445. https://doi.org/10.1002/elps.201100531

Kodama, S., Yamamoto, A., Ohura, T., Matsunaga, A., Kanbe, T., 2003. Enantioseparation of imazalil residue in orange by capillary electrophoresis with 2-Hydroxypropyl-β-cyclodextrin as a chiral selector. *J. Agric. Food Chem.* 51, 6128–6131. https://doi.org/10.1021/jf030282x

Kodama, S., Yamamoto, A., Saitoh, Y., Matsunaga, A., Okamura, K., Kizu, R., Hayakawa, K., 2002. Enantioseparation of vinclozolin by γ-cyclodextrin-modified micellar electrokinetic chromatography. *J. Agric. Food Chem.* 50, 1312–1317. https://doi.org/10.1021/jf011238p

Konya, Y., Taniguchi, M., Fukusaki, E., 2017. Novel high-throughput and widely-targeted liquid chromatography–time of flight mass spectrometry method for d-amino acids in foods. *J. Biosci. Bioeng.* 123, 126–133. https://doi.org/https://doi.org/10.1016/j.jbiosc.2016.07.009

Kuang, H., Miao, H., Hou, X., Zhao, Y., Shen, J., Wu, Y., 2010. Determination of enantiomeric fractions of cypermethrin and cis-bifenthrin in Chinese teas by GC/ECD. *J. Sci. Food Agric.* 90, 1374–1379. https://doi.org/10.1002/jsfa.3934

Lämmerhofer, M., 2010. Chiral recognition by enantioselective liquid chromatography: mechanisms and modern chiral stationary phases. *J. Chromatogr. A* 1217, 814–56. https://doi.org/10.1016/j.chroma.2009.10.022

Langen, J., Fischer, U., Cavalar, M., Coetzee, C., Wegmann-Herr, P., Schmarr, H., 2016a. Enantiodifferentiation of 1,2-propanediol in various wines as phenylboronate ester with multidimensional gas chromatography-mass spectrometry. *Anal. Bioanal. Chem.* 408, 2425–2439. https://doi.org/10.1007/s00216-016-9379-1

Langen, J., Wegmann-Herr, P., Schmarr, H., 2016b. Quantitative determination of α-ionone, β-ionone, and β-damascenone and enantiodifferentiation of α-ionone in wine for authenticity control using multidimensional gas chromatography with tandem mass spectrometric detection. *Anal. Bioanal. Chem.* 408, 6483–6496. https://doi.org/10.1007/s00216-016-9767-6

Lasekan, O., 2018. Enantiomeric differentiation of three key volatile compounds in three different palm wines (Elaeis guineensis, Borassus flabellifer and Nypa fruticans). *J. Food* 16, 70–75. https://doi.org/10.1080/19476337.2017.1337813

Li, J., Dong, F., Cheng, Y., Liu, X., Xu, J., Li, Y., Chen, X., Kong, Z., Zheng, Y., 2012. Simultaneous enantioselective determination of triazole fungicide difenoconazole and its main chiral metabolite in vegetables and soil by normal-phase high-performance liquid chromatography. *Anal. Bioanal. Chem.* 404, 2017–2031. https://doi.org/10.1007/s00216-012-6240-z

Li, J., Dong, F., Xu, J., Liu, X., Li, Y., Shan, W., Zheng, Y., 2011. Enantioselective determination of triazole fungicide simeconazole in vegetables, fruits, and cereals using modified QuEChERS (quick, easy, cheap, effective, rugged, and safe) coupled to gas chromatography/tandem mass spectrometry. *Anal. Chim. Acta* 702, 127–135.

Li, R., Chen, Z., Dong, F., Xu, J., Liu, X., Wu, X., Pan, X., Tao, Y., Zheng, Y., 2018. Supercritical fluid chromatographic-tandem mass spectrometry method for monitoring dissipation of thiacloprid in greenhouse vegetables and soil under different application modes. *J. Chromatogr. B* 1081–1082, 25–32. https://doi.org/https://doi.org/10.1016/j.jchromb.2018.02.021

Li, S., Shi, W., Li, H., Xu, N., Zhang, R., Chen, X., Sun, W., Wen, D., He, S., Pan, J., He, Z., Fan, Y., 2018. Antibiotics in water and sediments of rivers and coastal area of Zhuhai city, pearl river estuary, South China. *Sci. Total Environ.* 636, 1009–1019. https://doi.org/10.1016/j.scitotenv.2018.04.358

Li, Y., Dong, F., Liu, X., Xu, J., Chen, X., Han, Y., Liang, X., Zheng, Y., 2013. Studies of enantiomeric degradation of the triazole fungicide hexaconazole in tomato, cucumber, and field soil by chiral liquid chromatography-tandem mass spectrometry. *Chirality* 25, 160–169.

Liberto, E., Cagliero, C., Sgorbini, B., Bicchi, C., Sciarrone, D., Zellner, B. D. A., Mondello, L., Rubiolo, P., 2008. Enantiomer identification in the flavour and fragrance fields by "interactive" combination of linear retention indices from enantioselective gas chromatography and mass spectrometry. *J. Chromatogr. A* 1195, 117–126. https://doi.org/10.1016/j.chroma.2008.04.045

Lin, E., Lin, K., Chang, C., Hsieh, M., 2013. On-line sample preconcentration by sweeping and poly(ethylene oxide)-mediated stacking for simultaneous analysis of nine pairs of amino acid enantiomers in capillary electrophoresis. *Talanta* 114, 297–303. https://doi.org/https://doi.org/10.1016/j.talanta.2013.05.039

Liu, H., Yi, X., Bi, J., Wang, P., Liu, D., Zhou, Z., 2019. The enantioselective environmental behavior and toxicological effects of pyriproxyfen in soil. *J. Hazard. Mater.* 365, 97–106. https://doi.org/10.1016/j.jhazmat.2018.10.079

Liu, N., Pan, X., Zhang, S., Ji, M., Zhang, Z., 2018. Enantioselective behaviour of tetraconazole during strawberry wine making. *Chirality 30(5)*, 1–9. https://doi.org/10.1002/chir.22845

Liu, W., Qin, S., Gan, J., 2005. Chiral stability of synthetic pyrethroid insecticides. *J. Agric. Food Chem.* 53, 3814–3820. https://doi.org/10.1021/jf048425i

Liu, W., Ye, J., Jin, M., 2009. Enantioselective phytoeffects of chiral pesticides. *J. Agric. Food Chem.* 57, 2087–2095. https://doi.org/10.1021/jf900079y

Lomenova, A., Hroboňová, K., 2018. Comparison of HPLC separation of phenylalanine enantiomers on different types of chiral stationary phases. *Food Anal. Methods* 11, 3314–3323. https://doi.org/10.1007/s12161-018-1308-9

Lu, Y., Shao, Y., Dai, S., Diao, J., Chen, X., 2016. Stereoselective behavior of the fungicide benalaxyl during grape growth and the wine-making process. *Chirality* 28, 394–398. https://doi.org/10.1002/chir.22589

Machonis, P. R., Jones, M. A., Schaneberg, B. T., Kwik-Uribe, C. L., 2012. Method for the determination of catechin and epicatechin enantiomers in cocoa-Based ingredients and products by High-Performance Liquid chromatography: Single-laboratory validation. *J. AOAC Int.* 95, 500–507. https://doi.org/10.5740/jaoacint.11-324

Malowicki, S. M. M., Martin, R., Qian, M. C., 2008a. Volatile composition in raspberry cultivars grown in the pacific northwest determined by stir bar sorptive extraction-gas chromatography-mass spectrometry. *J. Agric. Food Chem.* 56, 4128–4133. https://doi.org/10.1021/jf073489p

Malowicki, S. M. M., Martin, R., Qian, M. C., 2008b. Comparison of sugar, acids, and volatile composition in raspberry bushy dwarf virus-resistant transgenic raspberries and the wild type "Meeker" (*Rubus idaeus* L.). *J. Agric. Food Chem.* 56, 6648–6655. https://doi.org/10.1021/jf800253e

Martínez-Girón, A. B., García-Ruiz, C., Crego, A. L., Marina, M. L., 2009. Development of an in-capillary derivatization method by CE for the determination of chiral amino acids in dietary supplements and wines. *Electrophoresis* 30, 696–704. https://doi.org/10.1002/elps.200800481

Martinez, S. E., Davies, N. M., 2015. Stereospecific quantitation of 6-prenylnaringenin in commercially available H. lupulus-containing natural health products and dietary supplements. *Res. Pharm. Sci.* 10, 182–191.

McEachran, A. D., Blackwell, B. R., Hanson, J. D., Wooten, K. J., Mayer, G. D., Cox, S. B., Smith, P.N., 2015. Antibiotics, bacteria, and antibiotic resistance genes: Aerial transport from cattle feed yards via particulate matter. *Environ. Health Perspect.* 123, 337–343. https://doi.org/10.1289/ehp.1408555

Miao, Y., Liu, Q., Wang, W., Liu, L., Wang, L., 2017. Enantioseparation of amino acids by micellar capillary electrophoresis using binary chiral selectors and determination of D-glutamic acid and D-aspartic acid in rice wine. *J. Liq. Chromatogr. Relat. Technol.* 40, 783–789. https://doi.org/10.1080/10826076.2017.1364263

Mirasoli, M., Gotti, R., Di Fusco, M., Leoni, A., Colliva, C., Roda, A., 2014. Electronic nose and chiral-capillary electrophoresis in evaluation of the quality changes in commercial green tea leaves during a long-term storage. *Talanta* 129, 32–38. https://doi.org/10.1016/j.talanta.2014.04.044

Miyoshi, Y., Nagano, M., Ishigo, S., Ito, Y., Hashiguchi, K., Hishida, N., Mita, M., Lindner, W., Hamase, K., 2014. Chiral amino acid analysis of Japanese traditional Kurozu and the developmental changes during earthenware jar fermentation processes. *J. Chromatogr. B Anal. Technol. Biomed. Life Sci.* 966, 187–192. https://doi.org/10.1016/j.jchromb.2014.01.034

Mu, B., Zhu, Y., Lv, H., Yan, H., Peng, Q., Lin, Z., 2018. The enantiomeric distributions of volatile constituents in different tea cultivars. *Food Chem.* 265, 329–336. https://doi.org/https://doi.org/10.1016/j.foodchem.2018.05.094

Nakano, Y., Konya, Y., Taniguchi, M., Fukusaki, E., 2017. Development of a liquid chromatography-tandem mass spectrometry method for quantitative analysis of trace D-amino acids. *J. Biosci. Bioeng.* 123(1), 134–138. https://doi.org/10.1016/j.jbiosc.2016.07.008

Nakano, Y., Taniguchi, M., Fukusaki, E., 2019. High-sensitive liquid chromatography-tandem mass spectrometry-based chiral metabolic profiling focusing on amino acids and related metabolites. *J. Biosci. Bioeng.* 127, 520–527. https://doi.org/https://doi.org/10.1016/j.jbiosc.2018.10.003

Paik, M. J., Lee, J., Kim, K. R., 2008. N-Ethoxycarbonylation combined with (S)-1-phenylethylamidation for enantioseparation of amino acids by achiral gas chromatography and gas chromatography-mass spectrometry. *J. Chromatogr. A* 1214, 151–156. https://doi.org/10.1016/j.chroma.2008.10.068

Pan, X., Dong, F., Chen, Z., Xu, J., Liu, X., Wu, X., Zheng, Y., 2017. The application of chiral ultra-high-performance liquid chromatography tandem mass spectrometry to the separation of the zoxamide enantiomers and the study of enantioselective degradation process in agricultural plants. *J. Chromatogr. A* 1525, 87–95. https://doi.org/https://doi.org/10.1016/j.chroma.2017.10.016

Pan, X., Dong, F., Xu, J., Liu, X., Chen, Z., Zheng, Y., 2016. Stereoselective analysis of novel chiral fungicide pyrisoxazole in cucumber, tomato and soil under different application methods with supercritical fluid chromatography/tandem mass spectrometry. *J. Hazard. Mater.* 311, 115–124. https://doi.org/10.1016/j.jhazmat.2016.03.005

Pang, L., Zhou, J., Tang, J., Ng, S. C., Tang, W., 2014. Evaluation of perphenylcarbamated cyclodextrin clicked chiral stationary phase for enantioseparations in reversed phase high-performance liquid chromatography. *J. Chromatogr. A* 1363, 119–127. https://doi.org/10.1016/j.chroma.2014.08.040

Pasquini, B., Orlandini, S., Goodarzi, M., Caprini, C., Gotti, R., Furlanetto, S., 2016. Chiral cyclodextrin-modified micellar electrokinetic chromatography and chemometric techniques for green tea samples origin discrimination. *Talanta* 150, 7–13. https://doi.org/https://doi.org/10.1016/j.talanta.2015.12.003

Pažitná, A., Špánik, I., 2014. Enantiomeric distribution of major chiral volatile organic compounds in juniper-flavored distillates. *J. Sep. Sci.* 37, 398–403. https://doi.org/10.1002/jssc.201301151

Perez-Fernandez, V., Garcia, M. A., Marina, M. L., 2011a. Chiral separation of agricultural fungicides. *J. Chromatogr. A* 1218, 6561–6582.

Perez-Fernandez, V., Garcia, M. A., Marina, M. L., 2011b. Chiral separation of agricultural fungicides. *J. Chromatogr. A* 1218, 6561–6582.

Pérez-Mayán, L., Cobo-Golpe, M., Ramil, M., Cela, R., Rodríguez, I., 2020. Evaluation of supercritical fluid chromatography accurate mass spectrometry for neonicotinoid compounds determination in wine samples. *J. Chromatogr. A* 1620, 460963. https://doi.org/https://doi.org/10.1016/j.chroma.2020.460963

Pérez-Míguez, R., Marina, M. L., Castro-Puyana, M., 2016. Enantiomeric separation of non-protein amino acids by electrokinetic chromatography. *J. Chromatogr. A* 1467, 409–416. https://doi.org/10.1016/j.chroma.2016.06.058

Pons, A., Lavigne, V., Darriet, P., Dubourdieu, D., 2016. Identification and analysis of piperitone in red wines. *Food Chem.* 206, 191–196. https://doi.org/https://doi.org/10.1016/j.foodchem.2016.03.064

Pons, A., Lavigne, V., Landais, Y., Darriet, P., Dubourdieu, D., 2010. Identification of a sotolon pathway in dry white wines. *J. Agric. Food Chem.* 58, 7273–7279. https://doi.org/10.1021/jf100150q

Pons, A., Lavigne, V., Landais, Y., Darriet, P., Dubourdieu, D., 2008. Distribution and organoleptic impact of sotolon enantiomers in dry white wines. *J. Agric. Food Chem.* 56, 1606–1610. https://doi.org/10.1021/jf072337r

Prado-Cabrero, A., Beatty, S., Stack, J., Howard, A., Nolan, J. M., 2016. Quantification of zeaxanthin stereoisomers and lutein in trout flesh using chiral high-performance liquid chromatography-diode array detection. *J. Food Compos. Anal.* 50, 19–22. https://doi.org/10.1016/j.jfca.2016.05.004

Přibylka, A., Švidrnoch, M., Ševčík, J., Maier, V., 2015. Enantiomeric separation of 1,3-dimethylamylamine by capillary electrophoresis with indirect UV detection using a dual-selector system. *Electrophoresis* 36, 2866–2873. https://doi.org/10.1002/elps.201500182

Qi, P., Wang, Xiangyun, Zhang, H., Wang, Xinquan, Xu, H., Wang, Q., 2015. Rapid enantioseparation and determination of isocarbophos enantiomers in orange pulp, peel, and kumquat by chiral HPLC-MS/MS. *Food Anal. Methods* 8, 531–538. https://doi.org/10.1007/s12161-014-9922-7

Qi, P., Yuan, Y., Wang, Z., Wang, Xiangyun, Xu, H., Zhang, H., Wang, Q., Wang, Xinquan, 2016. Use of liquid chromatography- quadrupole time-of-flight mass spectrometry for enantioselective separation and determination of pyrisoxazole in vegetables, strawberry and soil. *J. Chromatogr. A* 1449, 62–70. https://doi.org/10.1016/j.chroma.2016.04.051

Qin, S., Budd, R., Bondarenko, S., Liu, W., Gan, J., 2006. Enantioselective degradation and chiral stability of pyrethroids in soil and sediment. *J. Agric. Food Chem.* 54, 5040–5045. https://doi.org/10.1021/jf060329p

Raimbault, A., Dorebska, M., West, C., 2019. A chiral unified chromatography–mass spectrometry method to analyze free amino acids. *Anal. Bioanal. Chem.* 411, 4909–4917. https://doi.org/10.1007/s00216-019-01783-5

Ravid, U., Elkabetz, M., Zamir, C., Cohen, K., Larkov, O., Aly, R., 2010. Authenticity assessment of natural fruit flavour compounds in foods and beverages by auto-HS–SPME stereoselective GC–MS. *Flavour Fragr. J.* 25, 20–27. https://doi.org/10.1002/ffj.1953

Ribeiro, A. R., Castro, P. M. L., Tiritan, M. E., 2012a. Environmental fate of chiral pharmaceuticals: Determination, degradation and toxicity, in: Lichtfouse, E., Schwarzbauer, J., Robert, D. (Eds.), *Environmental Chemistry for a Sustainable World*. Springer Netherlands, Dordrecht, pp. 3–45. https://doi.org/10.1007/978-94-007-2439-6_1

Ribeiro, A. R., Castro, P. M. L., Tiritan, M. E., 2012b. Chiral pharmaceuticals in the environment. *Environ. Chem. Lett.* 10, 239–253.

Rocco, A., Aturki, Z., Fanali, S., 2013. Chiral separations in food analysis. *Trends Anal. Chem.* 52, 206–225. https://doi.org/10.1016/j.trac.2013.05.022

Ruiz del Castillo, M. L., Caja, M. M., Herraiz, M., 2003. Use of the enantiomeric composition for the assessment of the authenticity of fruit beverages. *J. Agric. Food Chem.* 51, 1284–1288. https://doi.org/10.1021/jf025711q

Sánchez-Hernández, L., Castro-Puyana, M., García-Ruiz, C., Crego, A. L., Marina, M. L., 2010. Determination of l- and d-carnitine in dietary food supplements using capillary electrophoresis–tandem mass spectrometry. *Food Chem.* 120, 921–928. https://doi.org/https://doi.org/10.1016/j.foodchem.2009.11.004

Sanganyado, E., 2019. Comments on "Chiral pharmaceuticals: Environment sources, potential human health impacts, remediation technologies and future perspective." *Environ. Int.* 122, 412–415. https://doi.org/10.1016/j.envint.2018.11.032

Sanganyado, E., Lu, Z., Fu, Q., Schlenk, D., Gan, J., 2017. Chiral pharmaceuticals: A review on their environmental occurrence and fate processes. *Water Res.* 124, 527–542. https://doi.org/10.1016/j.watres.2017.08.003

Sanganyado, E., Lu, Z., Gan, J., 2014. Mechanistic insights on chaotropic interactions of liophilic ions with basic pharmaceuticals in polar ionic mode liquid chromatography. *J. Chromatogr. A* 1368, 82–88. https://doi.org/10.1016/j.chroma.2014.09.054

Sanganyado, E., Lu, Z., Liu, W., 2020. Application of enantiomeric fractions in environmental forensics: Uncertainties and inconsistencies. *Environ. Res.* 184, 109354. https://doi.org/10.1016/j.envres.2020.109354

Sardella, R., Lisanti, A., Marinozzi, M., Ianni, F., Natalini, B., Blanch, G. P., Ruiz del Castillo, M. L., 2013. Combined monodimensional chromatographic approaches to monitor the presence of d-amino acids in cheese. *Food Control.* 34, 478–487. https://doi.org/10.1016/j.foodcont.2013.05.026

Schaffrath, M., Weidmann, V., Maison, W., 2014. Enantioselective high-performance liquid chromatography and supercritical fluid chromatography separation of spirocyclic terpenoid flavor compounds. *J. Chromatogr. A* 1363, 270–277. https://doi.org/10.1016/j.chroma.2014.07.001

Schipilliti, L., Tranchida, P. Q., Sciarrone, D., Russo, M., Dugo, P., Dugo, G., Mondello, L., 2010. Genuineness assessment of mandarin essential oils employing gas chromatography-combustion-isotope ratio MS (GC-C-IRMS). *J. Sep. Sci.* 33, 617–625. https://doi.org/10.1002/jssc.200900504

Scriba, G. K. E., 2016. Chiral recognition in separation science—an update. *J. Chromatogr. A* 1467, 56–78. https://doi.org/10.1016/j.chroma.2016.05.061

Shaw, S. D., Harris, J. H., Berger, M. L., Subedi, B., Kannan, K., 2014. Brominated flame retardants and their replacements in food packaging and household products: Uses, human exposure, and health effects, in: Snedeker, S. M. (Ed.), *Molecular and Integrative Toxicology, Molecular and Integrative Toxicology.* Springer London, London, pp. 61–93. https://doi.org/10.1007/978-1-4471-6500-2_3

Simó, C., Barbas, C., Cifuentes, A., 2003. Chiral electromigration methods in food analysis. *Electrophoresis* 24, 2431–2441. https://doi.org/10.1002/elps.200305442

Simó, C., Rizzi, A., Barbas, C., Cifuentes, A., 2005. Chiral capillary electrophoresis-mass spectrometry of amino acids in foods. *Electrophoresis* 26, 1432–1441. https://doi.org/10.1002/elps.200406199

Soldo, T., Blank, I., Hofmann, T., 2003. (+)-(S)-alapyridaine—A general taste enhancer? *Chem. Senses* 28, 371–379. https://doi.org/10.1093/chemse/28.5.371

Špánik, I., Pažitná, A., Šiška, P., Szolcsányi, P., 2014. The determination of botanical origin of honeys based on enantiomer distribution of chiral volatile organic compounds. *Food Chem.* 158, 497–503. https://doi.org/https://doi.org/10.1016/j.foodchem.2014.02.129

Stamatopoulos, P., Brohan, E., Prevost, C., Siebert, T. E., Herderich, M., Darriet, P., 2016. Influence of chirality of lactones on the perception of some typical fruity notes through perceptual interaction phenomena in Bordeaux dessert wines. *J. Agric. Food Chem.* 64, 8160–8167. https://doi.org/10.1021/acs.jafc.6b03117

Stanley, J. K., Brooks, B. W., 2009. Perspectives on ecological risk assessment of chiral compounds. *Integr. Environ. Assess. Manag.* 5, 364–373. https://doi.org/10.1897/IEAM_2008-076.1

Stanley, J. K., Ramirez, A. J., Chambliss, C. K., Brooks, B. W., 2007. Enantiospecific sublethal effects of the antidepressant fluoxetine to a model aquatic vertebrate and invertebrate. *Chemosphere* 69, 9–16. https://doi.org/10.1016/j.chemosphere.2007.04.080

Stanley, J. K., Ramirez, A. J., Mottaleb, M., Chambliss, C. K., Brooks, B. W., 2006. Enantiospecific toxicity of the beta-blocker propranolol to *Daphnia magna* and *Pimephales promelas*. *Environ. Toxicol. Chem.* 25, 1780–6.

Sun, J., Yu, J., Zhang, P. C., Yue, Y. De, 2015. Enantiomeric determination of four diastereoisomeric oxyneolignans from bambusa tuldoides munro. *Phytochem. Anal.* 26, 54–60. https://doi.org/10.1002/pca.2536

Sun, M., Liu, D., Zhou, G., Li, J., Qiu, X., Zhou, Z., Wang, P., 2012. Enantioselective degradation and chiral stability of malathion in environmental samples. *J. Agric. Food Chem.* 60, 372–9. https://doi.org/10.1021/jf203767d

Tanaka, K., Muraoka, T., Otubo, Y., Takahashi, H., Ohnishi, A., 2016. HPLC enantioseparation on a homochiral MOF-silica composite as a novel chiral stationary phase. *RSC Adv.* 6, 21293–21301. https://doi.org/10.1039/c5ra26520g

Tang, Y., 1996. Significance of mobile phase composition in enantioseparation of chiral drugs by HPLC on a cellulose-based chiral stationary phase. *Chirality* 8, 136–142. https://doi.org/10.1002/(SICI)1520-636X(1996)8:1<136::AID-CHIR20>3.0.CO;2-N

Tao, Y., Dong, F., Xu, J., Liu, X., Cheng, Y., Liu, N., Chen, Z., Zheng, Y., 2014. Green and sensitive supercritical fluid chromatographic-tandem mass spectrometric method for the separation and determination of flutriafol enantiomers in vegetables, fruits, and soil. *J. Agric. Food Chem.* 62, 11457–11464. https://doi.org/10.1021/jf504324t

Tao, Y., Zheng, Z., Yu, Y., Xu, J., Liu, X., Wu, X., Dong, F., Zheng, Y., 2018. Supercritical fluid chromatography–tandem mass spectrometry-assisted methodology for rapid enantiomeric analysis of fenbuconazole and its chiral metabolites in fruits, vegetables, cereals, and soil. *Food Chem.* 241, 32–39. https://doi.org/https://doi.org/10.1016/j.foodchem.2017.08.038

Tezcan, F., Uzaşçı, S., Uyar, G., Öztekin, N., Erim, F. B., 2013. Determination of amino acids in pomegranate juices and fingerprint for adulteration with apple juices. *Food Chem.* 141, 1187–1191. https://doi.org/https://doi.org/10.1016/j.foodchem.2013.04.017

Tian, M., Zhang, Q., Shi, H., Gao, B., Hua, X., Wang, M., 2015. Simultaneous determination of chiral pesticide flufiprole enantiomers in vegetables, fruits, and soil by high-performance liquid chromatography. *Anal. Bioanal. Chem.* 407, 3499–3507. https://doi.org/10.1007/s00216-015-8543-3

Todoroki, K., Goto, K., Nakano, T., Ishii, Y., Min, J. Z., Inoue, K., Toyo'oka, T., 2014. 4-(4,6-Dimethoxy-1,3,5-triazin-2-yl)-4-methylmorpholinium chloride as an enantioseparation enhancer for fluorescence chiral derivatization-liquid chromatographic analysis of DL-lactic acid. *J. Chromatogr. A* 1360, 188–195. https://doi.org/10.1016/j.chroma.2014.07.077

Tominaga, T., Niclass, Y., Frérot, E., Dubourdieu, D., 2006. Stereoisomeric distribution of 3-Mercaptohexan-1-ol and 3-Mercaptohexyl acetate in dry and sweet white wines made from *Vitis vinifera* (var. Sauvignon Blanc and Semillon). *J. Agric. Food Chem.* 54, 7251–7255. https://doi.org/10.1021/jf061566v

Tomiyama, K., Aoki, H., Oikawa, T., Sakurai, K., Kasahara, Y., Kawakami, Y., 2012. Characteristic volatile components of Japanese sour citrus fruits: Yuzu, Sudachi and Kabosu. *Flavour Fragr. J.* 27, 341–355. https://doi.org/10.1002/ffj.3104

Vyviurska, O., Zvrškovcová, H., Špánik, I., 2017. Distribution of enantiomers of volatile organic compounds in selected fruit distillates. *Chirality* 29, 14–18. https://doi.org/10.1002/chir.22669

Wang, Q., Sánchez-López, E., Han, H., Wu, H., Zhu, P., Crommen, J., Marina, M. L., Jiang, Z., 2016. Separation of N-derivatized di- and tri-peptide stereoisomers by micro-liquid chromatography using a quinidine-based monolithic column - Analysis of l-carnosine in dietary supplements. *J. Chromatogr. A* 1428, 176–184. https://doi.org/10.1016/j.chroma.2015.09.016

Wang, X., Wu, H., Luo, R., Xia, D., Jiang, Z., Han, H., 2017. Separation and detection of free d- and l-amino acids in tea by off-line two-dimensional liquid chromatography. *Anal. Methods* 9, 6131–6138. https://doi.org/10.1039/C7AY01569K

Wang, X. X., Qi, P., Zhang, H., Xu, H., Wang, X. X., Li, Z., Wang, Z., Wang, Q., 2014. Enantioselective analysis and dissipation of triazole fungicide penconazole in vegetables by liquid chromatography-tandem mass spectrometry. *J. Agric. Food Chem.* 62, 11047–11053. https://doi.org/10.1021/jf5034653

Wang, Xiangyun, Li, Z., Zhang, H., Xu, J., Qi, P., Xu, H., Wang, Q., Wang, Xinquan, 2013. Environmental behavior of the chiral organophosphorus insecticide acephate and its chiral metabolite methamidophos: Enantioselective transformation and degradation in soils. *Environ. Sci. Technol.* 47, 9233–9240. https://doi.org/10.1021/es401842f

Wang, Z. L., Zhang, J. L., Zhang, Y. N., Zhao, Y., 2016. Mass spectrometric analysis of residual clenbuterol enantiomers in swine, beef and lamb meat by liquid chromatography tandem mass spectrometry. *Anal. Methods* 8, 4127–4133. https://doi.org/10.1039/c6ay00606j

Ward, T. J., Ward, K. D., 2012. Chiral separations: A review of current topics and trends. *Anal. Chem.* 84, 626–35. https://doi.org/10.1021/ac202892w

Xie, J., Zhao, J., Liu, K., Guo, F., Liu, W., 2016. Enantioseparation of four amide herbicide stereoisomers using high-performance liquid chromatography. *J. Chromatogr. A* 1471, 145–154. https://doi.org/https://doi.org/10.1016/j.chroma.2016.10.029

Xinzhong, Z., Yuechen, Z., Xinyi, C., Xinru, W., Huishan, S., Zongmao, C., Chun, H., Meruva, N., Li, Z., Fang, W., Luchao, W., Fengjian, L., 2018. Application and enantiomeric residue determination of diniconazole in tea and grape and apple by supercritical fluid chromatography coupled with quadrupole-time-of-flight mass spectrometry. *J. Chromatogr. A*.

Xu, G., Jia, X., Wu, X., Xu, J., Liu, X., Pan, X., Li, R., Li, X., Dong, F., 2018. Enantioselective monitoring of chiral fungicide famoxadone enantiomers in tomato, apple, and grape by chiral liquid chromatography with tandem mass spectrometry. *J. Sep. Sci.* 41, 3871–3880. https://doi.org/10.1002/jssc.201800681

Yan, X., Zhou, Y., Tang, J., Ji, M., Lou, H., Fan, P., 2016. Diketopiperazine indole alkaloids from hemp seed. *Phytochem. Lett.* 18, 77–82. https://doi.org/10.1016/j.phytol.2016.09.001

Ye, J., Wu, J., Liu, W., 2009. Enantioselective separation and analysis of chiral pesticides by high-performance liquid chromatography. *Trends Anal. Chem.* 28, 1148–1163. https://doi.org/10.1016/j.trac.2009.07.008

Zeng, H., Xie, X., Huang, Y., Chen, J., Liu, Y., Zhang, Y., Mai, X., Deng, J., Fan, H., Zhang, W., 2019. Enantioseparation and determination of triazole fungicides in vegetables and fruits by aqueous two-phase extraction coupled with online heart-cutting two-dimensional liquid chromatography. *Food Chem.* 301. https://doi.org/10.1016/j.foodchem.2019.125265

Zhang, H., Wang, X., Qian, M., Wang, X., Xu, H., Xu, M., Wang, Q., 2011. Residue analysis and degradation studies of fenbuconazole and mycobutanil in strawberry by chiral high-performance liquid chromatography-tandem mass spectrometry. *J. Agric. Food Chem.* 59, 12012–12017.

Zhang, H., Wang, Xiangyun, Wang, Xinquan, Qian, M., Xu, M., Xu, H., Qi, P., Wang, Q., Zhuang, S., 2014. Enantioselective determination of carboxyl acid amide fungicide mandipropamid in vegetables and fruits by chiral LC coupled with MS/MS. *J. Sep. Sci.* 37, 211–218. https://doi.org/10.1002/jssc.201301080

Zhang, Q., Gao, B., Tian, M., Shi, H., Hua, X., Wang, M., 2016. Enantioseparation and determination of triticonazole enantiomers in fruits, vegetables, and soil using efficient extraction and clean-up methods. *J. Chromatogr. B* 1009–1010, 130–137. https://doi.org/10.1016/j.jchromb.2015.12.018

Zhang, Q., Tian, M., Wang, Meiyun, Shi, H., Wang, Minghua, 2014. Simultaneous enantioselective determination of triazole fungicide flutriafol in vegetables, fruits, wheat, soil, and water by reversed-phase high-performance liquid chromatography. *J. Agric. Food Chem.* 62, 2809–2815. https://doi.org/10.1021/jf405689n

Zhang, Q., Xiong, W., Gao, B., Cryder, Z., Zhang, Z., Tian, M., Sanganyado, E., Shi, H., Wang, M., 2018a. Enantioselectivity in degradation and ecological risk of the chiral pesticide ethiprole. *L. Degrad. Dev.* 29, 4242–4251. https://doi.org/10.1002/ldr.3179

Zhang, Q., Zhang, Z., Tang, B., Gao, B., Tian, M., Sanganyado, E., Shi, H., Wang, M., 2018b. Mechanistic insights into stereospecific bioactivity and dissipation of chiral fungicide triticonazole in agricultural management. *J. Agric. Food Chem.* 66, 7286–7293. https://doi.org/10.1021/acs.jafc.8b01771

Zhang, X., Wang, X., Luo, F., Sheng, H., Zhou, L., Zhong, Q., Lou, Z., Sun, H., Yang, M., Cui, X., Chen, Z., 2019. Application and enantioselective residue determination of chiral pesticide penconazole in grape, tea, aquatic vegetables and soil by ultra performance liquid chromatography-tandem mass spectrometry. *Ecotoxicol. Environ. Saf.* 172, 530–537. https://doi.org/10.1016/j.ecoenv.2019.01.103

Zhang, Z., Gao, B., Li, L., Zhang, Q., Xia, W., Wang, M., 2018. Enantioselective degradation and transformation of the chiral fungicide prothioconazole and its chiral metabolite in soils. *Sci. Total Environ.* 634, 875–883. https://doi.org/10.1016/j.scitotenv.2018.03.375

Zhao, L., Chen, F., Guo, F., Liu, W., Liu, K., 2019. Enantioseparation of chiral perfluorooctane sulfonate (PFOS) by supercritical fluid chromatography (SFC): Effects of the chromatographic conditions and separation mechanism. *Chirality* 31, 870–878. https://doi.org/10.1002/chir.23120

Zhao, P., Li, S., Chen, X., Guo, X., Zhao, L., 2019a. Simultaneous enantiomeric analysis of six chiral pesticides in functional foods using magnetic solid-phase extraction based on carbon nanospheres as adsorbent and chiral liquid chromatography coupled with tandem mass spectrometry. *J. Pharm. Biomed. Anal.* 175, 112784. https://doi.org/10.1016/j.jpba.2019.112784

Zhao, P., Wang, Z., Gao, X., Guo, X., Zhao, L., 2019b. Simultaneous enantioselective determination of 22 chiral pesticides in fruits and vegetables using chiral liquid chromatography coupled with tandem mass spectrometry. *Food Chem.* 277, 298–306. https://doi.org/10.1016/j.foodchem.2018.10.128

Zhu, Y., Shao, C., Lv, H., Zhang, Y., Dai, W., Guo, L., Tan, J., Peng, Q., Lin, Z., 2017. Enantiomeric and quantitative analysis of volatile terpenoids in different teas (Camellia sinensis). *J. Chromatogr. A* 1490, 177–190. https://doi.org/https://doi.org/10.1016/j.chroma.2017.02.013

Zor, E., Bingol, H., Ersoz, M., 2019. Chiral sensors. *Trends Anal. Chem.* 121, 115662. https://doi.org/10.1016/j.trac.2019.115662

9 Occurrence of Chiral Pesticides in Food and Their Effects on Food Processing and the Health of Humans

Herbert Musarurwa and Nikita Tawanda Tavengwa

CONTENTS

9.1 Introduction 227
9.2 Occurrence of Chiral Pesticides in Food 228
    9.2.1 Food Products of Plant Origin 228
    9.2.2 Food Products of Animal Origin 233
9.3 Food Processing and Its Impact on Chiral Pesticide Residues in Food 235
9.4 Health Risks Associated with Chiral Pesticides 237
9.5 Challenges and Future Prospects 241
9.6 Conclusion 241
Acknowledgment 241
Conflicts of Interest...241
References 241

9.1 INTRODUCTION

The use of pesticides to control pests is a widespread practice in the agricultural sector (Birolli et al., 2019; Buerge et al., 2019; Xie et al., 2018). Most of the pesticides in current use have one or more asymmetric centers, and consequently, they are chiral (Caballo et al., 2013; Chen et al., 2016; Cui et al., 2018; Gao et al., 2019). The enantiomers of chiral pesticides have different bioactivities and toxicities in chiral environments. Their bioactivities are, therefore, enantioselective (Kaziem et al., 2020; Tong et al., 2019; Zhang et al., 2019). Despite this, most chiral pesticides are currently being sold and applied to crops as racemates (Asad et al., 2017; Maia et al., 2017). Thus, the inactive enantiomers are discharged into the environment where they cause toxicity to non-target organisms as well as contaminating food sources of human beings.

Many types of food consumed by humans have trace amounts of the enantiomers of chiral pesticides (Lin et al., 2018; Pirsaheb et al., 2018). Researchers have published papers showing the presence of chiral pesticide residues in different food products (Giroud et al., 2019; Hamadamin and Hassan, 2020; Li et al., 2019; do Lago et al., 2020). For instance, Zhang et al. (2016) performed a study to determine the presence of the enantiomers of triticonazole pesticide in cucumber samples. Their results showed that the concentration of S-triticonazole in cucumber was 124 µg kg^{-1}, while the R-enantiomer had a concentration of 112 µg kg^{-1}. Researchers have also shown that some fruits also contain some traces of the enantiomers of chiral pesticides (Qi et al., 2015; Zeng et al., 2019; Zhao et al., 2019). For instance, Li et al. (2019) detected the concentration of R-tebuconazole in grape samples as 28.9 µg kg^{-1} while the S-enantiomers had a concentration of 35.8 µg kg^{-1}. Some cereals are also contaminated with chiral pesticides. Pirsaheb et al. (2018), for example, determined the concentration of chiral deltamethrin residues in wheat samples, and they found out that its concentration ranged from 23.5 to 73.6 µg kg^{-1}. Thus, some food derived from plants that are consumed by humans is contaminated with chiral pesticides.

Chiral pesticides are not found only in food derived from plants. Foods derived from animals are equally prone to contamination by the enantiomers of chiral pesticides. Dallegrave et al. (2016), for instance, found out that the concentration of 17 chiral pyrethroids in eggs, fish, and bovine milk samples ranged from 0.03 to 270 ng g^{-1}. Ye et al. (2018) found out that the concentration of paclobutrazol in honey ranged from 21.6 to 94.0 ng kg^{-1}, while Mendes et al. (2019) found out that the concentration of DDT in fish muscles ranged from 11.58 to 48.40 ng g^{-1}. Some animal products, therefore, are contaminated with chiral pesticides.

Most types of food are subjected to some form of processing before consumption by humans. These food-processing techniques include fermentation (Yang et al., 2020), pickling (Guo et al., 2020), canning (Yigit and Velioglu, 2019), pasteurization (Yigit and Velioglu, 2019), soaking (Kiwango et al., 2020), boiling, and washing (Luo et al., 2020). Some chiral pesticides undergo degradation and dissipation during processing, and this results in the reduction of the amount of their residues in food (Bian et al., 2020; Luo et al., 2020). Some researchers, however, have observed that the concentrations of some chiral pesticides remain unchanged throughout the processing period (Guo et al., 2020; Lu et al., 2016). Such chiral pesticides would remain in large quantities in the final product and have the potential of causing serious health problems to humans. Food processing, generally, can facilitate the reduction of significant proportions of chiral pesticide residues from food before humans consume it.

The concentrations of chiral pesticide residues should be carefully monitored so that they do not reach toxic levels in food. If the concentrations of chiral pesticides in food exceed the maximum allowable limits, usually serious health problems would occur in humans. Various health issues are associated with consuming large quantities of chiral pesticides (Chang et al., 2018; Qian et al., 2019; Zhang et al., 2019). Zhao et al. (2019), for instance, found that chiral bifenthrin residues have endocrine disrupting effects in the human body while Carrão et al. (2019) studies showed that fipronil is an inhibitor of some enzymes in the body. Liu et al. (2010), on the other hand, found out that isocarbophos is a cytotoxin in the body. Thus, chiral pesticide residues in food cause a myriad of health problems in the human body if their concentrations are not kept within the acceptable limits.

This chapter gives a detailed account of the occurrence of chiral pesticides in a wide range of food samples derived from both plants and animals. The various processing techniques food is subjected to before consumption by humans are discussed with emphasis on their impact on chiral pesticide residues. Lastly, a detailed account of the health issues associated with the consumption of food contaminated with chiral pesticides is given.

9.2 OCCURRENCE OF CHIRAL PESTICIDES IN FOOD

9.2.1 Food Products of Plant Origin

There is a wide array of chiral pesticides that have been developed to control pests in the agro-industry (Asad et al., 2017; Li et al., 2019; Zhang et al., 2017). The use of chiral pesticides is associated with many benefits such as efficient food production and effective control of vectors of diseases (Chen et al., 2016). Pesticide use, however, has its own drawbacks. One of them is that chiral pesticides discharged into the environment may contaminate food sources used by human beings. Many researchers have found out that a lot of food products derived from plants are contaminated with chiral pesticides such as herbicides (Saito-Shida et al., 2020; Xie et al., 2018), fungicides (Li et al.,

2019; Zeng et al., 2019), organophosphates (Alaboudi et al., 2019; do Lago et al., 2020), and pyrethroids (Hamadamin and Hassan, 2020; Zhao et al., 2019).

Fungicides are chemicals that are used in the agro-industry to control fungi. Researchers have shown that many types of food derived from plants contain a relatively high proportion of chiral fungicides (Table 9.1) (Li et al., 2012; Li et al., 2019; Zeng et al., 2019). For instance, Zeng et al. (2019) determined the concentrations of chiral fungicides (paclobutrazol, myclobutanil, tebuconazole, and bitertanol) in fruits and vegetables using two-dimensional liquid chromatography (2D-LC). They found out that the concentration of racemic paclobutrazol in apples was 0.228 µg kg^{-1}

TABLE 9.1
Chiral Pesticides Detected in Some Food Products of Plant Origin

Food Matrix	Chiral Pesticide Studied	Extraction Technique	Analytical Instrument	Concentration of Racemate (µg kg^{-1})	Concentration of Enantiomers (µg kg^{-1})	Reference
Orange	Isocarbophos	QuEChERS	HPLC	ng	(R)-(-)-isomer: 6.6, 18 and 290 (S)-(+)-isomer: 7.7, 20 and 320	Qi et al. (2015)
Kumquat	Isocarbophos	QuEChERS	HPLC	ng	(R)-(-)-isomer: 26, 43 and 103 (S)-(+)-isomer: 33, 50 and 125	Qi et al. (2015)
Apple, cucumber, pepper, tomato, eggplant, and Chinese cabbage	Paclobutrazol, myclobutanil, tebuconazole and bitertanol	Aqueous two-phase extraction	2D-LC	0.228 (Paclobutrazol in apples), 0.470 (myclobutanil in cucumber), 0.851 (tebuconazole in pepper) and 0.940 (bitertanol in tomato)	ng	Zeng et al. (2019)
Rice, tomato, and apple	Dinotefuran	SPE	HPLC	ng	(+)-d-isomer: 2.23 (rice), 2.52 (apple) and 2.39 (tomato) (-)-d-isomer: 2.53 (rice), 2.40 (apple) and 2.38 (tomato)	Chen et al. (2012)
Cucumber, tomato, apple, pear, wheat, and rice	Simeconazole	QuEChERS	GC-MS/MS	6.52 – 10.06	ng	Li et al. (2011)
Apple	Bifenthrin and permethrin	MSPE	GC-MS	4.53**(bifentrin) and 7.49** (permethrin)	ng	Zhao et al. (2019)
Tomato, cucumber, pear, and apple	Tebuconazole	QuEChERS	LC-MS/MS	3.5 – 14.6	ng	Li et al. (2012)

(Continued)

TABLE 9.1 *(Continued)*
Chiral Pesticides Detected in Some Food Products of Plant Origin

Food Matrix	Chiral Pesticide Studied	Extraction Technique	Analytical Instrument	Concentration of Racemate (µg kg⁻¹)	Concentration of Enantiomers (µg kg⁻¹)	Reference
Cucumber	Triticonazole	QuEChERS	HPLC-UV	ng	124* (S-triticonazole) and 112* ((R)-triticonazole)	Zhang et al. (2016)
Wheat	Malathion, deltamethrin and permethrin	DLLME-SFO	HPLC-UV	25.5 – 167.4 (malathion) and 23.5 -73.6 (deltamethrin)	ng	Pirsaheb et al. (2018)
Tomato and kidney bean	Bifenthrin, cyhalothrin and cypermethrin	QuEChERS	GC-MS/MS	8 (bifenthrin in tomato), 54 (cyhalothrin in tomato) and 30 (cypermethrin in kidney bean)	ng	Lin et al. (2018)
Grapes, apples, lettuces, rape, and cucumber	Triticonazole, hexaconazole, tebuconazole, triadimefon, metalaxyl and benalaxyl	QuEChERS	LC-MS/MS	ng	0.712 ((R)-hexaconazole in grapes), 0.774 ((S)-hexaconazole in grapes), 28.9 ((R)-tebuconazole in grapes), 35.8 ((S)-tebuconazole in grapes), 2.73 ((R)-tebuconazole in apples), 2.90 ((S)-tebuconazole in apples), 1.39 ((R)-triadimefon in apples) and 1.14 ((S)-triadimefon in apples)	Li et al. (2019)
Cucumber, tomato, and apple	Cyflumetofen	LLE and SPE	HPLC	ng	100* - 150*	Chen et al. (2013)
Cucumber, spinach, tomato, apple, and peach	Ethiprole	SPE	HPLC	20* - 171* (tomato and cucumber)	ng	Zhang et al. (2016)
Garlic, onion, and sugar beet	Ethofumesate	LLE	GC-MS/MS	Concentration less than 0.5* (LOD)	ng	Saito-Shida et al. (2020)

Note: ng – not given.
* Converted from mg kg⁻¹ to µg kg⁻¹.

Occurrence of Chiral Pesticides in Food

while the concentration of racemic myclobutanil was 0.470 μg kg^{-1} in cucumber and tebuconazole had a concentration of 0.851 μg kg^{-1} in pepper. Bitertranol was also detected in tomatoes and its concentration was 0.940 μg kg^{-1}. In a similar study, Li et al. (2012) determined the concentration of racemic tebuconazole in tomato, cucumber, pear, and apple samples using LC-MS/MS. The concentration of racemic tebuconazole in these fruit and vegetable samples ranged from 3.5 to 14.6 μg kg^{-1}. In the same vein, Li et al. (2011) determined the concentrations of racemic simeconazole in cucumber, tomato, pear, apple, wheat, and rice samples using GC-MS/MS and they were found to range from 6.52 to 10.06 μg kg^{-1}. Figure 9.1A gives the chromatograms obtained in some of the rice samples. Thus, these findings show that chiral fungicides contaminate some types of food derived from plants that are used by humans.

Some researchers determined the concentrations of the individual enantiomers of some chiral fungicides found in fruits and vegetables (Table 9.1) (Chen et al., 2012; Li et al., 2019; Qi et al., 2015; Zhang et al., 2016). For instance, Zhang et al. (2016) determined the concentration of the enantiomers of triticonazole in cucumber samples using HPLC-UV. They found the concentration of (S)-triticonazole to be 124 μg kg^{-1} while that of the R-enantiomer was 112 μg kg^{-1}. The results revealed that the S-enantiomer was more stable in cucumber as compared to its R-isomer. In another study, Li et al. (2019) determined the concentrations of enantiomers of six chiral fungicides (triticonazole, hexaconazole, tebuconazole, triadimefon, metalaxyl, and benalaxyl) in fruits and vegetables. They found the concentration of R-hexaconazole in grapes to be 0.712 μg kg^{-1} while the concentration of (S)-hexaconazole was 0.774 μg kg^{-1}. Tebuconazole was also detected in grapes and the concentration of its R-enantiomer was 28.9 μg kg^{-1} while that of the S-isomer was 35.8 μg kg^{-1}. The slight differences between the concentrations of the R- and S-enantiomers show that enantioselective degradation was a less pronounced phenomenon. During the same study, triadimefon was detected in apple samples and the concentration of the R-triadimefon was found to be 1.39 μg kg^{-1} while its S-enantiomer had a concentration of 1.14 μg kg^{-1}. The other fungicides, however, had concentrations below the LOQ of the analytical instrument, and as a result, were not detected. Thus, enantiomers of some chiral fungicides are found in food consumed by humans and monitoring their concentrations in food product is, therefore, an imperative.

Pyrethroids are among the widely used chiral pesticides in the agro-industry. Researchers have shown that some types of food derived from plants are contaminated with pyrethroid pesticide residues (Table 9.1) (Lin et al., 2018; Pirsaheb et al., 2018; Zhao et al., 2019). For instance, Zhao et al. (2019) detected the presence of bifenthrin and permethrin pesticides in apple samples. They found the concentration of racemic bifenthrin in apple samples to be 4.53 μg kg^{-1} while the concentration of racemic permethrin was 7.49 μg kg^{-1}. In a similar study, Pirsaheb et al. (2018) determined the concentration of the racemate of deltamethrin in wheat using HPLC-UV. They found out that its concentration ranged from 23.5 to 73.6 μg kg^{-1} in the wheat samples used. In another study, Lin et al. (2018) determined concentrations of the racemic mixtures of bifenthrin, cyhalothrin, and cypermethrin in tomato and kidney bean samples. Their results showed that the concentration of bifenthrin in the tomato samples was 8 μg kg^{-1} while cyhalothrin had a concentration of 54 μg kg^{-1}. The concentration of cypermethrin in kidney bean samples, on the other hand, was found to be 30 μg kg^{-1}. Thus, some chiral pyrethroid pesticide residues are found in some types of food consumed by humans.

Some researchers detected chiral organophosphate pesticide residues in some types of food derived from plants (Table 9.1) (Pirsaheb et al., 2018; Qi et al., 2015). For instance, Qi et al. (2015) investigated the presence of the enantiomers of isocarbophos in orange and kumquat samples using HPLC, and they obtained good resolutions of the enantiomers (Figure 9.1B). Their results indicated that the concentration of (R)-isocarbophos in orange samples ranged from 6.6 to 290 μg kg^{-1} while the concentration of the S-enantiomer ranged from 7.7 to 320 μg kg^{-1}. They also observed a significant level of the enantiomers of isocarbophos in kumquat samples they used. The concentration of (R)-isocarbophos in the kumquat samples they used ranged from 26.43 to 103 μg kg^{-1} while the concentration of the S-enantiomer ranged from 33.50 to 125 μg kg^{-1}. In another study, Pirsaheb

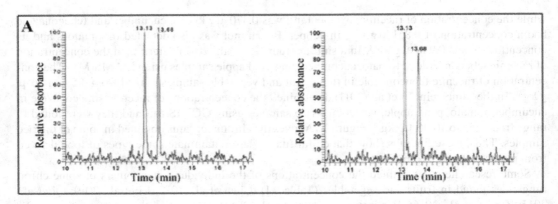

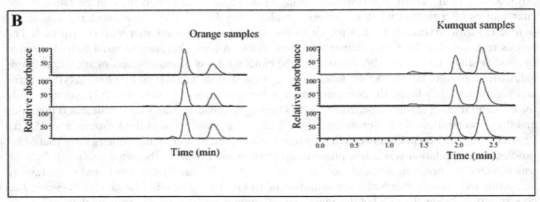

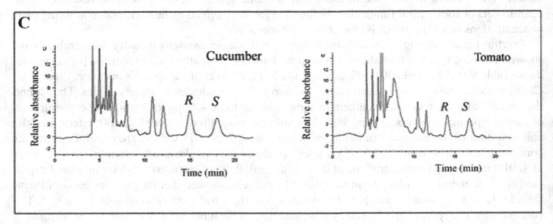

FIGURE 9.1 Chromatograms showing the presence of some chiral pesticide residues in food of plant origin. (A) Simeconazole residues in rice samples. (From Li et al., 2011. Used with permission of Elsevier.) (B) Residues of isocarbophos enantiomers in orange and kumquat samples. (Reprinted by permission from Springer Nature, *Food Analytical Methods*, Rapid Enantioseparation and Determination of Isocarbophos Enantiomers in Orange Pulp, Peel, and Kumquat by Chiral HPLC-MS/MS, Peipei Qi et al., Copyright 2015.) (C) Residues of enantiomers of ethiprole in cucumber and tomato samples. (From Zhang et al. (2016). Used with permission of Elsevier.)

et al. (2018) determined the concentration of racemic malathion in wheat samples. They observed that its concentration ranged from 25.5 to 167.4 µg kg^{-1} in the wheat samples they used. These findings confirm the presence of organophosphate residues in some food consumed by humans.

The residues of other classes of pesticides are also found in food. For instance, Chen et al. (2012) investigated the presence of the enantiomers of dinotefuran pesticide residues in rice, tomato, and apple samples using HPLC. Dinotefuran is a neonicotinoid pesticide with broad spectrum and systemic insecticidal activities. They detected both (+)-d-dinotefuran and (-)-d-dinotefuran in the food samples they used. The concentrations of the (+)-d-enantiomer were 2.23 µg kg^{-1} in rice, 2.52 µg kg^{-1} in apple, and 2.39 µg kg^{-1} in tomato while the concentrations of the (-)-d-enantiomer were 2.53 µg kg^{-1} in rice, 2.40 µg kg^{-1} in apple, and 2.38 µg kg^{-1} in tomato samples. In another study, Chen et al. (2013) investigated the presence of racemic cyflumetofon in cucumber, tomato, and apple samples. Cyflumetofon is an acaricide pesticide. Their results indicated that the concentration of racemic cyflumetofon ranged from 100 to 150 µg kg^{-1} in the food samples they used. Zhang et al. (2016), on the other hand, investigated the presence of ethiprole pesticide in cucumber, spinach, and tomato samples using HPLC and obtained good resolutions of its enantiomers (Figure 9.1C). They observed that the concentration of total ethiprole in tomato and cucumber ranged from 20 to 171 µg kg^{-1}. Ethiprole residues, however, were not detected in spinach. Thus, residues of chiral neonicotinoid and acaricides are also found in some food consumed by humans.

9.2.2 Food Products of Animal Origin

Pesticides are extensively used to control pests in the agro-industry as well as in residential and business settings (Hu et al., 2020). Animals are exposed to these pesticides mainly through food, water, air, and skin contact (Mehrpour et al., 2014). Thus, a lot of food products from animal origins such as chicken eggs (Hatta et al., 2019; Nardelli et al., 2019), bovine milk (Aydin et al., 2019; Tripathy et al., 2019), honey (Iturbe-Requena et al., 2020; Mohebbi et al., 2020), and fish (Martyniuk et al., 2020; Ojemaye et al., 2020) are contaminated with pesticide residues.

Many people around the world use bovine milk as an integral part of their diet every day. Bovine milk is a rich source of proteins and other nutrients required by the body. Unfortunately, research has shown that some of the bovine milk is contaminated with agro-chemicals such as chiral pesticides (Table 9.2) (do Lago et al., 2020; Dallegrave et al., 2016; Koleini et al., 2020). For instance, Koleini et al. (2020) investigated the presence of 10 chiral pesticides in bovine milk. Their results showed that the concentrations of the chiral pesticides studied ranged from 2.49 to 10.48 mg mL^{-1}. In a similar study, do Lago et al. (2020) identified trace amounts of chlorpyriphos, malathion, and disulfoton chiral organophosphates in bovine milk. Dallegrave et al. (2016), on the other hand, investigated the presence of 17 chiral pyrethroids in multi-class food products derived from animals such as beef, chicken eggs, fish, and bovine milk. The concentrations of the racemic pyrethroids determined in these foods ranged from 0.03 to 270 ng g^{-1}. These results were further confirmation that foods derived from animals are sometimes contaminated with chiral pesticide residues.

Chicken products such as eggs and meat are among the widely used food products from animal origin. These food products from chicken are not free from contamination by chiral pesticide residues and many studies have confirmed this (Alaboudi et al., 2019; Dallegrave et al., 2016; Zhang et al., 2019) (Table 9.2). For instance, Zhang et al. (2019) studied the presence of 58 chiral pesticide residues in chicken eggs. Among the chiral pesticides that were studied only three (acetamiprid, pyrimethanil, and fipronil sulfone) were found to be present in the chicken eggs used during the investigation. The concentration of racemic acetamiprid ranged from 0.68 to 1.56 µg kg^{-1} in the chicken egg sample while racemic pyrimethanil had a concentration of 4.94 µg kg^{-1} and concentrations of racemic fipronil sulfone ranged from 1.75 to 7.50 µg kg^{-1}. In the same vein, Li et al. (2019) determined fipronil and fipronil sulfone residues in chicken eggs using GC-MS. The concentrations of fipronil in the egg samples used ranged from 3.61 to 15.18 ng g^{-1} while fipronil sulfone concentrations ranged from 69.58 to 323.01 ng g^{-1}. These results are an indication that chicken products in general and chicken eggs are sometimes contaminated with chiral pesticide residues, and these may cause detrimental effects on human health.

Honey is one of the popular agro-based food products globally. It is nutritionally rich as it contains sugars, amino acids, proteins, and vitamins (Choi et al., 2020). The use of chiral pesticides in

TABLE 9.2
Chiral Pesticides Detected in Some Food Products of Animal Origin

Food Matrix	Chiral Pesticide Studied	Extraction Technique	Analytical Instrument	Concentration of Racemate (ng g^{-1})	Reference
Milk	10 chiral pesticides	QuEChERS-DLLME	GC-FID	2.49[+] – 10.48[+]	Koleini et al. 2020
Honey	Paclobutrazol, myclobutanil, diniconazole, and epoxiconazole	SPE-DLLME	LC-MS/MS	0.0216[*] – 0.094.0[*] (both enantiomers of paclobutrazol. Concentrations of others were below LOQ	Ye et al. 2018
Cattle fats, goat fats, and sheep fats	Deltamethrin	QuEChERS	GC-MS	248[**] (deltamethrin in sheep fats) and 122[**] (deltamethrin in goat fats)	Hamadamin and Hassan, 2020
Bovine milk	Clorpyriphos, malathion, and disulfoton	Dispersive solid phase extraction	LC	Concentrations below LODs (0.36[++] to 0.95[++])	do Lago et al. 2020
Eggs	Acetamiprid, pyrimethanil, and fipronil sulfone	QuEChERS	LC-MS/MS	0.68[***]-1.56[***] (acetamiprid), 4.94[***] (pyrimethanil) and 1.75[***] – 7.50[***] (fipronil sulfone)	Zhang et al. 2019
Fish muscles	DDT	Ng	GC	11.58 -48.4	Mendes et al. 2019
Beef, chicken, eggs, fish and milk	17 pyrethroids	LLE	GC-MS/MS	0.03 – 270	Dallegrave et al. 2016
Chicken eggs	Malathion and aldrin	LLE	HPLC-UV	75[**] (malathion in egg yolk), 67[**] (Aldrin in egg yolk), 49[**] (malathion in egg white) and 58[**] (Aldrin in egg white)	Alaboudi et al. 2019
Chicken eggs	Fipronil and fipronil sulfone	SPE	GC-MS	3.61 - 15.18 (fipronil) and 69.58 – 323.01 (fipronil sulfone)	Li et al. 2019
Honey	Thiamethoxan and clothianidin	QuEChERS	LC-MS/MS	0.15 – 0.25 (thiamethoxan) and clothianidin concentration was below LOQ	Giroud et al. 2019

Note: ng – not given.
* converted from ng kg^{-1} to ng g^{-1}.
** converted from mg kg^{-1} to ng g^{-1}.
*** converted from μg kg^{-1} to ng g^{-1}.
\+ units in mg mL^{-1}.
++ units in mu g L^{-1}.

Occurrence of Chiral Pesticides in Food

the agro-industry may result in the contamination of honey by these chemicals. For instance, honeybees may carry chiral pesticide residue from the flowers of a plant to their hive resulting in contamination of the honey. Recent research findings have confirmed the presence of chiral pesticide residues in honey (Table 9.2) (Cervera-Chiner et al., 2020; Ruiz et al., 2020; Villalba et al., 2020). For instance, Ye et al. (2018) investigated the presence of chiral triazole pesticides (paclobutrazol, myclobutanil, diniconazole, and epoxiconazole) in honey samples using LC-MS/MS. The concentrations of myclobutanil, diniconazole, and epoxiconazole were found to be below the limit of quantification (LOQ) of the LC-MS/MS used. However, the concentrations of paclobutrazol in the honey samples used ranged from 21.6 to 94.0 ng kg^{-1}. In the same vein, Giroud et al. (2019) studied the presence of chiral thiamethoxam and clothianidin pesticides in honey samples using LC-MS/MS. The concentrations of thiamethoxan in the honey samples used ranged from 0.15 to 0.25 ng g^{-1} while the concentrations of clothianidin were found to be below LOQ in all the honey samples used. Thus, chiral pesticide residues are sometimes found in honey samples and this may adversely affect the health of human beings.

Agro-chemicals such as chiral pesticides are washed into water bodies such as rivers and dams by rain where they place aquatic organisms such as fish at high health risk. Water contaminated with chiral pesticides may be absorbed into the fish's body through gills, skin, and their digestive system (Ghelichpour et al., 2020). These toxic chemicals are then dispersed by blood to the various tissues of fish, exposing humans to danger when they eat fish. The problem is compounded by the bioaccumulation of some of the chiral pesticides in the adipose tissues of fish, leading to the amplification of their concentrations to toxic levels. Research has indicated the presence of some chiral pesticides in fish tissues (Table 9.2) (Da Cuña et al., 2020; Ojemaye et al., 2020). For instance, Olisah et al. (2019) studied the distribution of organochlorine pesticides in fresh fish carcass using GC-ECD. They found that the total concentration of organochlorine pesticides in the gills of *Mugil cephalus* fish ranged from 238 to 18 043 ng g^{-1}. In another study, Mendes et al. (2019) found that the concentrations of DDT in fish muscle samples ranged from 11.58 to 48.40 ng g^{-1}. Thus, fish products, just like other animal-based food products, may be contaminated with chiral pesticide residues and, therefore, close monitoring of their concentrations is imperative.

9.3 FOOD PROCESSING AND ITS IMPACT ON CHIRAL PESTICIDE RESIDUES IN FOOD

The amount of chiral pesticide residues in food products depends on the nature of processing it is subjected to before its consumption. The processing of food can facilitate the degradation and dissipation of some chiral pesticides (Table 9.3) (Lu et al., 2011; Yang et al., 2020). In some cases, the presence of chiral pesticide residues may cause preferential growth of a certain species of microorganisms, which in turn may enhance processing or work against it by way of reducing the quality of food (Guo et al., 2020). The amount of chiral pesticides in the food consumed by humans is, in most cases, reduced by most of the processing techniques.

Pickling is a food processing technique that is done to preserve vegetables to increase their shelf-life. It can be done using natural fermentation or by the direct addition of salt and vinegar to the food (Yigit and Velioglu, 2019). Many researchers performed some studies to find out the effect of pickling on chiral pesticide residues in food (Table 9.3) (Bian et al., 2020; Guo et al., 2020; Zhao et al., 2019). For instance, Zhao et al. (2019) investigated the effect of the cabbage pickling process on the chiral paclobutrazol pesticide residues. They observed enantioselective dissipation of the enantiomers of paclobutrazol from the cabbage. The dissipation of (*R*)-paclobutrazol was faster than that of (*S*)-paclobutrazol. In the same vein, Guo et al. (2020) studied the impact of paclobutrazol residues on cucumber pickling process. They detected that (2*R*,3*R*)-enantiomer of paclobutrazol caused microbial disturbance during the pickling process. Thus, paclobutrazol residues derailed the pickling process by killing some of the microorganisms involved, thereby affecting the quality of

TABLE 9.3
Effects of Food Processing on the Amount of Chiral Pesticide Residues in Food Products

Process	Food Product	Chiral Pesticide(s)	Effect of Chiral Pesticides on Food Processing	Effect of Food Processing on Chiral Pesticides	Reference
Cucumber pickling process	Cucumber pickle	Paclobutrazol	(2R,3R)-paclobutrazol caused microbial disturbance during pickling	Degradation and interconversion of the isomers occurred	Guo et al. (2020)
Grape and sucrose fermentation	Wine and ethanol	Diclofop-methyl	ng	(S)-(-)-diclofop-methyl degraded faster than (R)-(+)-diclofop-methyl. Both enantiomers stable	Lu et al. (2011)
Household processing (washing, boiling, soaking, and drying)	Okra fruits	Tebufenozide	ng	Degradation of tebufenozide	Luo et al. (2020)
Washing, peeling, and pickling	Cucumber	Azoxystrobin and meptyldinocap	ng	Dissipation of azoxystrobin and meptyldinocap	Bian et al. (2020)
Soy sauce brewing process	Soy sauce	Diclofop-methyl and diclofop	ng	(S)-(-)-diclofop-methyl degraded faster than (R)-(+)-diclofop-methyl. Diclofop relatively stable	Lu et al. (2016)
Gaseous ozone fumigation	Dried basil leaves	Cypermethrin and diclofol	ng	99.9% reduction in chiral pesticide residue	Chanrattanayothin et al. (2019)
Wine fermentation	Wine	Paclobutrazol	Disturbance of microbial activities during fermentation	Enantiomers stable under fermentation conditions	Guo et al. (2020)
Cabbage pickling	Cabbage pickle	Paclobutrazol	ng	The dissipation of (R)-paclobutrazol was faster than the (S)-paclobutrazol	Zhao et al. (2019)
Winemaking process	Wine	Benalaxyl	Little effect on growth of yeast and production of wine	Both (R)- and (S)-benalaxyl stable under fermentation conditions	Lu et al. (2016)
Winemaking process	Wine	Cyazofamid	Ng	Degradation occurred during fermentation	Yang et al. (2020)

Note: ng – not given.

the pickle. They also observed the occurrence of degradation of the enantiomers of paclobutrazol as well as their interconversion. From their findings, it can be inferred that pickling generally, cause dissipation of some chiral pesticide residues in food resulting in the reduction of their concentration.

Fermentation is another technique that is used during processing of food. It involves the microbial breakdown of food into other substances. It is the main process used in the wine industry during the conversion of grapes into wine. The effect of chiral pesticide residues during wine fermentation have been studied by many researchers (Table 9.3) (Guo et al., 2020; Lu et al., 2016; Lu et al., 2011; Yang et al., 2020). Some chiral pesticide residues undergo degradation during the wine fermentation process. For instance, Yang et al. (2020) investigated the effect of the enantiomers of cyazofamid during the wine making process. Their results indicated that the enantiomers of cyazofamid residues in grapes undergo degradation under the fermentation conditions. In another study, Lu et al. (2011) studied the effect of diclofop-methyl during grape fermentation. Their results revealed that diclofop-methyl residues in grapes undergo enantioselective degradation during the fermentation process. They observed that (S)-(-)-diclofop-methyl degraded faster than (R)-(+)-diclofop-methyl under the fermentation conditions. However, the two enantiomers of diclofop-methyl did not undergo interconversion during the fermentation process. Thus, the fermentation process used during winemaking assists to reduce the concentrations of some chiral pesticide residues in grapes, thereby minimizing health risks to the consumers.

Research has shown that other chiral pesticides do not undergo degradation under the fermentation conditions used during winemaking (Table 9.3) (Guo et al., 2020; Lu et al., 2016). The concentrations of their enantiomers remain unchanged throughout the fermentation process. For instance, Lu et al. (2016) performed a study to find out the behavior of benalaxyl pesticides during wine fermentation. Their results revealed that the two enantiomers of benalaxyl were stable under the fermentation conditions and their concentrations remained unchanged until the end of the process. In a similar study, Guo et al. (2020) investigated the effects of the enantiomers of paclobutrazol pesticide during the fermentation process. They observed that the enantiomers of paclobutrazol were stable under the fermentation conditions they used. Their results also revealed that enantiomers of paclobutrazol interfered with the microbial activities during the fermentation process. Thus, these stable enantiomers of some chiral pesticides would remain in the wine until they reach the consumers where they may cause serious health problems.

Household processing plays an integral part in reducing the amount of pesticide residues consumed by humans in food. It involves simple but effective processes such as washing, boiling, soaking, peeling, and drying of food (Table 9.3) (Bian et al., 2020; Luo et al., 2020). Researches have been conducted to verify the effectiveness of these household processing techniques in reducing the amount of chiral pesticide residues in food. For instance, Luo et al. (2020) investigated the effects of household food processing on the amount of tebufenozide pesticide residue in okra fruits. Their results indicated that simple processes like washing, boiling, soaking, and drying of food facilitated the degradation of tebufenozide pesticide residues in okra fruits. In the same vein, Bian et al. (2020) investigated the effects of washing, peeling, and pickling on the amount of azoxystrobin and meptyldinocap residues in cucumber. They also observed that these simple processes promoted the degradation of these two chiral pesticides. Drying of food is a common processing technique and should be done cautiously as it may end up concentrating the chiral pesticide residues. Some researchers, however, subject dried food to ozone fumigation to ensure efficient removal of chiral pesticide residues (Chanrattanayothin et al., 2019).

9.4 HEALTH RISKS ASSOCIATED WITH CHIRAL PESTICIDES

Human exposure to chiral pesticides in food can lead to both acute and chronic health effects (Table 9.4). Acute exposure to chiral pesticides in food can lead to several symptoms that include loss of coordination, headache, fatigue, slow pulse, central nervous system depression, nausea, sweating, diarrhea, confusion, and coma (Md Meftaul et al., 2020; Sang et al., 2020). Some chiral

TABLE 9.4
Health Effects of Chiral Pesticides

Class of Chiral Pesticide	Chiral Pesticide	Matrix	Type of Study Used	Health Effect(s)	Enantioselective Toxicity	Reference
Triazole fungicide	Prothioconazole	ng	In vitro	Disruption of the endocrine system	(S)-prothioconazole had greater endocrine-disrupting effect than (R)-prothioconazole	Zhang et al. (2019)
Insecticide	Fipronil	CYP$_{450}$ enzymes	In vitro	Inhibition of CYP$_2$D$_6$ enzymes in humans	Negligible enantioselective toxicity	Carrão et al. (2019)
Insecticide	Fipronil	Embryonic and larval zebrafish	In vivo	Epigenetic modifications such as DNA methylation	(R)-fipronil caused greater epigenetic modifications	Qian et al. (2019)
Pyrethroid	Lambda-cyhalothrin	Eremias argus	In vivo	Thyroid endocrine disruption	(+)-Lambda-cyhalothrin had greater effect on thyroid gland	Chang et al. (2018)
Fungicide	Tebuconazole	Human liver microsomes	In vitro			Habenschus et al. (2019)
Fungicide	Myclobutanil	CYP$_{450}$ enzymes	In vitro	Inhibition of CYP3A and CYP2G9 enzymes	Both enantiomers strongly caused inhibition	Fonseca et al. (2019)
Herbicide	Acetochlor	Zebrafish embryo-larvae	In vivo	Disruption of the activities of the thyroid gland	(S)-(+)-acetochlor had greater thyroid disruptive effects	Xu et al. (2019)
Insecticide	Fipronil	Zebrafish	In vivo	Transgenerational toxicity	Preferential accumulation of (S)-fipronil in both adult and young zebrafish	Xu et al. (2019)
Herbicide	Diclofop-methyl	Human serum albumin	In vitro	Disruption of the secondary structure of albumin	Human serum albumin had greater affinity for S-enantiomer than the R-isomer	Zhang et al. (2014)
Pyrethroid	Bifenthrin	Tropoblast cells	In vitro	Endocrine disrupting effect	(S)-bifenthrin had greater Endocrine disrupting effect than R-isomer	Zhao et al. (2014)
Organophosphate	Pyraclofos	Zebrafish	In vivo	Immunotoxicity	(R)-pyraclofos more toxic than (S)-pyraclofos	Zhuang et al. (2015)

TABLE 9.4 (Continued)
Health Effects of Chiral Pesticides

Class of Chiral Pesticide	Chiral Pesticide	Matrix	Type of Study Used	Health Effect(s)	Enantioselective Toxicity	Reference
Pyrethroid	Cis-bifenthrin	Zebrafish	In vivo	Caused reproductive impairment	(1S)-cis-bifenthrin caused stronger reproductive impairment	Xiang et al. (2019)
Herbicide	Diclofop	Human serum albumin	In vitro	Alteration of the secondary structure of albumin	(S)-diclofop had greater binding with albumin than (R)-diclofop	Zhang et al. (2013)
Organochlorine	Acetofenate	MCF-7 and JEG-3 cells	In vitro	Endocrine disruption	(S)-(+)-acetofenate had greater endocrine disrupting effect than R-isomer	Chen et al. (2012)
Pyrethroid	Beta-cypermethrin	Zebrafish embryo cells	In vivo	Causes york edema, pericardial edema and crooked body	(1R-cis-αS)- and (1R-trans-αS)-enantiomers showed greater developmental toxicity	Xu et al. (2010)
Organophosphate	Isocarbophos	Hep G$_2$ cells	In vitro	Hepatocyte toxicity (cytotoxicity)	(-)-isocarbopho had greater cytotoxicity than (+)-isocarbophos	Liu et al. 2010
Pyrethroids	Lambda-cyhalothrin and cis-bifenthrin	Macrophage cells and human breast carcinoma cells	In vitro	Endocrine disruption and immunotoxicity	(-)-lambda-cyhalothrin and (S)-bifenthrin had greater estrogenic potential and immunocytoxicity	Zhao et al. (2010)
Organochlorine	Acetofenate	Macrophage cells	In vitro	Immunotoxicity	(S)-(+)-acetofenate more toxic than (R)-(-)-acetofenate	Zhao and Liu (2009)
Pyrethroid	Cis-Bifenthrin	Human amnion epithelial cells	In vitro	Toxicity to human amnion epithelial cells	(S)-bifenthrin more toxic than (R)-bifenthrin	Liu et al. (2008)

Note: ng – not given.

pesticides have been associated with, through *in vivo* studies, adverse reproductive effects (Xiang et al., 2019), hormonal imbalance (Xu et al., 2019), mutations (Qian et al., 2019), and disruption of the immune system (Zhuang et al., 2015). Consumption of food contaminated with chiral pesticide residues can also damage important organs and systems of the body such as the cardiovascular system, digestive system, respiratory system, and endocrine system as well as damaging the liver and the kidneys (Md Meftaul et al., 2020).

Some chiral pesticides may cause endocrine disruption in the human body (Table 9.4). This causes the derailment of all the hormone-based activities of the body resulting in detrimental effects to human health. Many researchers have performed both *in vivo* and *in vitro* studies to show the endocrine disrupting effects of chiral pesticides (Chang et al., 2018; Fonseca et al., 2019; Habenschus et al., 2019; Xu et al., 2019; Xiang et al., 2019). For instance, Zhao et al. (2014) investigated the endocrine effects of chiral bifenthrin pesticide using trophoblast cells. Their results showed that there was enantioselective endocrine disruption with S-bifenthrin having greater toxicity than the R-enantiomer. In another study, Chen et al. (2012) performed an in vitro study on the endocrine disrupting effects of chiral acetofenate pesticide using MCF-7 and JFG-3 cells. Enantioselectivity was also observed with (S)-(+)-acetofenate being more toxic than R-(-)-acetofenate. The findings of the *in vitro* studies were supported by *in vivo* studies performed by some researchers. For instance, Xu et al. (2019) studied the endocrine disrupting effects of chiral acetochlor pesticide using zebrafish embryo larvae. Their results showed enantioselectivity with (S)-(+)-acetochlor having greater thyroid disrupting effects than (R)-(-)-acetochlor.

Proteins play vital roles in the human body and their ability to execute their functions depends on their three-dimensional conformations. Some chiral pesticides may disrupt the secondary and tertiary structures of proteins, thereby disrupting their three-dimensional configurations and adversely affecting their functions especially as enzymes. Many researchers have performed some studies that showed the potential of chiral pesticides as inhibitors of enzymes (Table 9.4) (Carrão et al., 2019; Fonseca et al., 2019; Zhao et al., 2014). For instance, Carrão et al. (2019) performed *in vitro* studies that showed that fipronil could cause inhibition of CYP2D6 enzymes in human. Their results showed the enantiomers of fipronil were equally toxic during the inhibition and, therefore, there was negligible enantioselective toxicity. In a similar *in vitro* study, Fonseca et al. (2019) showed that myclobutanil was a potential inhibitor of CYP3A and CYD2G9 enzymes in the human body. Their results also showed that the enantiomers of myclobutanil were equally toxic to the enzymes and therefore no enantioselective toxicity was observed. On the other hand, Zhang et al. (2014) studied the effect of diclofop-methyl on human albumin protein. Their results showed that diclofop-methyl disrupted the secondary structure of albumin and the S-enantiomers showed greater toxicity than the R-enantiomers. In the same vein, Zhang et al. (2013) performed an *in vivo* study on the effects of diclofop on human albumin. They observed that S-diclofop caused greater disruption of the secondary structure of albumin than R-diclofop.

The immune system is very essential for the general well-being of the human body as it plays protective functions against infections. Research has shown that some chiral pesticides can cause immunotoxicity, rendering the immune system ineffective (Table 9.4) (Zhuang et al., 2015; Zhao and Liu, 2009; Zhao et al., 2010). Zhao and Liu (2009) performed an *in vitro* study to show the effects of acetofenate on macrophage cells. Their results showed that acetofenate interfered with the functioning of macrophage cells causing the immune system to malfunction. Enantioselectivity was observed with S-(+)-acetofenate being more toxic than R-(-)-acetofenate. In the same vein, Zhao et al. (2010) showed that lambda-cyhalothrin and cis-bifenthrin had immunotoxic effects in the human body. Their results showed that (-)-lambda cyhalothrin and S-bifenthrin were more toxic to the macrophage cells than (+)-lambda cyhalothrin and R-bifenthrin. Similar results were obtained using the *in vivo* studies. For instance, Zhuang et al. (2015) performed an *in vivo* study on the effect of pyraclofos pesticide on zebrafish. They found out that chiral pyraclofos was an immunotoxin to zebrafish with the R-enantiomer being more toxic than the S-enantiomer.

Some chiral pesticides interfere with the growth of cells and they are, therefore, cytotoxins in the body. Some researchers have performed studies to verify the cytotoxicity of some chiral pesticides in the human body (Liu et al., 2010, 2008). For instance, Liu et al. (2008) investigated the effect of cis-bifenthrin on human amnion epithelial cells. Their results showed that chiral cis-bifenthrin retarded the growth of amnion epithelial cells. Their results also revealed that (S)-bifenthrin was more toxic than R-bifenthrin. In another study, Liu et al. (2010) studied the effects of isocarbophos on hepatic cells. Their results revealed that isocarbophos caused hepatocyte toxicity. Enantioselective toxicity was also observed with (-)-isocarbophos being more toxic than (+)-isocarbophos.

9.5 CHALLENGES AND FUTURE PROSPECTS

The bioactivities of chiral pesticides are enantioselective in chiral environments. Despite this, most chiral pesticides are currently being manufactured and sold as racemates. When racemates are used in pest management, only the active enantiomer would be involved in destroying the pest. The other enantiomers would be inactive and would be discharged into the environment. The entry of these inactive enantiomers into the environment is a serious challenge, as they would cause contamination of the food sources used by human beings. The problem would be further compounded by the persistent nature of some of the chiral pesticides, which make them remain in the environment for long periods. In addition, some of these chiral pesticides undergo bioaccumulation in plants and animals that are used as food by humans. Bioaccumulation may cause the elevation of the concentrations of these chiral pesticides to toxic levels. Humans usually subject food to some form of processing before consumption. Some chiral pesticides undergo degradation during processing. The challenge associated with the degradation of food during processing is the production of degradation products that might be more toxic than the starting materials. Moreover, some chiral pesticides might be stable under the conditions used during processing and, as a result, they enter the final product unchanged putting the health of humans in danger. The problem of contamination of food with chiral pesticides may be minimized by employing slow-release techniques and nanomaterial encapsulation during their application to crops.

9.6 CONCLUSION

Different classes of chiral pesticides are used in the agricultural sector to control pests. Trace amounts of the enantiomers of these chiral pesticides usually find their way in food products consumed by humans if the pesticides are properly applied. Humans should take some measures, therefore, to minimize the amount of chiral pesticides they consume in food. Most of the time, simple food processing techniques such as pickling, fermentation, boiling, washing, and soaking assist to remove a significant amount of chiral pesticide residues in food. Processing techniques just reduce but do not eliminate chiral pesticide residues from food. The consumption of chiral pesticide residues in food is, therefore, inevitable as long as they are still being used in pest management. If the quantities of chiral pesticides consumed exceed the maximum allowable limits, humans usually encounter serious health problems.

Acknowledgment

Authors are grateful to the Research Center at the University of Venda, for financial support.

Conflicts of Interest

The authors declare no conflict of interest.

REFERENCES

Alaboudi, Akram R., Tareq M. Osaili, and Arwa Alrwashdeh. 2019. "Pesticides (hexachlorocyclohexane, aldrin, and malathion) residues in home-grown eggs: Prevalence, distribution, and effect of storage and heat treatments." *Journal of Food Science* 84 (12): 3383–90. doi:10.1111/1750-3841.14918.

Asad, Muhammad Asad Ullah, Michel Lavoie, Hao Song, Yujian Jin, Zhengwei Fu, and Haifeng Qian. 2017. "Interaction of chiral herbicides with soil microorganisms, algae and vascular plants." *Science of the Total Environment* 580: 1287–99. doi:10.1016/j.scitotenv.2016.12.092.

Aydin, Senar, Mehmet Emin Aydin, Fatma Beduk, and Arzu Ulvi. 2019. "Organo-halogenated pollutants in raw and UHT cow's milk from Turkey: A risk assessment of dietary intake." *Environmental Science and Pollution Research* 26 (13): 12788–97. doi:10.1007/s11356-019-04617-0.

Bian, Yanli, Gang Guo, Fengmao Liu, Xiaochu Chen, Zongyi Wang, and Tongyao Hou. 2020. "Meptyldinocap and azoxystrobin residue behaviours in different ecosystems under open field conditions and distribution on processed cucumber." *Journal of the Science of Food and Agriculture* 100 (2): 648–55. doi:10.1002/jsfa.10059.

Birolli, Willian G., Marylyn S. Arai, Marcia Nitschke, and André L. M. Porto. 2019. "The pyrethroid (±)-lambda-cyhalothrin enantioselective biodegradation by a bacterial consortium." *Pesticide Biochemistry and Physiology* 156: 129–37. doi:10.1016/j.pestbp.2019.02.014.

Buerge, Ignaz J., Astrid Bächli, Roy Kasteel, Reto Portmann, Rocío López-Cabeza, Lars F. Schwab, and Thomas Poiger. 2019. "Behaviour of the chiral herbicide imazamox in soils: pH-dependent, enantioselective degradation, formation and degradation of several chiral metabolites." *Environmental Science & Technology* 53 (10): 5725–32. doi:10.1021/acs.est.8b07209.

Caballo, C., M. D. Sicilia, and S. Rubio. 2013. "Stereoselective quantitation of mecoprop and dichlorprop in natural waters by supramolecular solvent-based micro-extraction, chiral liquid chromatography and tandem mass spectrometry." *Analytica Chimica Acta* 761: 102–8. doi:10.1016/j.aca.2012.11.044.

Carrão, Daniel Blascke, Maísa Daniela Habenchus, Nayara Cristina Perez de Albuquerque, Rodrigo Moreira da Silva, Norberto Peporine Lopes, and Anderson Rodrigo Moraes de Oliveira. 2019. "*In vitro* inhibition of human CYP2D6 by the chiral pesticide fipronil and its metabolite fipronil sulfone: Prediction of pesticide-drug Interactions." *Toxicology Letters* 313: 196–204. doi:10.1016/j.toxlet.2019.07.005.

Cervera-Chiner, Lourdes, Carmen March, Antonio Arnau, Yolanda Jimenez, and Angel Montoya. 2020. "Detection of DDT and carbaryl pesticides in honey by means of immune-sensors based on high fundamental frequency quartz crystal microbalance (HFF-QCM)." *Journal of the Science of Food and Agriculture* 100 (6): 2468–72. doi:10.1002/jsfa.10267.

Chang, Jing, Weiyu Hao, Yuanyuan Xu, Peng Xu, Wei Li, Jianzhong Li, and Huili Wang. 2018. "Stereoselective degradation and thyroid endocrine disruption of lambda-cyhalothrin in lizards (*Eremias argus*) following oral exposure." *Environmental Pollution* 232: 300–309. doi:10.1016/j.envpol.2017.09.072.

Chanrattanayothin, Phatthanawan, Danuwat Peng-Ont, Anchalee Masa-Ad, Tinnakrit Warisson, Rotjapun Nirunsin, and Hathaithip Sintuya. 2019. "Degradation of cypermethrin and dicofol pesticides residue in dried basil leave by gaseous ozone fumigation." *Ozone: Science & Engineering* 0 (0): 1–8. doi:10.1080/01919512.2019.1708699.

Chen, Fang, Quan Zhang, Cui Wang, Yingchong Lu, and Meirong Zhao. 2012. "Enantioselectivity in estrogenicity of the organochlorine insecticide acetofenate in human trophoblast and MCF-7 cells." *Reproductive Toxicology* 33 (1): 53–59. doi:10.1016/j.reprotox.2011.10.016.

Chen, Xiu, Fengshou Dong, Xingang Liu, Jun Xu, Jing Li, Yuanbo Li, Minmin Li, Yunhao Wang, and Yongquan Zheng. 2013. "Chiral determination of a novel acaricide cyflumetofen enantiomers in cucumber, tomato, and apple by HPLC." *Journal of Separation Science* 36 (2): 225–31. doi:10.1002/jssc.201200636.

Chen, Xiu, Fengshou Dong, Xingang Liu, Jun Xu, Jing Li, Yuanbo Li, Yunhao Wang, and Yongquan Zheng. 2012. "Enantioselective separation and determination of the dinotefuran enantiomers in rice, tomato and apple by HPLC." *Journal of Separation Science* 35 (2): 200–205. doi:10.1002/jssc.201100823.

Chen, Xixi, Fengshou Dong, Jun Xu, Xingang Liu, Zenglong Chen, Na Liu, and Yongquan Zheng. 2016. "Enantioseparation and determination of isofenphos-methyl enantiomers in wheat, corn, peanut and soil with supercritical fluid chromatography/tandem mass spectrometric method." *Journal of Chromatography B* 1015–1016: 13–21. doi:10.1016/j.jchromb.2016.02.003.

Choi, Yi-Ching, Tsz-Tsun Ng, Bin Hu, Rong Li, and Zhong-Ping Yao. 2020. "Rapid detection of pesticides in honey by solid-phase micro-extraction coupled with electrospray ionization mass spectrometry." *Journal of Mass Spectrometry* 55 (2): e4380. doi:10.1002/jms.4380.

Cui, Ning, Haoyu Xu, Shijie Yao, Yiwen He, Hongchao Zhang, and Yunlong Yu. 2018. "Chiral triazole fungicide tebuconazole: Enantioselective bioaccumulation, bioactivity, acute toxicity, and dissipation in soils." *Environmental Science and Pollution Research* 25 (25): 25468–75. doi:10.1007/s11356-018-2587-9.

Da Cuña, Rodrigo Hérnan, Fabiana Laura Lo Nostro, Valeria Shimabukuro, Paola Mariana Ondarza, and Karina Silvia Beatriz Miglioranza. 2020. "Bioaccumulation and distribution behavior of endosulfan on a cichlid fish: Differences between exposure to the active ingredient and a commercial formulation." *Environmental Toxicology and Chemistry* 39 (3): 604–11. doi:10.1002/etc.4643.

Dallegrave, Alexsandro, Tânia Mara Pizzolato, Fabiano Barreto, Ethel Eljarrat, and Damià Barceló. 2016. "Methodology for trace analysis of 17 pyrethroids and chlorpyrifos in foodstuff by gas chromatography–tandem mass spectrometry." *Analytical and Bioanalytical Chemistry* 408 (27): 7689–97. doi:10.1007/s00216-016-9865-5.

Fonseca, Franciele S., Daniel B. Carrão, Nayara C. P. de Albuquerque, Viviani Nardini, Luís G. Dias, Rodrigo M. da Silva, Norberto P. Lopes, and Anderson R. M. de Oliveira. 2019. "Myclobutanil enantioselective risk assessment in humans through *in vitro* CYP450 reactions: Metabolism and inhibition studies." *Food and Chemical Toxicology* 128: 202–11. doi:10.1016/j.fct.2019.04.009.

Gao, Beibei, Zhaoxian Zhang, Lianshan Li, Amir E. Kaziem, Zongzhe He, Qianwen Yang, Qing Peiyang, Qing Zhang, and Minghua Wang. 2019. "Stereoselective environmental behavior and biological effect of the chiral organophosphorus insecticide isofenphos-methyl." *Science of the Total Environment* 648: 703–10. doi:10.1016/j.scitotenv.2018.08.182.

Ghelichpour, Melika, Ali Taheri Mirghaed, Seyyed Morteza Hoseini, and Amalia Perez Jimenez. 2020. "Plasma antioxidant and hepatic enzymes activity, thyroid hormones alterations and health status of liver tissue in common carp (*Cyprinus carpio*) Exposed to Lufenuron." *Aquaculture* 516: 734634. doi:10.1016/j.aquaculture.2019.734634.

Giroud, Barbara, Selina Bruckner, Lars Straub, Peter Neumann, Geoffrey R. Williams, and Emmanuelle Vulliet. 2019. "Trace-level determination of two neonicotinoid insecticide residues in honey bee royal jelly using ultra-sound assisted salting-out liquid liquid extraction followed by ultra-high-performance liquid chromatography-tandem mass spectrometry." *Microchemical Journal* 151: 104249. doi:10.1016/j.microc.2019.104249.

Guo, Hao-Ming, Yue Zhao, Mei-Nan Ou Yang, and Zhong-Hua Yang. 2020. "Enantiomeric effect of paclobutrazol on the microorganism composition during wine fermentation." *Chirality* 32 (4): 489–99. doi:10.1002/chir.23183.

Guo, Hao-Ming, Yue Zhao, Mei-Nan Ou Yang, Zhong-Hua Yang, and Jian-Hong Li. 2020. "The enantioselective effects and potential risks of paclobutrazol residue during cucumber pickling process." *Journal of Hazardous Materials* 386: 121882. doi:10.1016/j.jhazmat.2019.121882.

Habenschus, Maísa Daniela, Viviani Nardini, Luís Gustavo Dias, Bruno Alves Rocha, Fernando Barbosa, and Anderson Rodrigo Moraes de Oliveira. 2019. "*In vitro* enantioselective study of the toxicokinetic effects of chiral fungicide tebuconazole in human liver microsomes." *Ecotoxicology and Environmental Safety* 181: 96–105. doi:10.1016/j.ecoenv.2019.05.071.

Hamadamin, Ahmad Yaseen, and Khulod Ibraheem Hassan. 2020. "Gas chromatography-mass spectrometry based sensitive analytical approach to detect and quantify non-polar pesticides accumulated in the fat tissues of domestic animals." *Saudi Journal of Biological Sciences* 27 (3): 887–93. doi:10.1016/j.sjbs.2019.12.029.

Hatta, Sumito, Yohei Odagawa, Jun Izumi, and Satoshi Nakamae. 2019. "Study on fipronil and other pesticide residues in chicken eggs produced and sold in Japan." *Food Hygiene and Safety Science* 60 (5): 154–58.

Hu, Yi, Yan Zhang, Angela Vinturache, Yiwen Wang, Rong Shi, Limie Chen, Kaili Qin, Ying Tian, and Yu Gao. 2020. "Effects of environmental pyrethroids exposure on semen quality in reproductive-age men in Shanghai, China." *Chemosphere* 245: 125580. doi:10.1016/j.chemosphere.2019.125580.

Iturbe-Requena, Sandra L., Maria G. Prado-Ochoa, Marco A. Munoz-Guzman, Liborio Carrillo-Miranda, Ana M. Velazquez-Sanchez, Enrique Angeles, and Fernando Alba-Hurtado. 2020. "Acute oral and contact toxicity of new ethyl-carbamates on the mortality and acetylcholinesterase activity of honey bee (*Apis mellifera*)." *Chemosphere* 242: UNSP 125293. doi:10.1016/j.chemosphere.2019.125293.

Kaziem, Amir E., Beibei Gao, Lianshan Li, Zhaoxian Zhang, Zongzhe He, Yong Wen, and Ming-hua Wang. 2020. "Enantioselective bioactivity, toxicity, and degradation in different environmental mediums of chiral fungicide epoxiconazole." *Journal of Hazardous Materials* 386: 121951. doi:10.1016/j.jhazmat.2019.121951.

Kiwango, Purificator A., Neema Kassim, and Martin E. Kimanya. 2020. "Household vegetable processing practices influencing occurrence of pesticide residues in ready-to-eat vegetables." *Journal of Food Safety* 40 (1): e12737. doi:10.1111/jfs.12737.

Koleini, Farnoosh, Parvaneh Balsini, and Hadi Parastar. 2020. "Evaluation of partial least-squares regression with multivariate analytical figures of merit for determination of 10 pesticides in milk." *International Journal of Environmental Analytical Chemistry* 0 (0): 1–11. doi:10.1080/03067319.2020.1745198.

Lago, Ayla Campos do, Marcello Henrique da Silva Cavalcanti, Mariana Azevedo Rosa, Alberto Thalison Silveira, Cesar Ricardo Teixeira Tarley, and Eduardo Costa Figueiredo. 2020. "Magnetic restricted-access carbon nanotubes for dispersive solid phase extraction of organophosphates pesticides from bovine milk samples." *Analytica Chimica Acta* 1102: 11–23. doi:10.1016/j.aca.2019.12.039.

Li, Jing, Fengshou Dong, Jun Xu, Xingang Liu, Yuanbo Li, Weili Shan, and Yongquan Zheng. 2011. "Enantioselective determination of triazole fungicide simeconazole in vegetables, fruits, and cereals using modified QuEChERS (Quick, Easy, Cheap, Effective, Rugged and Safe) coupled to gas chromatography/tandem mass spectrometry." *Analytica Chimica Acta* 702 (1): 127–35. doi:10.1016/j.aca.2011.06.034.

Li, Liang, Hui Wang, Yazhou Shuang, and Laisheng Li. 2019. "The preparation of a new 3,5-dichlorophenyl-carbamated cellulose-bonded stationary phase and its application for the enantioseparation and determination of chiral fungicides by LC-MS/MS." *Talanta* 202: 494–506. doi:10.1016/j.talanta.2019.05.011.

Li, Runan, Xinglu Pan, Qinqin Wang, Yan Tao, Zenglong Chen, Duoduo Jiang, Chi Wu, et al. 2019. "Development of S-fluxametamide for bioactivity improvement and risk reduction: Systemic evaluation of the novel insecticide fluxametamide at the enantiomeric level." *Environmental Science & Technology* 53 (23): 13657–65. doi:10.1021/acs.est.9b03697.

Li, Xianjiang, Hongmei Li, Wen Ma, Zhen Guo, Xiuqin Li, Shanjun Song, Hua Tang, Xiaomin Li, and Qinghe Zhang. 2019. "Development of precise GC-EI-MS method to determine the residual fipronil and its metabolites in chicken egg." *Food Chemistry* 281 (May): 85–90. doi:10.1016/j.foodchem.2018.12.041.

Li, Yuanbo, Fengshou Dong, Xingang Liu, Jun Xu, Jing Li, Zhiqiang Kong, Xiu Chen, and Yongquan Zheng. 2012. "Enantioselective determination of triazole fungicide tebuconazole in vegetables, fruits, soil and water by chiral liquid chromatography/tandem mass spectrometry." *Journal of Separation Science* 35 (2): 206–15. doi:10.1002/jssc.201100674.

Lin, Xiao-Yan, Ren-Xiang Mou, Zhao-Yun Cao, Zhen-Zhen Cao, and Ming-Xue Chen. 2018. "Analysis of pyrethroid pesticides in Chinese vegetables and fruits by GC–MS/MS." *Chemical Papers* 72 (8): 1953–62. doi:10.1007/s11696-018-0447-1.

Liu, Huigang, Jing Liu, Lihong Xu, Shanshan Zhou, Ling Li, and Weiping Liu. 2010. "Enantioselective cytotoxicity of isocarbophos is mediated by oxidative stress-induced JNK activation in human hepatocytes." *Toxicology* 276 (2): 115–21. doi:10.1016/j.tox.2010.07.018.

Liu, Huigang, Meirong Zhao, Cong Zhang, Yun Ma, and Weiping Liu. 2008. "Enantioselective cytotoxicity of the insecticide bifenthrin on a human amnion epithelial cell line." *Toxicology*, This issue includes: Proceedings of the annual congress of the British Toxicology Society, 253 (1): 89–96. doi:10.1016/j.tox.2008.08.015.

Lu, Yuele, Jinling Diao, Xu Gu, Yanfeng Zhang, Peng Xu, Peng Wang, and Zhiqiang Zhou. 2011. "Stereoselective degradation of diclofop-methyl during alcohol fermentation process." *Chirality* 23 (5): 424–28. doi:10.1002/chir.20946.

Lu, Yuele, Yihua Shao, Songjun Dai, Jinling Diao, and Xiaolong Chen. 2016. "Stereoselective behaviour of the fungicide benalaxyl during grape growth and the wine-making process." *Chirality* 28 (5): 394–98. doi:10.1002/chir.22589.

Lu, Yuele, Dong Zhang, Yahui Liao, Jinling Diao, and Xiaolong Chen. 2016. "Stereoselective behaviour of the chiral herbicides diclofop-methyl and diclofop during the soy sauce brewing process." *Chirality* 28 (1): 78–84. doi:10.1002/chir.22545.

Luo, Xiaoshuang, Xinxian Qin, Dan Chen, Zhengyi Liu, Kankan Zhang, and Deyu Hu. 2020. "Determination, residue analysis, risk assessment and processing factors of tebufenozide in okra fruits under field conditions." *Journal of the Science of Food and Agriculture* 100 (3): 1230–37. doi:10.1002/jsfa.10134.

Maia, Alexandra S., Ana R. Ribeiro, Paula M. L. Castro, and Maria Elizabeth Tiritan. 2017. "Chiral analysis of pesticides and drugs of environmental concern: Biodegradation and enantiomeric fraction." *Symmetry-Basel* 9 (9): 196–225. doi:10.3390/sym9090196.

Martyniuk, Christopher J., Alvine C. Mehinto, and Nancy D. Denslow. 2020. "Organochlorine pesticides: Agrochemicals with potent endocrine-disrupting properties in fish." *Molecular and Cellular Endocrinology* 507: 110764. doi:10.1016/j.mce.2020.110764.

Md Meftaul, Islam, Kadiyala Venkateswarlu, Rajarathnam Dharmarajan, Prasath Annamalai, and Mallavarapu Megharaj. 2020. "Pesticides in the urban environment: A potential threat that knocks at the door." *Science of the Total Environment* 711: 134612. doi:10.1016/j.scitotenv.2019.134612.

Mehrpour, Omid, Parissa Karrari, Nasim Zamani, Aristides M. Tsatsakis, and Mohammad Abdollahi. 2014. "Occupational exposure to pesticides and consequences on male semen and fertility: A review." *Toxicology Letters, Environmental contaminants and target organ toxicities*, 230 (2): 146–56. doi:10.1016/j.toxlet.2014.01.029.

Mendes, Rosivaldo A., Marcelo O. Lima, Ricardo J. A. de Deus, Adaelson C. Medeiros, Kelson C. F. Faial, Iracina M. Jesus, Kleber R. F. Faial, and Lourivaldo S. Santos. 2019. "Assessment of DDT and mercury levels in fish and sediments in the Iriri river, Brazil: Distribution and ecological risk." *Journal of Environmental Science and Health Part B-Pesticides Food Contaminants and Agricultural Wastes* 54 (12): 915–24. doi:10.1080/03601234.2019.1647060.

Mohebbi, Ali, Saeid Yaripour, Mir Ali Farajzadeh, Mohammad Reza Afshar Mogaddam, and Hassan Malekinejad. 2020. "Control of organophosphorus pesticides residues in honey samples using a miniaturized tandem pre-concentration technique coupled with high-performance liquid chromatography." *Pharmaceutical Sciences* 26 (1): 52–60. doi:10.34172/PS.2019.63.

Nardelli, Valeria, Valeria D'Amico, Francesco Casamassima, Giuseppe Gesualdo, Donghao Li, Wadir M. V. Marchesiello, Donatella Nardiello, and Maurizio Quinto. 2019. "Development of a screening analytical method for the determination of non-dioxin-like polychlorinated biphenyls in chicken eggs by gas chromatography and electron capture detection." *Food Additives & Contaminants: Part A* 36 (9): 1393–1403. doi:10.1080/19440049.2019.1627002.

Ojemaye, C. Y., C. T. Onwordi, and L. Petrik. 2020. "Herbicides in the tissues and organs of different fish species (Kalk bay harbour, South Africa): Occurrence, levels and risk assessment." *International Journal of Environmental Science and Technology* 17 (3): 1637–48. doi:10.1007/s13762-019-02621-y.

Olisah, Chijioke, Omobola O. Okoh, and Anthony I. Okoh. 2019. "Distribution of organochlorine pesticides in fresh fish carcasses from selected estuaries in Eastern Cape province, South Africa, and the associated health risk assessment." *Marine Pollution Bulletin* 149: 110605. doi:10.1016/j.marpolbul.2019.110605.

Pirsaheb, Meghdad, Nazir Fattahi, Mohammad Karami, and Hamid Reza Ghaffari. 2018. "Simultaneous determination of deltamethrin, permethrin and malathion in stored wheat samples using continuous sample drop flow micro-extraction followed by HPLC–UV." *Journal of Food Measurement and Characterization* 12 (1): 118–27. doi:10.1007/s11694-017-9622-2.

Qi, Peipei, Xiangyun Wang, Hu Zhang, Xinquan Wang, Hao Xu, and Qiang Wang. 2015. "Rapid enantioseparation and determination of isocarbophos enantiomers in orange pulp, peel, and kumquat by chiral HPLC-MS/MS." *Food Analytical Methods* 8 (2): 531–38. doi:10.1007/s12161-014-9922-7.

Qian, Yi, Chenyang Ji, Siqing Yue, and Meirong Zhao. 2019. "Exposure of low-dose fipronil enantioselectively induced anxiety-like behaviour associated with DNA methylation changes in embryonic and larval zebrafish." *Environmental Pollution* 249: 362–71. doi:10.1016/j.envpol.2019.03.038.

Ruiz, Paola, Ana M. Ares, Silvia Valverde, Maria T. Martin, and Jose Bernal. 2020. "Development and validation of a new method for the simultaneous determination of spinetoram in honey from different botanical origins employing solid-phase extraction with a polymeric sorbent and liquid chromatography coupled to quadrupole time-of-flight mass spectrometry." *Food Research International* 130: 108904. doi:10.1016/j.foodres.2019.108904.

Saito-Shida, Shizuka, Nao Kashiwabara, Kouji Shiono, Satoru Nemoto, and Hiroshi Akiyama. 2020. "Development of an analytical method for determination of total ethofumesate residues in foods by gas chromatography-tandem mass spectrometry." *Food Chemistry* 313: 126132. doi:10.1016/j.foodchem.2019.126132.

Sang, Chenhui, Peter Borgen Sørensen, Wei An, Jens Hinge Andersen, and Min Yang. 2020. "Chronic health risk comparison between china and denmark on dietary exposure to chlorpyrifos." *Environmental Pollution* 257: 113590. doi:10.1016/j.envpol.2019.113590.

Tong, Zhou, Xu Dong, Shasha Yang, Mingna Sun, Tongchun Gao, Jinsheng Duan, and Haiqun Cao. 2019. "Enantioselective effects of the chiral fungicide tetraconazole in wheat: Fungicidal activity and degradation behaviour." *Environmental Pollution* 247: 1–8. doi:10.1016/j.envpol.2019.01.013.

Tripathy, Vandana, Krishan Kumar Sharma, Rajbir Yadav, Suneeta Devi, Amol Tayade, Khushbu Sharma, Priya Pandey, et al. 2019. "Development, validation of QuEChERS-based method for simultaneous determination of multiclass pesticide residue in milk, and evaluation of the matrix effect." *Journal of Environmental Science and Health, Part B* 54 (5): 394–406. doi:10.1080/03601234.2019.1574169.

Villalba, A., M. Maggi, P. M. Ondarza, N. Szawarski, and K. S. B. Migliorana. 2020. "Influence of land use on chlorpyrifos and persistent organic pollutant levels in honey bees, bee bread and honey: Beehive exposure assessment." *Science of the Total Environment* 713: 136554. doi:10.1016/j.scitotenv.2020.136554.

Xiang, Dandan, Kun Qiao, Zhuoying Song, Shuyuan Shen, Mengcen Wang, and Qiangwei Wang. 2019. "Enantioselectivity of toxicological responses induced by maternal exposure of cis-bifenthrin enantiomers in zebrafish (*Danio rerio*) larvae." *Journal of Hazardous Materials* 371: 655–65. doi:10.1016/j.jhazmat.2019.03.049.

Xie, Jingqian, Lu Zhao, Kai Liu, Fangjie Guo, and Weiping Liu. 2018. "Enantioselective effects of chiral amide herbicides napropamide, acetochlor and propisochlor: The more efficient R-enantiomer and its environmental Friendly." *Science of the Total Environment* 626: 860–66. doi:10.1016/j.scitotenv.2018.01.140.

Xu, Chao, Lili Niu, Jinsong Liu, Xiaohui Sun, Chaonan Zhang, Jing Ye, and Weiping Liu. 2019. "Maternal exposure to fipronil results in sulfone metabolite enrichment and transgenerational toxicity in zebrafish offspring: Indication for an overlooked risk in maternal transfer?" *Environmental Pollution* 246: 876–84. doi:10.1016/j.envpol.2018.12.096.

Xu, Chao, Xiaohui Sun, Lili Niu, Wenjing Yang, Wenqing Tu, Liping Lu, Shuang Song, and Weiping Liu. 2019. "Enantioselective thyroid disruption in zebrafish embryo-larvae via exposure to environmental concentrations of the chloroacetamide herbicide acetochlor." *Science of the Total Environment* 653: 1140–48. doi:10.1016/j.scitotenv.2018.11.037.

Xu, Chao, Wenqing Tu, Chun Lou, Yingying Hong, and Meirong Zhao. 2010. "Enantioselective separation and zebrafish embryo toxicity of insecticide beta-cypermethrin." *Journal of Environmental Sciences* 22 (5): 738–43. doi:10.1016/S1001-0742(09)60171-6.

Yang, Qingxi, Shiwei Wei, Na Liu, and Zumin Gu. 2020. "The dissipation of cyazofamid and its main metabolite CCIM during wine-making process." *Molecules* 25 (4): 777. doi:10.3390/molecules25040777.

Ye, Xiu, Shuping Ma, Lianjun Zhang, Pengfei Zhao, Xiaohong Hou, Longshan Zhao, and Ning Liang. 2018. "Trace enantioselective determination of triazole fungicides in honey by a sensitive and efficient method." *Journal of Food Composition and Analysis* 74: 62–70. doi:10.1016/j.jfca.2018.09.005.

Yigit, Nuran, and Yakup Sedat Velioglu. 2019. "Effects of processing and storage on pesticide residues in foods." *Critical Reviews in Food Science and Nutrition* 0 (0): 1–20. doi:10.1080/10408398.2019.1702 501.

Zeng, Huiyun, Xiujuan Xie, Yejing Huang, Jieming Chen, Yingtao Liu, Yi Zhang, Xiaoman Mai, Jianchao Deng, Huajun Fan, and Wei Zhang. 2019. "Enantioseparation and determination of triazole fungicides in vegetables and fruits by aqueous two-phase extraction coupled with online heart-cutting two-dimensional liquid chromatography." *Food Chemistry* 301: 125265. doi:10.1016/j.foodchem.2019.125265.

Zhang, Aiqian, Yunsong Mu, and Fengchang Wu. 2017. "An enantiomer-based virtual screening approach: Discovery of chiral organophosphates as acetyl cholinesterase inhibitors." *Ecotoxicology and Environmental Safety* 138: 215–22. doi:10.1016/j.ecoenv.2016.12.035.

Zhang, Ping, Zhe Li, Xinru Wang, Zhigang Shen, Yao Wang, Jin Yan, Zhiqiang Zhou, and Wentao Zhu. 2013. "Study of the enantioselective interaction of diclofop and human serum albumin by spectroscopic and molecular modeling approaches *in vitro*." *Chirality* 25 (11): 719–25. doi:10.1002/chir.22204.

Zhang, Ping, Donghui Liu, Zhe Li, Zhigang Shen, Peng Wang, Meng Zhou, Zhiqiang Zhou, and Wentao Zhu. 2014. "Multispectroscopic and molecular modeling approach to investigate the interaction of diclofop-methyl enantiomers with human serum albumin." *Journal of Luminescence* 155: 231–37. doi:10.1016/j.jlumin.2014.06.040.

Zhang, Qing, Beibei Gao, Mingming Tian, Haiyan Shi, Xiude Hua, and Minghua Wang. 2016. "Enantioseparation and determination of triticonazole enantiomers in fruits, vegetables, and soil using efficient extraction and clean-up methods." *Journal of Chromatography B* 1009–1010 (January): 130–37. doi:10.1016/j.jchromb.2015.12.018.

Zhang, Qing, Haiyan Shi, Beibei Gao, Mingming Tian, Xiude Hua, and Minghua Wang. 2016. "Enantioseparation and determination of the chiral phenylpyrazole insecticide ethiprole in agricultural and environmental samples and its enantioselective degradation in soil." *Science of the Total Environment* 542: 845–53. doi:10.1016/j.scitotenv.2015.10.132.

Zhang, Xining, Yue Song, Qi Jia, Lin Zhang, Wei Zhang, Pengqian Mu, Yanbo Jia, Yongzhong Qian, and Jing Qiu. 2019. "Simultaneous determination of 58 pesticides and relevant metabolites in eggs with a multifunctional filter by ultra-high-performance liquid chromatography-tandem mass spectrometry." *Journal of Chromatography A* 1593: 81–90. doi:10.1016/j.chroma.2019.01.074.

Zhang, Zhaoxian, Guizhen Du, Beibei Gao, Kunming Hu, Amir E. Kaziem, Lianshan Li, Zongzhe He, Haiyan Shi, and Minghua Wang. 2019. "Stereoselective endocrine-disrupting effects of the chiral triazole fungicide prothioconazole and its chiral metabolite." *Environmental Pollution* 251: 30–36. doi:10.1016/j.envpol.2019.04.124.

Zhao, Bihong, Dan Wu, Huiyuan Chu, Chaozhan Wang, and Yinmao Wei. 2019. "Magnetic mesoporous nanoparticles modified with poly(ionic liquids) with multi-functional groups for enrichment and determination of pyrethroid residues in Apples." *Journal of Separation Science* 42 (10): 1896–1904. doi:10.1002/jssc.201900038.

Zhao, Meirong, Fang Chen, Cui Wang, Quan Zhang, Jianying Gan, and Weiping Liu. 2010. "Integrative assessment of enantioselectivity in endocrine disruption and immunotoxicity of synthetic pyrethroids." *Environmental Pollution* 158 (5): 1968–73. doi:10.1016/j.envpol.2009.10.027.

Zhao, Meirong, and Weiping Liu. 2009. "Enantioselectivity in the immunotoxicity of the insecticide acetofenate in an *in vitro* model." *Environmental Toxicology and Chemistry* 28 (3): 578–85. doi:10.1897/08-246.1.

Zhao, Meirong, Ying Zhang, Shulin Zhuang, Quan Zhang, Chengsheng Lu, and Weiping Liu. 2014. "Disruption of the hormonal network and the enantioselectivity of bifenthrin in trophoblast: Maternal–foetal health risk of chiral pesticides." *Environmental Science & Technology* 48 (14): 8109–16. doi:10.1021/es501903b.

Zhao, Yue, Meinan Ouyang, Yabing Xiong, Dandan Wang, Haoming Guo, and Zhonghua Yang. 2019. "The different dissipation behaviour of chiral pesticide paclobutrazol in the brine during Chinese cabbage pickling process." *Chirality* 31 (3): 230–35. doi:10.1002/chir.23051.

Zhuang, Shulin, Zhisheng Zhang, Wenjing Zhang, Lingling Bao, Chao Xu, and Hu Zhang. 2015. "Enantioselective developmental toxicity and immunotoxicity of pyraclofos toward zebrafish (*Danio rerio*)." *Aquatic Toxicology* 159: 119–26. doi:10.1016/j.aquatox.2014.12.006.

10 Chiral Pharmaceuticals
Source, Human Risk, and Future Studies

Christina Nannou and Dimitra Lambropoulou

CONTENTS

10.1 Introduction 249
10.2 Chiral Pharmaceuticals 251
10.3 Analytical Aspects for Determination of Chiral Pharmaceuticals.................................... 251
10.4 Sources of Pharmaceuticals in Food Chain and Uptake Processes 252
    10.4.1 Sources of Pharmaceuticals for Human Consumption in Food Crops.................... 253
    10.4.2 Sources of Veterinary Drugs in Food Crops 254
    10.4.3 Uptake and Translocation of Chiral Pharmaceuticals in Food Crops..................... 255
        10.4.3.1 Plant Uptake Processes 256
        10.4.3.2 Chemical-Specific Factors Influencing Plant Uptake of Pharmaceuticals.. 256
    10.4.4 Stereoselective Uptake, Translocation, and Metabolism 257
    10.4.5 Stereoselective Sorption 257
10.5 Enantioselective Occurrence of Selected Pharmaceuticals in Food Matrices 257
    10.5.1 Antibiotics.. 260
    10.5.2 Non-Steroidal Anti-Inflammatory Drugs (NSAIDs)... 262
    10.5.3 Psychoactive Drugs... 263
    10.5.4 Anthelmintic Drugs 265
    10.5.5 Other Drugs 266
10.6 Human Risk 267
10.7. Enantiospecific Toxicity 269
10.8 Concluding Remarks and Future Challenges 270
Acknowledgments... 272
References... 272

10.1 INTRODUCTION

Chirality, or stereoisomerism, or enantiomerism, is a geometric property of certain compounds, which are not superimposable on their mirror images (Alvarez-Rivera et al., 2020). Technically, molecules having at least one asymmetric carbon, thus one or more stereogenic centers, are chiral, and the two non-superimposable mirror-image forms of chiral molecules are called enantiomers (Fanali et al., 2019). Chiral compounds are further divided into enantiomers and diastereomers. Since the distinctive feature of the enantiomers of a chiral molecule is that they are mirror-images, they are of identical chemical structure. However, enantiomers differ in the spatial arrangement of the atoms around the stereogenic center, they differ substantially in their optical activity, and especially their ability to rotate polarized light, and they exhibit rather similar physicochemical properties (Ribeiro et al., 2013; Sanganyado et al., 2017). On the other hand, diastereomers are stereoisomers but not mirror images, and they show moderately to considerably different physicochemical properties. Sometimes, apart from carbon, sulfur, phosphorus, and nitrogen can form chiral molecules (i.e., omeprazole, cyclophosphamide, and methaqualone, respectively). The phenomenon of chirality is widespread in nature.

Different conventions apply to name after and distinct chiral molecules. For instance, the *R/S* convention is used to name the enantiomers. An alternative identification of the enantiomers is defined by whether they rotate a polarized light clockwise, or counterclockwise, where the symbols (+) and (−) are used, respectively. It is noteworthy that in the case of racemic mixtures (equimolar mixtures of enantiomers) there is no rotation of the polarized light, thus the symbol (±) is used (Nguyen et al., 2006).

The chirality of compounds can be of utmost importance for food control and food quality, not only because it is proved that the chirality of molecules highly affects their biological activity, but also because individual enantiomers or enantiomeric ratios of various natural and synthetic compounds can affect the food appropriateness for consumption. There is a misconception that food analysis is limited to the analysis of nutrients, fatty acids, lipids, amines, phenols and proteins, enzymes, amino acids, carbohydrates, nucleosides, and a few alkaloids and hormones that may be chiral compounds. Nevertheless, food analysis aims to an integrated food quality control and it is conducted for a broad-spectrum of substances including pesticides, mycotoxins, and emerging contaminants, with pharmaceuticals being the most representative, focusing on antibiotics and veterinary drugs, as well as their metabolites or transformation products (Ribeiro et al., 2020). The massive introduction of emerging contaminants in the food chain imposes their simultaneous analysis to ensure an integral food control, although pesticides used to be the typical markers to guarantee food safety until recently, due to their persistence and potential bioaccumulation.

Regarding food components, a variety of them show chirality; it is remarkable that individual enantiomers or enantiomeric ratios in food components may be the origin for different effects, such as taste and odor. During food processing, some natural chiral components that are endogenous in food matrix can racemize, shifting the enantiomer distribution (Nguyen et al., 2006). In addition, different enantiomeric forms of contaminants may show different toxicity. For example, almost half of the authorized pesticides are chiral, hence many of them end up in the environment and then in the food chain as racemates (Carlsson et al., 2014). Those racemate(s) exert different toxicities and degradation rates of the enantiomers. Consequently, the impact on human health may vary importantly.

Based on the aforementioned comments, apart from the determination of chiral pesticides, it is also crucial to study exhaustively the behavior of chiral drugs in all fields of their production, from pharmaceutical and agricultural industry to food quality control. For instance, S(−)-propranolol and S(+)-citalopram are 100 times more active than their respective enantiomers, as well as (S)-ibuprofen is 100 times more active than (R)-ibuprofen, meaning that they provoke different toxicities to nontarget organisms. Hence, it is obvious that chirality is a top-class subject for research and largely affects the life of plants, animals, and humans and so does for the food industry. Therefore, determination of enantiomeric drug ratios in food and beverages can provide valuable information for food safety. The necessity of studying these compounds is indirectly associated with human health, as underlined by the fact that food safety control institutions strictly inspect the established maximum residues levels (MRL) for pesticides and drugs in food (Masiá et al., 2016). Consequently, it is challenging to study the occurrence of chiral drugs to food matrices in order to assess the potential risks for human health.

In this context, this chapter attempts to be a critical appraisal for the most interesting findings about enantioselective determination of pharmaceuticals in food matrices, as well as their potential sources and human risk, with an emphasis on antibiotics and veterinary drugs that have gained most of the scientific attention about this topic. Lastly, the perspectives for future research as well as some recommendations in this burning issue are discussed after a substantial investigation and identification of the gaps in the most recent (2015–2020) relevant literature.

10.2 CHIRAL PHARMACEUTICALS

The broad class of pharmaceuticals and personal care products (PPCPs) include prescribed and over-the-counter (OTC) drugs, antibiotics, veterinary drugs, illicit drugs, and ingredients of personal care products such as fragrances, sunscreens, cosmetics, and detergents (Kosma et al., 2014). Whilst pharmaceuticals are applied in both humans and animals for medical purposes, personal care products are typically applied to the human body (Daughton and Ternes, 1999). According to data from pharmaceuticals industry, over 50% of the licensed pharmaceuticals in use are chiral (Nguyen et al., 2006). In addition, 90% of them are being released into the market as racemates, consisting of an equimolar mixture of two enantiomers. Currently, chiral drugs are often developed and employed as single enantiomers in the pharmaceutical industry, and more enantiopure pharmaceutical preparations are being approved each year (Calcaterra and D'Acquarica, 2018). Therefore, chiral pharmaceuticals are consumed as racemates or as enantiomerically pure forms. Regarding chiral drugs, they are categorized in the following three classes: (i) drugs having one major bioactive form; for example mirtazapine's isomers, (R)-(–)-mirtazapine and (S)-(+)-mirtazapine exhibit different pharmacokinetic and pharmacodynamic properties, since (S)-(+)-mirtazapine is a more prone receptor antagonist than its mirror at almost all binding sites (Muth-Selbach et al., 2009), (ii) drugs having equally bioactive enantiomers, and (iii) drugs having a single bioactive form where the other enantiomer might be converted into its bioactive enantiomer in the body through chiral inversion (Patel et al., 2019). A nice example is that of ibuprofen which undergoes unidirectional enantioselective inversion, from the R- to S-enantiomer; the latter will not be reversed to the R-enantiomer again (Sanganyado et al., 2017).

Enantiomers of chiral drugs have the same physicochemical properties but differ in biological activities, which can be similar, opposite, null, or different (Kasprzyk-Hordern, 2010). Given that each enantiomer is submitted to metabolism, other stereoisomeric compounds can occur or be excreted unchanged. The differences in enantioselectivity in the metabolism and/or in the biodegradation in WWTPs are the reason why chiral pharmaceuticals end in the final receivers exhibiting different enantiomeric fractions. Apart from negative environmental impact, pharmaceuticals and consequently chiral pharmaceuticals as well, have indirect detrimental effects on several organisms via the food chain web (Zhou et al., 2018).

Notably, some achiral drugs may undergo metabolism generating chiral metabolites, by introducing a chiral center. Typical examples, where the thioether function is oxidized to a sulfoxide group, are that of the chiral metabolites of albendazole (veterinary drug), or cimetidine (H2-receptor antagonist). Then, the sulfur is the new chiral center (Patel et al., 2019). Hence, their enantioselective absorption, distribution, and metabolism occur *in vivo*, and finally, the corresponding enantiomers are excreted into the environment either as the parent compound or as metabolites by human or animal body (Li et al., 2020).

10.3 ANALYTICAL ASPECTS FOR DETERMINATION OF CHIRAL PHARMACEUTICALS

Since a lot of interest has been occasioned by the chiral recognition, there is an increasing number of analytical methods targeting the reliable identification and/or quantification of any enantiomeric forms and monitoring the stereoselective behavior of enantiomers on several applications (Alvarez-Rivera et al., 2020; D'Orazio, 2020). Method development for chiral analysis is demanding and time-consuming given the insufficiently clarified mechanism of chiral separation, as well as complex and unpredictable enantioselectivity. The sample pretreatment preceding the determination of enantiomeric composition in food, environmental, and biological samples is similar to that for non-chiral analysis, except an extra step for the derivatization of polar enantiomers that sometimes is necessary (Caballo et al., 2015). The process applied to two separate isomers of a racemic mixture is known with the term chiral separation or chiral resolution. It is a challenging analytical process,

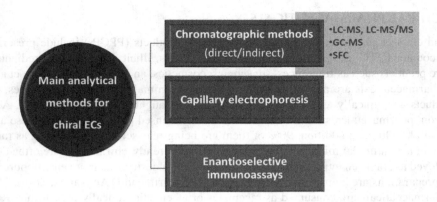

FIGURE 10.1 The main analytical approaches of chiral analysis in environmental matrices.

as regards both preparative and analytical steps. The main analytical approaches of chiral analysis in various matrices are depicted in Figure 10.1.

For the analysis of racemic drugs, two kinds of analytical methods have been developed: (a) the physical methods and (b) the enantioselective immunoassays. The main physical methods include gas or liquid chromatography coupled to mass spectrometry (GC–MS, LC–MS, respectively) and gas or liquid chromatography-tandem mass spectrometry (GC–MS/MS, LC–MS/MS), high-performance liquid chromatography (HPLC), supercritical fluid chromatography (SFC), and capillary electrophoresis (CE), with LC techniques being the most popular among them. The indirect LC uses chiral derivatization reagent(s) with the formation of two diastereomers and it is often performed in bioanalysis for its high sensitivity. However, the existence of a functional group in the drug is required (amine, hydroxyl, carboxyl, etc.). A chiral derivatizing reagent (a pure single enantiomer) added in the sample will react with these functions to form two diastereomers that can be separated by a classical reversed-phase column (C_{18} or C_8) (Alvarez-Rivera et al., 2020). Direct chiral separations using chiral stationary phases (CSPs) are the most used because of its simplicity and its rapidity. In general, the scarcity in commercially available chiral columns inhibits chiral analysis. On the other hand, enantioselective immunoassays are a helpful alternative tool, enabling fast separation with a high sample throughput (Buchberger, 2011), where a preliminary extraction is not necessary. Furthermore, enantioselective immunoassays are more sensitive and require smaller sample amounts.

10.4 SOURCES OF PHARMACEUTICALS IN FOOD CHAIN AND UPTAKE PROCESSES

Chiral pharmaceuticals play a major role in the field of the pharmaceutical industry and its applications. Their worldwide extensive use and consumption for veterinary and/or human purposes makes them notorious contaminants in the environment and related matrices, such as plants, animals, food, and feed. They follow the main pathways as pharmaceuticals in general, and the main sources for their unintended occurrence include wastewater effluents, improper storage and disposal, emissions from manufacturing, and so on (Huang et al., 2015). Humans are exposed to chiral pharmaceuticals mainly via "contaminated" food or drinking water, breathing, or surgical treatment. Contamination of food matrices by pharmaceuticals is related with the factors for environment contamination, since environment is the first receiver (Figure 10.2). Hence, it depends on source strengths, geographical distributions of source types, climatic conditions, etc. (Patel et al., 2019). The impact of food on human health is an incentive for the scientific community to pay attention to the sources, and human risk of chiral pharmaceuticals in food, as well as to define and set the future challenges.

Chiral Pharmaceuticals

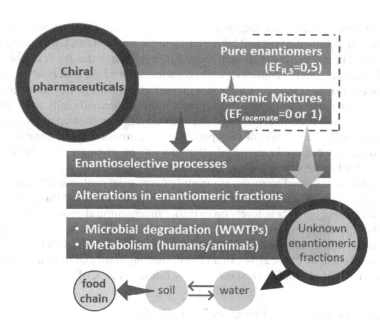

FIGURE 10.2 Chiral pharmaceuticals: from production to food chain. (Modified from Maia et al., 2017.)

Some of the most well-studied human pharmaceuticals, such as antibiotics, non-steroidal anti-inflammatory drugs (NSAIDs), β-blockers, and psychoactive substances are chiral. Chirality characterizes also some antibiotics, such as sulfonamides, tetracyclines, and β-lactams that are used on a large scale as veterinary drugs in aquaculture, husbandry, and many other fields apart from human health (Wang et al., 2017; Yanovych et al., 2018). Consumption of antibiotics in animals exceeds by far their human consumption (Sarmah et al., 2006). They are administered to animals either orally (through water, feed) or topically (by injection, paste, drench, and more rarely implants), for the prophylaxis or treatment of infectious and noninfectious diseases. Other pharmaceuticals used as veterinary drugs are non-steroidal anti-inflammatory drugs, psychotropic substances, anthelmintics, and sometimes hormones and estrogens, in order to manage animals' reproductive systems; typical examples are steroids, progesterone, prostaglandins, oxytocin, ergonovine, HCG, GnRH, and FSH (Patel et al., 2019).

10.4.1 Sources of Pharmaceuticals for Human Consumption in Food Crops

There is increasing evidence that food crops can uptake and accumulate pharmaceuticals (Fu et al., 2019; Malchi et al., 2014), either chiral or not. The main pathways followed by pharmaceuticals to enter agroecosystems are: (a) irrigation with treated wastewater, (b) land application of biosolids, and (c) excretion of livestock containing administered veterinary pharmaceuticals (Christou et al., 2017).

Indeed, in several arid or semi-arid regions worldwide, treated wastewater is used as an alternative for agricultural irrigation to face water scarcity (Chuang et al., 2019) and to improve soil conditions by the presence of (micro)nutrients and organic matter that remain in wastewater after treatment. This practice results in dissipation of pharmaceuticals in agricultural soil, which are afterwards taken up by certain crops (Grossberger et al., 2014). This is partially attributed to their medium-high to high polarity—hence water solubility and low affinity to soil—and presence in ground water without filtration which results in accumulation in plants via plant uptake. Among the most frequently detected pharmaceuticals and personal care products in treated wastewater are paracetamol, caffeine, atenolol, carbamazepine, sulfamethoxazole, diclofenac, fluoxetine, and N,N-diethyl-metatoluamide (DEET), triclosan, and triclocarban, respectively (Wu et al., 2014).

In many areas, sewage biosolids generated after treatment and livestock manure are applied to agriculture for the enrichment of the soil by recycling urban nutrients and organic matter into crops for the improvement of soil quality (Christou et al., 2017). Although the use of biosolids is a sustainable practice, it requires careful consideration and anticipation of the potential environmental and human risk, since it can be a source of microbes and organic or inorganic contaminants (Sabourin et al., 2012). Such risk may be mitigated if this practice is regulated officially and includes pretreatment methods to decrease the content in pathogens. Notably, the application of biosolids is regulated by certain criteria, such the location of the application, the frequency and duration of application, and the offset time between application and allowance off livestock to graze or the harvesting of crops, if destined for human consumption (Tasho and Cho, 2016).

Pharmaceuticals have gained attention among other emerging contaminants and are scrutinized in biosolids (Carter et al., 2018; Thomaidi et al., 2016). It is reported that this practice may lead to accumulation of pharmaceuticals at levels up to µg/kg in agricultural products (Calderón-Preciado et al., 2013). These levels are rarely capable of posing risks to human health, since they are much lower than the usual therapeutic dosage. However, they are harmful for plant health if changing the plant hormone levels (Carter et al., 2014). In addition, the long-term exposure may entail synergistic or antagonistic effects of pharmaceutical mixtures (Chitescu et al., 2013).

10.4.2 Sources of Veterinary Drugs in Food Crops

Veterinary medicines including antibiotics, hormones, and parasiticides end up to land either directly by feces or urine or indirectly through the application of manure as a fertilizer; they are detected in soils, surface waters, and groundwaters and incidentally are taken up by crops (Boxall et al., 2006). Due to hydrological, climatic, and land-use settings, such contaminants are detected throughout the year, with some of them being persistent enough, such as oxytetracycline (Chitescu et al., 2013; Chuang et al., 2019).

Antibiotics have been used for many decades in order to protect public health from bacterial diseases. In addition, they have been used for veterinary purposes for more than 60 years, as therapy, disease prevention and control, and growth promotion (Tasho and Cho, 2016). By then, antibiotics have been used increasingly as veterinary drugs and in animal feed, especially after the growing trend to use industrial farming to produce food. The repeated manure application results in high concentrations of veterinary antibiotics in the terrestrial environment, hence in plants. Several studies have been carried out in order to investigate the uptake of veterinary antibiotics, either in plant cultures in a medium/soil artificially spiked with antibiotic or plants grow up in soil manured with animal effluents, containing antibiotics (Boxall et al., 2004; Chung et al., 2017; Tasho and Cho, 2016).

The annual use of antibiotics for either medical or veterinary use is estimated to be 200,000 tons worldwide (Tang et al., 2017), although data about specific quantities of certain antibiotics used in food animals are not publicly available. The extensive and improper use of antibiotics in animals and humans is an issue of concern and has introduced the worrisome global phenomenon of antimicrobial resistance and the potential transmission of resistant bacterial strains from animals to humans (Christou et al., 2017; Rana et al., 2019). For this reason, maximum residual limits (MRLs) setting the maximum acceptable level of residue in the tissue of animal-origin food products have been established worldwide.

In Europe, there was a lack of legislation framework controlling the veterinary drugs for years. Even though the environmental assessment of veterinary drugs was established in the United States in 1980, it was 10 years later that Europe (EU) set the legal limit for veterinary drugs in food matrices for the first time. This was the limit of 4–1500 µg/kg for antibiotic residues in milk, while for other food products of animal origin the limit ranged from 25–6000 µg/kg (Ramatla et al., 2017). Due to the growing concern of antimicrobial resistance globally, promotors are also banned in many regions (such as Europe) already since 2006 and phasing out strategies have been recommended to all countries by WHO in their recent documents on combating antimicrobial resistance (FAO, 2008).

The "International Cooperation on Harmonization of Technical Requirements for the Registration of Veterinary Medical Products" has also set a limit for the total soil concentration in veterinary pharmaceuticals (≤100 μg/kg), based on data derived from ecotoxic effects on earthworms, plants, and bacteria (Patel et al., 2019).

Among the antibiotic classes used for veterinary purposes, the lion's share belongs to tetracyclines, followed by sulfonamides and macrolides—altogether comprising 90% of the total antibiotics. According to a critical review, the most commonly used antibiotics in livestock industry are oxytetracycline, chlortetracycline, penicillin, sulfamethazine, neomycin, monensin, tylosin, virginiamycin, and bacitracin (Tasho and Cho, 2016). Depending on the type and age of the animal, the antibiotic's dosage varies from a few gram to 220 g/Mg of feed (Kumar and Gupta, 2016), while most of them are not digested completely by the animal system and are excreted out into the environment. The unsafe and improper disposal of huge amounts of animal excreta produced every day is another problem to be faced, while it is observed noteworthy discrepancy for the residue levels between both the sampling sites and animal species (Zhao et al., 2010). For instance, in animal manures from industrial livestock farms are reported higher concentration of antibiotic residues than in the farmer's households. It is interesting that in most areas, there is no legal limitation for the livestock waste treatment, hence a large portion of such waste ends up to land and may be used later as fertilizer, as it is rich in nutrients and organic matter. The application of liquid manure is shown to produce equivalent crop yields as the chemical fertilizers (Tasho and Cho, 2016). During storage, many veterinary drugs exhibit slow degradation, thus there is a potential release of active biosolids from land enriched with manure to the environment, and then plants and the food chain.

It is noteworthy that animal feed may be also contaminated by various substances, such as pesticides and their TPs, polychlorinated biphenyls (PCBs), dioxins, mycotoxins, heavy metals as well as multi-class chiral pharmaceuticals and intentionally applied veterinary drugs. Besides, all veterinary drugs are submitted to tests for approval before being released in the market, in order to identify the probability of human health concerns (carcinogenicity, mutagenesis) (Bamidele Falowo and Festus Akimoladun, 2020). As a result, public health risk associated with residues remains low. However, the undesirable carryover of veterinary drugs in feed or its ingredients during production is underestimated, despite the potential to lead to unexpected residues in animal tissues.

10.4.3 Uptake and Translocation of Chiral Pharmaceuticals in Food Crops

It has been found that plant organs and tissues respond differently depending on the antibiotic concentration and exposure time. For example, a study on *Lythrum salicaria* showed that the phytotoxic effect of sulfadimethoxine varied from organ to organ (Migliore et al., 2010). Namely, roots, cotyledons, and cotyledon petioles showed a toxic effect, while internodes and leaf length exhibited an increased growth at lower drug concentrations and toxicity at a higher level (Migliore et al., 2010).

The uptake and translocation of pharmaceuticals in food crops can take place in roots as well as in the edible parts of the plants imposing unintended effects on food safety (Sanganyado et al., 2017). According to relevant research, the amounts of pharmaceuticals detected in food crops are rather low (Colon and Toor, 2016). However, these low levels should not considered as reassuring, since the toxicity and potential long-term effects posed to humans remains poorly documented (Caballo et al., 2015; Colon and Toor, 2016). Uptake of such contaminants by plants is highly dependent on the physicochemical properties of pharmaceuticals, namely, polarity, expressed by the octanol-water partition coefficient, K_{ow}, water solubility, vapor pressure, and molecular weight (Fu et al., 2019). In addition, there are some environmental factors that have influence on the process, such as soil type and its water content, temperature, and agricultural practices (Colon and Toor, 2016; Fatta-Kassinos et al., 2011). Also, the plant characteristics such as the root system, shape and size of leaves, and lipid content could affect to some extent the uptake (Kumar and Gupta, 2016). Furthermore, the high number of both cultivated species (over 7000 for human consumption) (Bharucha and Pretty, 2010) and released pharmaceuticals into the environment pose an extra difficulty in the exploration of the process and make it even more complicated.

10.4.3.1 Plant Uptake Processes

It is of primary importance to anticipate the main pathways and processes involved with the uptake and translocation of organic contaminants. First, the main interaction occurs between the root system and soil-soil solution. The simplest type of passive transport for pharmaceuticals and similar contaminants to move across a cell membrane is passive diffusion (Calderón-Preciado et al., 2013). There is also the active transport that requires energy to move nutrients and contaminants across the cell membrane. Once the contaminants reach the root system, they can translocate to the aerial parts of the plant through the xylem of the vascular system, which oversees transferring water and nutrients upwards from the roots to the upper plant parts (Colon and Toor, 2016). Apart from roots, plants can also uptake contaminants as free gas molecules from the atmosphere, through small pores on the surface of leaves, called stomata. After entering the stomata, phloem translocates the contaminants into the other plants of the plants, thanks to the root system. Furthermore, when contaminants are in the vapor phase, they are translocated to plant roots through the soil (O'Connor, 1996) or diffuse into the plant after dissolvement in water droplets or sorption to particles that are deposited on plant surfaces (Zhang et al., 2017). It is obvious from this process that pharmaceuticals with Henry's law constant affect the uptake of contaminants from the air. It is noteworthy that irrigation with reclaimed water is gaining more and more ground worldwide, thus, in the field of pharmaceuticals uptake it is much more crucial to focus on their uptake through the root system, which is also the object of the most available related studies.

10.4.3.2 Chemical-Specific Factors Influencing Plant Uptake of Pharmaceuticals

As mentioned above, the uptake of pharmaceuticals in plants can vary depending on the species. A nice example is that of root vegetables (beet roots, carrots, onions, etc.), in comparison with tree fruit (orange, apple, peach). In the first case, the uptake taking place through soil and roots is much higher than in the second one, due to the proximity of the crops to soil. In contrast, any volatile contaminants are more probable to reach on the plant trees. Other plant-specific parameters are the root system, shape and size of leaves, lipid and water content, and transpiration date. However, it seems that the chemical-specific factors are more crucial in the process of plant uptake of pharmaceuticals. For neutral pharmaceuticals, the most critical factor is that of the tendency of an organic compound to adsorb to soil, expressed by the octanol-water partition coefficient (K_{ow}) (Carter et al., 2018). This constant shows a wide range, since there are extremely polar (hydrophilic) to highly lipophilic pharmaceuticals. According to the Gaussian distribution, where the maximum translocation of chemicals is observed at a log K_{ow} of ~1.8 (Carter et al., 2014), this means that an extremely hydrophilic contaminant cannot pass through lipid membranes of roots. On the contrary, hydrophobic compounds tend to bind strongly to root tissues and are not translocated. This is in accordance with several studies reporting the uptake of organic contaminants by plants, where the translocation takes place when the log K_{ow} is between 1 to 4 (Limmer and Burken, 2014). For instance, carbamazepine has positive relationship between the root uptake and log K_{ow} (Carter et al., 2014), underlining that it is hydrophobicity that mostly affects its uptake. Unlike neutral pharmaceuticals, the ionic ones, being more polar and water soluble, exhibit a different behavior, due to different mechanisms of action that affect the accumulation (electrical attraction or repulsion, and ion trap) (Dodgen et al., 2015). As a result, there is no reliable correlation between hydrophobicity and plant uptake. For instance, caffeine, a non-charged compound is easily taken up by aquatic plants (Colon and Toor, 2016). On the other hand, diclofenac which is negatively charged is not proved to be taken up by such plants (Calderón-Preciado et al., 2013). According to Calderón-Preciado et al. (2013) this is due to the negative electrical potential at the cell membrane of the plants that repulse the negatively charged anions. Lastly, the stability of pharmaceuticals into the soil system for a period is necessary so that degradation does not occur and to allow uptake by plants to take place. Technically, this is expressed by the half-life time of the potential contaminant; it is reported that contaminants with $t_{1/2} < 14$ days are more likely to be taken up by plants (Carter et al., 2014). An array of studies proves this claim; for example, diclofenac, with a total half-life (including metabolites) of 25–33 h, has not been detected in vegetable crops irrigated with treated wastewater (Carter et al., 2014).

10.4.4 Stereoselective Uptake, Translocation, and Metabolism

Numerous studies raise the issue of biodegradation in soil as a process that regulates the inevitable pollution including runoff, leaching, and transfer to terrestrial organisms, but other equally important processes, including wetland removal, soil degradation, and plant uptake and metabolism are ignored, even lesser for chiral pharmaceuticals (Dodgen et al., 2015; Koba et al., 2017). The partitioning of chiral compounds is poorly documented compared to achiral compounds. It is reported that selective sorption to soil and sludge occurs with chiral compounds, but this process is almost unexplored yet, as well as the factors that affect it (Grossberger et al., 2014). In an excellent review, Sanganyado et al. (2017) report that the main cause for stereoselective sorption in soil or sludge is probably organic matter with chiral structures. The same authors underline that the chiral surfaces of soil-forming minerals should also be taken into consideration, since it has been proved that chiral surfaces, such as calcite, quartz, alkali feldspar, leonardite, and clinopyroxene, are involved in enantioselective sorption (Sanganyado et al., 2017).

Stereoselective adsorption is also affected by environmental conditions or the presence of environmental components that affect the polarity of the chiral pharmaceuticals. Such parameters are pH, temperature, ionic strength, and the presence of co-solutes, and their study along with the stereoselective sorption could be of great help in order to understand the mechanism of stereoselective adsorption and to predict the distribution of chiral pharmaceuticals among different environmental media (Sanganyado et al., 2017).

10.4.5 Stereoselective Sorption

Experimental evidence for stereoselective adsorption of chiral pharmaceuticals in food matrices either of plant or animal origin remains scarce. Stereoselectivity in adsorption of chiral pharmaceuticals is a complex issue, given that adsorption is supposed to be a non-stereoselective process. This may explain the fact that the stereoselectivity in adsorption of chiral pharmaceuticals is poorly investigated. Such evidence has been demonstrated in several studies for pesticides in agricultural soils, mainly after its irrigation with treated wastewater, and in few studies about foodstuff, while for pharmaceuticals, most of chirality studies focus on drugs themselves (tablets, products), or WWTPs and aqueous bodies. This is attributed to the fact that wastewater treatment is the key point for the removal of (a)chiral pharmaceuticals.

However, substituents with chiral structures such as minerals and organic matter create a chiral environment and probably foster stereoselectivity (Camacho-Muñoz et al., 2019). However, there is evidence that during the adsorption of pharmaceuticals onto sludge, enantiomer enrichment may happen (Camacho-Muñoz et al., 2019).

10.5 ENANTIOSELECTIVE OCCURRENCE OF SELECTED PHARMACEUTICALS IN FOOD MATRICES

There are relatively few studies on stereoselective occurrence of chiral pharmaceuticals in the food matrices. Table 10.1 compiles the published studies within the years 2015–2020, on the condition that enantioselective analysis was conducted in parallel with the measurement of the pharmaceutical's concentration in food/feed matrices. Although this table includes the most recent publications,

TABLE 10.1
Recent Literature Concerning Enantioselective Occurrence of Pharmaceuticals in Food Matrices

Analytes	Matrix	Sample Preparation	Instrumentation	Type of Study	Concentration in Real Samples (μg/kg)	Reference
Antibiotics						
Chloramphenicol	Honey (sunflower, blossom, blossom with)	Multi-step liquid-liquid extraction	ELISA, chiral LC-MS/MS	Monitoring	0.1-5.9 (racemate) <0.1-5.5 (RR-chloramphenicol <0.1-2.2 (SS-chloramphenicol)	Rimkus et al. (2020)
Chloramphenicol	Honey	Multi-step liquid-liquid extraction	Chiral LC-MS	Monitoring	0.3-5.9 (racemate) <0.1-5.5 (RR-chloramphenicol <0.1-2.2 (SS-chloramphenicol) <0.1 (RS-chloramphenicol) <0.1 (SS-chloramphenicol)	Rimkus and Hoffmann (2017)
Chloramphenicol	Honey	Solvent extraction	ELISA, chiral LC-MS/MS	Monitoring	n.d. - 0.4	Yanovych et al. (2018)
Chloramphenicol	Beehive strips	Solid phase extraction	ELISA, chiral LC-MS/MS	Monitoring	0.03-150 (racemate) 2.7-300,000 (RR-chloramphenicol 14-5000 (SS-chlorampenicol)	Yanovych et al. (2018)
Dextramycin (isomer of chloramphenicol)	Honey	Solid phase extraction	ELISA, chiral LC-MS/MS	Monitoring	0.3-3.1	Yanovych et al. (2018)
Chlortetracycline	Pig muscle	Solvent extraction	LC-MS/MS	Spiked samples, comparison of peak profile		Gaugain et al. (2015)
Flumequine	Milk, yogurt, honey, chicken, beef	Solvent extraction	Chiral LC-MS	Application to real samples	n.d.	Li et al. (2019)
Norfloxacin	Milk, chicken, pork, and fish	Solvent extraction	Capillary Electrophoresis Immunoassay with Laser-Induced Fluorescence Detector	Recovery study of spiked samples		Liu et al. (2015)
Ofloxacin	Milk	Home-made ionization device for extraction	Chiral LC-MS	Application of the method in eye drops		Wang et al. (2017)
NSAIDs						
Ibuprofen, naproxen and ketoprofen	Freshwater fish (O. mykiss, A. alburnus, L. gibbosus, M. salmoides, C. carpio)	Supramolecular solvent-based extraction	Chiral LC-MS	Recovery study of spiked samples/ Application to real samples	n.d.	Caballo et al. (2015)
Ibuprofen	Breast milk	Vortex-assisted matrix solid-phase dispersion	Chiral LC-MS	Recovery study of spiked samples		León-González and Rosales-Conrado (2017)
Ibuprofen, indoprofen, pranoprofen, flurbiprofen, ketoprofen, carprofen, naproxen, loxoprofen, etodolac	Fish tissues	Ultrasound-assisted extraction/ SPE for clean up	Chiral LC-MS	Application to real samples	n.d.	Li et al. (2020)

TABLE 10.1 (Continued)
Recent Literature Concerning Enantioselective Occurrence of Pharmaceuticals in Food Matrices

Analytes	Matrix	Sample Preparation	Instrumentation	Type of Study	Concentration in Real Samples (µg/kg)	Reference
Naproxen, ibuprofen, ketoprofen flurbiprofen	Fish			Ecotoxicity bioassay		Neale et al. (2019)
Psychotropic drugs						
Cyamemazine, promethazine, fluoxetine, oxazepam, mirtazapine, bupropion, phencynoate	Pork, beef and lamb muscles	Ultrasound-assisted extraction/ SPE for clean up	Chiral LC-MS	Application to real samples	n.d.	Zhu et al. (2019a)
Citalopram, desmethylcitalopram	Breast milk	Protein precipitation, solid phsa extraction	Chiral LC-MS	Monitoring	*(Foremilk) 30.0 ng/mL, S-(+)-citalopram 37.9 ng/mL, R-(−)-citalopram 9.9 ng/mL, S-(+)-desmethylcitalopram 9.0 ng/mL, R-(−)-desmethylcitalopram (hindmilk) 39.0 ng/mL, S-(+)-citalopram 49.1 ng/mL, R-(−)-citalopram 13.2 ng/mL, S-(+)-desmethylcitalopram 12.4 ng/mL, R-(−)-desmethylcitalopram	Weisskopf et al. (2016)
Fluoxetine, norfluoxetine	Human milk	Direct sample injection	2-dimensional LC-MS/MS	Application to real samples	*22.1-44.0 ng/mL, S-fluoxetine 9.7-18.7 ng/mL, R-fluoxetine 29.6-59.8 ng/mL, S-fluoxetine 12.9-23.5 ng/mL, R-fluoxetine	Alvim et al. (2016)
Steroid-like drugs						
Clenbuterol	Meat	RIKILT SOP A1172 method (LLE/SPE)	Chiral LC-MS	Spiked samples, human uptake		Parr et al., (2017)
Clenbuterol	Animal feed	QuEChERS/ Solid phsa extraction	LC-MS	Spiked sample		Rosales-Conrado et al. (2015)
Clenbuterol	Swine, beef and lamb	Solvent extraction/ Solid phase extraction	HPLC-MS/MS	Spiked samples, study of the distribution of R/S ratio		Wang et al. (2016)
4-Hydroxypraziquantel metabolites	Aquaculture fish (perch, tilapia, and ricefield-eel muscles)	Solvent extraction	LC-MS/MS	Evaluation of the depletion profiles		Zhang et al. (2019)

it represents a big fraction of the total articles handling chiral analysis of pharmaceuticals in food. This is partially attributed to the fact that most research articles and reviews discussing chiral contaminants in food emphasize on chiral pesticides rather than drugs. In addition, food is far less documented than environmental samples regarding PPCPs that are considered as contaminants fewer years than pesticides, which are the most popular contamination markers in such matrices. Enantioselective analysis of pharmaceuticals has been applied mostly in WWTPs (Kasprzyk-Hordern and Baker, 2012; Zhou et al., 2018), and little attention has been paid to surface water, sludge, soil, sediment, or biological samples, and incidentally in food (D'Orazio et al., 2017). In addition, it is noteworthy and at the same time challenging for future research that most studies describe the development/application of single-methods (the enantiomers of a single compound is monitored), or more rarely few compounds of the same therapeutic class—at the most nine compounds (Li et al., 2020). Moreover, almost all those studies are limited to an application of the selective method in real samples so that the applicability and feasibility of the method is tested, without a further monitoring study (Gaugain et al., 2015; Li et al., 2020; Liu et al., 2015). Therefore, there is an urgent need to develop and validate enantioselective multiresidue methods and carry out more monitoring studies. The lack of such data entails the inconclusiveness for any synergistic effects about the co-occurrence of various chiral pharmaceuticals in food products.

10.5.1 Antibiotics

Among the plethora of antibiotics, fluoroquinolones are one of the major groups having a broad activity spectrum against mycoplasma, Gram-positive, and Gram-negative bacteria (Pham et al., 2019). They comprise a class of synthetic antibiotics, being characterized by an extensive usage in both veterinary medicine and farming (Pham et al., 2019). Hence, their residual monitoring is of vital importance to ensure the quality and appropriateness of foodstuff of animal origin, and to conform with the regulation 37/2010 of the European Union. Human consumption of animal-derived food with quinolone residues above the regular limits may pose threats to human health, since those substances can indirectly be absorbed by the human body for a long time, and once accumulated in the human body will harm human health. As such, MRLs for animal-derived food have been established for quinolones in many European countries (Guidi et al., 2018).

Ofloxacin, a second-generation fluoroquinolone, used to be administered widely to treat mild-to-moderate bacterial infections. Today, it has been replaced by more potent and less toxic fluoroquinolones. It is frequently studied in milk and honey, and rarely in meat or other food matrices (Wang et al., 2017). The on-line identification of chiral ofloxacin in milk with an extraction/ionization device coupled to ESI-MS is described in a recent study (Wang et al., 2017). Therein, a homemade apparatus was used to simplify sample pretreatment. Milk sample was directly injected and chiral ofloxacin in the sample was extracted at PTFE membrane for further ionization, in a total time of ten seconds. By means of reaction thermodynamics method, trimeric cluster ion [NiII(ref)$_2$Ofloxacin-H]$^+$ was formed and dissociated to get chiral resolution of levofloxacin and dextrofloxacin, due to the different relative stabilities of the two diastereomeric clusters produced through the dissociation of NiII bound trimeric clusters. The application of that method enabled the qualitative and quantitative chiral analysis of ofloxacin in milk (Wang et al., 2017). However, the application of the method included ofloxacin and levofloxacin eye drops instead of human milk, since the method detection limit (MDL) was higher (μg/mL) than the expected concentration (ng/mL). This finding is a nice example of the need to develop more enantioselective methods that achieve lower limits, in order to detect trace levels of such compounds.

Norfloxacin is a second-generation fluoroquinolone antibacterial which is broadly used in animals as well; as a result, residues may occur in animal-derived foods and subsequently lead to a public health hazard. Norfloxacin shows striking potency against many pathogenic Gram-negative and Gram-positive bacteria by the inhibition of different kinds of enzymes (Liu et al., 2015). Regarding its enantioselective determination in food matrices, a capillary electrophoresis immunoassay method with laser-induced fluorescence (CEIA-LIF) for the quantitative analysis of norfloxacin in milk, chicken, pork, and fish samples was developed (Liu et al., 2015). The analytical performance was assessed by the recovery studies of the spiked matrices and gave very satisfactory results. The authors reported that CEIA-LIF combines the effective separation power of CE, the ligand specificity of IA, and the high sensitivity of LIF detection that can be further applied for enantioselective determination. However, this study did not expand further for application to real samples.

Flumequine is a fluoroquinolone of first generation that acts by the selective inhibition of type II topoisomerase and bacterial DNA gyrase, two essential enzymes for the bacterial DNA replication and transcription in the process of cell growth and division (Zhao et al., 2007). The high antimicrobial activity against a wide range of Gram-negative and Gram-positive pathogens renders flumequine a necessary component of the treatment of infections in food-producing animals such as cattle, turkey, pig, and poultry (Li et al., 2019). In a recent study, the enantioseparation and determination of flumequine enantiomers was successfully attempted in milk, yogurt, chicken, beef, egg, and honey samples by chiral liquid chromatography-tandem mass spectrometry (Li et al., 2019). The method can separate (R) and (S) enantiomers, while the relative recoveries for the S-enantiomer of flumequine varied from 83.7 to 110.5% in all studied matrices, and for R-enantiomer of flumequine varied from 82.3 to 108.3% in all studied matrices. The application of the method in 36 real samples of all six matrices obtained from different local markets in Benxi, China, revealed that no flumequine residues were detected in any of the randomly purchased food.

Among the limited studies focusing on the enantioselective determination of antibiotics in food matrices, chloramphenicol is one of the most well-studied. Chloramphenicol is a broad-spectrum antibiotic, biosynthesized by the soil organism *Streptomyces venezuelae* and several other actinomycetes while it is chemically synthesized for commercial use both in human and in veterinary medication. Since 1994, chloramphenicol has been banned from use in food-producing animals in Europe and several countries worldwide due to several adverse effects (Rimkus and Hoffmann, 2017). Despite the prohibition, residual monitoring for this compound is still carried out worldwide, due to the high entailed risk of any positive findings. Analytical methods for the determination of chloramphenicol should follow strict standards and ensure high quality assurance.

In a very recent study, chiral and achiral LC–MS/MS, ELISA, and a radioimmunoassay under the brand name "Charm® II Chloramphenicol test" were employed after a multistep liquid extraction, for the enantioselective determination of chloramphenicol in honeys from different origin (sunflower, blossom). The reported levels ranged from below 0.1 up to 5.9 μg/kg in samples collected from Vietnam, Ukraine, Southeastern Asia, Bulgaria, and Eastern Europe (Rimkus et al., 2020). In a previous study of the same research group, 40 honey samples were analyzed by chiral LC–MS/MS to investigate the four chloramphenicol stereoisomers. Of 40 samples, only in 9 honey samples the bioactive RR-chloramphenicol was analyzed as anticipated. In the remaining 31 samples the non-bioactive SS-chloramphenicol was identified and quantified unambiguously. In addition, in 10 of these samples, mixtures of RR-chloramphenicol and SS-chloramphenicol were analyzed, in a R/S ratio from 0.1 to 2.2, with 2 samples with an approximately ratio of 1. RS-chloramphenicol or SR-chloramphenicol could not be detected in any sample (Rimkus and Hoffmann, 2017).

A similar research for chloramphenicol and its optical isomer, dextramycin, in honey and beehive strips, with the aid of revealed chiral LC–MS/MS, revealed the presence of dextramycin in honey samples. Interestingly, discrepancies between the immuno-chemical methods and confirmatory RP LC–MS/MS methods were observed. It is remarkable that using LC-MS/MS methods, chloramphenicol was detected in unregistered drugs used for treating bees in violation of national regulations, as well as in medicinal preparations. The high dextramycin concentrations (up to 300,000 μg/kg) in some of the strips suspected for wastes of chloramphenicol chemical synthesis in the production (Yanovych et al., 2018).

Chloramphenicol stereoisomers need to be distinguished in every step, from production to application. The statement is based on an experience during a proficiency test including prawns as a matrix (Sykes et al., 2017). Due to the large discrepancies between the results, the reference standards under the same CAS number were put in doubt, addressing the fact that the stereoisomeric forms were not distinguishable by conventional LC–MS/MS or NMR. There is evidence that important variation is the result of the different stereoisomeric forms of chloramphenicol. The authors believe that the potential impact of incompletely specified stereoisomers in veterinary drug residues analysis should be taken seriously into consideration. In addition, it is documented that there is a stereoisomeric dependency on antimicrobial activity and immunoaffinity determination, and as such, routine analysis laboratories are not actually aware of the in vivo antimicrobial activity of the chloramphenicol residues.

Tetracyclines are a broad-spectrum antibiotic class, claimed to be the most commonly used in veterinary medicine. This is attributed not only to their activity, but also to low toxicity and cost. Tetracyclines are used as powders, solutions, and premixes. The first tetracycline, chlortetracycline, was isolated in 1948 by *Streptomycin aureofacien* (Chopra and Roberts, 2001). It is generally used for both preventive and curative treatment of pulmonary and digestive infections in poultry, calves, lambs, rabbits, and, particularly, in breeding pigs (Gaugain et al., 2015). If human consumption is intended, then the residues of chlortetracycline should be checked in animal tissues, in order to conform with the MRL of 100 µg/kg (for the sum of tetracycline and 4-epi-chlortetracycline, according to European Commission directive 37/2010; European Commission, 2010). There is an interesting study in which the distinction of the different isobaric forms of chlortetracycline in pig muscle tissue samples was attempted. In particular, the four compounds that finally were successfully differentiated were stereoisomers (4-epimers), structural isomers (keto-enol forms) and metabolites (6-iso-chlortetracycline and 4-epi-6-iso-chlortetracycline). Although there was no application to real samples, method development and validation revealed that the ratio between the keto-enol forms is the same in spiked or in incurred muscle samples, confirming the validity of the quantification method (Gaugain et al., 2015).

10.5.2 Non-Steroidal Anti-Inflammatory Drugs (NSAIDs)

Non-steroidal anti-inflammatory drugs (NSAIDs) are by far the therapeutic class of pharmaceuticals that most people are familiar with, given not only their widespread usage but also the over the counter (OTC) availability in the market. Their analgesic, antipyretic, and anti-inflammatory activity is mainly based on the inhibition of cyclooxygenase or 5-lipoxidase and the suppression of biosynthesis of prostaglandin (Attiq et al., 2018). The common characteristic in the structure of NSAIDs is the 2-arylpropionic acids (2-APAs), well-known as "profens." To the best of our knowledge, most profens are commonly administered as racemates for economic reasons, however, their therapeutic effects are attributed almost entirely to (S)-enantiomers (Caballo et al., 2015; Li et al., 2020); however, their therapeutic effects are attributed almost entirely to (S)-enantiomers. Ibuprofen, ketoprofen, fenoprofen, diclofenac, mefenamic acid, indomethacin, and naproxen are the most popular NSAIDs (Shah et al., 2017). Notably, NSAIDs are the most studied micropollutants in the environment, because of their potential occurrence at high concentrations, a fact that may entrain the presence in food matrices. This probability is higher when reclaimed wastewater is reused for irrigation, thus the impact of NSAIDs on consumers is increased.

Ibuprofen is a chiral drug, derivative of the propionic acid, with analgesic, antipyretic, and anti-inflammatory action. It is sold over the counter and as a result it is one of the most famous pharmaceutically active compounds worldwide. Thousands of tons of ibuprofen have been synthesized around the world (Qu, 2019). The anti-inflammatory action of ibuprofen is mainly attributed the (S)-(+)-ibuprofen, which is 160 times more active than its antipode. Ibuprofen under-goes stereoselective metabolism, resulting in stereoselective pharmacokinetics parameters, with higher plasma and urinary concentrations for the (S)-(+)-isomer.

Ibuprofen and its enantiomers were also studied with a single residue method in breast milk, using vortex-assisted matrix solid-phase dispersion and direct chiral liquid chromatography, employing an experimental design to optimize the extraction parameters (León-González and Rosales-Conrado,

2017). The application of the method in real samples revealed enantiomer levels between 0.15 and 6.0 mg/kg. The novelty of that study was based (a) on the fact that ibuprofen is rarely determined in milk from lactating women—usually cow and bovine milk is monitored (Arroyo et al., 2011; Azzouz et al., 2011; Gentili et al., 2012; Ghorbani et al., 2016), (b) on the introduction of enantioselective analysis in order to distinguish (S)- and (R)-ibuprofen.

Like ibuprofen, ketoprofen is a derivative of propionic acid as well, belonging to the aryl propionic acid class of NSAIDs. The inhibitory spectrum of the drug is cyclooxygenase (COX-1 and COX-2 isoforms). This is an essential enzyme for the biosynthesis of prostaglandins. COX-1 is involved in the physiological production of prostaglandins, expressed in the kidneys and the gastrointestinal tract. COX-2 is induced in the process of inflammation. Side effects of ketoprofen include gastrointestinal bleeding, and gastrointestinal side effects mainly come from inhibition of COX-1 (Zhou et al., 2018). Ketoprofen is more specifically a COX-2 inhibitor and therefore has fewer ulcer gene effects (Al Mekhlafi et al., 2015). Ketoprofen is appropriate for treating mild to moderate pain connected with dental pain, migraine, dysmenorrhea, postoperative, and headache. Furthermore, it is also a conventional pharmaceutical that suppresses inflammation and rheumatic diseases (Guo et al., 2011). Similar is the action of naproxen, for which it is found that (S)-form is more potent than (R)-form.

Ketoprofen along with naproxen were studied in fish muscle, by means of microextraction with a supramolecular liquid made up of inverted hexagonal aggregates of decanoic acid, and enantiomeric separation by liquid chromatography onto a (R)-1-naphthylglycine and 3,5-dinitrobenzoic acid stationary phase and quantification by tandem mass spectrometry (Caballo et al., 2015). The application to real samples tested the suitability of the method. For this reason, fish belonging to five different species, namely *O. mykiss, A. alburnus, L. gibbosus, M. salmoides,* and *C. carpio*, were analyzed. No residues of NSAIDs were found, while at the same time, fortified samples from those species were spiked with the three NSAIDs, in order to calculate the enantiomeric fractions that were close to the value 0.5.

Nine non-steroidal anti-inflammatory drugs including ibuprofen, indoprofen, pranoprofen, flurbiprofen, ketoprofen, carprofen, naproxen, loxoprofen, and etodolac were investigated in fish tissues by LC–MS/MS, separated on four polysaccharide-based chiral stationary phases (Chiralpak IA, Chiralpak IB, Chiralpak IC, and Chiralpak ID) (Li et al., 2020). The aim of that work was to develop and validate a method, and not to monitor NSAIDs in many sets of samples. However, the validated method was finally applied to determine the residues of four target profens enantiomers in twenty randomly collected fish samples from a district market in China (Shenyang, Benxi, and Fuxin cities). The analysis revealed no residues of profen enantiomers in any samples (Li et al., 2020).

10.5.3 Psychoactive Drugs

There is no doubt that most veterinary drugs involved in food analysis are antibiotics. However, there are some other classes of pharmaceuticals that are underestimated while they may utterly affect human health. One of these are the class of psychoactive drugs.

It is intriguing that apart from human psychological disorders, psychotropic drugs are extensively applied in modern farming, although it is not legal. Among the various purposes of such usage, the enhancement of the feed conversion ratio by reducing animal activity in animal husbandry is listed. Moreover, some antipsychotics are given to animals such as pig, bovine, and lamb to avoid or to reduce stress during the transportation to slaughterhouse, since it is observed that this could make the meat of poor quality or even lead to premature death. According to forensic toxicological cases, many symptoms appear after the administration of antipsychotics, including excessive sedation, anesthesia, coma, overdose, dependence, and death (Zhu et al., 2019b). As a result, if an animal has been illegally administered such drugs, especially before its slaughter, it is likely that drug residues can be found in the edible animal parts and end up to humans via food chain. The worst-case scenario, if there is an extremely high concentration in the animal tissue, is to provoke side effects to humans, but this is not of high probability. However, there is a risk for consumers. Given that despite the ban of certain antipsychotics (chlorpromazine and propionylpromazine) by the joint FAO/WHO Expert Committee on Food Additives (JECFA) for three decades, in the meantime such residues have been detected in animals, the absence of antipsychotics residues in animal productions is deemed necessary to ensure food safety.

Very few studies illustrate their chiral determination in food matrices (Caballo et al., 2015; Gaugain et al., 2015; León-González and Rosales-Conrado, 2017; Li et al., 2020, 2019; Rimkus et al., 2020; Rimkus and Hoffmann, 2017; Wang et al., 2017; Yanovych et al., 2018). However, given that pork, beef, and lamb are the main meat in our daily live, residues of antipsychotics and their potential enantiomers merit investigation of any highly active, toxic, or even dangerous stereoisomers that can be harmful to consumers. Several common antipsychotics comprise chiral centers and are administrated as racemates (Nageswara Rao and Guru Prasad, 2015). The determination of chiral antipsychotics in animal foodstuff is paramount in importance about the quality control of the products, product authenticity, and safety as well as compliance with food and trade laws, while the field of the toxicity of the enantiomers is not yet well-documented and fully characterized. Up to now, there is not enough literature on the enantioselective determination of antipsychotics in meat samples. Regarding the psychoactive drugs, very few studies illustrate their chiral determination, even lesser in food matrices. However, it is proved that there are important differences between their enantiomers.

Fluoxetine is a selective serotonin reuptake inhibitor belonging to antidepressants, and it is designed to treat or to mediate major depression, compulsive, premenstrual dysphoric, panic, and post-traumatic stress disorders. It has also been used "off-label" for cataplexy, obesity, and alcohol dependence, as well as binge eating disorder (Henry et al., 2004). It is an optical active pharmaceutical used as medicine as a racemic mixture, R- and S-fluoxetine and its interactions with the serotonin-reuptake carrier and its biotransformation is highly associated with stereospecificity (Sahu et al., 2009). It is N-demethylated to R- and S-norfluoxetine, which is also pharmacologically active. The latter occurs after metabolization in the liver via N-demethylation, hence the biological activity of the prodrug is prolonged (Mandrioli et al., 2006). While R-, S-fluoxetine and S-norfluoxetine are equally potent selective serotonin reuptake inhibitors, R-norfluoxetine is 20-fold less potent. Given that it is the most prescribed antidepressant drug worldwide, it is considered a "blockbuster" drug and the "breakthrough" drug in the modern treatment of depression (Hancu et al., 2018). Yet, there is evidence in literature claiming that the serotonin-reuptake carrier is associated with stereospecific interactions. This applies also for the parent drug norfluoxetine. The enantiomers of the parent compound exhibit similar potencies as inhibitors of serotonin reuptake. Nevertheless, the potencies of the active metabolite are different, with S-norfluoxetine being the more potent one (Hancu et al., 2018). In addition, the two enantiomers show different metabolic rates (R-fluoxetine is about four times greater than that of S-fluoxetine; consequently, the half-life of S-fluoxetine is 25% that of R-fluoxetine (Fuller et al., 1992).

An interesting study by Alvim et al. (2016) presents the enantiomeric quantification of fluoxetine and norfluoxetine in human milk, by means of a two-dimensional LC system coupled to a triple quadrupole tandem mass spectrometer (2D LC/MS/MS), after direct injection. The first step of analysis aimed at the depletion of the milk proteins, with the aid of a restricted access media of bovine serum albumin C18 column. In the second step, an antibiotic-based chiral column was employed. In an older study of Kim et al. (2004) the concentration of both the parent compound and metabolite was measured, while at the same time the stereoselective disposition of fluoxetine and norfluoxetine in sheep milk was observed. At the same time, milk and serum of newborn and fetus were studied to better correlate the stereoselectivity with the animal's age. The S/R ratios in the fetus and newborn were higher than in the serum or milk of the mothers. As a result, infants and fetus are more prone to the biologically active enantiomers, particularly S-norfluoxetine. This study underlines the importance to study the stereoselective disposition of fluoxetine in pregnancy and the postpartum period in both mother and infant and in breast milk as well (Kim et al., 2006).

Fluoxetine is the most studied of the chiral psychoactive drugs in food matrices. Due to the fact that fluoxetine is also administered during pregnancy in case of depression, and afterwards to treat postpartum depression thanks to its effectiveness and lower incidence of maternal side effects (Wong et al., 1995), it has been studied in breast milk. The stereoselective disposition of fluoxetine has been studied again (Kim et al., 2006), but there is a recent interesting article about this issue. Alvim et al. describe the simultaneous enantioselective quantification of fluoxetine and its metabolite norfluoxetine in human milk by direct sample injection using two-dimensional liquid chromatography–tandem mass spectrometry (Alvim et al., 2016). This work included the development and validation of the enantioselective method as well as its application to a nursing mother who was consuming 40 mg of the drug daily. According to the authors, the results were in accordance with data for the metabolism of fluoxetine that indicate that the elimination half-lives of the enantiomers of both the precursor drug (1.1–9.5 days) and metabolite (5.5–17.4 days) are different and highly dependent on the patient's metabolism (Fjordsid et al., 1999). Consequently, it is justified that S-norfluoxetine was found at concentrations 2.5 times higher than the average concentration of the R enantiomer. Similar results were reported for the precursor drug, since the average concentration of S-fluoxetine was also 2.3 times higher than the average concentration of R-fluoxetine. Comparing the concentration of the S-metabolite with the S precursor, a ratio of 1.4 was calculated.

A similar observational study aimed at the simultaneous, stereoselective quantification of the antidepressant citalopram and its active metabolite desmethylcitalopram in breast milk and human plasma (Weisskopf et al., 2016). No matter that the latter is not a food matrix, we need to closely observe the results, since there is an interesting conclusion by the authors. The measured concentrations in hindmilk were 39.0 ng/mL for S-(+)-citalopram, 49.1 ng/mL for R-(−)-citalopram, 13.2 ng/mL for S-(+)- desmethylcitalopram, and 12.4 ng/mL for R-(−)- desmethylcitalopram. On the one hand, the respective concentrations for plasma were 22.2 ng/mL, 25.5 ng/mL, 6.7 ng/mL, and 6.5 ng/mL. From the interpretation of the results, it is concluded that in both matrices, R-(−)-citalopram was found at higher concentrations than the S-(+) enantiomer, but not for its metabolite, S/R=0.8–0.9 in breast milk for citalopram, while 1.0–1.1 for desmethylcitalopram. In addition, milk levels of substances were higher than in plasma, suggesting a possible drug accumulation in breast milk.

Apart from the above-mentioned studies for fluoxetine and citalopram, there is an innovative article of 2019, presenting a multi residue method for enantioselective determination of seven psychoactive drugs enantiomers in multi-species animal tissues, including pork, beef, and lamb muscles (Zhu et al., 2019b). The drugs were R/S(±)-cyamemazine, R/S(±)-promethazine, R/S(±)-fluoxetine, R/S(±)-oxazepam, R/S(±)-mirtazapine, R/S(±)-bupropion, and R/S(±)-phencynonate. After validation the method was applied to 30 meat samples (leg of the animals) bought from different local markets in China. Meat samples (a portion from the leg of slaughtered animals) were analyzed by chiral LC–MS/MS and the results showed that no antipsychotic was detected in any of the animal tissue samples.

10.5.4 Anthelmintic Drugs

Anthelmintic drugs are of pivotal importance for both human tropical medicine and for veterinary medicine. They are designated to treat infections with parasitic worms (flukes, tapeworms, worms). According to WHO (2004) 2 billion people are hosts of parasitic worm infections. Livestock and crops are not insusceptible in parasites, and consequently their infections may affect produce. The infection of domestic pets is of similar importance.

Praziquantel is a chiral pyrazino-isoquinoline compound that serves as a broad-spectrum anthelmintic drug, administered for the prophylaxis and treatment of trematode and cestode infections, mainly in food-producing animals. Praziquantel, which is extensively used in aquaculture, is administered as a racemate, but its enantiomers show different pharmacodynamics, pharmacokinetics, and toxicities, as applies for most chiral drugs. As far the functions of the two enantiomers, the R-praziquantel is used to treat schistosomiasis, while the S-form causes the bitter taste of the racemate (Meister et al., 2016).

Despite its huge necessity, praziquantel is rarely studied, especially its enantiomeric forms. Such a work has been published by Zhang et al. (2019), discussing the selective depletion of the chiral 4-hydroxypraziquantel and its metabolites in three types of aquaculture fish, namely perch, tilapia, and rice field eel. The interesting findings revealed (a) species-specific differences in the metabolism of the drug isomers, and (b) new metabolites, as 3-hydroxypraziquantel diastereomers (Zhang et al., 2019).

10.5.5 Other Drugs

Apart from antibiotics, NSAIDs, psychoactive drugs, as well as anthelmintic drugs that are the main classes of pharmaceuticals whose chirality exhibits relevance in food analysis, clenbuterol is extensively studied, even if it does not belong to the above-mentioned classes. To the best of the authors' knowledge, clenbuterol is the only representative of other pharmaceutical classes for which chirality and food safety has been connected.

Clenbuterol is a potent chiral long-acting β2-sympathomimetic drug. All currently approved pharmaceuticals contain the racemic mixture of R-(–) and S-(+)-clenbuterol. It has been proven that R-(-)-clenbuterol form is active at the b2-adrenergic receptor, while the S-(+)-clenbuterol form is inactive. When used legally, clebuterol is applied to treat bronchial asthma and pulmonary disease, or as tocolytic/mucolytic agent in the symptomatic treatment of respiratory diseases in both humans and animals (Wang et al., 2016). However, it can induce the muscular tissue growth and body fat loss, and as a result, its illicit usage as a nutrient-repartitioning agent in meat-producing animals has been observed (Rosales-Conrado et al., 2015), despite the ban by the World Anti-Doping Agency due to anabolic action. In fact, sometimes it is used even 5–10 times higher than recommended in the therapeutic dose to accelerate animal's growth and increase the famers' benefits (Wang et al., 2016). Eventually, the presence and potential accumulation of clenbuterol in foodstuffs can have a negative impact on human health, if "contaminated" meat products and liver are consumed (Ramos et al., 2011). It is reported that the accumulation of clenbuterol in edible tissues of swine is enriched enantiomerically towards the dextrorotatory (+)-clenbuterol, the S-(+)-clenbuterol is enriched in tissues while R-(-)-clenbuterol is depleted (Wang et al., 2016). In addition, it is also misused by humans for muscle building and weight-loss purposes, found as ingredient of unapproved and black market (Parr et al., 2017).

An LC–MS/MS method for the determination of residual clenbuterol enantiomers in swine, beef, and lamb meat was developed by Wang et al. (2016), hoping to be the starting point for future studies to monitor the enantiomeric residues in livestock production and exploit the calculation of the difference in the R/S ratio to ensure food safety and enhance doping control as well. To be more specific, it was found that the time course of the drug administration to the animal can affect the ratio of clenbuterol enantiomers in edible tissue. Furthermore, enantiomers were depleted significantly in swine, beef, and lamb. Whilst the S-(+)-clenbuterol accumulation in swine muscle and liver was time-dependent, the R-(-)-clenbuterol accumulated in beef and lamb muscle wasn't.

Another work described a controlled administration trial for the distinction of clenbuterol intake from drug or contaminated food of animal origin, carrying out an enantiomeric separation for doping control purposes (Parr et al., 2017). However, the findings are relevant since "contaminated" meat intended to be consumed by athletes (volunteers) was analyzed. More specifically, cattle meat samples were analyzed for their enantiomeric composition. For the trial needs, meat samples contained clenbuterol at concentrations 0.04–2.82 μg/kg. R-(–)-clenbuterol was mainly found enriched in cattle meat resulting in proportions of S-clenbuterol (pS) = 0.509 ± 0.004 (total amount of clenbuterol 1.67 μg/kg) and pS = 0.478 ± 0.001 (total amount of clenbuterol 1.93 μg/kg). The liver of these animals was found enriched in S-clenbuterol. In addition, trial meatballs were analyzed, and the calculated concentration of the total drug was 2.38 μg/kg.

Clenbuterol was also the objective of a study by Rosales-Conrado (Rosales-Conrado et al., 2015) who developed and validated an analytical method for the chiral determination of clenbuterol in animal feed, albeit not applied in real samples. Good recoveries (64 to 101%) were achieved for both enantiomers for the spiked feed samples. In addition, differences between the recoveries obtained for each clenbuterol enantiomer were also evident, suggesting a different enantiomer behavior and interaction with the feed matrix.

10.6 HUMAN RISK

The potential human risk associated with human exposure to chiral pharmaceuticals is an issue of concern, although it is considered to be relatively low. Up to now, there is a lack of information about the potential risk of exposure to chiral drugs or their mixtures and consequently it is uncertain if there are any long-term risks from the low-level exposure. The existing data are limited, and it is supposed that allows representative estimations (Zhou et al., 2018). The main limitations in the study of human risk are the variability in the applied methods as well as the lack of studies conducted in real, field conditions (Calderón-Preciado et al., 2013). According to toxic studies in non-target organisms, there is evidence that certain diseases, including cancer, may be correlated to exposure to chiral pharmaceuticals. Therefore, the long-term risk may vary from negligible to potentially carcinogenic or with lethal effects. Enantioselectivity also affects the potency of different drugs on the human body, e.g., the (R)-(+)-thalidomide is responsible for sedative effects, while the (S)-(–)enantiomer and its derivatives are teratogenic (Tokunaga et al., 2018).

It is already discussed that the main sources of chiral pharmaceuticals are from industrial processes, such as the discharge of raw or insufficiently treated wastewaters containing chiral pharmaceuticals, incorrect storage and use, animal waste, and biosolids. The exposure of animals to various chiral and achiral pharmaceuticals may result in poisoning, low fertility, or even worse, alteration in gene expression. Human risk is entailed by exposure to such chiral pharmaceuticals via food chain and drinking water, apart from disease treatment process, skin contact, and inhalation (Rather et al., 2017).

Regarding veterinary drugs, human exposure occurs through several routes, namely, the consumption of crops produced in "contaminated" soil which accumulated substances as a result of exposure to manure and slurry; consumption of food of animal origin that have accumulated veterinary medicines via the food chain web or consumption of aquaculture fish exposed to treatments; abstracted groundwater and surface waters containing veterinary medicines. Rarely, humans are exposed to veterinary drugs through the inhalation of dust emitted from intensively reared livestock facilities or as a result of contact with contaminated fleece from treated sheep (Boxall et al., 2006).

Although veterinary medicines are routinely monitored in food materials from treated animals to ensure that concentrations are below the maximum residue limits, the magnitude of the exposure via many of the routes listed and the health impacts of such exposure have not been extensively quantified. Recent work has assessed the potential risks arising from exposure to veterinary medicines in fish and water. Studies have demonstrated the presence of medicines in water bodies (Boxall et al., 2004; Charuaud et al., 2019) and the accumulation of veterinary medicines from surface waters by fish, shellfish, and crustacea (Boxall et al., 2004; Hoff et al., 2016). Figure 10.3 summarizes the possible routes that pharmaceuticals and veterinary drugs follow before ending up to humans.

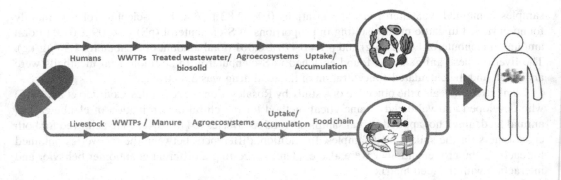

FIGURE 10.3 Human and veterinary pharmaceuticals: the entry into food chain web and human uptake.

Specific classes that may be significantly harmful to human health through the enrichment of the food chain are some beta-blockers, NSAIDs, antibiotics, and psychotropic substances. Some typical examples of such "suspect" compounds were compiled and are discussed herein.

The concentration of sulfamethoxazole and trimethoprim follows the trophic magnification, since it is associated with trophic levels. They show trophic magnification factors (TMF) 2.19 and 2.40, respectively (Liu et al., 2017). The HQ value of enrofloxacin (HQ=0.07) in seafood (2014 and 2015) in the Laizhou Bay exceeds the standard of obvious risk (HQ > 0.05). The HQ values of ciprofloxacin, clarithromycin, and enrofloxacin were 0.01, 0.01, and 0.04 in urban residents while in rural residents are higher than the considerable risk (HQ≥0.01) (Liu et al., 2017).

As implied by their name, beta-blockers act thanks to the competitive inhibition of β-adrenergic receptors (Kasprzyk-Hordern, 2010), slowing down the heart rate, decreasing the blood pressure and smoothing the contracted muscle. Many of the β-blockers are available in the market as racemates (Agustian et al., 2010), while some of the exceptions are penbutolol, timolol, and propranolol, for which the single l-isomer is available. The main representative of this significant pharmaceutical class are atenolol, propranolol, and metoprolol. Comprising the main agents for the treatment of cardiovascular diseases, they are widely prescribed and consumed globally (Ali et al., 2017), to treat cardiovascular diseases, such as hypertension, angina, arrhythmias, and glaucoma, coronary heart disease, while there is proof for their effectiveness against anxiety and neurosis (Zhou et al., 2018). Among the health effects on humans, disturbed peripheral circulations and bronchial contractions have been reported. If combined with the β-adrenergic receptor, they are highly stereoselective in the body. Unfortunately, there is a lack of studies issuing directly the occurrence of b-blocker in food chain.

The enantiomers of another commonly used b-blocker, metoprolol, have also different potency in the human body. The (S)- form is found to be more than 35 times more potent than the (R)- form (Patel et al., 2019). Continuing with b-blockers, sotalol has also (R)- and (S)± enantiomers, and it is reported that the (R)- form possesses most of the pharmacological b-blocking activity. D-sotalol caused mortality in patients with left ventricular dysfunction after recent and remote myocardial infarction. Propranolol, maybe the most famous of the chiral b-blockers, has (S)(-) and (R)-(+) enantiomers. The first is 100 times more potent than the latter. Propranolol is one of the most prescribed β-blockers worldwide, thus it is frequently detected as a toxic pollutant in the environment and thus in food chain (Azzouz and Ballesteros, 2012). Initially, propranolol was designed to treat human diseases, but there is growing concern that it poses threats to aquatic organisms (Ding et al., 2015). Likewise, it may cause adverse effects in humans, including infant hemangioma, apoptosis induction, insomnia, and nightmares (Ali et al., 2017). Now it is thought to be one of the most toxic drugs, if not the most toxic one, in terms of chronic toxicity, since the acute toxicity for the class of b-blockers is not extensively studied so far (Kasprzyk-Hordern, 2010). It is proved that the S-(–)-propranolol is 100 times more active than its R-(+)-enantiomer (Nguyen et al., 2006).

Atenolol is another chiral b-blocker, which is a globally prescribed b-blocker used to treat human cardiovascular diseases, including hypertension, arrythmia, angina, and other ischaemic heart diseases (Tsujimoto et al., 2017). It is well-documented that the activity of atenolol is primarily related to its S-enantiomers (Mehvar and Brocks, 2001; Sanganyado et al., 2017). Although the toxicity of atenolol itself is negligible, in the presence of other β-blockers synergistic effects may occur (Zhou et al., 2018). Atenolol is considered as a marker of efficient treatment of sewage treatment plants and endocrine disrupting compounds (Sanganyado et al., 2017), since it is excreted nearly unchanged, ending up in WWTPs.

Regarding the NSAIDs, for ibuprofen it is reported that the (S)-ibuprofen inhibits COX1 and COX2 equally. In addition, the (R)-enantiomer weakly inhibits COX1 and has no effects on COX2. (R)-ibuprofenoyl-CoA, the (R)-metabolite of ibuprofen, exerts greater COX2 inhibition. In general, the (S)- form is >100 times more potent than the (R)- form. It is reported that side effects of NSAIDs include severe problems in kidney (Ungprasert et al., 2015). A more specific human risk includes problems in bone healing, especially after bone-related operations (Marongiu et al., 2020).

10.7 ENANTIOSPECIFIC TOXICITY

As it is obvious from the evidence presented in the previous section, human risk associated with the exposure to chiral compounds has not been studied extensively. Sometimes, the study of toxicity in marker organisms is helpful for the comprehension of the mechanisms that exert human risk. Studies on enantiospecific toxicity of chiral pharmaceuticals are rare, despite their already known and well-studied pharmacological properties and the fact that there is evidence that chiral pharmaceuticals pose enantiospecific toxicity on different organisms (Patel et al., 2019). There are significant differences between the toxicological response and pharmaceuticals' potency of enantiomers in humans. In general, most of the toxicological studies of chiral pharmaceuticals employed racemates, confirming the urgent need for new research on this topic that will be laborious. In order to have a holistic image of the potential stereoselective toxicity, extensive data on transport, fate, persistence, and bioaccumulation of the enantiomers should precede.

An example of such study about the stereospecific toxicity of atenolol revealed that (S)-atenolol inhibited the growth of *Daphnia magna* more in comparison with the (R)-atenolol (de Andrés et al., 2009). Observations in the toxicological responses of the chiral enantiomers of propranolol showed higher chronic toxicity of the (S)- form in *P. promelas*, while this was equal for *D. magna*. Regarding the acute toxicity in *D. magna*, similar acute response was observed for both species. Studies indicate that propranolol is detected in the fish's reproductive tissue, liver, and heart, causing chronic toxicity in both the reproductive and cardiovascular systems (Kasprzyk-Hordern, 2010). According to a number of excellent studies, it is found that propranolol has a negative impact on hatchability, heart rate, embryo-larval development, and gene expression in fish (Ding et al., 2015).

Likewise, the enantiomers of fluoxetine caused different toxicities in *Daphnia magna* exposed to them (de Andrés et al., 2009; Andrés-Costa et al., 2017). A chronic toxicity study for fluoxetine demonstrated that it caused sublethal and behavioral effects stereoselectively on *Pimephales promelas*, with (S)-fluoxetine more adversely affecting feeding behavior and growth than (R)-fluoxetine (Stanley et al., 2007).

Diclofenac causes severe visceral gout and renal failure, leading to high mortality (5–86%) on oriental white-backed vulture adults and subadults in the Indian subcontinent. This indirect effect has caused a 95% population decline. Diclofenac bioaccumulates in vultures when they feed on dead livestock treated with diclofenac. *Gammarus pulex*, a European fresh water shrimp exhibits a clear feeding preference for leaves that were not conditioned with sulfadiazine and oxytetracycline (Patel et al., 2019). For ketoprofen, a toxicity evaluation has been carried out for racemic ketoprofen and its enantiomer S(+)-ketoprofen (dexketoprofen) for both acute and chronic toxicity tests using 3 representative model organisms (*Vibrio fischeri, Pseudokirchneriella subcapitata*, and

Ceriodaphnia dubia). Toxicity data from dexketoprofen exposure revealed higher sensitivity through inhibition of bioluminescence and algal growth and through increased mortality/immobilization compared to racemic ketoprofen exposure. The growth inhibition test showed that racemic ketoprofen and dexketoprofen exhibited different effect concentration values (240.2 and 65.6 mg/L, respectively) (Mennillo et al., 2018). Induced bacterial toxicity was observed when the enantiomers of naproxen, ibuprofen, ketoprofen, and flurbiprofen was evaluated in bioassays with bacteria, algae, and fish cells. (R)-naproxen was more toxic than (S)-naproxen (EC50 0.75 vs 0.93 mg/L) and (S)-flurbiprofen more toxic than (R)-flurbiprofen (EC_{50} 1.22 vs 2.13 mg/L). Both (R)-flurbiprofen and (S)-flurbiprofen induced photosystem II inhibition in green algae, with (R)-flurbiprofen having a greater effect in the assay after 24 h (EC_{10} 5.47 vs 9.07 mg/L). Only the (R)-enantiomers of flurbiprofen and ketoprofen induced ethoxyresorufin-O-deethylase (EROD) activity in fish cells, while (S)-naproxen was 2.5 times more active than (R)-naproxen in the EROD assay. These findings were proof that (R)-enantiomer of the commonly used NSAID ketoprofen is imperative as it was at least six times more potent in the EROD assay than the inactive (S)-ketoprofen (Neale et al., 2019). Such data could be of high value to improve the prediction models for human risk. Most test organisms are aquatic (biota, invertebrates, etc.), and the concentrations in aquatic bodies are in almost all cases below the aquatic toxicity levels. Therefore, it arises that the study of chronic toxicity is more urgent to be investigated.

For sulfamethazine, which is one of the most common antibiotics used both for human and veterinary purposes, it is reported that it can suppress epigenetic silencing in *Arabidopsis thaliana* mutant, where the transgenes are transcriptionally silenced, by impairing folate synthesis (Zhang et al., 2012). According to Zhang et al., 2012 sulfamethazine was released RdDm (RNA-directed DNA methylation)-dependent and -independent transcriptional gene silencing (TGS) of two transgenes and the silencing of endogenous loci. Sulfamethazine causes methyl deficiency by decreasing the plant folate pool size which leads to a reduction in DNA methylation and repressive histone mark (H3K9me2).

Lastly, future toxicological studies of chiral drugs should focus on any molecular interactions of enantiomers with the primary, because they may have additive, antagonistic, or synergistic action and should be treated as mixtures and not individually (Küster and Adler, 2014). Since pharmacological data often include biological activities of the enantiomers, such data could be used to predict enantiospecific toxicity.

10.8 CONCLUDING REMARKS AND FUTURE CHALLENGES

Despite the great importance and massive study of emerging contaminants in food, among other matrices, it is obvious that stereochemistry is frequently underestimated in food analysis because of the misconception that enantiomers are a unique molecular entity. Several enantioselective methods and biodegradation experiments have been developed during the last decades, regarding mainly pesticides and more recently pharmaceuticals, illicit drugs, personal care products, industrial chemicals, surfactants, etc., while this field is yet to be enriched. The prospect of the current chapter was to shed light on the state-of-the-art of the sources, human risk, and future challenges of chiral pharmaceuticals in foodstuffs, presenting data derived from the period 2015–2020. The main sources of chiral pharmaceuticals are discussed, together with the potential human risk, while the importance of studying such contaminants in food is highlighted in the future challenges. Many articles are involved with the analysis, occurrence, and human risk of chiral emerging contaminants in food, discussing also any practical limitations or obstacles mainly because of demanding separation and determination techniques. However, it is obvious that the issue of chiral ECs in food is a challenging issue for further studies and further work is needed in the search of more advantageous techniques and more in-depth studies of human risk. A major concern is that stereoselectivity can occur during the biotransformation pathways of enantiomers, hence there is a potential bioaccumulation in the food web. This can be proved by the relative accumulation of the enantiomers in biota and variations in EF.

Despite the above concluding remarks, there are some recommendations and challenges which should be applied in the future studies. In general, promising research in the field of chiral pharmaceuticals in food analysis could include various aspects. Undoubtedly, any recommendation demands comprehensive knowledge and expertise on the topic of chirality, which is characterized by diversity and interdisciplinarity, the key feature for efficient and conclusive research.

It is of primary importance to anticipate that it is not only (chiral) pesticides residues that worsen food quality, and as such entail human risks; hence the focus of attention should be allotted to several contaminants. First, a "blacklist" of chiral pharmaceuticals and other emerging contaminants along with their TPs should be established. The included substances should be further prioritized according to the extent that they affect food quality and consumer health. This could be the first step to formulate a series of standards to study further the risk posed by such contaminants.

To this point, it is critical to invest in high-resolution mass spectrometry techniques which enable ultra-trace analysis in matrices such as foodstuff; conventional instrumentation is rather unreliable to discriminate ultra-low concentrations of chiral compounds in too complex matrices. Moreover, special attention should be paid on sample pretreatment of chiral pharmaceuticals. Pretreatment is not only the first but also the determining step of analysis, it can entail fatal errors that unintentionally forge the analysis and lead to false positive or false negative results. For instance, it is necessary to execute preliminary experiments investigating the conditions of sample pretreatment that affect the behavior of chiral substances, such as pH, extraction time, operating temperatures, etc. Besides, the complexity of food as a matrix in combination with the complex interaction of chiral drugs with such samples demand detailed research of enantioselectivity and the potential effects to non-target organisms (animals, plants) involved.

Like for the achiral pharmaceuticals, there is an urgent need to seek remediation technologies in order to accelerate the repairing process of chiral pharmaceuticals occurrence in food chain. The principles of those methods should be based on environmental remediation technologies, either biological (microbial participation, etc.) that are demanding and time consuming, or non-biological (sonolysis, photocatalysis, advance oxidation processes, etc.). Both approaches require specific and expensive equipment, without promising at the same time a flawless remediation. For instance, all these processes may generate metabolites and transformation products that are eventually more toxic and resistant to biotransformation, not to mention the difficulty in their handling due to their unknown behavior.

Another challenge that should be the next trend in chirality study of pharmaceuticals is associated with the study of enantioselectivity involved with the formation and degradation of chiral metabolites and TPs of chiral compounds. In particular, it is highly challenging and demanding to study and evaluate any inherent enantioselectivity of metabolites and TPs, apart from that of the chiral parent compounds themselves. In addition, the potential human risk posed by that phenomenon merit research and further assessment.

The comprehensive study of stereoselectivity and to what extent it affects the degradation and transformation of emerging contaminants in food can be a useful tool to enhance the human risk assessment of those substances, since their toxicological effects are still unknown and unexpected threats can be revealed. Focusing on chiral transformation products of pesticides, the understanding of enantioselective processes in food matrices could contribute to the evolution of agrochemical industry and redefine the massive production of chiral substances applied to enantiomeric formulations. This could result in the reduction of the pollutant load of racemic mixtures into the agriculture and hence agricultural products.

ACKNOWLEDGMENTS

This research is co-financed by Greece and the European Union (European Social Fund-ESF) through the Operational Programme "Human Resources Development, Education and Lifelong

Learning" in the context of the project "Reinforcement of Postdoctoral Researchers—2nd Cycle" (MIS-5033021), implemented by the State Scholarships Foundation (IKY).

Operational Programme
Human Resources Development,
Education and Lifelong Learning
Co-financed by Greece and the European Union

REFERENCES

Agustian, J., Kamaruddin, A.H., Bhatia, S., 2010. Single enantiomeric β-blockers—The existing technologies. *Process Biochem.* 45(10), 1587–1604. https://doi.org/10.1016/j.procbio.2010.06.022

Al Mekhlafi, S., Alkadi, H., Kotb, M.-I., Sayed, E.-, 2015. Synthesis and anti-inflammatory activity of novel Ketoprofen and Ibuprofen derivatives. *J. Chem. Pharm. Res.* 7, 503–510.

Ali, I., Alothman, Z. A., Alwarthan, A., 2017. Uptake of propranolol on ionic liquid iron nanocomposite adsorbent: Kinetic, thermodynamics and mechanism of adsorption. *J. Mol. Liq.* 236, 205–213. https://doi.org/10.1016/j.molliq.2017.04.028

Alvarez-Rivera, G., Bueno, M., Ballesteros-Vivas, D., Cifuentes, A., 2020. Chiral analysis in food science. *Trends Anal. Chem.* 123. https://doi.org/10.1016/j.trac.2019.115761

Alvim, J., Lopes, B. R., Cass, Q. B., 2016. Simultaneous enantioselective quantification of fluoxetine and norfluoxetine in human milk by direct sample injection using 2-dimensional liquid chromatography-tandem mass spectrometry. *J. Chromatogr. A* 1451, 120–126. https://doi.org/10.1016/j.chroma.2016.05.022

Andrés-Costa, M. J., Proctor, K., Sabatini, M. T., Gee, A. P., Lewis, S. E., Pico, Y., Kasprzyk-Hordern, B., 2017. Enantioselective transformation of fluoxetine in water and its ecotoxicological relevance. *Sci. Rep.* 7. https://doi.org/10.1038/s41598-017-15585-1

Arroyo, D., Ortiz, M. C., Sarabia, L. A., 2011. Optimization of the derivatization reaction and the solid-phase microextraction conditions using a D-optimal design and three-way calibration in the determination of non-steroidal anti-inflammatory drugs in bovine milk by gas chromatography-mass spectrometry. *J. Chromatogr. A* 1218, 4487–4497. https://doi.org/10.1016/j.chroma.2011.05.010

Attiq, A., Jalil, J., Husain, K., Ahmad, W., 2018. Raging the war against inflammation with natural products. *Front. Pharmacol.* https://doi.org/10.3389/fphar.2018.00976

Azzouz, A., Ballesteros, E., 2012. Combined microwave-assisted extraction and continuous solid-phase extraction prior to gas chromatography–mass spectrometry determination of pharmaceuticals, personal care products and hormones in soils, sediments and sludge. *Sci. Total Environ.* 419, 208–215. https://doi.org/10.1016/j.scitotenv.2011.12.058

Azzouz, A., Jurado-Sánchez, B., Souhail, B., Ballesteros, E., 2011. Simultaneous determination of 20 pharmacologically active substances in cow's milk, goat's milk, and human breast milk by gas chromatography-mass spectrometry. *J. Agric. Food Chem.* 59, 5125–5132. https://doi.org/10.1021/jf200364w

Bamidele Falowo, A., Festus Akimoladun, O., 2020. Veterinary drug residues in meat and meat products: Occurrence, detection and implications, in: *Veterinary Medicine and Pharmaceuticals.* IntechOpen. https://doi.org/10.5772/intechopen.83616

Bharucha, Z., Pretty, J., 2010. The roles and values of wild foods in agricultural systems. *Philos. Trans. R. Soc. B Biol. Sci.* https://doi.org/10.1098/rstb.2010.0123

Boxall, A. B., Fogg, L. A., Blackwell, P. A., Kay, P., Pemberton, E. J., Croxford, A., 2004. Veterinary medicines in the environment. *Rev. Environ. Contam. Toxicol.* 180.

Boxall, A. B. A., Johnson, P., Smith, E. J., Sinclair, C. J., Stutt, E., Levy, L. S., 2006. Uptake of Veterinary Medicines from Soils into Plants. *J. Agric. Food Chem.* 54, 2288–2297. https://doi.org/10.1021/jf053041t

Buchberger, W. W., 2011. Current approaches to trace analysis of pharmaceuticals and personal care products in the environment. *J. Chromatogr. A* 1218, 603–18. https://doi.org/10.1016/j.chroma.2010.10.040

Caballo, C., Sicilia, M. D., Rubio, S., 2015. Enantioselective analysis of non-steroidal anti-inflammatory drugs in freshwater fish based on microextraction with a supramolecular liquid and chiral liquid chromatography-tandem mass spectrometry. *Anal. Bioanal. Chem.* 407, 4721–4731. https://doi.org/10.1007/s00216-015-8675-5

Calcaterra, A., D'Acquarica, I., 2018. The market of chiral drugs: Chiral switches versus de novo enantiomerically pure compounds. *J. Pharm. Biomed. Anal.* 147, 323–340. https://doi.org/10.1016/j.jpba.2017.07.008

Calderón-Preciado, D., Matamoros, V., Savé, R., Muñoz, P., Biel, C., Bayona, J. M., 2013. Uptake of microcontaminants by crops irrigated with reclaimed water and groundwater under real field greenhouse conditions. *Environ. Sci. Pollut. Res.* 20, 3629–3638. https://doi.org/10.1007/s11356-013-1509-0

Camacho-Muñoz, D., Petrie, B., Lopardo, L., Proctor, K., Rice, J., Youdan, J., Barden, R., Kasprzyk-Hordern, B., 2019. Stereoisomeric profiling of chiral pharmaceutically active compounds in wastewaters and the receiving environment – A catchment-scale and a laboratory study. *Environ. Int.* 127, 558–572. https://doi.org/10.1016/j.envint.2019.03.050

Carlsson, P., Herzke, D., Kallenborn, R., 2014. Enantiomer-selective and quantitative trace analysis of selected persistent organic pollutants (pop) in traditional food from Western Greenland, in: *Journal of Toxicology and Environmental Health - Part A: Current Issues.* Taylor and Francis Inc., pp. 616–627. https://doi.org/10.1080/15287394.2014.887425

Carter, L. J., Harris, E., Williams, M., Ryan, J. J., Kookana, R. S., Boxall, A. B. A., 2014. Fate and uptake of pharmaceuticals in soil-plant systems. *J. Agric. Food Chem.* 62, 816–825. https://doi.org/10.1021/jf404282y

Carter, L. J., Williams, M., Martin, S., Kamaludeen, S. P. B., Kookana, R. S., 2018. Sorption, plant uptake and metabolism of benzodiazepines. *Sci. Total Environ.* 628–629, 18–25. https://doi.org/10.1016/j.scitotenv.2018.01.337

Charuaud, L., Jarde, E., Jaffrezic, A., Thomas, M. F., Le Bot, B., 2019. Veterinary pharmaceutical residues from natural water to tap water: Sales, occurrence and fate. *J. Hazard. Mater.* 361, 169–186. https://doi.org/10.1016/j.jhazmat.2018.08.075

Chitescu, C. L., Nicolau, A. I., Stolker, A. A. M., 2013. Uptake of oxytetracycline, sulfamethoxazole and ketoconazole from fertilised soils by plants. *Food Addit. Contam. Part A* 30, 1138–1146. https://doi.org/10.1080/19440049.2012.725479

Chopra, I., Roberts, M., 2001. Tetracycline antibiotics: Mode of action, applications, molecular biology, and epidemiology of bacterial resistance. *Microbiol. Mol. Biol. Rev.* 65, 232–260. https://doi.org/10.1128/mmbr.65.2.232-260.2001

Christou, A., Agüera, A., Bayona, J. M., Cytryn, E., Fotopoulos, V., Lambropoulou, D., Manaia, C. M., Michael, C., Revitt, M., Schröder, P., Fatta-Kassinos, D., 2017. The potential implications of reclaimed wastewater reuse for irrigation on the agricultural environment: The knowns and unknowns of the fate of antibiotics and antibiotic resistant bacteria and resistance genes – A review. *Water Res.* https://doi.org/10.1016/j.watres.2017.07.004

Chuang, Y. H., Liu, C. H., Sallach, J. B., Hammerschmidt, R., Zhang, W., Boyd, S. A., Li, H., 2019. Mechanistic study on uptake and transport of pharmaceuticals in lettuce from water. *Environ. Int.* 131, 104976. https://doi.org/10.1016/j.envint.2019.104976

Chung, H. S., Lee, Y.-J., Rahman, M. M., Abd El-Aty, A. M., Lee, H. S., Kabir, M. H., Kim, S. W., Park, B.-J., Kim, J.-E., Hacımüftüoğlu, F., Nahar, N., Shin, H.-C., Shim, J.-H., 2017. Uptake of the veterinary antibiotics chlortetracycline, enrofloxacin, and sulphathiazole from soil by radish. *Sci. Total Environ.* 605–606, 322–331. https://doi.org/10.1016/J.SCITOTENV.2017.06.231

Colon, B., Toor, G. S., 2016. A Review of uptake and translocation of pharmaceuticals and personal care products by food crops irrigated with treated wastewater. *Adv. Agron.* 140, 75–100. https://doi.org/10.1016/bs.agron.2016.07.001

D'Orazio, G., 2020. Chiral analysis by nano-liquid chromatography. *Trends Anal. Chem.* 125. https://doi.org/10.1016/j.trac.2020.115832

D'Orazio, G., Fanali, C., Asensio-Ramos, M., Fanali, S., 2017. Chiral separations in food analysis. *TrAC - Trends Anal. Chem.* 96, 151–171. https://doi.org/10.1016/j.trac.2017.05.013

Daughton, C. G., Ternes, T. A., 1999. Pharmaceuticals and personal care products in the environment: Agents of subtle change? *Environ. Health Perspect.* 107(Suppl 6), 907–938. https://doi.org/10.2307/3434573

de Andrés, F., Castañeda, G., Ríos, Á., 2009. Use of toxicity assays for enantiomeric discrimination of pharmaceutical substances. *Chirality.* 21, 751–759. https://doi.org/10.1002/chir.20675

Ding, J., Lu, G., Li, S., Nie, Y., Liu, J., 2015. Biological fate and effects of propranolol in an experimental aquatic food chain. *Sci. Total Environ.* 532, 31–39. https://doi.org/10.1016/j.scitotenv.2015.06.002

Dodgen, L. K., Ueda, A., Wu, X., Parker, D. R., Gan, J., 2015. Effect of transpiration on plant accumulation and translocation of PPCP/EDCs. *Environ. Pollut.* 198, 144–153. https://doi.org/10.1016/j.envpol.2015.01.002

European Commission, 2010. Commission Regulation (EU) 37/2010 of 22 December 2009 on pharmacologically active substances and their classification regarding maximum residue limits in foodstuffs of animal origin. *Off. J. Eur. Union L15*, 1–72. https://doi.org/2004R0726 - v.7 of 05.06.2013

Fanali, C., D'Orazio, G., Gentili, A., Fanali, S., 2019. Analysis of enantiomers in products of food interest. *Molecules*. 24. https://doi.org/10.3390/molecules24061119

FAO, 2008. Biological Hazards in Animal Feed 10–12.

Fatta-Kassinos, D., Kalavrouziotis, I. K., Koukoulakis, P. H., Vasquez, M. I., 2011. The risks associated with wastewater reuse and xenobiotics in the agroecological environment. *Sci. Total Environ.* 409. https://doi.org/10.1016/j.scitotenv.2010.03.036

Fjordsid, L., Jeppesen, U., Eap, C. B., Powell, K., Baumann, P., Brøsen, K., 1999. The stereoselective metabolism of fluoxetine in poor and extensive metabolizers of sparteine. *Pharmacogenetics* 9, 55–60. https://doi.org/10.1097/00008571-199902000-00008

Fu, Q., Malchi, T., Carter, L. J., Li, H., Gan, J., Chefetz, B., 2019. Pharmaceutical an personal care products: From wastewater treatment into Agro-Foo systems. *Environ. Sci. Technol.* 53, 14083–14090. https://doi.org/10.1021/acs.est.9b06206

Fuller, R. W., Snoddy, H. D., Rushinski, J. H., et al., 1992. Comparison of norfluoxetine enantiomers as serotonin uptake inhibitors in vivo. *Neuropharmacology* 31:997–1000. https://doi.org/10.1016/0028-3908(92)90100-4

Gaugain, M., Gautier, S., Bourcier, S., Jacques, A. M., Laurentie, M., Abjean, J. P., Hurtaud-Pessel, D., Verdon, E., 2015. 6-Iso-chlortetracycline or keto form of chlortetracycline? Need for clarification for relevant monitoring of chlortetracycline residues in food. *Food Addit. Contam. - Part A Chem. Anal. Control. Expo. Risk Assess.* 32, 1105–1115. https://doi.org/10.1080/19440049.2015.1036381

Gentili, A., Caretti, F., Bellante, S., Mainero Rocca, L., Curini, R., Venditti, A., 2012. Development and validation of two multiresidue liquid chromatography tandem mass spectrometry methods based on a versatile extraction procedure for isolating non-steroidal anti-inflammatory drugs from bovine milk and muscle tissue. *Anal. Bioanal. Chem.* 404, 1375–1388. https://doi.org/10.1007/s00216-012-6231-0

Ghorbani, M., Chamsaz, M., Rounaghi, G. H., 2016. Ultrasound-assisted magnetic dispersive solid-phase microextraction: A novel approach for the rapid and efficient microextraction of naproxen and ibuprofen employing experimental design with high-performance liquid chromatography. *J. Sep. Sci.* 39, 1082–1089. https://doi.org/10.1002/jssc.201501246

Grossberger, A., Hadar, Y., Borch, T., Chefetz, B., 2014. Biodegradability of pharmaceutical compounds in agricultural soils irrigated with treated wastewater. *Environ. Pollut.* 185, 168–177. https://doi.org/10.1016/j.envpol.2013.10.038

Guidi, L. R., Santos, F. A., Ribeiro, A. C. S. R., Fernandes, C., Silva, L. H. M., Gloria, M. B. A., 2018. Quinolones and tetracyclines in aquaculture fish by a simple and rapid LC-MS/MS method. *Food Chem.* 245, 1232–1238. https://doi.org/10.1016/j.foodchem.2017.11.094

Guo, C.-C., Tang, Y.-H., Hu, H.-H., Yu, L.-S., Jiang, H.-D., Zeng, S., 2011. Analysis of chiral non-steroidal anti-inflammatory drugs flurbiprofen, ketoprofen and etodolac binding with HSA. *J. Pharm. Anal.* 1, 184–190. https://doi.org/10.1016/j.jpha.2011.06.005

Hancu, G., Cârcu-Dobrin, M., Budău, M., Rusu, A., 2018. Analytical methodologies for the stereoselective determination of fluoxetine: An overview. *Biomed. Chromatogr.* 32, 1–9. https://doi.org/10.1002/bmc.4040

Henry, T. B., Kwon, J. W., Armbrust, K. L., Black, M. C., 2004. Acute and chronic toxicity of five selective serotonin reuptake inhibitors in *Ceriodaphnia dubia*. *Environ. Toxicol. Chem.* 23, 2229–2233. https://doi.org/10.1897/03-278

Hoff, R., Pizzolato, T. M., Diaz-Cruz, M. S., 2016. Trends in sulfonamides and their by-products analysis in environmental samples using mass spectrometry techniques. *Trends Environ. Anal. Chem.* 9, 24–36. https://doi.org/10.1016/j.teac.2016.02.002

Huang, S., Zhu, F., Jiang, R., Zhou, S., Zhu, D., Liu, H., Ouyang, G., 2015. Determination of eight pharmaceuticals in an aqueous sample using automated derivatization solid-phase microextraction combined with gas chromatography–mass spectrometry. *Talanta* 136, 198–203. https://doi.org/10.1016/J.TALANTA.2014.11.071

Kasprzyk-Hordern, B., 2010. Pharmacologically active compounds in the environment and their chirality. *Chem. Soc. Rev.* 39, 4466–4503. https://doi.org/10.1039/c000408c

Kasprzyk-Hordern, B., Baker, D. R., 2012. Enantiomeric profiling of chiral drugs in wastewater and receiving waters. *Environ. Sci. Technol.* 46, 1681–1691. https://doi.org/10.1021/es203113y

Kim, J., Riggs, K. W., Misri, S., Kent, N., Oberlander, T. F., Grunau, R. E., Fitzgerald, C., Rurak, D. W., 2006. Stereoselective disposition of fluoxetine and norfluoxetine during pregnancy and breast-feeding. *Br. J. Clin. Pharmacol.* 61, 155–163. https://doi.org/10.1111/j.1365-2125.2005.02538.x

Koba, O., Golovko, O., Kodešová, R., Fér, M., Grabic, R., 2017. Antibiotics degradation in soil: A case of clindamycin, trimethoprim, sulfamethoxazole and their transformation products. *Environ. Pollut.* 220, 1251–1263. https://doi.org/10.1016/j.envpol.2016.11.007

Kosma, C. I., Lambropoulou, D. A., Albanis, T. A., 2014. Investigation of PPCPs in wastewater treatment plants in Greece: Occurrence, removal and environmental risk assessment. *Sci. Total Environ.* 466–467, 421–438. https://doi.org/10.1016/j.scitotenv.2013.07.044

Kumar, K., Gupta, S. C., 2016. A framework to predict uptake of trace organic compounds by plants. *J. Environ. Qual.* 45, 555–564. https://doi.org/10.2134/jeq2015.06.0261

Küster, A., Adler, N., 2014. Pharmaceuticals in the environment: Scientific evidence of risks and its regulation. *Philos. Trans. R. Soc. Lond. B. Biol. Sci.* 369, 20130587-. https://doi.org/10.1098/rstb.2013.0587

León-González, M. E., Rosales-Conrado, N., 2017. Determination of ibuprofen enantiomers in breast milk using vortex-assisted matrix solid-phase dispersion and direct chiral liquid chromatography. *J. Chromatogr. A* 1514, 88–94. https://doi.org/10.1016/j.chroma.2017.07.072

Li, M., Liang, X., Guo, X., Di, X., Jiang, Z., 2020. Enantiomeric separation and enantioselective determination of some representative non-steroidal anti-inflammatory drug enantiomers in fish tissues by using chiral liquid chromatography coupled with tandem mass spectrometry. *Microchem. J.* 153, 104511. https://doi.org/10.1016/j.microc.2019.104511

Li, S., Liu, B., Xue, M., Yu, J., Guo, X., 2019. Enantioseparation and determination of flumequine enantiomers in multiple food matrices with chiral liquid chromatography coupled with tandem mass spectrometry. *Chirality*. 31, 968–978. https://doi.org/10.1002/chir.23125

Limmer, M. A., Burken, J. G., 2014. Plant translocation of organic compounds: Molecular and physicochemical predictors. *Environ. Sci. Technol. Lett.* 1, 156–161. https://doi.org/10.1021/ez400214q

Liu, C., Feng, X., Qian, H., Fang, G., Wang, S., 2015. Determination of norfloxacin in food by capillary electrophoresis immunoassay with laser-induced fluorescence detector. *Food Anal. Methods* 8, 596–603. https://doi.org/10.1007/s12161-014-9936-1

Liu, S., Zhao, H., Lehmler, H. J., Cai, X., Chen, J., 2017. Antibiotic pollution in marine food webs in Laizhou bay, North China: Trophodynamics and human exposure implication. *Environ. Sci. Technol.* 51, 2392–2400. https://doi.org/10.1021/acs.est.6b04556

Maia, A. S., Ribeiro, A. R., Castro, P. M. L., Tiritan, M. E., 2017. Chiral analysis of pesticides and drugs of environmental concern: Biodegradation and enantiomeric fraction. *Symmetry (Basel)*. https://doi.org/10.3390/sym9090196

Malchi, T., Maor, Y., Tadmor, G., Shenker, M., Chefetz, B., 2014. Irrigation of root vegetables with treated wastewater: Evaluating uptake of pharmaceuticals and the associated human health risks. *Environ. Sci. Technol.* 48, 9325–9333. https://doi.org/10.1021/es5017894

Mandrioli, R., Forti, G., Raggi, M., 2006. Fluoxetine metabolism and pharmacological interactions: The role of cytochrome P450. *Curr. Drug Metab.* 7, 127–133. https://doi.org/10.2174/138920006775541561

Marongiu, G., Dolci, A., Verona, M., Capone, A., 2020. The biology and treatment of acute long-bones diaphyseal fractures: Overview of the current options for bone healing enhancement. *Bone Reports*. https://doi.org/10.1016/j.bonr.2020.100249

Masiá, A., Suarez-Varela, M. M., Llopis-Gonzalez, A., Picó, Y., 2016. Determination of pesticides and veterinary drug residues in food by liquid chromatography-mass spectrometry: A review. *Anal. Chim. Acta* 936, 40–61. https://doi.org/10.1016/j.aca.2016.07.023

Mehvar, R., Brocks, D. R., 2001. Stereospecific pharmacokinetics and pharmacodynamics of beta-adrenergic blockers in humans. *J. Pharm. Sci.* 4, 185–200.

Meister, I., Kovac, J., Duthaler, U., Odermatt, P., Huwyler, J., Vanobberghen, F., Sayasone, S., Keiser, J., 2016. Pharmacokinetic study of praziquantel enantiomers and its main metabolite R-trans-4-OH-PZQ in plasma, blood and dried blood spots in opisthorchis viverrini-infected patients. *PLoS Negl. Trop. Dis.* 10, e0004700. https://doi.org/10.1371/journal.pntd.0004700

Mennillo, E., Arukwe, A., Monni, G., Meucci, V., Intorre, L., Pretti, C., 2018. Ecotoxicological properties of ketoprofen and the S(+)-enantiomer (dexketoprofen): Bioassays in freshwater model species and biomarkers in fish PLHC-1 cell line. *Environ. Toxicol. Chem.* 37, 201–212. https://doi.org/10.1002/etc.3943

Migliore, L., Rotini, A., Cerioli, N. L., Cozzolino, S., Fiori, M., 2010. Phytotoxic antibiotic sulfadimethoxine elicits a complex hormetic response in the weed Lythrum salicaria L. *Dose-Response* 8, 414–427. https://doi.org/10.2203/dose-response.09-033.Migliore

Muth-Selbach, U., Hermanns, H., Driehsen, C., Lipfert, P., Freynhagen, R., 2009. Racemic intrathecal mirtazapine but not its enantiomers acts anti-neuropathic after chronic constriction injury in rats. *Brain Res. Bull.* 79, 63–68. https://doi.org/10.1016/j.brainresbull.2008.12.015

Nageswara Rao, R., Guru Prasad, K., 2015. Stereo-specific LC and LC-MS bioassays of antidepressants and psychotics. *Biomed. Chromatogr.* 29, 21–40. https://doi.org/10.1002/bmc.3356

Neale, P. A., Branch, A., Khan, S. J., Leusch, F. D. L., 2019. Evaluating the enantiospecific differences of non-steroidal anti-inflammatory drugs (NSAIDs) using an ecotoxicity bioassay test battery. *Sci. Total Environ.* 694, 133659. https://doi.org/10.1016/j.scitotenv.2019.133659

Nguyen, L. A., He, H., Pham-Huy, C., 2006. Chiral drugs: an overview. *Int. J. Biomed. Sci.* 2, 85–100.

O'Connor, G. A., 1996. Organic compounds in sludge-amended soils and their potential for uptake by crop plants. *Sci. Total Environ.* 185(1–3):71–81. https://doi.org/10.1016/0048-9697(95)05043-4

Parr, M. K., Blokland, M. H., Liebetrau, F., Schmidt, A. H., Meijer, T., Stanic, M., Kwiatkowska, D., Waraksa, E., Sterk, S. S., 2017. Distinction of clenbuterol intake from drug or contaminated food of animal origin in a controlled administration trial–the potential of enantiomeric separation for doping control analysis. *Food Addit. Contam. - Part A Chem. Anal. Control. Expo. Risk Assess.* 34, 525–535. https://doi.org/10.1080/19440049.2016.1242169

Patel, M., Kumar, R., Kishor, K., Mlsna, T., Pittman, C. U., Mohan, D., 2019. Pharmaceuticals of emerging concern in aquatic systems: Chemistry, occurrence, effects, and removal methods. *Chem. Rev.* 119, 3510–3673. https://doi.org/10.1021/acs.chemrev.8b00299

Pham, T. D. M., Ziora, Z. M., Blaskovich, M. A. T., 2019. Quinolone antibiotics. *Medchemcomm.* https://doi.org/10.1039/c9md00120d

Qu, R., 2019. Earth and Environmental Science Evaluation and Improvement of Synthesis Method for Ibuprofen Evaluation and Improvement of Synthesis Method for Ibuprofen. *IOP Conf. Ser.* https://doi.org/10.1088/1755-1315/242/5/052001

Ramatla, T., Ngoma, L., Adetunji, M., Mwanza, M., 2017. Evaluation of antibiotic residues in raw meat using different analytical methods. *Antibiotics* 6. https://doi.org/10.3390/antibiotics6040034

Ramos, F., Lurdes Baeta, M., Reis, J., Noronha Da Silveira, M. I., Noronha, M. I., Evalua, S., 2011. Evaluation of the illegal use of clenbuterol in feed, drinking water, urine, and hair samples collected in bovine farms in Portugal. https://doi.org/10.1080/02652030902729908ï

Rana, M. S., Lee, S. Y., Kang, H. J., Hur, S. J., 2019. Reducing veterinary drug residues in animal products: A review. *Food Sci. Anim. Resour.* 39, 687–703. https://doi.org/10.5851/kosfa.2019.e65

Rather, I. A., Koh, W. Y., Paek, W. K., Lim, J., 2017. The sources of chemical contaminants in food and their health implications. *Front. Pharmacol.* https://doi.org/10.3389/fphar.2017.00830

Ribeiro, A. R., Afonso, C. M., Castro, P. M. L., Tiritan, M. E., 2013. Enantioselective biodegradation of pharmaceuticals, alprenolol and propranolol, by an activated sludge inoculum. *Ecotoxicol. Environ. Saf.* 87, 108–114. https://doi.org/10.1016/j.ecoenv.2012.10.009

Ribeiro, A. R. L., Maia, A. S., Ribeiro, C., Tiritan, M. E., 2020. Analysis of chiral drugs in environmental matrices: Current knowledge and trends in environmental, biodegradation and forensic fields. *TrAC - Trends Anal. Chem.* https://doi.org/10.1016/j.trac.2019.115783

Rimkus, G. G., Hoffmann, D., 2017. Enantioselective analysis of chloramphenicol residues in honey samples by chiral LC-MS/MS and results of a honey survey. *Food Addit. Contam. - Part A Chem. Anal. Control. Expo. Risk Assess.* 34, 950–961. https://doi.org/10.1080/19440049.2017.1319073

Rimkus, G. G., Huth, T., Harms, D., 2020. Screening of stereoisomeric chloramphenicol residues in honey by ELISA and CHARM ® II test–the potential risk of systematically false-compliant (false negative) results. *Food Addit. Contam. - Part A Chem. Anal. Control. Expo. Risk Assess.* 37, 94–103. https://doi.org/10.1080/19440049.2019.1682685

Rosales-Conrado, N., de León-González, M. E., Polo-Díez, L. M., 2015. Development and validation of analytical method for clenbuterol chiral determination in animal feed by direct liquid chromatography. *Food Anal. Methods.* 8, 2647–2659. https://doi.org/10.1007/s12161-015-0146-2

Sabourin, L., Duenk, P., Bonte-Gelok, S., Payne, M., Lapen, D. R., Topp, E., 2012. Uptake of pharmaceuticals, hormones and parabens into vegetables grown in soil fertilized with municipal biosolids. *Sci. Total Environ.* 431, 233–236. https://doi.org/10.1016/j.scitotenv.2012.05.017

Sahu, P. K., Wang, C. H., Lee, S. L., 2009. Interaction of serotonin and fluoxetine: Toward understanding the importance of the chirality of fluoxetine (S form and R form). *J. Phys. Chem. B* 113, 14529–14535. https://doi.org/10.1021/jp907151n

Sanganyado, E., Lu, Z., Fu, Q., Schlenk, D., Gan, J., 2017. Chiral pharmaceuticals: A review on their environmental occurrence and fate processes. *Water Res.* 124, 527–542. https://doi.org/10.1016/j.watres.2017.08.003

Sarmah, A. K., Meyer, M. T., Boxall, A. B. A., 2006. A global perspective on the use, sales, exposure pathways, occurrence, fate and effects of veterinary antibiotics (VAs) in the environment. *Chemosphere.* https://doi.org/10.1016/j.chemosphere.2006.03.026

Shah, K., Gupta, J. K., Chauhan, N. S., Upmanyu, N., Shrivastava, S. K., Mishra, P., 2017. Prodrugs of NSAIDs: A review. *Open Med. Chem. J.* 11, 146–195. https://doi.org/10.2174/1874104501711010146

Stanley, J. K., Ramirez, A. J., Chambliss, C. K., Brooks, B. W., 2007. Enantiospecific sublethal effects of the antidepressant fluoxetine to a model aquatic vertebrate and invertebrate. *Chemosphere* 69, 9–16. https://doi.org/10.1016/j.chemosphere.2007.04.080

Sykes, M., Czymai, T., Hektor, T., Sharman, M., Knaggs, M., 2017. Chloramphenicol stereoisomers need to be distinguished: Consequences observed from a proficiency test. *Food Addit. Contam. - Part A Chem. Anal. Control. Expo. Risk Assess.* 34, 536–541. https://doi.org/10.1080/19440049.2016.1270469

Tang, K. L., Caffrey, N. P., Nóbrega, D. B., Cork, S. C., Ronksley, P. E., Barkema, H. W., Polachek, A. J., Ganshorn, H., Sharma, N., Kellner, J. D., Ghali, W. A., 2017. Restricting the use of antibiotics in food-producing animals and its associations with antibiotic resistance in food-producing animals and human beings: a systematic review and meta-analysis. *Lancet Planet. Heal.* 1, e316–e327. https://doi.org/10.1016/S2542-5196(17)30141-9

Tasho, R. P., Cho, J. Y., 2016. Veterinary antibiotics in animal waste, its distribution in soil and uptake by plants: A review. *Sci. Total Environ.* 563–564, 366–376. https://doi.org/10.1016/j.scitotenv.2016.04.140

Thomaidi, V. S., Stasinakis, A. S., Borova, V. L., Thomaidis, N. S., 2016. Assessing the risk associated with the presence of emerging organic contaminants in sludge-amended soil: A country-level analysis. *Sci. Total Environ.* 548–549, 280–288. https://doi.org/10.1016/j.scitotenv.2016.01.043

Tokunaga, E., Yamamoto, T., Ito, E., Shibata, N., 2018. Understanding the thalidomide chirality in biological processes by the self-disproportionation of enantiomers. *Sci. Rep.* 8, 1–7. https://doi.org/10.1038/s41598-018-35457-6

Tsujimoto, T., Sugiyama, T., Shapiro, M. F., Noda, M., Kajio, H., 2017. Risk of cardiovascular events in patients with diabetes mellitus on β-blockers. *Hypertens.* (Dallas, Tex. 1979) 70, 103–110. https://doi.org/10.1161/HYPERTENSIONAHA.117.09259

Ungprasert, P., Cheungpasitporn, W., Crowson, C. S., Matteson, E. L., 2015. Individual non-steroidal anti-inflammatory drugs and risk of acute kidney injury: A systematic review and meta-analysis of observational studies. *Eur. J. Intern. Med.* 26, 285–291. https://doi.org/10.1016/j.ejim.2015.03.008

Wang, J., Jiang, X. X., Zhao, W., Hu, J., Guan, Q. Y., Xu, J. J., Chen, H. Y., 2017. On-line Identification of chiral ofloxacin in milk with an extraction/ionization device coupled to electrospray mass spectrometry. *Talanta* 171, 190–196. https://doi.org/10.1016/j.talanta.2017.04.071

Wang, Z. L., Zhang, J. L., Zhang, Y. N., Zhao, Y., 2016. Mass spectrometric analysis of residual clenbuterol enantiomers in swine, beef and lamb meat by liquid chromatography tandem mass spectrometry. *Anal. Methods* 8, 4127–4133. https://doi.org/10.1039/c6ay00606j

Weisskopf, E., Panchaud, A., Nguyen, K. A., Grosjean, D., Hascoët, J. M., Csajka, C., Eap, C. B., Ansermot, N., 2016. Stereoselective determination of citalopram and desmethylcitalopram in human plasma and breast milk by liquid chromatography tandem mass spectrometry. *J. Pharm. Biomed. Anal.* 131, 233–245. https://doi.org/10.1016/j.jpba.2016.08.014

WHO, 2004. ACTION AGAINST WORMS ARE YOU DEWORMING?

Wong, D. T., Bymaster, F. P., Engleman, E. A., 1995. Prozac (fluoxetine, lilly 110140), the first selective serotonin uptake inhibitor and an antidepressant drug: Twenty years since its first publication. *Life Sci.* https://doi.org/10.1016/0024-3205(95)00209-O

Wu, C., Huang, X., Witter, J. D., Spongberg, A. L., Wang, K., Wang, D., Liu, J., 2014. Occurrence of pharmaceuticals and personal care products and associated environmental risks in the central and lower Yangtze river, China. *Ecotoxicol. Environ. Saf.* 106, 19–26. https://doi.org/10.1016/j.ecoenv.2014.04.029

Yanovych, D., Berendsen, B., Zasadna, Z., Rydchuk, M., Czymai, T., 2018. A study of the origin of chloramphenicol isomers in honey. *Drug Test. Anal.* 10, 416–422. https://doi.org/10.1002/dta.2234

Zhang, C., Feng, Y., Liu, Y. Wang, Cangh, H. Qing, Li, Z. jun, Xue, J. ming, 2017. Uptake and translocation of organic pollutants in plants: A review. *J. Integr. Agric.* https://doi.org/10.1016/S2095-3119(16)61590-3

Zhang, H., Deng, X., Miki, D., Cutler, S., La, H., Hou, Y. J., Oh, J. E., Zhu, J. K., 2012. Sulfamethazine suppresses epigenetic silencing in Arabidopsis by impairing folate synthesis. *Plant Cell* 24, 1230–1241. https://doi.org/10.1105/tpc.112.096149

Zhang, Y., He, L., Li, X., Wang, Y., Xie, J., Lee, J. T., Armstrong, D. W., 2019. Selective depletion of chiral 4-hydroxypraziquantel metabolites in three types of aquaculture fish by LC-MS/MS. *J. Agric. Food Chem.* 67, 4098–4104. https://doi.org/10.1021/acs.jafc.8b04728

Zhao, L., Dong, Y. H., Wang, H., 2010. Residues of veterinary antibiotics in manures from feedlot livestock in eight provinces of China. *Sci. Total Environ.* 408, 1069–1075. https://doi.org/10.1016/j.scitotenv.2009.11.014

Zhao, S., Jiang, H., Li, X., Mi, T., Li, C., Shen, J., 2007. Simultaneous determination of trace levels of 10 quinolones in swine, chicken, and shrimp muscle tissues using HPLC with programmable fluorescence detection, in: *Journal of Agricultural and Food Chemistry*. pp. 3829–3834. https://doi.org/10.1021/jf0635309

Zhou, Y., Wu, S., Zhou, H., Huang, H., Zhao, J., Deng, Y., Wang, H., Yang, Y., Yang, J., Luo, L., 2018. Chiral pharmaceuticals: Environment sources, potential human health impacts, remediation technologies and future perspective. *Environ. Int.* 121, 523–537. https://doi.org/10.1016/j.envint.2018.09.041

Zhu, B., Li, S., Zhou, L., Li, Q., Guo, X., 2019a. Simultaneous enantioselective determination of seven psychoactive drugs enantiomers in multi-specie animal tissues with chiral liquid chromatography coupled with tandem mass spectrometry. *Food Chem.* 300. https://doi.org/10.1016/j.foodchem.2019.125241

Zhu, B., Li, S., Zhou, L., Li, Q., Guo, X., 2019b. Simultaneous enantioselective determination of seven psychoactive drugs enantiomers in multi-specie animal tissues with chiral liquid chromatography coupled with tandem mass spectrometry. *Food Chem.* 300. https://doi.org/10.1016/j.foodchem.2019.125241

11 Food Authenticity and Adulteration

Leo M.L. Nollet

CONTENTS

11.1 Authenticity and Traceability 279
11.2 Phenols 281
11.3 Seafood 285
11.4 Olive Oil 286
11.5 Honey 287
11.6 Fruit Beverages 288
11.7 Dairy Products 289
11.8 Essential Oils 289
11.9 Flavors 291
References 294

11.1 AUTHENTICITY AND TRACEABILITY

Food authentication is the process that verifies that a food is in compliance with its label description (1). This may include, among others, the origin (species, geographical, or genetic), production method (conventional, organic, traditional procedures, free range), or processing technologies (irradiation, freezing, microwave heating). The declaration of specific quality attributes in high-value products is of particular interest since these products are often the target of fraudulent labeling. Proof of provenance is an important topic for food safety, food quality, and consumer protection, as well as the compliance with national legislation, international standards, and guidelines. Due to the globalization of food markets and the resulting increase in variability and availability of food products from other countries, consumers are increasingly interested in knowing the geographical origin along with the assumed quality of the products they eat and drink.

Authenticity has been a major concern of consumers, producers, and regulators since ancient times. Modern instrumentation and advances in basic sciences and in information and communication technologies provide means for precise measurement and elucidation of origin of foods. Since the beginning of the 20th century, organizations that set standards for and control the origin of ingredients and the production process have appeared all over the world, e.g., the French "Institut National des Appellations d'Origine (INAO)," Italy's "Denominazione di Origine Controllata," Spain's "Denominación de Origen," South Africa's "Wine of Origin," or the United States' "American Viticultural Areas." The production of consumer goods according to these standardized procedures normally results in better products and is rewarded with higher prices at the point of sale. Unfortunately, these financial benefits attract the production of counterfeit food and illegal food trades (2).

In Europe, origin is one of the main authenticity issues concerning food. European Union legislation reserving specific names for foods and beverages of a particular quality or reputation has been abundant since the dawn of the European integration process (Council Regulation, EEC No.

2081/92). These legislations introduced regulatory framework for wines and spirits and quality schemes for food products including PDO (Protected Designation of Origin) that links products to the defined geographical area where they are produced, PGI (Protected Geographical Indication) that links products to a geographical area where at minimum one production step occurred, and TSG (Traditional Specialities Guaranteed) that protects traditional methods of production. Furthermore, recently defined optional quality terms (OQT) such as "mountain product" and "product of island farming" were defined (1151/2012 EU Regulation). The purpose of these EU schemes is to protect the reputation of the regional foods and to promote good practices in rural and agricultural activity. Such practices help producers obtain premium prices for authentic products and minimize the unfair and misleading competition from non-genuine products, usually of inferior quality or of different flavor (1151/2012 EU Regulation). The information includes a characterization of the geographic region and reinforces the consumer perception of special quality attributed to mountain and island products. In the case of cultivated species, EU indicates that a reference should be made to the country in which the food undergoes the final production stage. Vigorous research activities in EU are supported by the coordination actions of the European Union, including "Food Integrity," "MoniQa," and "TRACE" within HORIZON 2020 (2).

After a detailed investigation of the true definition of traceability, Olsen and Borit (3) came with a compendium of all the definitions saying that traceability is "The ability to access any or all information relating to that which is under consideration, throughout its entire life cycle, by means of recorded identifications."

Historically, food scares have been with human beings for many years. In the modern livestock production sector, long distance animal transport is increasing (4). This in turn not only has increased the potential for infection and spread of diseases related to livestock, but also has exposed the sector for bioterrorism attacks. These challenges have triggered the importance of animal identification and certification processes. Another risk of food such as contamination with radioactive materials disturbs the FSC (food chain supply). After the release of radioactive releases from damaged nuclear plants due to the earthquake in Japan in 2011 (WHO, 2011), many countries have implemented intensive food control measures concerning their food trade relationship with Japan while some countries suspended transporting food from Japan. A study in Finland (5) has indicated that, as proactive strategy, effective training on food logistics is important. Such training should include cleaning warehouses and retails, controlling and transporting non-contaminated food items to retail storage facilities after the radioactive materials have passed over a region.

In addition to the risk to public health, food crises lead to economic crises due to direct and indirect (damage to reputation and brand name) costs of product recall. The indirect cost dominates the recall cost as the loss of market value and reputation could lead to the total bankruptcy of the brand name (6). Therefore, traceability is an important component of contemporary supply chains in the production industry in general and in the food sector in particular as the food sector is sensitive from the human and animal health point of view.

The terms "tracing" and "tracking" are generally discussed in the traceability. Tracing is the backward process where the origin is identified by history or records in supply chain and tracking is the forward process where end users and trading partners are identified by location in the supply chain, while both terms provide the visibility to the supply chain (7).

Current traceability systems are characterized by the inability to link food chains records, inaccuracy and errors in records, and delays in obtaining essential data, which are fundamental in the case of food outbreak disease (8). These systems should address the recall and withdrawal of non-consumable products, however to date there are still recent reports covering the implementation of food assurance systems that do not mention the traceability question although they are highly related to food traceability.

11.2 PHENOLS

Numerous papers on enantioseparation of phenols, and especially catechin and epicatechin may be found. Catechins are a type of phenolic compound very abundant in tea, cocoa, and berries to which are ascribed a potent antioxidant activity. It is a plant secondary metabolite. It belongs to the group of flavan-3-ols (or simply flavanols), part of the chemical family of flavonoids.

The enantiomers of catechin and epicatechin were separated by chiral capillary electrophoresis using modified cyclodextrins as chiral selectors (9). Various conditions for the separation system were optimized, including the pH value and the concentration of the running buffer. A baseline separation of the catechin and epicatechin enantiomers could be achieved by using 0.1 mol l^{-1} borate buffer (pH 8.5) with 12 mmol l^{-1} (2-hydroxypropyl)-γ-cyclodextrin as chiral selector, a fused-silica capillary with 40 cm effective length (75 μm I.D.), +18 kV applied voltage, a temperature of 20°C, and direct UV detection at 280 nm. The method was applied to different plant food samples. (+)-Catechin and (–)-epicatechin could be verified as the most common flavan-3-ols. In the case of guaraná, however, it was possible to identify all four enantiomers, both (+)- and (–)-catechin and (+)- and (–)-epicatechin, as naturally occurring compounds. This finding was verified by further isolation and purification of the flavan-3-ols and subsequent LC–MS analysis. This method allowed for the identification of the authenticity of guaraná through the analysis of the catechin and epicatechin enantiomers.

Cocoa contains high levels of different flavonoids. An enantioseparation of catechin and epicatechin in cocoa and cocoa products by chiral capillary electrophoresis (CCE) was performed (10). A baseline separation of the catechin and epicatechin enantiomers was achieved by using 0.1 mol·L^{-1} borate buffer (pH 8.5) with 12 mmol·L^{-1} (2-hydroxypropyl)-γ-cyclodextrin as chiral selector, a fused-silica capillary with 50 cm effective length (75 μm I.D.), +18 kV applied voltage, a temperature of 20°C and direct UV detection at 280 nm. To avoid comigration or coelution of other similar substances, the flavan-3-ols were isolated and purified using polyamide-solid-phase-extraction and LC-MS analysis. As expected, (-)-epicatechin and (+)-catechin in unfermented, dried, unroasted cocoa beans were found. In contrast, roasted cocoa beans and cocoa products additionally contained the atypical flavan-3-ol (-)-catechin. This is generally formed during the manufacturing process by an epimerization which converts (-)-epicatechin to its epimer (-)-catechin. High temperatures during the cocoa bean roasting process and particularly the alkalization of the cocoa powder are the main factors inducing the epimerization reaction. In addition to the analysis of cocoa and cocoa products, peak ratios were calculated for a better differentiation of the cocoa products.

When catechins are found in plant extracts, they are almost always identified as catechin and/or epicatechin probably due to stereoselectivity of the enzymes involved in the biosynthesis of these substances. However, the lack of reports regarding to *ent*-catechin as well as *ent*-epicatechin does not necessarily mean that these compounds have not been produced. In fact, most of the previous reports used chromatographic conditions not suitable for such separation. Rinaldo D. et al. (11) describe a simple and reliable analytical HPLC-PAD-CD method for simultaneous determination of catechin diastereomers both in infusions and extracts from the leaves of *Byrsonima* species. The direct separation of catechin, *ent*-catechin, epicatechin, and *ent*-epicatechin was obtained in normal phase by HPLC-PAD-CD using Chiralcel OD-H as chiral stationary phase and *n*-hexane/ethanol with 0.1% of TFA as mobile phase.

A single-laboratory validation study was performed for an HPLC method to identify and quantify the flavanol enantiomers (+)- and (–)-epicatechin and (+)- and (–)-catechin in cocoa-based ingredients and products (12). These compounds were eluted isocratically with an ammonium acetate–methanol mobile phase applied to a modified β-cyclodextrin chiral stationary phase and detected using fluorescence. Spike recovery experiments using appropriate matrix blanks, along with cocoa extract, cocoa powder, and dark chocolate, were used to evaluate accuracy, repeatability, specificity, LOD, LOQ, and linearity of the method as performed by a single analyst on multiple days. In all samples analyzed, (–)-epicatechin was the predominant flavanol and represented 68–91% of the total monomeric flavanols detected. For the cocoa-based products, within-day (intraday) precision for (–)-epicatechin was between 1.46–3.22%, for (+)-catechin between 3.66–6.90%, and for (–)-catechin between 1.69–6.89%; (+)-epicatechin was not detected in these samples. Recoveries for the three sample types investigated ranged from 82.2 to 102.1% at the 50% spiking level, 83.7 to 102.0% at the 100% spiking level, and 80.4 to 101.1% at the 200% spiking level.

Single-laboratory validation data previously published in the *Journal of AOAC INTERNATIONAL* 95(2), 500–507 (2012) was reviewed by the Stakeholder Panel on Strategic Food Analytical Methods Expert Review Panel (ERP) at the AOAC INTERNATIONAL Mid-Year Meeting held on March 12–14, 2013 in Rockville, MD. The ERP determined the data presented met the established standard method performance requirement and approved the method as AOAC Official First Action on March 14, 2013 (13). Using high-performance liquid chromatography (HPLC), flavanol enantiomers, (+)- and (-)-epicatechin and (+)- and (-)-catechin, are eluted isocratically using ammonium acetate and methanol mobile phase. The mobile phase is applied to a modified β-cyclodextrin chiral stationary phase and the flavanols detected by fluorescence. Using several cocoa-based matrices, recoveries for the four enantiomers ranged from 82.2–102.1% at a 50% spike level, and 80.4–101.1% at a 100% spike level. Precision was determined to be 1.46–3.22% for (-)-epicatechin, 3.66–6.90% for (+)-catechin, 1.69–6.89% for (-)-catechin. (+)-Epicatechin was not detected in any of the samples used for this work, so precision could not be determined for this molecule.

Assessment of the flavanol composition of 41 commercial chocolates was performed by HPLC-DAD (14). Among individual flavonols ranged from 0.095 to 3.264 mg g^{-1}, epicatechin was the predominant flavanol accounting for 32.9%. Contrary to catechin, epicatechin was a reliable predictive value of the polyphenol content. Conversely the percentage of theobromine used as a proxy measure for nonfat cocoa solids (NFCS) was not a good predictor of epicatechin or flavanol content. In a further chiral analysis, the naturally occurring forms of cocoa flavanols, (–)-epicatechin and (+)-catechin, was determined joint the occurrence of (+)-epicatechin and (–)-catechin due to the epimerization reactions produced in chocolate manufacture. (–)-Epicatechin, the most bioactive compound and predominant form accounted for 93%. However, no positive correlation was found with percentage of cocoa solids, the most significant quality parameter.

The separation of enantiomers of flavanone and 5 chiral flavanone derivatives was studied on two experimental polysaccharide-based chiral columns by high-performance liquid chromatography with methanol, ethanol, acetonitrile, *n*-hexane/ethanol, and *n*-hexane/2-propanol as mobile phases with emphasis on the effect of analyte and chiral selector structure on separation parameters (15). Since chiral selectors were covalently immobilized onto the surface of silica, it was possible to use a wider variety of mobile phases in order to optimize enantioseparations. Together with the backbone of the polysaccharide, the significant role of substituents in the phenyl moiety of chiral selectors, as well as the structure of the analyte, was observed on retention and enantioselectivity.

In the past decade, macrocyclic antibiotics have proved to be an exceptionally useful class of chiral selectors for the separation of enantiomers of biological and pharmacological importance by means of HPLC, TLC, and electrophoresis (16). More chiral analytes have been resolved through the use of glycopeptides than with all the other macrocyclic antibiotics combined (ansamycins, thiostrepton, aminoglycosides, etc.). The glycopeptides avoparcin, teicoplanin, ristocetin A, and vancomycin have been extensively used as chiral selectors in the form of chiral bonded phases in HPLC, and HPLC stationary phases based on these glycopeptides have been commercialized. Teicoplanin, vancomycin, their analogs, and ristocetin A seem to be the most useful glycopeptide HPLC bonded phases for the enantioseparation of proteins and unusual native and derivatized amino acids. In fact, the macrocyclic glycopeptides are to some extent complementary to one another: where partial enantioresolution is obtained with one glycopeptide, there is a high probability that baseline or better separation can be obtained with another. This review sets out to characterize the physicochemical properties of these antibiotics and their application in the enantioseparations of amino acids.

The mechanism of separation, the sequence of elution of the stereoisomers, and the relation to the absolute configuration are also discussed.

Most HPLC enantiomer separations are performed with columns packed with a chiral stationary phase (CSP) operated with an achiral mobile phase. The intrinsically limited chemical selectivity of most CSPs to the simultaneous resolution of several pairs of enantiomers means that complex mixtures of diverse pairs of enantiomers cannot be resolved in a single run due to peak overlapping. Moreover, some drawbacks remain when the analyte is present in very complex samples containing other achiral compounds which can co-elute with the enantiomer peaks. Multidimensional chromatography becomes an option to increase peak capacity and resolve these samples.

An online fully comprehensive 2D-LC mode utilizing a combination of a chiral column in the first dimension and an achiral column in the second dimension was evaluated (17). The 2D-LC system was built with an active flow splitter pump in order to easily adjust the volume of sample transferred into the second dimension and to independently optimize the flow rate in the first dimension.

The present LC × LC method was optimized for the separation of amino acids present in honey samples, taking into account key parameters that influence the bidimensional peak capacity (orthogonality, sampling frequency, etc.). The amino acids have been pre-concentrated on a cation-exchange column followed by derivatization. Several amino acids present in different honey samples have been identified and the data generated has been analyzed by principal component analysis.

Cyclodextrin-modified micellar electrokinetic chromatography was applied to the enantioseparation of catechin and epicatechin using 6-O-α-D-glucosyl-β-cyclodextrin together with sodium dodecyl sulfate and borate-phosphate buffer (18). Factors affecting chiral resolution and migration time of catechin and epicatechin were studied. The optimum running conditions were found to be 200 mM borate-20 mM phosphate buffer (pH 6.4) containing 25 mM 6-O-α-D-glucosyl-β-cyclodextrin and 240 mM sodium dodecyl sulfate with an effective voltage of +25 kV at 20°C using direct detection at 210 nm. Under these conditions, the resolution (R_s) of *racemic* catechin and epicatechin were 4.15 and 1.92, respectively. With this system, catechin and epicatechin enantiomers along with other four catechins ((−)-catechin gallate, (−)-epicatechin gallate, (−)-epigallocatechin, (−)-epigallocatechin gallate) and caffeine in tea samples were analyzed successfully.

Besides catechins, L-Theanine (γ-glutamylethylamide), a characteristic amino acid in tea leaves, has become a further focus of the phytochemical research for the reported beneficial effects mainly on cognitive performance, emotional state, and sleep quality. In the study of Fiori J. et al. (19) has been developed a CD-MEKC method based on sodium dodecyl sulfate (SDS) and Heptakis (2,6-di-O-methyl)-β-cyclodextrin for the separation of six major green tea catechins and enantiomers of theanine. The latter, because of the poor detectability was derivatized prior analysis by o-phthaldialdehyde in the presence of *N*-acetyl-L-cysteine which, under mild conditions (neutral pH, in two minutes) allowed two diastereomers isoindole derivatives to be obtained. The derivatization reaction was directly carried out on tea infusion and derivatized samples were analyzed by CD-MEKC involving 65 mM SDS and 28 mM cyclodextrin in acidic buffer (pH 2.5). The separation of six major green tea catechins including enantioresolution of (±)-Catechin and D/L-Theanine was obtained in about 5 min allowing D-Theanine to be quantified at least at 0.5% m/m level with respect to L-Theanine. Since (−)-Catechin and D-Theanine can be considered as non-native enantiomers (distomers), their presence in real samples provides an indication of tea leaves treatments (thermal treatment, fermentation, etc.) and could represent an opportunity for grading tea. The obtained results were confirmed by a RP-HPLC approach; even though the chromatography was developed in achiral conditions, the derivatization approach applied to theanine (diastereomers formation), allowed for D/L-Theanine chiral analysis.

The reliable detection of catechin and epicatechin by spectrophotometry is limited by the interference with substances that absorb light at similar wavelengths; thus, an effective alternative is to detect the oxidation products of these compounds at more specific wavelengths. The study of Dias T. et al. (20) aimed to develop an analytical method to quantify total catechin and epicatechin in some food samples (cocoa beans, guarana powder, apples, and tea leaves) by enzymatic oxidation with tyrosinase. The samples were extracted in an ultrasonic bath and the purified extracts were oxidized with tyrosinase solution. The quinones formed in the reaction were detected by spectrophotometry. The method was selective and presented linearity between 1.21 and 7.26 mg g^{-1}, limit of detection of 0.48 mg g^{-1}, and limit of quantification of 1.61 mg g^{-1}. The accuracy of the proposed method was reasonable, with recoveries between 84.3 and 90.7% in the fortified matrices. High precision was observed, with low coefficients of variation for repeatability (3.17 to 3.68%) and intermediate precision (3.27%). In cocoa beans, the total catechin and epicatechin level was 1.65 mg g^{-1} (3.58 mg g^{-1}). The successful application of the method was also demonstrated in guarana powder.

The aim of the study of Rockenbach I. et al. was to evaluate the profile of galloylated and non-galloylated flavan-3-ols and the presence of enantiomers of catechin and epicatechin in seeds of pomace from the vinification of different grape varieties with a view to their exploitation as a source of natural antioxidants (21). Chiral capillary electrophoretic analysis showed that the fermentation process of winemaking did not give rise to (–)-catechin and (+)-epicatechin enantiomers. Only (+)-catechin and (–)-epicatechin were detected. High-performance liquid chromatography coupled with a diode array detector and an ion trap mass spectrometer (HPLC/DAD-MSn) showed the presence of several galloylated and non-galloylated flavan-3-ol compounds and the presence of condensed products of catechin with acetaldehyde. Fourier-transform ion cyclotron resonance mass spectrometry (LC-ESI-FTICR-MS) enabled the assignment of elemental compositions to 251 different flavan-3-ol compounds in Cabernet Sauvignon variety, including isomers of 28 different molecular classes. The data obtained may support the exploitation of grape seeds as a source of natural antioxidants.

The detailed phenolic composition of five single-cultivar (Baboso Negro, Listán Negro, Negramoll, Tintilla, and Vijariego Negro) young and aged (vintages 2005–2009) red wines of the Canary Islands has been determined by HPLC-DAD-ESI-MSn (22). Despite the total monomeric anthocyanin content decreasing for older wines in each set of single-cultivar wines, the corresponding anthocyanin profiles remained almost unchanged. Although all wine anthocyanin profiles were dominated by malvidin 3-glucoside, their differentiation by grape cultivar was possible, with the exception of Listán Negro. In contrast, the total content of non-anthocyanin phenolics did not appreciably change within vintages but polymerization, hydrolysis, and isomerization reactions greatly modified the phenolic profiles. Aglycone-type flavonol profiles offered the best results for differentiation of the wines according to grape cultivar (Listán Negro and Negramoll; Baboso Negro and Vijariego Negro; and Tintilla). Within flavan-3-ols, the B-ring trihydroxylated monomers ((–)-epigallocatechin and (–)-gallocatechin) and also (–)-epicatechin provided additional cultivar differentiation. Hydroxycinnamic acid derivatives and stilbene profiles were very heterogeneous with regard to both grape cultivar and vintage and did not significantly contribute to wine differentiation, even when structure-type profiles were obtained, with the exception of Tintilla, which always appeared as the most different single-cultivar wines. Finally, most Canary Islands wines showed characteristic high contents of stilbenes, especially trans-resveratrol.

The composition of robusta and arabica coffee species in terms of their amino acid enantiomers in the green and roasted states was reported (23). The analyses were conducted for the free amino acids, as well as for the amino acids obtained after acid hydrolysis. The amino acids were extracted/hydrolyzed and isolated by SPE on strong cation exchange columns, derivatized to their N-ethoxycarbonylheptafluorobutyl esters, and analyzed by gas chromatography/FID on a Chirasil l-Val column. Multivariate analyses applied to the results showed that the free amino acids can be used as a tool for discrimination between coffee species, with a special reference to l-glutamic acid, l-tryptophan, and pipecolic acid. There is also some evidence that these compounds can be used for discrimination between green coffees subjected to different postharvest processes. It is also shown that the amino acid levels observed after acid hydrolysis can be used for the same purposes, although displaying less discriminatory power.

A micellar electrokinetic chromatography (MEKC) method was developed for the quantitation of polyphenols (+)-catechin and (–)-epicatechin (catechin monomers) and the methylxanthine theobromine in *Theobroma cacao* beans (24). Owing to the poor stability of catechin monomers in alkaline conditions, a 50 mM Britton–Robinson buffer at a pH 2.50 was preferred as the background electrolyte. Under these conditions, the addition of hydroxypropyl-β-cyclodextrin (HP-β-CD) at a concentration of 12 mM to the SDS micellar solution (90 mM), resulted in a cyclodextrin-modified micellar electrokinetic chromatography (CD-MEKC) endowed with two peculiar advantages compare to the conventional MEKC: (i) strong improvement of separation of the most important phytomarkers of *T. cacao* and (ii) enantioselectivity toward (±)-catechin. In particular, separation of methylxanthines (theobromine and caffeine), procyanidin dimers B1 and B2, and catechins (epicatechin and catechin) was obtained simultaneously to the enantioseparation of racemic catechin within 10 min. The enantioselectivity of the method makes it suitable in evaluation of possible epimerization at the C-2 position of epicatechin monomer potentially occurring during heat processing and storage of *T. cacao* beans. The extraction procedure of the phytomarkers from the beans was approached using ultrasonic bath under mild conditions optimized by a multivariate strategy. The method was validated for robustness, selectivity, sensitivity, linearity, range, accuracy, and precision and it was applied to *T. cacao* beans from different countries; interestingly, the native enantiomer (+)-catechin was found in the beans whereas for the first time we reported that in chocolate, predominantly (−)-catechin is present, probably yielded by epimerization of (−)-epicatechin occurred during the manufacture of chocolate.

11.3 SEAFOOD

To develop a practical methodology for the authentication of organic salmonid products, 130 fillet samples of trout and salmon originating from organic and conventional aquaculture as well as wild stocks (salmon) were collected from the German market over one year (25). Combined stable isotope analysis of $\delta^{15}N$ and $\delta^{13}C$ in defatted dry matter allowed differentiation of organically farmed from conventionally farmed salmon and brown trout, whether raw, smoked, or graved. For the additional distinction of organic and wild salmon, a second analysis of $\delta^{13}C$ in fish lipids was required. Fatty acid analysis completely differentiated the three production types of salmon just by the linoleic acid content in the fish lipids, which was lowest in wild and highest in conventional salmon. Moreover, the elevated myristic acid content allowed organic to be distinguished from wild and conventional salmon. Furthermore, organic and conventional brown trout could be distinguished by combining the oleic acid and gondoic acid contents. Analysis of the free astaxanthin isomeric pattern allowed a clear distinction of conventional and wild salmon, but organic salmon showed variable patterns that did not consistently allow the authentication of their origin. While a special feed composition is required in organic aquaculture, the composition of conventional aquaculture feed has changed considerably within the last decade. Consequently, the percentages of animal and vegetable components, which clearly vary between the production types, result in distinctive features in terms of stable isotope or fatty acid composition that are utilizable for the authentication of organic salmonid products.

The presence of carotenoids in animal tissue reflects their sources along the food chain. Astaxanthin, the main carotenoid used for salmonid pigmentation, is usually included in the feed as a synthetic product. However, other dietary sources of astaxanthin such as shrimp or krill wastes, algae meal or yeasts are also available on the market (26). Astaxanthin possesses two identical asymmetric atoms at C-3 and C-3' making possible three optical isomers with all-*trans* configuration of the chain: $3S,3'S$, $3R,3'S$, and $3R,3'R$. The distribution of the isomers in natural astaxanthin differs from that of the synthetic product. This latter is a racemic mixture, with a typical ratio of 1:2:1 ($3S,3'S:3R,3'S:3R,3'R$), while astaxanthin from natural sources has a variable distribution of the isomers deriving from the different biological organism that synthesized it. A HPLC analysis of all-*trans* isomers of astaxanthin was performed in different pigment sources, such as red yeast *Phaffia rhodozyma*, alga meal *Haematococcus pluvialis*, krill meal and oil, and shrimp meal. With the aim to investigate astaxanthin isomer ratios in flesh of fish fed different carotenoid sources, three groups of rainbow trout were fed for 60 days diets containing astaxanthin from synthetic source, *H. pluvialis* algae meal and *P. rhodozyma* red yeast. Moreover, the distribution of optical isomers of astaxanthin in trout purchased on the Italian market was investigated. A characteristic distribution of astaxanthin stereoisomers was detected for each pigment source and such distribution was reproduced in the flesh of trout fed with that source. Color values measured in different sites of fillet of rainbow trout fed with different pigment sources showed no significant differences. Similarly, different sources of pigment (natural or synthetic) produced color values of fresh fillet with no relevant or significant differences. The coefficient of distance computed amongst the feed ingredient and the trout fillet astaxanthin stereoisomers was a useful tool to identify the origin of the pigment used on farm.

A method of separating eggs and alevins from wild and farmed Atlantic salmon (*Salmo salar*) is described, using HPLC to trace optical isomers of astaxanthin commonly occurring in the diet of both wild and farmed salmon (27). The proportions of isomers in eggs and alevins of farmed salmon, fed synthetic astaxanthin, differ highly from those of wild fish. The method can be used as a tool to determine the spawning success of escaped farmed female Atlantic salmon.

11.4 OLIVE OIL

The presence or absence of filbertone in 21 admixtures of olive oil with virgin and refined hazelnut oils obtained using various processing techniques from different varieties and geographical origins was evaluated by solid phase microextraction and multidimensional gas chromatography (SPME–MDGC) (28). The obtained results showed that the sensitivity achievable with the proposed procedure was enough to detect filbertone and, hence, to establish the adulteration of olive oil of different varieties with virgin hazelnut oils in percentages of up to 7%. The very low concentrations in which filbertone occurs in some refined hazelnut oils made difficult its detection in specific admixtures. In any case, the minimum adulteration level to be detected depends on the oil varieties present in the adulterated samples. In the present study, the presence of *R*- and *S*-enantiomers of filbertone could be occasionally detected in olive oils adulterated with 10–20% of refined hazelnut oil.

The optimization of the interface performance in the on-line coupling of reversed phase liquid chromatography and gas chromatography was intended to improve the sensitivity achievable in the direct analysis of olive oils adulterated with virgin and refined hazelnut oils (29). The efficient elimination of the eluent coming from the pre-separation was achieved by considering some experimental variables (i.e., transfer volume, interface temperature during transfer, helium flow during both transfer and purge, and purge time) affecting the operation of a vertically positioned programmed temperature vaporizer which acted as the interface of the system. The obtained results demonstrated the possibility of evaluating the genuineness of olive and hazelnut oils as well as of detecting adulterations of olive oil with percentages of around 5% and 10% of virgin and refined hazelnut oils, respectively, in less than 30 min by the method proposed.

On-line coupling of reversed-phase liquid chromatography to gas chromatography (RPLC-GC) using a programmable temperature vaporizer (PTV) as an interface is used for detecting adulterations of hazelnut oil in olive oil on the basis of the determination of the enantiomeric composition of (E)-5-methylhept-2-en-4-one (filbertone) (30). Different variables (i.e., packing material used for trapping the LC–GC transferred solutes, desorption temperature, and eluent composition of the mobile phase during the LC preseparation) are investigated in terms of their influence on the sensitivity achievable in the RPLC-GC analysis. The method does not require any type of sample pretreatment, thus allowing the direct and rapid analysis (under 50 min) of sample oils. Working under the experimental conditions proposed, adulterations of olive oils with hazelnut oil percentages lower than 5% are detectable.

A new method was proposed for the determination of the enantiomeric composition of γ-lactones in different vegetable edible oils (i.e., olive oil, almond oil, hazelnut oil, peanut oil, and walnut oil), and its potential for authenticity control is underlined for a limited number of samples (31).

The method is based on the direct injection (i.e., without requiring a sample pretreatment step) in on-line coupled reversed phase liquid chromatography to gas chromatography (RPLC-GC) using a chiral stationary phase in the GC-step. Different experimental values for both speed of sample introduction into GC and volume of the transferred fraction are considered to improve the recoveries obtained. Relative standard deviations lower than 10% and detection limits ranging from 0.06 to 0.22 mg/L were achieved for the investigated γ-lactones.

11.5 HONEY

Assessment of the botanical origin of the unifloral honeys is of great concern in the context of consumer protection and quality control. In order to find alternatives to the time consuming and uncertain methods, a new analytical approach is proposed; it is based on the enantiomeric ratio investigation of chiral volatile constituents which are derived from the plants being visited by the bees (32). The method was applied to orange honeys; first, the volatile fraction of orange honey and flowers were studied by SPME-GC-MS; a large number of components were identified in orange honeys while linalool prevailed among orange flower volatiles. The enantiomeric ratios of linalool and its oxides were determined and analogous values between honey and flowers resulted. Even if a wide variability in the amount of typical volatile constituents of orange honeys emerged, the enantiomeric ratios of linalool and its oxides remained stable and thus less influenced by production period, conditioning, packaging, storage, etc. As a result the enantiomeric distribution of the honey volatile constituents that directly come from flowers could represent a rapid and easy method for floral origin authenticity.

The distribution of enantiomers of selected chiral volatile organic compounds in 45 monofloral honey samples was studied by GC (33). The volatile organic compounds were extracted from Slovakian rapeseed, acacia, sunflower basswood, and raspberry honeys by solid phase microextraction followed by GC-MS analysis. The chiral compounds present at higher contents were selected from 230 organic compounds found in studied samples for determination of their isomeric ratios. One-dimensional GC with chiral stationary phases provided excellent enantiomer separations; however, the resolved enantiomers often co-eluted with other non-chiral or already separated enantiomers of other organic compounds. Thus, two-dimensional GC with two independent thermostats and proper column setup was required to determine correct isomeric ratios. Finally, the isomeric ratios of linalool, cis- and trans-furanoid linalool oxides, hotrienol, and four isomers of lilac aldehydes were determined. The distribution of enantiomers in honey samples partially depended on their botanical origin. The differences in ratios of lilac aldehyde isomer B and hotrienol were observed for acacia honey that allowed us to distinguish this type of honey from others. Similarly, a different isomeric ratio of trans-furanoid linalool oxide was found for sunflower honeys.

The enantiomer ratios of chiral volatile organic compounds in rapeseed, chestnut, orange, acacia, sunflower, and linden honeys were determined by multi-dimensional gas chromatography using SPME as a sample pre-treatment procedure (34). Linalool oxides, linalool, and hotrienol were present at the highest concentration levels, while significantly lower amounts of α-terpineol, 4-terpineol and all isomers of lilac aldehydes were found in all studied samples. On the other hand, enantiomer distribution of some chiral organic compounds in honey depends on their botanical origin. The significant differences in enantiomer ratio of linalool were observed for rapeseed honey that allows us to distinguish this type of honey from the other ones. The enantiomer ratios of lilac aldehydes were useful for distinguishing of orange and acacia honey from other studied monofloral honeys. Similarly, different enantiomer ratio of 4-terpineol was found for sunflower honeys.

Volatile organic compounds in 14 honey samples (rosemary, eucalyptus, orange, thyme, sage, and lavender) were identified by Pažitná et al. (35). Volatile organic compounds were extracted using a solid phase microextraction method followed by gas chromatography connected with mass spectrometry analysis. The studied honey samples were compared based on their volatile organic compounds composition. In total, more than 180 compounds were detected in the studied samples.

The detected compounds belong to various chemical classes such as terpenes, alcohols, acids, aldehydes, ketones, esters, norisoprenoids, benzene, and furane derivatives, and organic compounds containing sulfur and nitrogen heteroatom. Ten chiral compounds (linalool, *trans*-linalool oxide, *cis*-linalool oxide, 4-terpineol, α-terpineol, hotrienol, and four stereoisomers of lilac aldehydes) were selected for further chiral separation.

Most HPLC enantiomer separations are performed with columns packed with a chiral stationary phase (CSP) operated with an achiral mobile phase. The intrinsically limited chemical selectivity of most CSPs to the simultaneous resolution of several pairs of enantiomers means that complex mixtures of diverse pairs of enantiomers cannot be resolved in a single run due to peak overlapping. Moreover, some drawbacks remain when the analyte is present in very complex samples containing other achiral compounds which can co-elute with the enantiomer peaks. Multidimensional chromatography becomes an option to increase peak capacity and resolve these samples.

The aim of the work of Acquaviva et al. (17) was to study an online fully comprehensive 2D-LC mode utilizing a combination of a chiral column in the first dimension and an achiral column in the second dimension. The 2D-LC system was built with an active flow splitter pump in order to easily adjust the volume of sample transferred into the second dimension and to independently optimize the flow rate in the first dimension.

The present LCxLC method was optimized for the separation of amino acids present in honey samples, taking into account key parameters that influence the bidimensional peak capacity (orthogonality, sampling frequency, etc.). The amino acids have been pre-concentrated on a cation-exchange column followed by derivatization. Several amino acids present in different honey samples have been identified and the data generated has been analyzed by principal component analysis.

11.6 FRUIT BEVERAGES

Enantiomeric compositions of chiral terpenes in commercial fruit beverages were examined by SPME-GC (36). Optimization of the method was accomplished on the basis of some parameters involved in the extraction, such as heating temperature and extraction time, that provided the highest peak areas, 60°C and 2 min being the optimal values. With the proposed method relative standard deviation values from three replicates ranging from 2 to 12% were obtained. The enantiomeric distribution of some terpenes remained constant, whereas other terpenes (linalool, terpinen-4-ol, and α-terpineol) exhibited a considerable variation among samples. This can be indicative of the eventual addition of aromas to some fruit beverages.

The enantiomer ratios of chiral volatile organic compounds in fruit distillates were determined by multidimensional gas chromatography using solid-phase microextraction (SPME) as a sample treatment procedure (37). Linalool and its oxides, limonene, α-terpineol, and nerolidol, were present at the highest concentration levels, while significantly lower amounts of β-citronellol and lactones were found in the studied samples. However, almost all terpenoids mainly occur as a racemic or near-racemic mixture; enantiomer distribution of some chiral organic compounds in fruit distillates correlated to a botanical origin. In particular, a significant enantiomeric excess of (R)-linalool and (S)-α-terpineol was found only for pear brandy, and likewise the dominance (R)-limonene and the second eluted enantiomer of nerolidol for *Sorbus domestica* and strawberry, respectively. The distribution of γ-lactones stereoisomers was more nonspecific, with a general excess of the R-enantiomer.

The content of α-hydroxy acids and their enantiomers can be used to distinguish authentic and adulterated fruit juices (38). The use of ligand exchange CE with two kinds of central metal ion in a BGE for the simultaneous determination of enantiomers of DL-malic, DL-tartaric, and DL-isocitric acids, and citric acid was investigated. Ligand exchange CE with 100 mM D-quinic acid as a chiral selector ligand and 10 mM Cu(II) ion as a central metal ion could enantioseparate DL-tartaric acid but not DL-malic acid or DL-isocitric acid. Addition of 1.8 mM Sc(III) ion to the BGE with 10 mM Cu(II) ion to create a dual central metal ion system permitted the simultaneous determination of these α-hydroxy acid enantiomers and citric acid. The proposed ligand exchange CE was thus well suited for detecting adulteration of fruit juices.

A method based on the on-line coupling of reversed phase liquid chromatography with gas chromatography/mass spectrometry (RPLC-GC-MS) for the chiral evaluation of characteristic constituents of fruit beverage aroma was investigated (39). The consideration of a variety of parameters involved in the transfer step allowed to achieve relative standard deviations ranging from 0.4 to 10% in most cases and detection limits from 0.2 to 2.5 mg/l. By applying the developed method to fruit beverages, racemic mixtures of ethyl 2-methylbutanoate and γ-nonalactone were found. This fact suggests the eventual addition of artificial aromas. The method proposed in the present work can be useful to assess reliably the authenticity of aqueous samples, such as fruit beverages.

Headspace–solid phase microextraction stereoselective gas chromatography–mass spectrometry (HS–SPME stereoselective GC–MS) was used to determine the enantiomeric ratios of chiral flavor and fragrance indicators present in the raw materials and in the products (40). γ-Decalactone in peach, nectarine, strawberry, passion fruit, and apricot, as well as (E)-α-ionone in raspberry and linalool and linalyl acetate in bergamot, were used as indicators in stereoselective analysis of commercial foods and beverages. In the fresh fruit aromas only one enantiomer was predominant or enantiomerically pure. In many commercial fruit products racemic mixtures of the above-mentioned chiral flavor compounds were detected.

Hesperidin (hesperetin-7-O-rutinoside), a flavonoid affecting vascular function, is abundant in citrus fruits and derived products such as juices (41). After oral administration, hesperidin is hydrolyzed by the colonic microbiota producing hesperetin-7-O-glucoside, the glucoside group is further cleaved and the resulting hesperetin is absorbed and metabolized. Flavanones have a chiral carbon generating (R)- and (S)-enantiomers, with potentially different biological activities. A rapid UPLC–MS/MS method for the analysis of (R)- and (S)-hesperetin enantiomers in human plasma and urine was developed and validated. Biological matrices were incubated with β-glucuronidase/sulfatase, and hesperetin was isolated by solid-phase extraction using 96-well plate mixed-mode cartridges having reversed-phase and anion-exchange functionalities. Racemic hesperetin was analyzed with a UPLC HSS T3 reversed phase column and hesperetin enantiomers with a HPLC Chiralpak IA-3 column using H_2O with 0.1% CHOOH as solvent A and acetonitrile with 0.1% CHOOH as solvent B. The method was linear between 50 and 5000 nM for racemic hesperetin in plasma and between 25 and 2500 nM for (S)- and (R)-hesperetin in plasma. Linearity was achieved between 100 and 10,000 nM for racemic hesperetin in urine and between 50 and 5000 nM for (S)- and (R)-hesperetin in urine. Values of repeatability and intermediate reproducibility for racemic hesperetin and enantiomers in plasma and urine were below 15% of deviation in general, and maximum 20% for the lowest concentrations. In addition, the method was applied for the quantification of total hesperetin and of hesperetin enantiomers in human plasma and urine samples, obtained after oral ingestion of purified hesperetin-7-O-glucoside.

11.7 DAIRY PRODUCTS

The review of Mariaca et al. (42) lists a total of 126 chiral compounds described in the literature for at least 21 different cheese varieties: 30 alcohols, 8 ketones, 2 aldehydes, 20 esters, 20 lactones, 27 carboxylic acids, 13 hydrocarbons (8 terpenes), 4 furanones, and 2 amines. Of these only a few authentic reference compounds are commercially available.

11.8 ESSENTIAL OILS

The review of König and Hochmuth describes current analytical technology for the analysis of chiral constituents in essential oils and other natural volatiles, flavor, and fragrance compounds (43). The technique of enantioselective gas chromatography is described and applied for assigning absolute configuration of chiral natural compounds, which is strongly connected to differences in odor properties of their enantiomers. In addition, some recent results to facilitate the handling of GC-mass spectrometry data of known and unknown plant volatiles are discussed.

The paper of Schubert and Mosandl (44) reports on the chiral evaluation of linalool as an additional criterion of authenticity of lavender oils, using enantioselective multidimensional gas chromatography with a column combination of Superox and heptakis(2,3,6-tri-O-ethyl)-β-cyclodextrin in OV-1701-vi. This method is extended to some other valuable spice oils-investigations on the enantiomeric distribution of their linalool content are discussed in terms of quality control of essential oil.

Enantioselective multidimensional gas chromatography using the column combination Carbowax 20 M/heptakis-(2,3-di-O-acetyl-6-O-tert-butyldimethylsilyl)-β-cyclodextrin in OV 1701-vinyl allows the simultaneous stereodifferentiation of trans- and cis-linalol oxides (furanoid), camphor, octan-3-ol, oct-1-en-3-ol, linalyl acetate, lavandulol, terpinen-4-ol, and linalool (53). The method is applied to commercially available as well as self-prepared essential oils of *Lavandula* species. Chirality evaluation is discussed in view of quality assessment of these essential oils.

Using enantioselective multidimensional gas chromatography (enantio-MDGC) and the column combination polyethylene glycol/heptakis (2,3-di-O-acetyl-6-O-tert-butyldimethylsilyl)-β-cyclodextrin in OV 1701-vi, the chiral monoterpenoids cis/trans-rose oxides, octan-3-ol, oct-1-en-3-ol, linalol, citronellal, citronellol, and citronellic acid methyl ester were stereoanalyzed (45). The method is applied to chirality evaluation of these compounds from balm oils. The enantiomeric distributions are discussed in order to assess the authenticity of this essential oil.

Linalool is a major volatile component of tea aroma. It occurs naturally as R-(−)- and S-(+)-linalool enantiomers, which exhibit entirely different sensorial properties. Using headspace solid-phase microextraction with chiral gas chromatography, R-(−)- and S-(+)-linalool was quantitated in teas (46). Optimal extraction conditions were as follows: CAR–DVB–PDMS fiber, 60 min at 60°C, and tea/water ratio of 1:6 (w(g)/v(mL)). The authors measured linalool levels in five different teas, in fresh leaves of 14 different cultivars, and in samples collected during processing of green and black teas. Enantiomeric distributions of linalool were significantly different among the different teas and among the different tea cultivars. R-(−)-Linalool and S-(+)-linalool reached maximum levels during the rolling of black tea, with the levels declining drastically during green tea processing. Although tea germplasm was an important factor in determining the enantiomeric distribution of linalool in tea products, the processing steps also had a large impact on linalool levels.

The sensitivity of SPME/GC/MS for the analysis of chiral volatile compounds in food matrices was intended to be improved (47). To that end, a new approach based on the freezing/defrosting of the sample prior to the extraction is described. The defrosting time was carefully optimized, obtaining different optimal values according to the matrix studied. Relative standard deviations from three replicates of the overall procedure were lower than 12% in all cases. By applying the method proposed, peak areas for the compounds of interest up to nine times higher were obtained in some cases. This improvement in the sensitivity allowed to increase the reliability in the identification of volatile components as well as to detect certain chiral minor compounds in foods.

Multidimensional enantioselective gas chromatography, employing heart-cutting from a nonchiral precolumn onto a selectivity adjusted, modified cyclodextrin coated chiral main column, permits the simultaneous stereo differentiation of the chiral major compounds of buchu leaf oil (48). Commercially available essential oil as well as laboratory prepared samples were directly analyzed. The enantiomeric distribution of limonene, menthone, isomenthone, and pulegone was investigated, and the characteristic flavor impact compounds, 3-oxo-p-menthane-8-thiol and its thiol acetate, were stereoanalyzed. Their chirality evaluation is discussed as an indicator of authenticity of "cassis" type fruit aromas and fragrances.

The review of Smelcerovic et al. (49) provides an overview of the emerging trends in analysis of essential oils and indicates future trends. Since early human history the essential oils have been used in folk medicine, food, and cosmetic industries in various parts of the world. GC-FID is the

Food Authenticity and Adulteration 291

traditional method for essential oils quantification while GC-MS is the most common analytical method for qualitative analysis. Chiral GC allowed the identification of a great number of chiral essential oil constituents. An alternative to GC analysis is HPLC chromatography. The use of hyphenated techniques, such as LC-MS-MS provides important information about the structure of essential oils constituents. However, only a comparatively small number of reports on essential oil analyses by HPLC can be found in the literature. Multidimensional chromatography is an approach capable of providing greater resolution. 13C-NMR spectroscopy is a complementary tool for analysis of essential oils. Advantage of this method compared to mass spectrometry is identification of stereoisomers and thermally unstable compounds while its disadvantage is inability to identify the minor oils constituents. 1H-NMR as an online tool for GC analysis showed promising results but requires further research to be applied on the analysis of essential oils. The range of information obtained from essential oils analysis enables the application of chemometrics.

Shellie (50) discusses the relationship between essential oil analysis and gas chromatography (GC). GC, particularly when combined with mass spectrometry (MS), contributes to the development of the science of essential oils and fragrances in the areas of phytochemistry, chemotaxonomy, olfactory research, biochemistry, plant-insect research, the search for new sources of odoriferous compounds for industry, and quality control. Essential oils are analyzed to determine the qualitative and/or quantitative composition of a product; control the quality and authenticity of the product; and detect the presence of adulteration or contamination. One of the major objectives of plant extract analysis by 2D gas chromatography (GCxGC) is to determine a sample composition. Characterization of volatile plant extracts relies heavily on the retention index-based identification of the separated components, and this is usually supported by MS. A simple procedure that permits the direct method translation of retention times in 2D for GCxGC–FID and GCxGC–ToF MS analyses is described in the chapter. The chapter discusses the use of retention indices in the GCxGC analyses of essential oils and fragrance compounds. Retention indices are extremely important in identification strategies of essential oil and fragrance components because mass spectral data alone are generally insufficient to provide positive identification.

11.9 FLAVORS

The acceptance of food strongly depends on their flavor and aroma impressions (50, 52). Consequently, authentication of genuine flavors is an important topic in view of quality assurance in the food industry and in consumer protection. Both phenomena, enantioselectivity as well as isotope discrimination during biosynthesis may serve as inherent parameters in the authenticity control of natural flavor compounds, provided that suitable methods and comprehensive data from authentic sources are available. Besides *site* specific *n*atural *i*sotope *f*ractionation, measured by NMR-spectroscopy (SNIF-NMR), enantio-selective *c*apillary *g*as *c*hromatography (enantio-cGC) and comparative *i*sotope *r*atio *m*ass spectrometry (IRMS), have proved to be highly efficient tools in the origin specific analysis. Nevertheless, analytical authentication of genuine food constituents is a permanent challenge, due to the complexity of food matrices. So far, enantioselective and/or IRMS online coupling techniques are the methods of choice in order to determine the authenticity of flavors. The benefits of *s*tir *b*ar *s*orptive *e*xtraction (SBSE)-enantio-MDGC/MS and multielement GC-IRMS techniques are outlined.

Chiral structure of proteins turned out to be significant in the perception of taste perceptions as well as the perception of aroma (53). The most characteristic phenomenon is chirality among odorants. Numerous compounds essential for aroma are found in nature in the form of two isomers, with considerable predominance of one of them, while the ratio of these isomers is specific and stable. Enantiomers of odorants may differ in the intensity and character of their smell. Only the development of methods facilitating the separation of optical isomers created a possibility to investigate its universality and importance in living organisms. The application of more and more advanced analytical techniques in the control of chiral nature of food ingredients may be used to detect adulterations and especially to determine authenticity of fragrances.

Native concentrations of α-ionone, β-ionone, and β-damascenone were studied in various authentic and commercial wines (54). In addition, the enantiomeric distribution of α-ionone was determined and its merits as a potential marker for aroma adulteration in wine were discussed. For extraction of volatiles, headspace solid-phase microextraction (HS-SPME) was applied, followed by heart-cut multidimensional gas chromatography coupled to tandem mass spectrometric detection for trace-level analysis. The enantioselective analysis of α-ionone was achieved with octakis(2,3-di-O-pentyl-6-O-methyl)-γ-cyclodextrin as the chiral selector in the separation column for gas chromatography (GC). In all the authentic wines studied, α-ionone showed a high enantiomeric ratio in favor of the (R)-enantiomer. Since an illegal addition of α-ionone in a racemic form changes the enantiomeric ratio, this ratio may serve as an adulteration marker. Concentrations varied between <LOD to 0.081 µg/L for α-ionone, <LOD to 1.0 µg/L for β-ionone, and 0.03–10.3 µg/L for β-damascenone. Commercial wines of suspiciously strong flavor yielded concentrations up to 4.6 µg/L for α-ionone, 3.6 µg/L for β-ionone, and 4.3 µg/L for β-damascenone. Elevated α- and β-ionone concentrations serve as additional indicators for a potential adulteration. In order to classify the concentrations of the analytes in the context of their odor activity in wine, odor thresholds were determined.

A fast and simple method for authenticating raspberry flavors from food products was developed (55). The two enantiomers of the compound (E)-α-ionone from raspberry flavor were separated on a chiral gas chromatographic column. Based on the ratio of these two enantiomers, the naturalness of a raspberry flavor can be evaluated due to the fact that a natural flavor will consist almost exclusively of the R enantiomer, while a chemical synthesis of the same compound will result in a racemic mixture. Twenty-seven food products containing raspberry flavors where investigated using SPME-chiral-GC-MS. The authors found raspberry jam, dried raspberries, and sodas declared to contain natural aroma all contained almost only R-(E)-α-ionone supporting the content of natural raspberry aroma. Six out of eight sweets tested did not indicate a content of natural aroma on the labeling which was in agreement with the almost equal distribution of the R and S isomer. Two products were labeled to contain natural raspberry flavors but were found to contain almost equal amounts of both enantiomers indicating a presence of synthetic raspberry flavors only. Additionally, two products that were labeled to contain both raspberry juice and flavor showed equal amounts of both enantiomers, indicating the presence of synthetic flavor.

Both phenomena, enantioselectivity as well as isotope discrimination during biosynthesis, may serve as "endogenous" parameters, provided that suitable methods and comprehensive data from authentic sources are available. The review of Mosandl A. (56) reports on enantioselective capillary gas chromatography and online methods of isotope-ratio mass spectrometry in the authentication of food flavor and essential oil compounds, referring to literature references published until 2002.

The authenticity of fruit flavors in foods and beverages is of great importance for the producers of jams, ice creams, soft drinks, yogurt, etc. Regulations regarding the authenticity of flavors in foods and beverages will help consumers choose the foods best suitable for them. In order to enhance this goal, a headspace–solid phase microextraction stereoselective gas chromatography–mass spectrometry (HS–SPME stereoselective GC–MS) was used to determine the enantiomeric ratios of chiral flavor and fragrance indicators present in the raw materials and in the products (57). γ-Decalactone in peach, nectarine, strawberry, passion fruit, and apricot, as well as (E)-α-ionone in raspberry and linalool and linalyl acetate in bergamot, were used as indicators in stereoselective analysis of commercial foods and beverages. In the fresh fruit aromas only one enantiomer was predominant or enantiomerically pure. In many commercial fruit products racemic mixtures of the above-mentioned chiral flavor compounds were detected. The results suggest that HS–SPME stereoselective GC–MS could be a useful tool for differentiating natural flavor compounds from synthetic ingredients in quality control of foods and beverages.

Solid-phase microextraction combined with gas chromatographic/mass spectrometric analysis and separation on a chiral cyclodextrin stationary phase was a rapid, reliable technique for profiling chiral aroma compounds in flavored alcoholic beverages (58). Several enantiomeric terpenes, esters, alcohols, norisoprenoids, and lactones were identified in berry-, peach-, strawberry-, and citrus-flavored wine and malt beverages (wine coolers). Using this technique, we were able to confirm the addition of synthetic flavoring to several beverages, consistent with label designations.

Stir bar sorptive extraction (SBSE), was applied for the enantioselective analysis of chiral flavor compounds in strawberries and their processed products using enantioselective multidimensional gas chromatography-mass spectrometry (enantio-MDGC-MS) (59).

Using this rapid method the usual preparation steps for complex matrices such as liquid/liquid extraction, solid phase extraction or distillation techniques can be omitted. The analyzed chiral compounds showed fruit-specific characteristic enantiomeric ratios, irrespective of food processing techniques employed. Hence, it is possible to differentiate the flavor of authentic and aromatized strawberry products.

Using SBSE the usual preparation steps for complex matrices such as liquid/liquid extraction, solid phase extraction or distillation techniques can be omitted (59). The analyzed chiral compounds showed fruit-specific characteristic enantiomeric ratios, irrespective of food processing techniques employed. Hence, it is possible to differentiate the flavor of authentic and aromatized strawberry products.

A new method to detect the existence of chiral amino acids in orange juice is presented (60). The method employs β-cyclodextrins and micellar electrokinetic chromatography with laser-induced fluorescence (MEKC–LIF) to separate and detect l- and d-amino acids (l-aa and d-aa) previously derivatized with fluorescein isothiocianate (FITC). A systematic optimization of the chiral-MEKC conditions is done bringing about in less than 20 min a good separation of the main amino acids found in orange juice (i.e., Pro, Asp, Ser, Asn, Glu, Ala, Arg, and the nonchiral GABA, i.e., γ-aminobutyric acid). Using this procedure, the analysis time reproducibility for the 15 standard compounds (l-aa, d-aa, and GABA) has been determined to be better than 0.2% (n = 5) for the same day and better than 0.7% (n = 15) for three different days. Corrected peak area reproducibility is somewhat lower, providing values better than 3.3% (n = 5) for the same day and 6.9% (n = 15) for three different days. The limit of detection using this procedure was determined to be 0.86 attomoles for l-Arg. The optimized FITC derivatization method allows the easy and straightforward detection of amino acids in orange concentrates and juices (i.e., only centrifugation of diluted samples for 5 min is needed prior to their derivatization). d-Ala, d-Asp, d-Arg, and d-Glu were determined in orange juices and orange concentrates from different geographical origins using this new method. Moreover, the effect of different temperature treatments (50, 92, and 150°C) on the content of d-aa in orange juice was evaluated.

The review of Giuffrida et al. (61) highlights recent progresses in the chiral recognition and separation of amino acid enantiomers obtained by capillary electromigration techniques, using different chiral selectors and especially cyclodextrins, covering the literature published from January 2010 to March 2014. Sections are dedicated to the use of derivatization reagents and to the possibility to enantioseparate underivatized amino acids by using either ligand exchange capillary electrophoresis (LECE) and capillary electrophoresis (CE) coupled on line with mass spectrometry. A short insight on frontier nanomaterials is also given.

The enantioselective analysis and the biogenesis of rose oxide, rose oxide ketone, nerol oxide, furanoid linalooloxide, pyranoid linalooloxide, dill ether, linden ether, menthofuran, and 3,6-dimethylcoumaran is reviewed (62). All these compounds are important chiral cyclic monoterpenoid ethers and can be analyzed enantioselectively using modified cyclodextrins as chiral stationary phases in capillary gas chromatography (cGC). The evaluation of the enantiomeric ratios of these compounds can be used in the authenticity control of flavors and fragrances. Moreover, the enantioselective cGC is also an efficient tool in elucidating the biogenesis of these compounds. Biomimetic studies as well as in vivo feeding experiments using stable isotope labelling have shown that polyhydroxylated or epoxidated monoterpenes are direct precursors of these cyclic ethers.

The variability of the enantiomeric distribution of biologically active chiral terpenes in *Mentha piperita* plants from different geographical origins was evaluated by solid phase microextraction–gas chromatography–mass spectrometry (63). The optimization of some parameters (i.e., exposure temperature, extraction time, and type of fiber) affecting SPME-extraction enabled relative standard deviations ranging from 4 to 13% to be achieved. The use of two different chiral stationary phases allowed to separate the identified chiral terpenes into their corresponding enantiomers as well as to verify the enantiomeric excesses of those compounds which were enantiomerically resolved on both phases. For all chiral terpenes, the enantiomeric composition varied within a very narrow range all over the samples. Consequently, it may be stated that the enantiomeric composition of chiral terpenes in *Mentha piperita* appears to be independent of the geographical origin of the plant and, thus, any alteration in the characteristic value may be related to an adulteration or inadequate sample handling.

The developments and applications of capillary electromigration methods coupled on-line with mass spectrometry for chiral analysis are discussed (64). The multiple enantiomeric applications of this hyphenated technology are covered including chiral analysis of drugs, food compounds, pesticides, natural metabolites, etc., in different matrices such as plasma, urine, medicines, foods, etc. This work intends to provide an updated overview (including works published till September 2009) on the principal chiral applications carried out by CZE-MS, CEC-MS, and MEKC-MS, discussing their main advantages and drawbacks in all their different areas of application as well as their foreseeable development in the nondistant future.

REFERENCES

1. Rizwanullah M., Khan N., Amin S., Ahmad J., Imam S.S., Fahkri K.U., Rizvi, M.M.A. Introduction to food authentication based on fingerprinting techniques. In *Fingerprinting Techniques in Food Authentication and Traceability*. Eds. K.S. Siddiqi and L.M.L. Nollet. CRC Press, Boca Raton Fl, 2018.
2. Danezis G.P., Tsagkaris A.S., Camin F., Brusic V., Georgiou, C.A. Food authentication: Techniques, trends & emerging approaches. *TrAC Trends Anal Chem*, 2016, 85, 123–32.
3. Olsen P., Borit, M. How to define traceability. *Trends Food Sci Tech*, 2013, 29, 142–50.
4. Rizwanullah M., Khan N., Amin S., Ahmad J., Imam S.S., Fahkri K.U., Rizvi, M.M.A. Introduction to food traceability based on fingerprinting techniques. In *Fingerprinting Techniques in Food Authentication and Traceability*. Eds. K.S. Siddiqi and L.M.L. Nollet. CRCPress, Boca Raton Fl, 2018.
5. Orre, K.. The logistics of food supply following radioactive fallout. *J Environ Radioact*, 2005, 83, 429–32.
6. Saltini R., Akkerman, R. Testing improvements in the chocolate traceability system: Impact on product recalls and production efficiency. *Food Control*, 2012, 23, 1, 221–26.
7. Van Dorp, K.J. Tracking and tracing: A structure for development and contemporary practices. *Logist Inform Manag*, 2002, 15, 1, 24–33.
8. Badia-Melis R., Mishra P., Ruiz-García, L. Food traceability: New trends and recent advances. A review. *Food Control*, 2015, 57, 393–401.
9. Kofink M., Papagiannopoulos M., Galensa, R. Enantioseparation of catechin and epicatechin in plant food by chiral capillary electrophoresis. *European Food Research and Technology*, 2007, 225, 569–577.
10. Kofink M., Papagiannopoulos M., Galensa, R. (-)-Catechin in cocoa and chocolate: Occurence and analysis of an atypical flavan-3-ol enantiomer. *Molecules*, 2007, 12, 7, 1274–1288.
11. Rinaldo D., Batista J.M., Rodrigues J., Befatti A.C., Rodrigues C.M., Dos Santos L.C., Furlan M., Vilegas, W. Determination of catechin diastereomers from the leaves of *Byrsonima* species using chiral HPLC-PAD-CD. *Chirality*, 2010, 22, 8, 726–733.
12. Machonis P.R., Jones M.A., Schaneberg B.T., Kwik-Uribe, C.L. Method for the determination of catechin and epicatechin enantiomers in cocoa-based ingredients and products by high-performance liquid chromatography: Single-laboratory validation. *Journal of AOAC International*, 2012, 95, 2, 500–507.

13. Machonis P., Jones M., Schaneberg B., Kwik-Uribe C., Dowell, D. Method for the determination of catechin and epicatechin enantiomers in cocoa-based ingredients and products by high-performance liquid chromatography: First action 2013.04. *Journal of AOAC International*, 2014, 97, 2, 506–509.
14. Alañón M.E., Castle S.M., Siswanto P.J., Cifuentes-Gómez T., Spencer, J.P.E. Assessment of flavanol stereoisomers and caffeine and theobromine content in commercial chocolates. *Food Chemistry*, 2016, 1, 177–184.
15. Fanali C., Fanali S., Chankvetadze, B. HPLC separation of enantiomers of some flavanone derivatives using polysaccharide-based chiral selectors covalently immobilized on silica. *Chromatographia*, 2016, 79, 119–124.
16. Ilisz I., Berkecz R., Péter, A. HPLC separation of amino acid enantiomers and small peptides on macrocyclic antibiotic-based chiral stationary phases: A review. *Journal of Separation Science*, 2006, 29, 10, 1305–1321.
17. Acquaviva A., Siano G., Quintas P., Filqueira M.R., Castells, C.B. Chiral x achiral multidimensional liquid chromatography. Application to the enantioseparation of dintitrophenyl amino acids in honey samples and their fingerprint classification. *Journal of Chromatography A*, 2020, 1614, 460729.
18. Kodama S., Yamamoto A., Matsunaga A., Yanai, H. Direct enantioseparation of catechin and epicatechin in tea drinks by 6-O-α-D-glucosyl-β-cyclodextrin-modified micellar electrokinetic chromatography. *Electrophoresis*, 2004, 25, 16, 2892–2898.
19. Fiori J., Pasquini B., Caprini C., Orlandini S., Furlanetto S., Gotti, R. Chiral analysis of theanine and catechin in characterization of green tea by cyclodextrin-modified micellar electrokinetic chromatography and high-performance liquid chromatography. *Journal of Chromatography A*, 2018, 1562, 115–122.
20. Dias T., Silva M.R., Damiani C., da Silva, F.A. Enzymatic-spectrophotometric method with tyrosinase. *Food Analytical Methods*, 2017, 10, 3914–3923.
21. Ismael I., Rockenbach I., Jungfer E., Ritter C., Santiago-Schübel B., Thiele B., Fett R., Galensa, R. Characterization of flavan-3-ols in seeds of grape pomace by CE, HPLC-DAD-MSn and LC-ESI-FTICR-MS
22. Pérez-Trujillo J.P., Hernández Z., López-Bellido F.J., Hermosín-Gutiérrez, I. Characteristic phenolic composition of single-cultivar red wines of the Canary Islands (Spain). *J. Agric. Food Chem.* 2011, 59, 11, 6150–6164
23. Casal S., Alves M.R., Mendes E., Oliveira M.B.P.P., Ferreira M.A. Discrimination between Arabica and Robusta coffee species on the basis of their amino acid enantiomers. *J. Agric. Food Chem.* 2003, 51, 22, 6495–6501.
24. Gotti R., Furlanette S., Pinzuti S., Cavrini, V. Analysis of catechins in *Theobroma cacao* beans by cyclodextrin-modified micellar electrokinetic chromatography. *Journal of Chromatography A*, 2006, 1112, 1-2, 345–352.
25. Molkentin J., Lehmann I., Ostermeyer U., Rehbein, H. Traceability of organic fish – Authenticating the production origin of salmonids by chemical and isotopic analyses. *Food Control*, 2015, 53, 55–66.
26. Moretti V.M., Mentasti T., Bellagamba F.,Luzzana U., Caprino F., Turchini, G.M. Determination of astaxanthin stereoisomers and colour attributes in flesh of rainbow trout (*Oncorhynchus mykiss*) as a tool to distinguish the dietary pigmentation source. *Food Additives & Contaminants*, 2006, 23, 11, 1056–1063.
27. Lura H., Saegrov, H. A method of separating offspring from farmed and wild atlantic salmon (*Salmo salar*) based on different ratios of optical isomers of astaxanthin. *Canadian Journal of Fisheries and Aquatic Sciences*, 1991, 48, 3, 429–433.
28. Flores G., Ruiz del Castillo M.L., Blanch G.P., Herraiz, M. Detection of the adulteration of olive oils by solid phase microextraction and multidimensional gas chromatography. *Food Chemistry*, 2006, 97, 2, 336–342.
29. Flores G., Ruiz del Castillo M.L., Herraiz M., Blanch, G.P. Study of the adulteration of olive oil with hazelnut oil by on-line coupled high-performance liquid chromatographic and gas chromatographic analysis of filbertone. *Food Chemistry*, 2006, 97, 4, 742–749.
30. Ruiz del Castillo M.L., del Mar Caja M., Herraiz M., Blanch, G.P. Rapid recognition of olive oil adulterated with hazelnut oil by direct analysis of the enantiomeric composition of filbertone. *J. Agric. Food Chem.*, 1998, 46, 12, 5128–5131.
31. Ruiz del Castillo M.L., Herraiz M., Blanch, G.P. Determination of the enantiomeric composition of γ-lactones in edible oils by on-line coupled high-performance liquid chromatography and gas chromatography. *J. Agric. Food Chem.*, 2000, 48, 4, 1186–1190
32. Verzera A., Tripodi G., Condurso C., Dima G., Marra, A. Chiral volatile compounds for the determination of orange honey authenticity. *Food Control*, 2014, 39, 237–243.

33. Pažitná A., Antónia Janáčová A., Špánik, I. The enantiomer distribution of major chiral volatile organic compounds in Slovakian monofloral honeys. *Journal of Food and Nutrition Research*, 2012, 51 (4), 236–241.
34. Špánik I., Pažitná A., Šiška P., Szolcsányi, P. The determination of botanical origin of honeys based on enantiomer distribution of chiral volatile organic compounds. *Food Chemistry*, 2014, 158, 497–503.
35. Pažitná A., Džúrová J., Špánik, I. Enantiomer distribution of major chiral volatile organic compounds in selected types of herbal honeys. *Chirality*, 2014, 26, 670–674.
36. Ruiz del Castillo M.L., Caja M.M., Herraiz, M. Use of the enantiomeric composition for the assessment of the authenticity of fruit beverages. *J. Agric. Food Chem.*, 2003, 51, 5, 1284–1288.
37. Vyviurska O., Zvrškovcová H., Špánik, I. Distribution of enantiomers of volatile organic compounds in selected fruit distillates. *Chirality*, 2017, 29, 1, 14–18.
38. Kodama S., Aizawa S., Taga A., Yamamoto A., Honda Y., Suzuki K., Kemmei T., Hayakawa K. Determination of α-hydroxy acids and their enantiomers in fruit juices by ligand exchange CE with a dual central metal ion system. *Electrophoresis*, 2013, 34, 9-10, 1327–1333.
39. Caja M.M., Blanch G.P., Herraiz M., Ruiz del Castillo, M.L. On-line reversed–phase liquid chromatography-gas chromatography coupled to mass spectrometry for enantiomeric analysis of chiral compounds in fruit beverages. *Journal of Chromatography A*, 2004, 1024, 1-2, 81–85.
40. Uzi R., Meital E., Zamir C., Cohen K., Larkov O., Aly, R. Authenticity assessment of natural fruit flavour compounds in foods and beverages by auto-HS–SPME stereoselective GC–MS. *Flavour and Fragrance Journal*, 2009, 25, 1, 20–27.
41. Lévèques A., Actis-Goretta L., Rein M.J., Williamson G., Dionisi F., Giuffrida, F. UPLC–MS/MS quantification of total hesperetin and hesperetin enantiomers in biological matrices. *Journal of Pharmaceutical and Biomedical Analysis*, 2012, 57, 1–6.
42. Mariaca R.G., Imhof M.I., Bosset, J.O. Occurrence of volatile chiral compounds in dairy products, especially cheese – A review. *European Food Research and Technology.*, 2001, 212, 253–261.
43. König W.A., Hochmuth, D.H. Enantioselective gas chromatography in flavor and fragrance analysis: Strategies for the identification of known and unknown plant volatiles. *Journal of Chromatographic Science*, 2004, 42, 8, 423–439.
44. Schubert V., Mosandl., A. Chiral compounds of essential oils. VIII: Stereodifferentiation of linalool using multidimensional gas chromatography. *Phytochemical Analysis*, 1991, 2, 4, 171–174.
45. Kreis P., Mosandl, A. Chiral compounds of essential oils. Part XVI. enantioselective multidimensional gas chromatography in authenticity control of balm oil (*Melissa officinalis* L.). *Flavour and Fragrance Journal*, 1994, 9, 5, 249–256.
46. Yang T., Zhu Y., Shao C.-Y., Zhang Y., Shi J., Lv H.-P., Lin, Z. Enantiomeric analysis of linalool in teas using headspace solid-phase microextraction with chiral gas chromatography. *Industrial Crops and Products*, 2016, 83, 17–23.
47. Flores G., Ruiz del Castillo M.L., Blanch G.P., Herraiz, M. Effect of sample freezing on the SPME performance in the analysis of chiral volatile compounds in foods. *Food Chemistry*, 2006, 96 2, 334–339.
48. Köpke T., Dietrich A., Mosandl, A. Chiral compounds of essential oils XIV: Simultaneous stereoanalysis of buchu leaf oil compounds. *Phytochemical Analysis*, 1994, 5, 2, 61–67.
49. Smelcerovic A., Djordjevic A., Lazarevic J., Stojanovic G. Recent advances in analysis of essential oils. *Current Analytical Chemistry*, 2013, 9, 1 61–70.
50. Zawirska-Wojtasiak R. Chirality and the nature of food authenticity of aroma. *Acta Scientarium Polonorum Technologia Alimentaria*, 2006, 5, 1, 21–36.
51. Shellie R.A. Chapter 9 volatile components of plants, essential oils, and fragrances. *Comprehensive Analytical Chemistry*, 2009, 55, 189–213.
52. Mosandl A., Kreck M., Jung J., Sewenig, S. Sophisticated Online Techniques in the Authenticity Assessment of Natural Flavors. 2006. Authentication of Food and Wine. Chapter 4pp 52-74. ACS Symposium Series Vol. 952.
53. Kreis P., Mosandl, A. Chiral compounds of essential oils. Part XI. Simultaneous stereoanalysis of lavandula oil constituents. *Flavour and Fragrance Journal*, 1992, 7, 4, 187–193.
54. Langen J., Wegman-Herr P., Schmarr, H.-G. Quantitative determination of α-ionone, β-ionone, and β-damascenone and enantiodifferentiation of α-ionone in wine for authenticity control using multidimensional gas chromatography with tandem mass spectrometric. Detection. *Analytical and Bioanalytical Chemistry*, 2016, 408, 6483–6496.
55. Hansen A.-M. S, Frandsen H.L., Fromberg, A. Authenticity of raspberry flavor in food products using SPME-chiral-GC-MS. *Food Science & Nutrition*, 2015, 4, 3, 348–354.

56. Mosandl, A. Authenticity assessment: A permanent challenge in food flavor and essential oil analysis. *Journal of Chromatographic Science*, 2004, 42, 8, 440–449.
57. Ravid U., Elkabetz M., Zamir C., Cohen K., Larkov O., Aly, R. Authenticity assessment of natural fruit flavour compounds in foods and beverages by auto-HS–SPME stereoselective GC–MS. *Flavour and Fragrance Journal*, 2010, 25, 1, 20–27.
58. Ebeler S.E., Sun G.M., Datta M., Stremple P., Vickers, A.K. Solid-phase microextraction for the enantiomeric analysis of flavors in beverages. *Journal of AOAC International*, 2001, 84, 2, 479–485.
59. Kreck M., Scharrer A., Bilke S., Mosandl, A. Stir bar sorptive extraction (SBSE)-enantio-MDGC-MS – a rapid method for the enantioselective analysis of chiral flavour compounds in strawberries. *European Food Research and Technology*, 2001, 213, 389–394.
60. Simó C., Barbas C., Cifuentes, A. Sensitive micellar electrokinetic chromatography–laser-induced fluorescence method to analyze chiral amino acids in orange juices. *J. Agric. Food Chem.* 2002, 50, 19, 5288–5293.
61. Giuffrida A., Maccarrone G., Cucinotta V., Orlandici S., Contino, A. Recent advances in chiral separation of amino acids using capillary electromigration techniques. *Journal of Chromatography A*, 2014, 1363, 41–50.
62. Wüst M., Mosand, A. Important chiral monoterpenoid ethers in flavours and essential oils – enantioselective analysis and biogenesis. *European Food Research and Technology*, 1999, 209, 3–11.
63. Ruiz del Castillo M.L., Blach G.P., Herraiz, M. Natural variability of the enantiomeric composition of bioactive chiral terpenes in *Mentha piperita*. *Journal of Chromatography A*, 2004, 1054, 87–93.
64. Simó C., García-Cañas V., Cifuentes, A. Chiral Capillary electrophoresis-mass spectrometry of amino acids in foods. *Electrophoresis*, 2005, 26, 7-8, 1432–1441.

Section III

Regulation and Remediation of Chiral Pollutants

12 Regulatory Perspectives and Challenges in Risk Assessment of Chiral Pollutants

Edmond Sanganyado

CONTENTS

12.1 Introduction 301
    12.1.1 Implications of Chirality on Risk Assessment 303
    12.1.2 Scope of Chapter 303
12.2 Regulation of Chiral Contaminants 303
    12.2.1 Types of Chemical Regulations 303
    12.2.2 Regulation of Chiral Contaminants 306
12.3 Risk Assessment of Chiral Pollutants 307
    12.3.1 Challenges in Risk Assessment 307
    12.3.2 Recent Advances in Risk Assessment 308
12.4 Case Study: Limitations of Pbt Models in Assessing Environmental Risk of Chiral Pharmaceuticals 309
    12.4.1 Challenges in Exposure Assessment 309
    12.4.2 Challenges in Toxicity Assessment 312
    12.4.3 Future Perspectives 313
12.5 Conclusion 313
Acknowledgments 313
References 313

12.1 INTRODUCTION

Chemicals have been recognized in the past centuries as essential for sustaining life. As humans abandoned the hunter-gatherer lifestyle thousands of years ago, chemicals played a significant role in paving the way for modern civilization (Johnson et al., 2020). Isolation of natural chemicals from plants for use in medicine, agriculture, and later the textile industry was critical in improving the quality of life. The risk posed by chemicals was known even then with Paracelsus postulating in the 15th century that it was the dose that makes a poison. In other words, Paracelsus posited that any chemical was potentially a poison at the right dose. Natural chemicals were also used for their adverse effects to kill or maim people. For example, Greek philosophers such as Theramenes and Socrates were killed by poisoning using hemlock extracts.

The chemical industry rapidly grew in the past century due to rapid population growth, improvements in the standard of life, and an increased global food demand (Sanganyado et al., 2018). It is estimated that at least 95% of all commercial and consumer products are heavily dependent on the chemical industry (Wang et al., 2020). A recent study found that there are more than 350 000 chemicals and mixtures registered for commercial production globally (Wang et al., 2020). In the United States, it is estimated that an additional 1000 chemicals are introduced into the market each year (Wang et al., 2020). Unfortunately, synthetic chemicals are released continuously into the environment or food products during production, distribution, use, and disposal of industrial or consumer

Chemical impacts on humans

In Michigan, USA, exposure to PFAS was linked to reproductive toxicity, immunotoxicity, and carcinogenicity in humans.

Chemicals produced during food processing such as acrylamide have been shown to be carcinogenic.

Neurological diseases were linked to eating fish contaminated with mercury in Minamata, Japan.

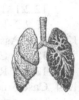

Polycyclic aromatic hydrocarbons are linked to malignant and non-malignant respiratory diseases.

Agrochemical residues in farm produce has been linked to developmental, reproductive, and immuno-toxicity.

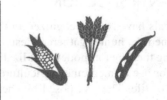

FIGURE 12.1 Examples of chemicals that have biological effects in humans following exposure.

products. Chemicals discharged from industries, hospitals, households, landfills, and wastewater treatment plants can enter the environment where they can bioaccumulate in organisms.

Since Rachel Carson's *Silent Spring*, chemical pollution has been recognized as a major threat to biodiversity, water quality, food security, and human health (Figure 12.1). For example, the use of neonicotinoids as an insecticide has been linked to population collapse of honeybees (Chen et al., 2019; Wang et al., 2013). Discharge of industrial effluent containing poly- and perfluoroalkyl substances (PFAS) in the environment has been linked to increased cancer risk, adverse effects on the immune system, and endocrine disruption (Buck et al., 2011; Gallen et al., 2018; Krafft and Riess, 2015). It was found that more than 1.9 million people in Michigan, USA were exposed to

PFAS through drinking water. In the 1950s, consumption of food contaminated with mercury in Minamata, Japan was shown to cause severe neurological disease (Sakamoto et al., 2010). Hence, synthetic chemicals may cause adverse human and environmental health effects.

12.1.1 Implications of Chirality on Risk Assessment

More than 50% and 30% of pharmaceuticals and agrochemicals are chiral compounds that exist as two or more enantiomers (Sanganyado et al., 2020). A chiral compound is a compound that can exist as two nonsuperimposable mirror-images (enantiomers) since it possesses an asymmetric center (Sanganyado et al., 2017). The asymmetric center can be a chiral center or axis. For example, pharmaceuticals such as fluoxetine, atenolol, venlafaxine, and chloroquine contain one or more chiral centers which are carbon atoms (Kasprzyk-Hordern, 2010; Ribeiro et al., 2013; Zhang et al., 2018a). Flame retardants such as polychlorinated biphenyls and polybrominated biphenyls, on the other hand, contain a biphenyl bond that acts as an axis of chirality (Ruan et al., 2019, 2018; von der Recke et al., 2005). Interestingly, molecules with helical structures such as deoxyribonucleic acid and some metal complexes (e.g., cobalt(III) metallocryptand) are inherently chiral even though they do not contain a chiral center (Rickhaus et al., 2016). However, planar and helical chirality are not common in chemical technology. Enantiomers of a chiral compound often have different properties in biological systems (Connors et al., 2013; Gonçalves et al., 2002; Khan et al., 2014; Sanganyado, 2019). For example, enantiomers of a chiral pharmaceutical often have different therapeutic and toxicological characteristics (Brocks, 2006; Shen et al., 2013; Smith, 2009). Several studies have shown that enantiomers of chiral pesticides also have different efficacies and toxicities to non-target organisms (Gámiz et al., 2018; Mueller and Buser, 1995; Zhang et al., 2018b). As a result, the risk posed by chiral contaminants to human and environmental health may be enantiomer specific.

12.1.2 Scope of Chapter

Research interest on the implications of chirality on the efficacy, toxicity, distribution, and fate of chiral chemicals grew significantly in the past three decades. This is probably because several studies showed that enantiomers of chiral compounds have different properties in biological systems. However, enantiomer-specific differences in the occurrence, transport, fate, and toxicity of chiral contaminants in food and the environment are generally overlooked by regulatory agencies across the globe. Unfortunately, current approaches on assessing risk ignore the implications of chirality on the biological effects of chemical contaminants, and this undermines the accuracy of the risk assessments (Figure 12.2). This chapter provides general backgrounds on chemical regulation and risk assessment. Current regulations, and the lack thereof, on chiral contaminants in food and the environment are also discussed.

12.2 REGULATION OF CHIRAL CONTAMINANTS

12.2.1 Types of Chemical Regulations

For the past five decades, identification of priority pollutants, restricting their use, and monitoring their presence in food and the environment has been a regulatory priority (Gago-Ferrero et al., 2018; Karahan Ozgun et al., 2016; Wang et al., 2012). There are several types of national and international chemical regulations; and these include substance control, pollution mitigation, chemical transport and storage, and chemical inventory regulations as well as international conventions (ECHA, 2017; Grisoni et al., 2016; Lillicrap et al., 2016). Table 12.1 offers brief descriptions of the several types of chemical regulations. It also shows examples of national regulations from USA and China as well as international conventions (Lorgeoux et al., 2016; Stanley and Brooks, 2009). A chemical regulation can fall under several different categories; for example, the Toxic Substances Control Act of

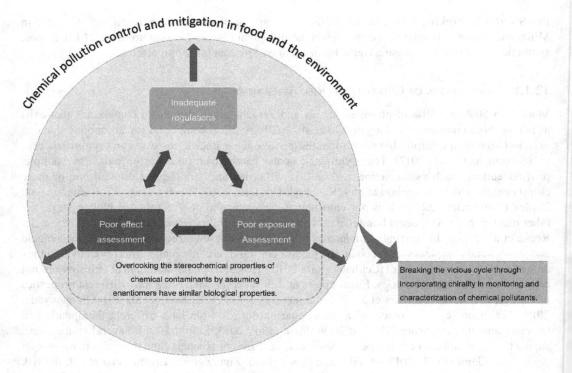

FIGURE 12.2 A vicious cycle demonstrating the implications of ignoring chirality in chemical regulations and risk assessment.

2016 contains regulations on control of toxic substances, new chemicals, and chemical inventories (Larsen et al., 2002).

Chemical regulations are sometimes focused primarily on addressing chemical contaminants in a single environmental compartment or commercial product. For example, the United States Environmental Protection Agency is involved in enforcing laws and regulations for preventing chemical pollution in the environment. The EPA enforces laws such as the Beaches Environmental Assessment and Coastal Health Act, Clean Air Act, Clean Water Act, and Safe Drinking Water Act that aim at protecting coastal environments, the atmosphere, aquatic systems, and drinking water, respectively (Guan et al., 2012; McCauley et al., 2000). Human exposure to chemical contaminants is further protected by commercial product specific regulations by the U.S. Food and Drug Administration which is responsible for enforcing the Federal Food, Drug, and Cosmetic Act, and Food Quality Protection Act (Calcaterra and D'Acquarica, 2018; U.S. Food and Drug Administration, 1992). Many countries have agencies that share the same mandate as the U.S. EPA and U.S. FDA. For example, there is Environment Canada and Canadian Food Inspection Agency and Chinese National Environmental Protection Agency and Chinese National Food and Drug Administration in Canada and China, respectively.

International conventions and regulations often address all the aspects of chemical regulations since they are often used to inform national regulations. In Europe, the Registration, Evaluation, Authorization, and Restriction of Chemicals (REACH) provides recommendations and guidelines for protecting human and environmental health by offering early detection of toxic substances. REACH identifies the biological and physicochemical properties of the chemicals throughout the chemical registration, evaluation, and authorization stages. When problematic chemicals are identified, REACH provides guidelines for restricting the production, distribution, usage, and disposal of the chemicals. REACH currently has regulations on persistent organic pollutants, fertilizer products,

TABLE 12.1
Brief Descriptions and Examples of National and International Chemical Regulations

Chemical Regulation	Description	Examples		
		United States	China	International Conventions
Substance Control	Guidelines for the production, processing, distribution, use, and disposal of chemicals.	Toxic Substances Control Act of 2016	Provisions on the First Import of Chemicals and the Import and Export of Toxic Chemicals (1994); The Regulations on Safe Management of Hazardous Chemicals (2011); The Measures for Environmental Management of New Chemical Substances	Montreal Protocol on Substances that Deplete the Ozone Layer; The Stockholm Convention on Persistent Organic Pollutants; International Convention on the Control of Harmful Antifouling Systems on Ships
Pollution mitigation	Recommendations for minimizing chemical pollution in the environment.	Pollution Prevention Act, Pesticide Chemicals Effluent Guidelines	Regulation on Pesticide Administration	International Convention for the Prevention of Pollution of Ships; The Stockholm Convention on Persistent Organic Pollutants; Montreal Guidelines for the Protection of the Marine Environment Against Pollution from Land-Based Sources
Food contamination	Recommendations for minimizing chemical pollution in food stuffs and packaging.	Federal Food, Drug, and Cosmetic Act of 1938; Kefauver-Harris Amendments of 1962	Regulations for Implementation of Food Safety Law; GB 2760-2011 Food Safety National Standards for the Usage of Food Additives	The Codex Alimentarius; Joint FAO/WHO Expert Committee on Food Additives
Chemical transport and storage	Regulations for the transport and storage of hazardous materials.	Pipeline and Hazardous Materials Safety Administration	GB 6944-2012 Classification and code of dangerous goods; GB 12268-2012 List of dangerous goods	Basel Convention on the Control of Transboundary Movements of Hazardous Wastes and Their Disposal; The United Nations Convention against Illicit Traffic in Narcotic Drugs and Psychotropic Substances
Chemical inventory		Toxic Substances Control Act Inventory	China Inventory of Existing Chemical Substances in China	Regulated Chemicals Listing (CHEMList);

detergents, explosives, and drug precursors. For example, the Regulation (EC) No 765/2008 of the European Parliament and of the Council provided guidelines for monitoring fertilizer products imported from non-EU countries. The regulations had guidelines for the maximum allowable concentrations of inorganic fertilizer ingredients and by-products such as cadmium. For organic pollutants, Regulation (EC) No 850/2004 of 29 April 2004 provided guidelines on persistent organic pollutants. This regulation emphasized eliminating the production and consumption of persistent organic pollutants that are recognized as priority contaminants internationally. The pollution mitigation regulations included restrictions on production, distribution, usage, and disposal of persistent organic pollutants as well as strategies to minimize unintentional release of the chemical during the product life cycle.

12.2.2 Regulation of Chiral Contaminants

Around 70 years ago, thalidomide was used widely as a drug for treating morning sickness in pregnant women. Many countries subsequently banned its use in 1961 as it was found to be teratogenic. Around 10,000 children in Europe, Japan, and Australia were born with birth defects such as phocomelia, congenital heart disease, and ocular abnormalities (Kim and Scialli, 2011; Smith, 2009; Tokunaga et al., 2018). Although thalidomide was commercially marketed as a racemic mixture, it is a chiral compound that contains a single chiral center. A study by Blaschke et al. (1979) found (R)-thalidomide had therapeutic properties while (S)-thalidomide was teratogenic. Blaschke et al. (1979) concluded that the thalidomide tragedy could have been averted if only (R)-(+)-enantiomer was used. However, recent studies have shown that (R)-(+)-thalidomide can undergo chiral inversion *in vivo* to yield the teratogenic (S)-(−)-enantiomer (Sanganyado, 2019). The thalidomide tragedy showed the importance of rigor in drug testing and approval. Following the thalidomide tragedy, the U.S. FDA began enforcing strict regulations on drug development and approval. Surprisingly, the U.S. FDA together with other regulatory agencies around the world continued to ignore the impact of stereochemistry on drug safety.

In 1992, 30 years after the thalidomide ban, the U.S. FDA issued a recommendation that encouraged drug manufacturers to consider the stereochemistry of a compound during drug development and manufacturing (U.S. Food and Drug Administration, 1992). The U.S. FDA recommended that manufacturers: (i) should put in place controls that ensure the enantiomeric composition of a drug were known and (ii) assume enantiomers have different pharmacokinetics and pharmacodynamics until experimental studies prove otherwise (Calcaterra and D'Acquarica, 2018; Stanley and Brooks, 2009; U.S. Food and Drug Administration, 1992). The U.S. FDA guidance on chiral drugs had several limitations which include: (i) it was not an enforceable legally binding document; (ii) it recommended evaluation of the racemic mixture only; and (iii) it did not take into account the fact that some enantiomers might have different therapeutic profiles (e.g., (R, R)-(+)-tramadol inhibits serotonin reuptake while (S, S)-(−)-tramadol inhibits noradrenalin reuptake (Ardakani et al., 2008)) (Stanley and Brooks, 2009). It is important to note that the FDA later issued corrections to the policy statement in 1997. In 1994, the European Union began enforcing the *Investigation of Chiral Active Substances* (CPMP/III/3501/91) guidance document on manufacturing and testing of chiral drugs (Branch, 2001). Canada began enforcing its comprehensive *Guidance for Industry: Stereochemical Issues in Chiral Drug Development* which even included recommendations for assessing the *in vivo* stability of single enantiomers (Minister of Public Works and Government Services Canada, 2000). However, these guidelines are limited to assessing risk of chiral drugs in humans and not on the environment.

Current guidelines on risk assessment of chiral pollutants in food and the environment are few and highly limited. For example, the *Interim Policy for Evaluation of Stereoisomeric Pesticides* by the U.S. EPA recommended that the environmental fate and effects of the commercially available pesticides marketed as racemic mixtures should be compared to the single enantiomer pesticides under development (Environmental Fate and Effects Division, 2000). The policy statement sought to establish whether single enantiomers posed higher environmental and human health risk than the commercially available racemic mixtures. First, it recognized the importance of accurate data by recommending the development of enantioselective techniques capable of determining the enantiomeric composition of chiral pesticides and their transformation products in various environmental matrices such as soil, water, and biota. This is because accurate environmental monitoring and characterization data is essential for determining the environmental risk of a chemical contaminant. Second, acute and chronic toxicity testing were recommended for assessing the toxicological differences between the racemic mixture and single enantiomer (Stanley and Brooks, 2009). Finally, the policy statement suggested that no additional fate data was essential when the biotransformation rates and products of the racemic mixture and the single enantiomer were similar following an aerobic soil metabolism study. However, these recommendations may not be adequate because they overlook that some chiral pesticides such as pyrethroids can undergo chiral inversion (Khan, 2014; Poiger et al., 2015; Sun et al., 2013). Furthermore, the improvement to risk assessment provided by the policy statement was limited because it did not require assessing the differences in fate and toxicities of single enantiomers. This is particularly important because the risk posed by chiral pesticides marketed as racemic mixtures should be evaluated as that of a chemical mixture. Each enantiomer needs to be treated as a distinct chemical which can interact with the other "chemicals" antagonistically, additively, or synergistically.

12.3 RISK ASSESSMENT OF CHIRAL POLLUTANTS

Risk assessment is a powerful tool for developing, implementing, evaluating, and amending chemical regulations. Regulations rely on accurate estimation of the risk posed by a chemical to humans or the environment. Risk assessment provides a formal framework for quantitatively assessing the effects of chemical exposure to a target species. Hence, risk assessment can be formally defined as a formal process for determining the probability that an adverse biological effect will occur following chemical exposure. This definition shows that risk assessment involves quantitatively assessing the effects of the chemical on humans or the environment (effect assessment) as well as assessing the degree of exposure (exposure assessment). Quantitative assessment of the exposure and effects of chemical contaminants is critical for providing evidence that can link chemical exposure to community and population level biological responses. By expressing contaminant risk quantitatively, it becomes easier for policy- and decision-makers to establish the extent of the pollution problem. Understanding the chemical pollution problem is essential for developing effective pollutant control and mitigation strategies as well as ecological restoration programs.

12.3.1 CHALLENGES IN RISK ASSESSMENT

Synthetic chemicals that are highly persistent (P), bioaccumulative (B), and toxic (T) have been recognized as high priority pollutants (Alonso et al., 2008; Pizzo et al., 2016). Highly persistent chemicals are difficult to regulate because their effects continue even after their environmental loading has stopped. As a result, the Stockholm Convention was established to provide recommendations and guidelines on priority persistent organic pollutants (Liu et al., 2016; Marvin et al., 2011; Raubenheimer and McIlgorm, 2018). Synthetic chemicals with high hydrophobicity can enter the food chain by bioaccumulating in organisms (Gramatica and Papa, 2003; Nfon and Cousins, 2007; Walters et al., 2016). The amount of the contaminant that bioaccumulates in the organisms may increase with increase in trophic level. This process whereby the concentration of the contaminant increases from primary producers up to the apex predator is called biomagnification. When trophic biomagnification occurs, contaminants pose greater risk in organisms that are at higher trophic levels. In addition, acute oral, dermal, and inhalation toxicity tests on species such as *Daphnia magna*, fish, and rats is traditionally used to identify priority pollutants (Crane et al., 2006; Creton et al., 2010; Lammer et al., 2009).

The PBT model is recognized by most national and international chemical regulators as a powerful tool for hazard assessment of chemical contaminants (Lillicrap et al., 2016; Pizzo et al., 2016). For example, the European Chemical Agency (ECHA) maintains an inventory of chemicals undergoing a PBT or a "very persistent and very bioaccumulative" (vPvB) assessment under the Registration, Evaluation, Authorization, and Restriction of Chemicals program (European Medicines Agency, 2015). As of March 2020, the ECHA PBT assessment list contains 176 substances including pyrene, chlorpyrifos, chlorinated paraffins, and tamoxifen. Table 12.2 summarizes the criteria set by ECHA for evaluating PBT and vPvB compounds.

TABLE 12.2
A Summary of PBT and vPvB Criteria Recommended by the European Union (Adapted from ECHA, 2017)

Property	PBT Criteria	vPvB Criteria	Indications
Persistence	Persistent contaminants have a half-life of at least: • 60 days in marine water. • 40 days in freshwater or estuarine waters. • 180 days in sediment. • 120 days in freshwater or estuarine sediment. • 120 days in soil.	Very persistent contaminants (vP) have a half-life of at least: • 60 days in marine water, freshwater or estuarine waters. • 180 days in marine water, freshwater or estuarine sediment. • 180 days in soil.	Results and other information are obtained from: • Ready biodegradability tests. • Enhanced ready and inherent biodegradability tests. • *In silico* methods (e.g., QSAR biodegradation models). • Field studies or monitoring studies. • Simulation testing in soil, surface water, and sediment.
Bioaccumulative	Bioaccumulative compounds have at least: • A bioconcentration factor higher than 2000 in aquatic. • Octanol-water partition coefficient, Log K_{OW} > 5	Very bioaccumulative compounds (vB) have at least: • A bioconcentration factor higher than 2000 in aquatic.	Results and other information are obtained from: • Octanol-water partitioning coefficient tests. • Estimated by QSAR models. • Bioaccumulation study. • Human biomonitoring studies. • Assessment of toxicokinetic behavior.
Toxicity	Toxic compounds have: • A long-term EC10 or no-observed effect concentration (NOEC) < 0.01 mg L^{-1} in aquatic species. • A classification of 1A or 1B carcinogen.[a] • A classification of 1A or 1B germ cell mutagen.[a] • A classification of 1A, IB, and 2 reproductive toxicity.[a] • A specific target organ toxicity after repeated exposure category of 1 or 2.[b]		Results and other information are obtained from: • Short-term aquatic toxicity testing in aquatic species. • Long-term toxicity testing in invertebrates and fish. • Reproductive toxicity with birds. • Growth inhibition study on aquatic plants. • Chronic toxicity study on animals

[a] Classification is according to REACH regulation EC No. 1272/2008.
[b] Classification is according to REACH Regulation EC No. 1272/2008.

12.3.2 RECENT ADVANCES IN RISK ASSESSMENT

A recent publication prepared by the European Food Safety Agency (EFSA) for members of the European Union provided critical guidelines for assessing the risk of chiral pesticides (Bura et al., 2019). For example, the guidance document: (i) succinctly stated that enantiomers should be treated

TABLE 12.3
A Summary of the European Food Safety Agency Recommendations on Dietary and Environmental Risk Assessment of Chiral Active Substances and Metabolites (Bura et al., 2019)

Issue	Stage	Recommendation
Analytical methods	Pre-approval	• Non-enantioselective analysis is sufficient when the pesticide is marketed as racemic mixture while enantioselective analysis is required for active substances non-racemic pesticides. • Quantification may be achieved using matrix-matched standards, although stable isotope labelled analogues are recommended.
	Post-approval	• Enantioselective analysis in the environment and human body fluids using stable isotope labelled standards is recommended.
Risk assessment	Pre-approval	• Stereoisomers should be treated separately even those considered inactive, impurities, or transformation products. • Racemization during storage, use, or in the environment should be considered.
	Post-approval	• Enantiomers with varying environmental fate and toxicities should probably have separate risk quotients. • Enantiomers are considered to have significantly different (eco)toxicological properties when they have (eco)toxicological endpoints that vary by a factor of three.
Toxicokinetics	Both	• Enantiomer specificity in mammalian absorption, distribution, metabolism, and excretion should be addressed.
Mammalian toxicity	Both	• Changes in enantiomeric composition are not expected to influence genotoxicity hazard identification given that all the enantiomers in the residue are sufficiently represented in the test mixture. • In general toxicity testing, changes in enantiomeric composition that do not exceed an enantiomeric excess of 10%, are not expected to influence the pesticide residue hazard and risk assessment. • Read-across approach can be used to establish whether there are enantiomer specific differences in the toxicological profile and/or toxicological potency.
Food residues	Post-approval	• Establish the effect of storage conditions on the chiral stability of the pesticide residues in food. • Determination of the enantiomeric composition of pesticide residues in food and agricultural produce is recommended.
Environmental fate and behavior	Post-approval	• Evaluate the enantiomeric composition of chiral pesticides in soils to which non-target species may be exposed. • Assumed that leaching and adsorption in soil or sediments and environmental behavior in air were non-enantioselective. • Preferential transformation and chiral inversion of racemic mixtures and single enantiomers should be investigated using aerobic mineralization or water/sediment studies.

as different compounds in risk assessments, as a general rule; (ii) recommended that manufacturers should consider the chirality of the compounds during pesticide development and approval; and (iii) provided guidelines for both dietary risk assessment and environmental risk assessment. Table 12.3 provides a summary of the recommendations of the guidance document.

12.4 CASE STUDY: LIMITATIONS OF PBT MODELS IN ASSESSING ENVIRONMENTAL RISK OF CHIRAL PHARMACEUTICALS

12.4.1 CHALLENGES IN EXPOSURE ASSESSMENT

Since there are more than 5,000 pharmaceuticals on the market, evaluating the environmental risk of all of them is overly ambitious and improbable. However, following the Stockholm protocol for

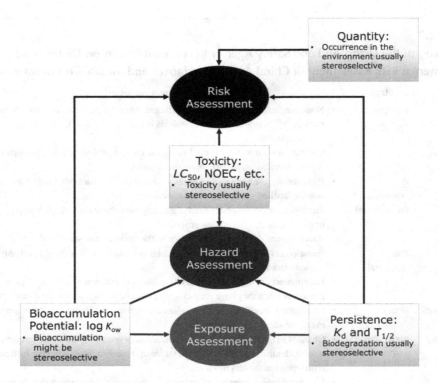

FIGURE 12.3 Limitations of persistence, bioaccumulation, and toxicity on establishing the risk of chiral pharmaceuticals in the environment.

persistent organic pollutants, environmental risk of pharmaceuticals are evaluated using PBT model (Figure 12.3) (Christen et al., 2010). There are two integrated approaches used in the evaluation of environmental risk of pollutants, namely establishing the exposure and the hazard factors. The hazard-based approach focuses on hazard identification using the intrinsic properties of the pollutant, such as volatility, octanol-water partition coefficients, half-life, sorption, polarity, and ionization. In contrast, exposure-based approaches identify risk using extrinsic behavior of the pollutants in the environment compartment. The exposure factor is mainly dependent on the magnitude, frequency, and duration of exposure, thus influencing the occurrence, transport, and partitioning of the pollutants in the environment. However, pharmaceuticals have different physicochemical and biological properties to POPs, and the PBT might require modifications to address the intrinsic and extrinsic properties of pharmaceuticals that make them a potential environmental risk (Table 12.4). Furthermore, the PBT model does not address the impacts of chirality on biological interactions between pharmaceuticals and non-target organisms.

The bioaccumulation potential of pharmaceuticals in non-target organisms is overlooked since pharmaceuticals are polar ionic compounds. However, most streams in the United States are effluent-dominated, implying aquatic organisms endure prolonged and frequent exposure to pharmaceuticals thus increasing the bioaccumulation potential (Du et al., 2014). Additionally, pharmaceuticals are designed for optimal absorption and distribution into the human body through hydrogen bonding, ionic interaction and, at times, high lipophilicity (Du et al., 2014). Hence, in the past decade, several studies reported the bioaccumulation of pharmaceuticals such as beta blockers and antidepressants in fish (Garcia et al., 2012; Valdés et al., 2014), algae, and crustaceans (Vernouillet et al., 2010). However, the effect of enantiomeric composition on exposure risk can be incorporated in exposure assessment if a read-across of pharmacological data is considered. For example, Williams et al. demonstrated a correlation between the K_d and V_D values for 13 pharmaceuticals with $r^2 = 0.62–0.72$

TABLE 12.4
Pharmacological and Ecotoxicological Behavior of Select Chiral Pharmaceuticals and Their Respective Primary Targets Obtained from DrugBank (www.drugbank.ca)

Drug	Class	Pharmacology Primary Drug Target	Stereoselectivity	Ecotoxicity	References
Atenolol	Beta-blocker	Beta-1 adrenergic receptor	Activity due to (S)-atenolol	(S)-atenolol was more toxic than the antipode to a microalga, but less toxic than (R)-atenolol to a protozoan.	De Andrés et al. (2009)
Betaxolol	Beta-blocker	Beta-1 adrenergic receptor	Activity due to (S)-betaxolol	Toxicity was estimated to be low using ECOSAR.	Sanderson et al. (2003)
Ibuprofen	NSAID	Prostaglandin G/H synthase 2	Activity due to (S)-ibuprofen	Ibuprofen was retarded development of zebrafish embryo at >10 µg/L.	David and Pancharatna (2009)
Fluoxetine	Antidepressant	Sodium-dependent serotonin transporter	Activity due to (S)-fluoxetine	(R)-fluoxetine was more toxic *P. subcapita* than the antipode.	De Andrés et al. (2009)
Metoprolol	Beta-blocker	Beta-1 adrenergic receptor	Activity due to (S)-metoprolol	No stereoselectivity in observed changes in heart rate, hatching rate or mortality to zebrafish.	Sun et al. (2014)
Propranolol	Beta-blocker	Beta-1 adrenergic receptor	Activity due to (S)-propranolol	(S)-propranolol was more chronically toxic to *P. promelas* than (R)-propranolol.	Stanley et al. (2006)
Pindolol	Beta-blocker	Beta-1 adrenergic receptor	Activity due to (S)-pindolol	—	Mehvar and Brocks (2001)
Salbutamol	Beta-blocker	Beta-2 adrenergic receptor	Activity due to (R)-salbutamol	—	Kasprzyk-Hordern (2010)
Sotalol	Beta-blocker	Beta-1 adrenergic receptor	Activity due to (R)-sotalol	Increased number of embryos in New Zealand mudsnail following exposure	Feiner et al. (2014)
Venlafaxine	Antidepressant	Serotonin re-uptake and noradrenergic re-uptake	—	Increased mortality to fathead minnows exposed to 305 ng/L venlafaxine.	Schultz et al. (2011)

in sediment (Williams et al., 2006) and soil systems (Williams et al., 2009). However, several pharmacological studies observed stereoselectivity in the V_D values of chiral pharmaceuticals (Brocks, 2006). Stereoselectivity in sorption of organic pollutants was previously reported in soil and sludge. Although no pharmaceutical studies on stereoselective bioaccumulation and plant uptake have been conducted this phenomenon has been observed in lizards (Wang et al., 2014), *Tubifex tubifex*

(Liu et al., 2014), and *Brassica campestris* (Wang et al., 2013) exposed to racemic pesticides. Thus, bioaccumulation and plant uptake of chiral pharmaceuticals may be stereoselective.

12.4.2 Challenges in Toxicity Assessment

Since pharmaceuticals are biologically active compounds, constant discharge of chiral pharmaceuticals by WWTP into the environment may result in adverse effects in aquatic organisms. Pharmaceuticals have short half-lives ranging from hours to months, but are considered pseudopersistent since they are continually discharged into the environment via wastewater effluent (Niemi et al., 2013). Thus, non-target organisms are chronically exposed to chiral pharmaceuticals. Several studies reported beta blockers, antidepressants, and other pharmaceuticals exhibited developmental (Airhart et al., 2007; David and Pancharatna, 2009), reproductive (Gust et al., 2009; Huggett et al., 2002), behavioral (Winder et al., 2009), and acute toxicity (Ferrari et al., 2004; Overturf et al., 2012) to various aquatic organisms. However, unlike pesticides, pharmaceuticals are designed for therapeutic purposes and not for their toxicity. Therefore, traditional endpoints, such acute toxicity, might not be suitable for establishing the environmental risk of pharmaceuticals. Aquatic organisms are exposed to chiral pharmaceuticals in surface water. Under environmental conditions, the dose of each enantiomer to aquatic species varies due to differences in magnitude of exposure. Numerous studies have demonstrated the disposition, metabolism, and toxicity of chiral pharmaceuticals is stereoselective. Enantiomers have similar physical and chemical properties, except optical activity, in a chiral environment. However, enantiomers and their racemic mixture have different biochemical and physicochemical properties in biological systems. Enantiomers usually engage in stereospecific interactions in chiral environments such as biological systems due to their three dimensional structure that results in stereoselective binding, catalysis, and stabilization of the chiral pharmaceuticals (Wong, 2006; Wong and Warner, 2010). When an enantiomer has high affinity for the receptor it is called a eutomer, and a distomer has less affinity (Ariens et al., 1988; Ariëns, 1984). Stoschitzky et al. (1993) showed (*S*)-atenolol was the eutomer since it contributed most of the beta-blocking effect and (*R*)-atenolol was the distomer (Table 12.4). However, sometimes eutomer-distomer classification is not that simple: trans-tramadol is an analgesic comprising a racemic mixture of (*R*,*R*)-(+)-tramadol and (*S*,*S*)-(−)-tramadol (Ardakani et al., 2008). However, most studies on ecotoxicity focus on acute toxicity ignoring that environmental exposure of non-target organisms to chiral pharmaceuticals is chronic thus potentially eliciting chronic toxicity. Therefore, understanding the stereoselective toxicity of chiral pharmaceuticals to non-target organisms is essential for a more accurate risk assessment.

There are a few studies that have demonstrated stereoselectivity in the toxicity of pharmaceuticals to non-target aquatic organisms. Previous studies on environmental risk of fluoxetine demonstrated that there is a need for incorporating chirality, exposure, and effects assessments for accurate pharmaceutical prioritization (Brooks et al., 2003; Scott et al., 2015; Stanley et al., 2007). The EC_{50} in growth inhibition of *Daphnia magna* by atenolol and fluoxetine ranged from 6.9 to 1450 mg/L (De Andrés et al., 2009). In this study, De Andres et al. (2009) observed stereoselectivity in the adverse effects of atenolol with (*S*)-enantiomer being almost more potent than the (*R*)-enantiomer, but not between enantiomers of fluoxetine. Even though environmental concentrations are below acute toxicity levels, pharmaceuticals are designed to effect physiological change at extremely low concentrations. Thus, pharmaceuticals may cause long-term effects, rather than short-term. After conducting a battery of traditional effects assessment, Brooks et al. (2003) found the environmental risk of fluoxetine was low, but they postulated that it could elicit adverse effects at non-traditional endpoints. However, fluoxetine was classified as a low priority pharmaceutical following a predicted exposure assessment using a corrected predicted environmental concentration (PEC)/predicted no effect concentration (PNEC) ratio (Oakes et al., 2010). A subsequent chronic toxicity study demonstrated that fluoxetine elicited sublethal and behavioral effects stereoselectively on *Pimephales promelas* (Stanley et al., 2007). These results agreed with the postulate made by Brooks et al. (2003)

thus showing knowledge of the evolutionary conservation of the primary target of the pharmaceuticals could help in their prioritization (Rand-Weaver et al., 2013).

12.4.3 FUTURE PERSPECTIVES

Since they are more than 5,000 pharmaceuticals in current use, long-term toxicity studies cannot be conducted for all of them, hence the need for prioritization. Pharmaceuticals are pollutants of environmental concern because they are designed to elicit physiological and behavioral responses at low concentrations. The biological responses are very specific because the pharmaceuticals compounds are designed to target specific molecular receptors, enzymes, metabolic pathways, cell signal pathways, and/or plasma proteins (Fabbri and Franzellitti, 2015; Fabbri and Moon, 2015). However, the primary drug targets are often conserved across species, thus occurrence of pharmaceuticals in the environment could potentially elicit biological effects in non-target organisms (LaLone et al., 2013). The read-across hypothesis is often described as the ability for a pharmaceutical to elicit a physiological or behavioral response on a non-target organism due to evolutionary conservation of the primary drug target (Berninger et al., 2016; Lalone et al., 2014; Rand-Weaver et al., 2013). Since several studies reported that some primary targets of pharmaceuticals are conserved in aquatic organisms such as zebra fish, mussels, and fathead minnow, the read-across hypothesis can be used for predicting ecotoxicity of pharmaceuticals and prioritization of pharmaceuticals in environmental risk assessments, through leveraging available pharmacological data (Berninger et al., 2016; Fabbri and Franzellitti, 2015; Lalone et al., 2014). It is interesting to note that the recent guidance document on chiral pesticides by EFSA recommended the use of read-across approaches in environmental risk assessment.

12.5 CONCLUSION

In a risk assessment program, both the exposure factor and effect factor are critical and need to be determined experimentally or using predictive models. There are currently few enforceable regulations that focus on the dietary and environmental risk of chiral contaminants. This is probably due to the lack of comprehensive studies on the ecological and human health effects of chiral chemicals. To curtail the shortage of (eco)toxicological data, the EFSA recommended the use of read-across models in assessing the risk of chiral pesticides. In this chapter, it was proposed that considering the availability of human toxicological data, read-across techniques can be used to prioritize pharmaceuticals. Incorporating stereoselectivity in pharmacological behavior can be used to predict environmental risk of chiral pharmaceuticals. Prediction of the effect ratio will help in the prioritization of chiral pharmaceuticals in environmental risk assessments. However, there is a need for additional studies on stereoselective chronic ecotoxicity of chiral pollutants.

ACKNOWLEDGMENTS

This study was supported by Shantou University Research Start-Up Program has been approved (Grant Number NTF20002).

REFERENCES

Airhart, M. J., Lee, D. H., Wilson, T. D., Miller, B. E., Miller, M. N., Skalko, R. G., 2007. Movement disorders and neurochemical changes in zebrafish larvae after bath exposure to fluoxetine (PROZAC). *Neurotoxicol. Teratol.* 29, 652–64.
Alonso, E., Tapie, N., Budzinski, H., Leménach, K., Peluhet, L., Tarazona, J. V., 2008. A model for estimating the potential biomagnification of chemicals in a generic food web: Preliminary development. *Environ. Sci. Pollut. Res.* 15, 31–40. https://doi.org/10.1065/espr2007.05.425

Ardakani, Y. H., Mehvar, R., Foroumadi, A., Rouini, M.-R., 2008. Enantioselective determination of tramadol and its main phase I metabolites in human plasma by high-performance liquid chromatography. *J. Chromatogr. B. Analyt. Technol. Biomed. Life Sci.* 864, 109–15.

Ariens, E. I., Wuis, E. W., Veringa, E. I., 1988. Stereoselectivity of bioactive xenobiotics: A pre-pasteur attitude in medicinal chemistry, pharmacokinetics and clinical pharmacology. *Biochem. Pharmacol.* 37, 9–18.

Ariëns, E. J., 1984. Stereochemistry: A basis for sophisticated nonsense in pharmacokinetics and clinical pharmacology. *Eur. J. Clin. Pharmacol.* 26, 663–668.

Berninger, J. P., LaLone, C. A., Villeneuve, D. L., Ankley, G. T., 2016. Prioritization of pharmaceuticals for potential environmental hazard through leveraging a large-scale mammalian pharmacological dataset. *Environ. Toxicol. Chem.* 35, 1007–1020. https://doi.org/10.1002/etc.2965

Blaschke, G., Kraft, H. P., Fickentscher, K., Köhler, F., 1979. Chromatographic separation of racemic thalidomide and teratogenic activity of its enantiomers. *Arzneimittelforschung.* 29, 1640–1642.

Branch, S. K., 2001. International regulation of chiral drugs, in: Subramanian, G. (Ed.), *Chiral Separation Techniques: A Practical Approach.* Wiley-VCH Verlag GmbH, Weinheim, FRG, pp. 317–341. https://doi.org/10.1002/3527600361.ch13

Brocks, D. R., 2006. Drug disposition in three dimensions: An update on stereoselectivity in pharmacokinetics. *Biopharm. Drug Dispos.* 27, 387–406. https://doi.org/10.1002/bdd.517

Brooks, B. W., Foran, C. M., Richards, S. M., Weston, J., Turner, P. K., Stanley, J. K., Solomon, K. R., Slattery, M., La Point, T. W., 2003. Aquatic ecotoxicology of fluoxetine. *Toxicol. Lett.* 142, 169–183.

Buck, R. C., Franklin, J., Berger, U., Conder, J. M., Cousins, I. T., Voogt, P. De, Jensen, A. A., Kannan, K., Mabury, S. A., van Leeuwen, S. P. J., 2011. Perfluoroalkyl and polyfluoroalkyl substances in the environment: Terminology, classification, and origins. *Integr. Environ. Assess. Manag.* 7, 513–541. https://doi.org/10.1002/ieam.258

Bura, L., Friel, A., Magrans, J. O., Parra-Morte, J. M., Szentes, C., 2019. Guidance of EFSA on risk assessments for active substances of plant protection products that have stereoisomers as components or impurities and for transformation products of active substances that may have stereoisomers. *EFSA J.* 17, 5804. https://doi.org/10.2903/j.efsa.2019.5804

Calcaterra, A., D'Acquarica, I., 2018. The market of chiral drugs: Chiral switches versus de novo enantiomerically pure compounds. *J. Pharm. Biomed. Anal.* 147, 323–340. https://doi.org/10.1016/j.jpba.2017.07.008

Chen, Z., Yao, X., Dong, F., Duan, H., Shao, X., Chen, X., Yang, T., Wang, G., Zheng, Y., 2019. Ecological toxicity reduction of dinotefuran to honeybee: New perspective from an enantiomeric level. *Environ. Int.* 130, 104854. https://doi.org/10.1016/j.envint.2019.05.048

Christen, V., Hickmann, S., Rechenberg, B., Fent, K., 2010. Highly active human pharmaceuticals in aquatic systems: A concept for their identification based on their mode of action. *Aquat. Toxicol.* 96, 167–181. https://doi.org/10.1016/j.aquatox.2009.11.021

Connors, K. A., Du, B., Fitzsimmons, P. N., Chambliss, C. K., Nichols, J. W., Brooks, B. W., 2013. Enantiomer-specific in vitro biotransformation of select pharmaceuticals in rainbow trout (Oncorhynchus mykiss). *Chirality* 25, 763–7. https://doi.org/10.1002/chir.22211

Crane, M., Watts, C., Boucard, T., 2006. Chronic aquatic environmental risks from exposure to human pharmaceuticals. *Sci. Total Environ.* 367, 23–41. https://doi.org/10.1016/j.scitotenv.2006.04.010

Creton, S., Dewhurst, I. C., Earl, L. K., Gehen, S. C., Guest, R. L., Hotchkiss, J. A., Indans, I., Woolhiser, M. R., Billington, R., 2010. Acute toxicity testing of chemicals - Opportunities to avoid redundant testing and use alternative approaches. *Crit. Rev. Toxicol.* 40, 50–83. https://doi.org/10.3109/10408440903401511

David, A., Pancharatna, K., 2009. Developmental anomalies induced by a non-selective COX inhibitor (ibuprofen) in zebrafish (*Danio rerio*). *Environ. Toxicol. Pharmacol.* 27, 390–5. https://doi.org/10.1016/j.etap.2009.01.002

De Andrés, F., Castañeda, G., Ríos, Á., 2009. Use of toxicity assays for enantiomeric discrimination of pharmaceutical substances. *Chirality* 21, 751–759. https://doi.org/10.1002/chir.20675

Du, B., Haddad, S. P., Luek, A., Scott, W. C., Saari, G. N., Kristofco, L. A., Connors, K. A., Rash, C., Rasmussen, J. B., Chambliss, C. K., Brooks, B. W., Brooks, B. W., 2014. Bioaccumulation and trophic dilution of human pharmaceuticals across trophic positions of an effluent-dependent wadeable stream. *Philos. Trans. R. Soc. B* 369, 1–10.

ECHA, 2017. Guidance on Information Requirements and Chemical Safety Assessment. Chapter R.11: PBT/vPvB Assessment, Guidance on Information Requirements and Chemical Safety Assessment. Helsinki, Finland. https://doi.org/10.2823/128621

Environmental Fate and Effects Division, 2000. Interim policy for evaluation of stereoisomeric pesticides [WWW Document]. *Pestic. Sci. Assess. Pestic. Risks.* https://www.epa.gov/pesticide-science-and-assessing-pesticide-risks/interim-policy-evaluation-stereoisomeric-pesticides (accessed 4.2.20).

European Medicines Agency, 2015. Guideline on the assessment of persistent, bioaccumulative and toxic (PBT) or very persistent and very bioaccumulative (vPvB) substances in veterinary medicinal products. London, UK.
Fabbri, E., Franzellitti, S., 2015. Human pharmaceuticals in the marine environment: Focus on exposure and biological effects in animal species. *Environ. Toxicol. Chem.* 35, 799–812. https://doi.org/10.1002/etc.3131
Fabbri, E., Moon, T. W., 2015. Adrenergic signaling in teleost fish liver, a challenging path. *Comp. Biochem. Physiol. Biochem. Mol. Biol.* https://doi.org/10.1016/j.cbpb.2015.10.002
Feiner, M., Laforsch, C., Letzel, T., Geist, J., 2014. Sublethal effects of the beta-blocker sotalol at environmentally relevant concentrations on the New Zealand mudsnail *Potamopyrgus antipodarum*. *Environ. Toxicol. Chem.* 33, 2510–2515. https://doi.org/10.1002/etc.2699
Ferrari, B., Mons, R., Vollat, B., Fraysse, B., Paxéus, N., Lo Giudice, R., Pollio, A., Garric, J., 2004. Environmental risk assessment of six human pharmaceuticals: Are the current environmental risk assessment procedures sufficient for the protection of the aquatic environment? *Environ. Toxicol. Chem.* 23, 1344–1354. https://doi.org/10.1897/03-246
Gago-Ferrero, P., Krettek, A., Fischer, S., Wiberg, K., Ahrens, L., 2018. Suspect screening and regulatory databases: A powerful combination to identify emerging micropollutants. *Environ. Sci. Technol.* 52, 6881–6894. https://doi.org/10.1021/acs.est.7b06598
Gallen, C., Eaglesham, G., Drage, D., Nguyen, T. H., Mueller, J. F., 2018. A mass estimate of perfluoroalkyl substance (PFAS) release from Australian wastewater treatment plants. *Chemosphere* 208, 975–983. https://doi.org/10.1016/j.chemosphere.2018.06.024
Gámiz, B., Hermosín, M. C., Celis, R., 2018. Appraising factors governing sorption and dissipation of the monoterpene carvone in agricultural soils. *Geoderma* 321, 61–68. https://doi.org/10.1016/j.geoderma.2018.02.005
Garcia, S. N., Foster, M., Constantine, L. A., Huggett, D. B., 2012. Field and laboratory fish tissue accumulation of the anti-convulsant drug carbamazepine. *Ecotoxicol. Environ. Saf.* 84, 207–211. https://doi.org/10.1016/j.ecoenv.2012.07.013
Gonçalves, P. V. B., Matthes, Â. D. C. S., Da Cunha, S. P., Lanchote, V. L., 2002. Enantioselectivity in the steady-state pharmacokinetics and transplacental distribution of pindolol at delivery in pregnancy-induced hypertension. *Chirality* 14, 683–687. https://doi.org/10.1002/chir.10124
Gramatica, P., Papa, E., 2003. QSAR modeling of bioconcentration factor by theoretical molecular descriptors. *QSAR Comb. Sci.* 22, 374–385. https://doi.org/10.1002/qsar.200390027
Grisoni, F., Consonni, V., Vighi, M., Villa, S., Todeschini, R., 2016. Expert QSAR system for predicting the bioconcentration factor under the REACH regulation. *Environ. Res.* 148, 507–512. https://doi.org/10.1016/j.envres.2016.04.032
Guan, Y.-F., Sun, J.-L., Ni, H.-G., Guo, J.-Y., 2012. Sedimentary record of polycyclic aromatic hydrocarbons in a sediment core from a maar lake, Northeast China: Evidence in historical atmospheric deposition. *J. Environ. Monit.* 14, 2475. https://doi.org/10.1039/c2em30461a
Gust, M., Buronfosse, T., Giamberini, L., Ramil, M., Mons, R., Garric, J., 2009. Effects of fluoxetine on the reproduction of two prosobranch mollusks : *Potamopyrgus antipodarum* and *Valvata piscinalis*. *Environ. Pollut.* 157, 423–429. https://doi.org/10.1016/j.envpol.2008.09.040
Huggett, D. B., Brooks, B.W., Peterson, B., Foran, C. M., Schlenk, D., 2002. Toxicity of select beta adrenergic receptor-blocking pharmaceuticals (B-blockers) on aquatic organisms. *Arch. Environ. Contam. Toxicol.* 43, 229–35.
Johnson, A. C., Jin, X., Nakada, N., Sumpter, J. P., 2020. Learning from the past and considering the future of chemicals in the environment. *Science (80-.).* 367, 384–387. https://doi.org/10.1126/science.aay6637
Karahan Ozgun, O., Basak, B., Eropak, C., Abat, S., Kirim, G., Girgin, E., Hanedar, A., Gunes, E., Citil, E., Görgün, E., Gomec, C. Y., Babuna, F. G., Ovez, S., Tanik, A., Ozturk, I., Kinaci, C., Karaaslan, Y., Gucver, S. M., Siltu, E., Orhon, A. K., 2016. Prioritization methodology of dangerous substances for water quality monitoring with scarce data. *Clean Technol. Environ. Policy* 1–18. https://doi.org/10.1007/s10098-016-1194-z
Kasprzyk-Hordern, B., 2010. Pharmacologically active compounds in the environment and their chirality. *Chem. Soc. Rev.* 39, 4466. https://doi.org/10.1039/c000408c
Khan, S. J., 2014. Biologically mediated chiral inversion of emerging contaminants, in: Lambropoulou, D. A., Nollet, L. M. L. (Eds.), *Transformation Products of Emerging Contaminants in the Environment: Analysis, Processes, Occurrence, Effects and Risks*. John Wiley & Sons, Chichester, UK, pp. 261–279.
Khan, S. J., Wang, L., Hashim, N. H., Mcdonald, J. A., 2014. Distinct enantiomeric signals of ibuprofen and naproxen in treated wastewater and sewer overflow. *Chirality* 26, 739–746. https://doi.org/10.1002/chir

Kim, J. H., Scialli, A. R., 2011. Thalidomide: The tragedy of birth defects and the effective treatment of disease. *Toxicol. Sci.* 122, 1–6. https://doi.org/10.1093/toxsci/kfr088

Krafft, M. P., Riess, J. G., 2015. Per- and polyfluorinated substances (PFASs): Environmental challenges. *Curr. Opin. Colloid Interface Sci.* 20, 192–212. https://doi.org/10.1016/j.cocis.2015.07.004

Lalone, C. A., Berninger, J. P., Villeneuve, D. L., Ankley, G. T., 2014. Leveraging existing data for prioritization of the ecological risks of human and veterinary pharmaceuticals to aquatic organisms. *Philos. Trans. R. Soc. B* 369, 1–10.

LaLone, C. A., Villeneuve, D. L., Burgoon, L. D., Russom, C. L., Helgen, H. W., Berninger, J. P., Tietge, J. E., Severson, M. N., Cavallin, J. E., Ankley, G. T., 2013. Molecular target sequence similarity as a basis for species extrapolation to assess the ecological risk of chemicals with known modes of action. *Aquat. Toxicol.* 144–145, 141–154. https://doi.org/10.1016/j.aquatox.2013.09.004

Lammer, E., Carr, G. J., Wendler, K., Rawlings, J. M., Belanger, S. E., Braunbeck, T., 2009. Is the fish embryo toxicity test (FET) with the zebrafish (*Danio rerio*) a potential alternative for the fish acute toxicity test? *Comp. Biochem. Physiol. C. Toxicol. Pharmacol.* 149, 196–209. https://doi.org/10.1016/j.cbpc.2008.11.006

Larsen, G. L., Beskid, C., Shirnamé-Moré, L., 2002. Environmental air toxics: Role in asthma occurrence? *Environ. Health Perspect.* 110 Suppl, 501–504. https://doi.org/sc271_5_1835[pii]

Lillicrap, A., Springer, T., Tyler, C. R., 2016. A tiered assessment strategy for more effective evaluation of bioaccumulation of chemicals in fish. *Regul. Toxicol. Pharmacol.* 75, 20–26. https://doi.org/10.1016/j.yrtph.2015.12.012

Liu, L.-Y., Ma, W.-L., Jia, H.-L., Zhang, Z.-F., Song, W.-W., Li, Y.-F., 2016. Research on persistent organic pollutants in China on a national scale: 10 years after the enforcement of the Stockholm Convention. *Environ. Pollut.* 217, 70–81. https://doi.org/10.1016/j.envpol.2015.12.056

Liu, T., Diao, J., Di, S., Zhou, Z., 2014. Stereoselective bioaccumulation and metabolite formation of triadimefon in *Tubifex tubifex*. *Environ. Sci. Technol.* 48(12), 6687–6693.

Lorgeoux, C., Moilleron, R., Gasperi, J., Ayrault, S., Bonté, P., Lefèvre, I., Tassin, B., 2016. Temporal trends of persistent organic pollutants in dated sediment cores: Chemical fingerprinting of the anthropogenic impacts in the Seine River basin, Paris. *Sci. Total Environ.* 541, 1355–1363. https://doi.org/10.1016/j.scitotenv.2015.09.147

Marvin, C. H., Tomy, G. T., Armitage, J. M., Arnot, J. A., McCarty, L., Covaci, A., Palace, V., 2011. Hexabromocyclododecane: Current understanding of chemistry, environmental fate and toxicology and implications for global management. *Environ. Sci. Technol.* 45, 8613–8623. https://doi.org/10.1021/es201548c

McCauley, D. J., Degraeve, G. M., Linton, T. K., 2000. Sediment quality guidelines and assessment: Overview and research needs. *Environ. Sci. Policy* 3, 133–144. https://doi.org/10.1016/S1462-9011(00)00040-X

Mehvar, R., Brocks, D. R., 2001. Stereospecific pharmacokinetics and pharmacodynamics of beta-adrenergic blockers in humans. *J. Pharm. Sci.* 4, 185–200.

Mehvar, R., Brocks, D. R., Vakily, M., 2002. Impact of stereoselectivity on the pharmacokinetics and pharmacodynamics of antiarrhythmic drugs. *Clin. Pharmacokinet.* 41, 533–58. https://doi.org/10.2165/00003088-200241080-00001

Minister of Public Works and Government Services Canada, 2000. *Guidance for Indusry: Stereochemical Issues in Chiral Drug Development, Therapeutic Products Programme*. Health Canada - Publications, Ottawa, Canada.

Mueller, M. D., Buser, H.-R., 1995. Environmental behavior of acetamide pesticide stereoisomers. 2. stereo- and enantioselective degradation in sewage sludge and soil. *Environ. Sci. Technol.* 29, 2031–2037. https://doi.org/10.1021/es00008a023

Nfon, E., Cousins, I. T., 2007. Modelling PCB bioaccumulation in a baltic food web. *Environ. Pollut.* 148, 73–82. https://doi.org/10.1016/j.envpol.2006.11.033

Niemi, L. M., Stencel, K. A., Murphy, M. J., Schultz, M. M., 2013. Quantitative determination of antidepressants and their select degradates by liquid chromatography/electrospray ionization tandem mass spectrometry in biosolids destined for land application. *Anal. Chem.* 85, 7279–86. https://doi.org/10.1021/ac401170s

Oakes, K. D., Coors, A., Escher, B. I., Fenner, K., Garric, J., Gust, M., Knacker, T., Küster, A., Kussatz, C., Metcalfe, C. D., Monteiro, S., Moon, T. W., Mennigen, J. A, Parrott, J., Péry, A. R. R., Ramil, M., Roennefahrt, I., Tarazona, J. V, Sánchez-Argüello, P., Ternes, T. A, Trudeau, V. L., Boucard, T., Van Der Kraak, G. J., Servos, M. R., 2010. Environmental risk assessment for the serotonin re-uptake inhibitor fluoxetine: Case study using the European risk assessment framework. *Integr. Environ. Assess. Manag.* 6 Suppl, 524–39. https://doi.org/10.1002/ieam.77

Overturf, M. D., Overturf, C. L., Baxter, D., Hala, D. N., Constantine, L., Venables, B., Huggett, D. B., 2012. Early life-stage toxicity of eight pharmaceuticals to the fathead minnow, *Pimephales promelas*. *Arch. Environ. Contam. Toxicol.* 62, 455–64. https://doi.org/10.1007/s00244-011-9723-6

Pizzo, F., Lombardo, A., Manganaro, A., Cappelli, C. I., Petoumenou, M. I., Albanese, F., Roncaglioni, A., Brandt, M., Benfenati, E., 2016. Integrated in silico strategy for PBT assessment and prioritization under REACH. *Environ. Res.* 151, 478–492. https://doi.org/10.1016/j.envres.2016.08.014

Poiger, T., Müller, M. D., Buser, H.-R. R., Buerge, I. J., 2015. Environmental behavior of the chiral herbicide haloxyfop. 1. Rapid and preferential interconversion of the enantiomers in soil. *J. Agric. Food Chem.* 63, 2583–2590. https://doi.org/10.1021/jf505241t

Rand-Weaver, M., Margiotta-Casaluci, L., Patel, A., Panter, G. H., Owen, S. F., Sumpter, J. P., 2013. The read-across hypothesis and environmental risk assessment of pharmaceuticals. *Environ. Sci. Technol.* 47, 11384–11395. https://doi.org/10.1021/es402065a

Raubenheimer, K., McIlgorm, A., 2018. Can the Basel and Stockholm Conventions provide a global framework to reduce the impact of marine plastic litter? *Mar. Policy* 96, 285–290. https://doi.org/10.1016/j.marpol.2018.01.013

Ribeiro, A. R., Afonso, C. M., Castro, P. M. L. L., Tiritan, M. E., 2013. Enantioselective biodegradation of pharmaceuticals, alprenolol and propranolol, by an activated sludge inoculum. *Ecotoxicol. Environ. Saf.* 87, 108–14. https://doi.org/10.1016/j.ecoenv.2012.10.009

Rickhaus, M., Mayor, M., Juríček, M., 2016. Strain-induced helical chirality in polyaromatic systems. *Chem. Soc. Rev.* 45, 1542–1556. https://doi.org/10.1039/c5cs00620a

Ruan, Y., Zhang, K., Lam, J. C. W., Wu, R., Lam, P. K. S., 2019. Stereoisomer-specific occurrence, distribution, and fate of chiral brominated flame retardants in different wastewater treatment systems in Hong Kong. *J. Hazard. Mater.* 374, 211–218. https://doi.org/10.1016/j.jhazmat.2019.04.041

Ruan, Y., Zhang, X., Qiu, J. W., Leung, K. M. Y., Lam, J. C. W., Lam, P. K. S., 2018. Stereoisomer-specific trophodynamics of the chiral brominated flame retardants HBCD and TBECH in a marine food web, with implications for human exposure. *Environ. Sci. Technol.* 52, 8183–8193. https://doi.org/10.1021/acs.est.8b02206

Sakamoto, M., Murata, K., Tsuruta, K., Miyamoto, K., Akagi, H., 2010. Retrospective study on temporal and regional variations of methylmercury concentrations in preserved umbilical cords collected from inhabitants of the Minamata area, Japan. *Ecotoxicol. Environ. Saf.* 73, 1144–1149. https://doi.org/10.1016/j.ecoenv.2010.05.007

Sanderson, H., Johnson, D. J., Wilson, C. J., Brain, R. A., Solomon, K. R., 2003. Probabilistic hazard assessment of environmentally occurring pharmaceuticals toxicity to fish, daphnids and algae by ECOSAR screening. *Toxicol. Lett.* 144, 383–395. https://doi.org/10.1016/S0378-4274(03)00257-1

Sanganyado, E., 2019. Comments on "Chiral pharmaceuticals: Environment sources, potential human health impacts, remediation technologies and future perspective." *Environ. Int.* 122, 412–415. https://doi.org/10.1016/j.envint.2018.11.032

Sanganyado, E., Lu, Z., Fu, Q., Schlenk, D., Gan, J., 2017. Chiral pharmaceuticals: A review on their environmental occurrence and fate processes. *Water Res.* 124, 527–542. https://doi.org/10.1016/j.watres.2017.08.003

Sanganyado, E., Lu, Z., Liu, W., 2020. Application of enantiomeric fractions in environmental forensics: Uncertainties and inconsistencies. *Environ. Res.* 184, 109354. https://doi.org/10.1016/j.envres.2020.109354

Sanganyado, E., Rajput, I. R., Liu, W., 2018. Bioaccumulation of organic pollutants in Indo-Pacific humpback dolphin: A review on current knowledge and future prospects. *Environ. Pollut.* 237. https://doi.org/10.1016/j.envpol.2018.01.055

Schultz, M. M., Painter, M. M., Bartell, S. E., Logue, A., Furlong, E. T., Werner, S. L., Schoenfuss, H. L., 2011. Selective uptake and biological consequences of environmentally relevant antidepressant pharmaceutical exposures on male fathead minnows. *Aquat. Toxicol.* 104, 38–47. https://doi.org/10.1016/j.aquatox.2011.03.011

Scott, W. C. C., Du, B., Haddad, S. P., Breed, C. S., Saari, G. N., Kelly, M., Broach, L., Chambliss, C. K., Brooks, B. W., 2015. Predicted and observed therapeutic dose exceedences of ionizable pharmaceuticals in fish plasma from urban coastal systems. *Environ. Toxicol. Chem.* 35, n/a-n/a. https://doi.org/10.1002/etc.3236

Shen, Q., Wang, L., Zhou, H., Jiang, H., Yu, L., Zeng, S., 2013. Stereoselective binding of chiral drugs to plasma proteins. *Acta Pharmacol. Sin.* 34, 998–1006. https://doi.org/10.1038/aps.2013.78

Smith, S. W., 2009. Chiral Toxicology: It's the Same Thing…Only Different. *Toxicol. Sci.* 110, 4–30. https://doi.org/10.1093/toxsci/kfp097

Stanley, J. K., Brooks, B. W., 2009. Perspectives on ecological risk assessment of chiral compounds. *Integr. Environ. Assess. Manag.* 5, 364. https://doi.org/10.1897/IEAM_2008-076.1

Stanley, J. K., Ramirez, A. J., Chambliss, C. K., Brooks, B. W., 2007. Enantiospecific sublethal effects of the antidepressant fluoxetine to a model aquatic vertebrate and invertebrate. *Chemosphere* 69, 9–16. https://doi.org/10.1016/j.chemosphere.2007.04.080

Stanley, J. K., Ramirez, A. J., Mottaleb, M., Chambliss, C. K., Brooks, B. W., 2006. Enantiospecific toxicity of the beta-blocker propranolol to *Daphnia magna* and *Pimephales promelas*. *Environ. Toxicol. Chem.* 25, 1780–6.

Stoschitzky, K., Egginger, G., Zernig, G., Klein, W., Lindner, W., 1993. Stereoselective features of (R)- and (S)-atenolol: Clinical pharmacological, pharmacokinetic, and radioligand binding studies. *Chirality* 5, 15–9.

Sun, D., Pang, J., Qiu, J., Li, L., Liu, C., Jiao, B., 2013. Enantioselective degradation and enantiomerization of indoxacarb in soil. *J. Agric. Food Chem.* 61, 11273–11277. https://doi.org/10.1021/jf4045952

Sun, L., Xin, L., Peng, Z., Jin, R., Jin, Y., Qian, H., Fu, Z., 2014. Toxicity and enantiospecific differences of two β-blockers, propranolol and metoprolol, in the embryos and larvae of zebrafish (*Danio rerio*). *Environ. Toxicol.* 29, 1367–1378. https://doi.org/10.1002/tox.21867

Tokunaga, E., Yamamoto, T., Ito, E., Shibata, N., 2018. Understanding the thalidomide chirality in biological processes by the self-disproportionation of enantiomers. *Sci. Rep.* 8, 6–12. https://doi.org/10.1038/s41598-018-35457-6

U.S. Food and Drug Administration, 1992. FDA's policy statement for the development of new stereoisomeric drugs. *Chirality* 4, 338–40.

Valdés, M. E., Amé, M. V., Bistoni, M. D. L. A., Wunderlin, D. A., 2014. Occurrence and bioaccumulation of pharmaceuticals in a fish species inhabiting the Suquía River basin (Córdoba, Argentina). *Sci. Total Environ.* 472, 389–396. https://doi.org/10.1016/j.scitotenv.2013.10.124

Vernouillet, G., Eullaffroy, P., Lajeunesse, A., Blaise, C., Gagné, F., Juneau, P., 2010. Toxic effects and bioaccumulation of carbamazepine evaluated by biomarkers measured in organisms of different trophic levels. *Chemosphere* 80, 1062–1068. https://doi.org/10.1016/j.chemosphere.2010.05.010

von der Recke, R., Mariussen, E., Berger, U., Götsch, A., Herzke, D., Vetter, W., 2005. Determination of the enantiomer fraction of PBB 149 by gas chromatography/electron capture negative ionization tandem mass spectrometry in the selected reaction monitoring mode. *Rapid Commun. Mass Spectrom.* 19, 3719–3723. https://doi.org/10.1002/rcm.2258

Walters, D. M., Jardine, T. D., Cade, B. S., Kidd, K. A., Muir, D. C. G., Leipzig-Scott, P., 2016. Trophic magnification of organic chemicals: A global synthesis. *Environ. Sci. Technol.* 50, 4650–4658. https://doi.org/10.1021/acs.est.6b00201

Wang, H., Yan, Z.-G., Li, H., Yang, N.-Y., Leung, K. M. Y., Wang, Y., Yu, R.-Z., Zhang, L., Wang, W.-H., Jiao, C.-Y., Liu, Z.-T., 2012. Progress of environmental management and risk assessment of industrial chemicals in China. *Environ. Pollut.* 165, 174–181. https://doi.org/10.1016/j.envpol.2011.12.008

Wang, H., Yang, Z., Liu, R., Fu, Q., Zhang, S., Cai, Z., Li, J., Zhao, X., Ye, Q., Wang, W., Li, Z., 2013. Stereoselective uptake and distribution of the chiral neonicotinoid insecticide, Paichongding, in Chinese pak choi (*Brassica campestris* ssp. chinenesis). *J. Hazard. Mater.* 262, 862–869. https://doi.org/10.1016/j.jhazmat.2013.09.054

Wang, Y., Yu, D., Xu, P., Guo, B., Zhang, Y., Li, J., Wang, H., 2014. Stereoselective metabolism, distribution, and bioaccumulation brof triadimefon and triadimenol in lizards. *Ecotoxicol. Environ. Saf.* 107, 276–283. https://doi.org/10.1016/j.ecoenv.2014.06.021

Wang, Z., Walker, G. W., Muir, D. C. G., Nagatani-Yoshida, K., 2020. Toward a global understanding of chemical pollution: A first comprehensive analysis of national and regional chemical inventories. *Environ. Sci. Technol.* 54, 2575–2584. https://doi.org/10.1021/acs.est.9b06379

Williams, M., Ong, P. L., Williams, D. B., Kookana, R. S., 2009. Estimating the sorption of pharmaceuticals based on their pharmacological distribution. *Environ. Toxicol. Chem.* 28, 2572–2579. https://doi.org/10.1897/08-587.1

Williams, M., Saison, C. L. A, Williams, D. B., Kookana, R. S., 2006. Can aquatic distribution of human pharmaceuticals be related to pharmacological data? *Chemosphere* 65, 2253–2259. https://doi.org/10.1016/j.chemosphere.2006.05.036

Winder, V. L., Sapozhnikova, Y., Pennington, P. L., Wirth, E. F., 2009. Effects of fluoxetine exposure on serotonin-related activity in the sheepshead minnow (*Cyprinodon variegatus*) using LC/MS/MS detection and quantitation. *Comp. Biochem. Physiol. C. Toxicol. Pharmacol.* 149, 559–65. https://doi.org/10.1016/j.cbpc.2008.12.008

Wong, C. S., 2006. Environmental fate processes and biochemical transformations of chiral emerging organic pollutants. *Anal. Bioanal. Chem.* 386, 544–58.

Wong, C. S., Warner, N. A., 2010. Chirality as an environmental forensics tool, in: Harrad, S. (Ed.), *Persistent Organic Pollutants*. John Wiley & Sons, Chichester, UK, pp. 71–136.

Zhang, Q., Xiong, W. W., Gao, B., Cryder, Z., Zhang, Z., Tian, M., Sanganyado, E., Shi, H., Wang, M., 2018a. Enantioselectivity in degradation and ecological risk of the chiral pesticide ethiprole. *L. Degrad. Dev.* 29, 4242–4251. https://doi.org/10.1002/ldr.3179

Zhang, Q., Zhang, Z., Tang, B., Gao, B., Tian, M., Sanganyado, E., Shi, H., Wang, M., 2018b. Mechanistic insights into stereospecific bioactivity and dissipation of chiral fungicide triticonazole in agricultural management. *J. Agric. Food Chem.* 66, 7286–7293. https://doi.org/10.1021/acs.jafc.8b01771

Section IV

Synthesis and Chiral Switching of Chiral Chemicals

13 Chiral Pesticides
Synthesis, Chiral Switching, and Absolute Configuration

Herbert Musarurwa, Mpelegeng Victoria Bvumbi, and Nikita Tawanda Tavengwa

CONTENTS

13.1 Introduction 323
13.2 Synthesis of Chiral Pesticides 324
    13.2.1 Catalytic Asymmetric Synthesis 325
    13.2.2 Chiron Approach 326
    13.2.3 Enzymatic Transformation 327
    13.2.4 Miscellaneous Methods of Synthesizing Chiral Pesticides 328
13.3 Molecularly Imprinted Polymers 330
13.4 Chiral Switching 332
13.5 Absolute Configuration of Chiral Pesticides 336
    13.5.1 X-ray Crystallography 336
    13.5.2 Circular Dichroism Spectroscopy 337
        13.5.2.1 ECD Spectroscopy 337
        13.5.2.2 VCD Spectroscopy 338
13.6 Challenges and Future Prospects 340
13.7 Conclusion 340
Acknowledgment 341
Conflicts of Interest 341
References 341

13.1 INTRODUCTION

The demand of pesticides in the world is always on the rise due to increasing food demands (Caballo et al., 2013). Most of the pesticides synthesized in the agro-industry have at least one asymmetric center and are therefore chiral (Cui et al., 2018; Chen et al., 2016). Most asymmetric centers in chiral pesticides are usually found on carbon, like what is found in metolachlor. However, some chiral pesticides such as profenofos have chiral centers on phosphorus while others, like fensulfothion, have them on sulfur.

Many methods have been developed for the synthesis of chiral pesticides. These include catalytic asymmetric synthesis (Sun et al., 2018; Yousefi et al., 2019), chiron approach (Aguiar et al., 2019; Yang et al., 2020), and enzymatic transformation (Zhang et al., 2018; Zheng et al., 2006). Catalytic asymmetric synthesis is widely used during production of chiral pesticides and it involves the use of chiral catalysts during the synthetic process (Jeschke, 2018). Most chiral catalysts used during pesticide synthesis are organometallic compounds. These catalysts are very efficient but they are generally very expensive. Enzymatic transformation, on the other hand, involves the use of enzymes during the synthesis of chiral pesticides (Zhang et al., 2018; Zheng et al., 2006). Enzymes are used as biocatalysts, and are relatively cheap and green as compared to organometallic catalysts used during catalytic asymmetric synthesis of chiral pesticides. Enzymatic transformation becomes more attractive and greener, from an environmental standpoint, when whole-cell catalysts are used during synthesis of chiral pesticides (Zhang et al., 2018). The chiron approach (chiral pool strategy) is yet another strategy that can be used during the synthesis of chiral pesticides. It involves the use of natural chiral compounds as chiral synthons for chiral pesticides (Aguiar et al., 2019).

Most chiral pesticides, after synthesis, are sold on the market as racemic mixtures. The enantiomers in the racemates do not exhibit the same bioactivity towards target organisms (Zhao et al., 2019). Bioactivity of chiral pesticides is enantioselective. Usually, there is preferential activity of one of the enantiomers and the rest would be inactive. These inactive enantiomers are discharged into the environment where they wreak havoc on non-target organisms (Cui et al., 2018; Xie et al., 2019). This serious environmental issue prompted researchers and environmentalists to advocate for the replacement of the racemates on the market with enantiopure chiral pesticides (chiral switching). Chiral switching has many benefits associated with it in pest management. For instance, it reduces the application rates of chiral pesticides (Wang et al., 2018). In addition, it reduces the amount of inactive enantiomers that are discharged into the environment during pest control (Hor et al., 2017). Thus, chiral switching is slowly gaining momentum in most countries in the world. For instance, racemic metolachlor has been replaced successfully with S-metolachlor while racemic metalaxyl is now being sold as enantiopure R-metalaxyl in most countries in the world.

The synthetic process of chiral pesticides is usually concluded by the determination of the absolute configurations of their enantiomers. The absolute configuration of the enantiomers of chiral pesticides may be determined using experimental techniques such as X-ray crystallography and nuclear magnetic resonance (Polavarapu, 2016). These techniques, although efficient, suffer from some drawbacks. For instance, during X-ray crystallography, good quality crystals must be obtained prior to analysis. Obtaining these quality crystals is not always easy with chiral pesticides. They are also time-consuming and use large quantities of the sample (Polavarapu, 2016; Ye et al., 2012). Consequently, these techniques are slowly giving way to methods like circular dichroism spectroscopy during the determination of the absolute configuration of chiral pesticides (Merten et al., 2019; Matsuo and Gekko, 2020; Ye et al., 2012). Circular dichroism spectroscopy is one of the popular techniques for assigning absolute configuration to chiral pesticides. It uses small volumes of the sample and is easy to implement (Yang et al., 2015).

This chapter discusses the synthesis of chiral pesticides using a variety of methods. After synthesis, the chiral pesticides are applied to crops in order to control pests. If they are not properly used, they end up polluting the environment. Some strategies, therefore, should be put in place to mitigate pollution of the environment by chiral pesticides. One such strategy is chiral switching. It is an important intervention for minimizing pollution of the environment by inactive enantiomers of chiral pesticides. This chapter gives a detailed account of the concept of chiral switching with appropriate examples, where necessary. The synthetic processes of chiral pesticides are concluded usually by the determination of their absolute configurations. The techniques that are used during the determination of the absolute configurations of chiral pesticides are also discussed in this chapter.

13.2 SYNTHESIS OF CHIRAL PESTICIDES

There are some basic strategies that are used to produce chiral pesticides that are pure enantiomerically. These include the use of naturally occurring optically active molecules and asymmetric synthesis. The latter is the most common method that is used to synthesize chiral compounds. Asymmetric syntheses can be performed in two ways, namely an enantioselective or diastereoselective way (Scheme 13.1). In enantioselective synthesis, the use of optically pure catalyst and enzymes are advantageous whereas in diastereoselective process a pro-chiral group in an enantiomerically pure compound, generally obtained from an achiral compound and a chiral auxiliary, is selectively converted with formation of a new chiral center (Feringa, 1988).

Chiral Pesticides

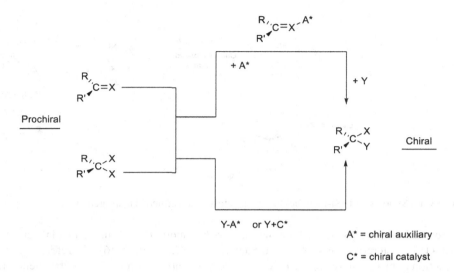

SCHEME 13.1 The enatioselective and diastereoselective ways of synthesizing chiral pesticides.

Furthermore, the use of chiral "synthons" in the preparation of new, optically pure pesticides is equally important (Feringa, 1988). The approach requires relative low molecular weight, enantiomerically pure and synthetically versatile compounds.

13.2.1 Catalytic Asymmetric Synthesis

Catalytic asymmetric synthesis involves the use of a chiral catalyst to transfer and amplify chirality during a chemical reaction (Jeschke, 2018). The synthesis is a very common way of producing chiral pesticides. For instance, metolachlor is generally synthesized from the hydrogenation reaction of 2-ethyl-6-methyl-aniline and methoxyacetone in the presence of platinum/carbon catalyst and the reaction afford precursor **1**. Furthermore, precursor **1** is then reacted with 2-chloroacetyl chloride to give rac-metolachlor (Scheme 13.2). Alternatively, a one pot reaction between 2-ethyl-6-methyl aniline, 2-bromo-1-methoxy-propane and chloroacetyl chloride can be used for the synthesis rac-metolachlor (Scheme 13.2).

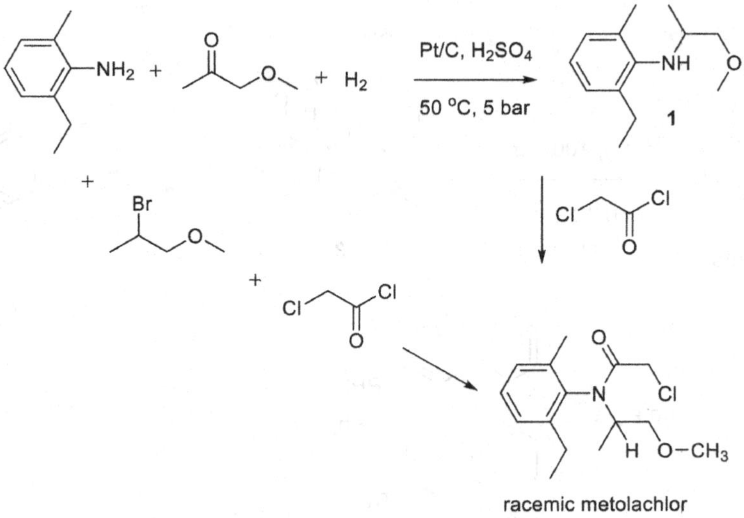

SCHEME 13.2 Catalytic asymmetric synthesis of rac-metolachlor.

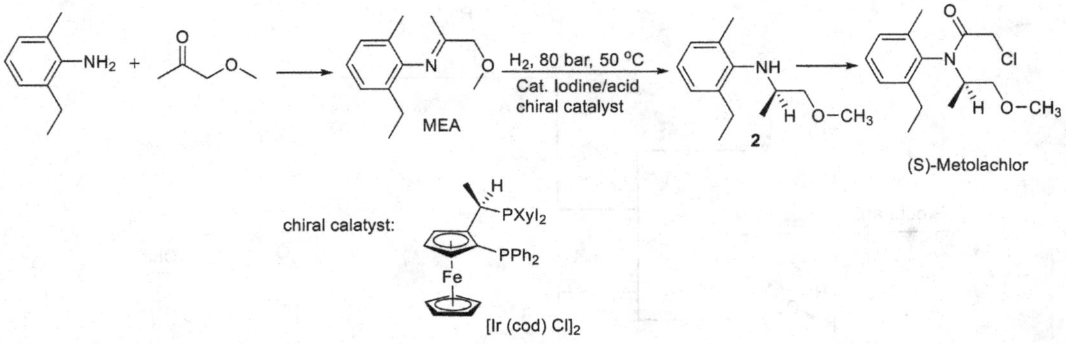

SCHEME 13.3 Synthesis of (S)-metolachlor using catalytic asymmetric hydrogenation.

Rac-metolachlor is a mixture of four enantiomers but about 95% of the herbicidal activity is due to the two (1S)-diastereoisomers (Rihs and Sauter, 1982; Xie et al., 2016). Therefore, the synthesis and production of enantiopure (S)-metolachlor is very essential to minimize environmental risks on non-target organisms by the inactive enantiomers. The synthetic process, just like for rac-metolachlor, starts with the formation of precursor **1** (Scheme 13.3). This is then followed by its reduction in the presence of a catalyst, which is generated *in situ* from [Ir(Cod)Cl]$_2$ and the ferrocenyldiphosphine ligand, xyliphos resulting in the formation compound 2. Finally, compound 2 is reacted with chloroacetyl chloride to form S-metolachlor (Scheme 13.3) (Müller, 2002).

Another chiral pesticide that can be synthesized via catalytic asymmetric hydrogenation is clozylacon (Spindler et al., 1998). It is a fungicide that is well suited for soil application against oomycetes. It exists as four stereoisomers and it can be synthesized via the enantioselective hydrogenation of an enamide precursor in the presence of Ruthenium/Binap catalyst to afford compound **3** which further undergoes chlorination reaction to give clozylacon (Scheme 13.4).

13.2.2 Chiron Approach

Chiron approach uses chiral compounds that occur naturally during the synthesis of chiral pesticides (Jeschke, 2018). Nature contains many chiral compounds that are cheap and commercially

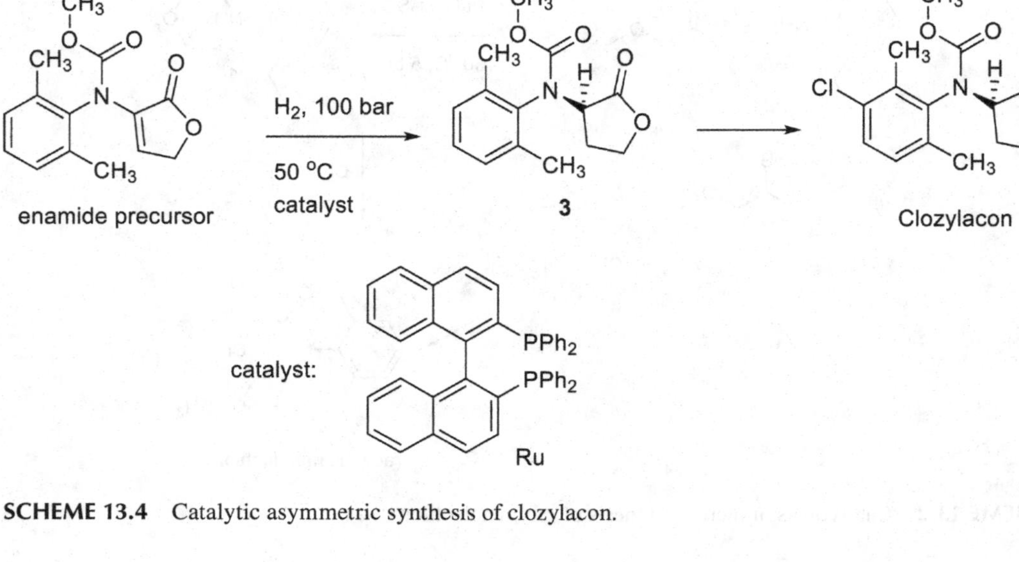

SCHEME 13.4 Catalytic asymmetric synthesis of clozylacon.

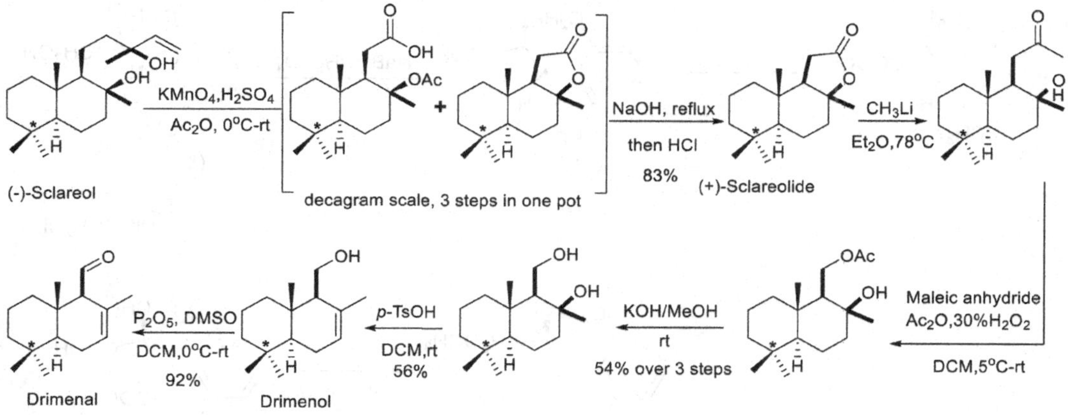

Chiral Pesticides

SCHEME 13.5 Synthesis of drimenal using the chiron approach.

available. Such compounds include carbohydrates, amino acids, alkaloids, and terpenes. These compounds can be used as precursors during synthesis of different types of chiral compounds including chiral pesticides. For instance, Li et al. (2018) used (-)-sclareol to synthesis a pesticide called drimenal (Scheme 13.5). Sclareol is a natural diterpene alcohol that is extracted from leaves and flowers of *Salvia sclarea*. It is a chiral natural compound and was successfully used by Li et al. (2018) to synthesis a chiral pesticide (Scheme 13.5).

Amino acids are among the most abundant chiral compounds that are found in nature. They can also be used as precursors during synthesis of chiral pesticides. For instance, Li et al. (2016) successfully used chiral amino acids to synthesis nicotamides (Scheme 13.6). These synthetic compounds can be used as pesticides since they have fungicidal properties.

13.2.3 Enzymatic Transformation

Enzymatic transformation involves the use of biocatalysts during the synthesis of chiral pesticides (Cheng et al., 2018; Jeschke, 2018). The catalysts that are used during catalytic asymmetric synthesis of chiral pesticides are usually made up of transition metals. These catalysts are generally expensive and make the production of chiral pesticides expensive. Thus, the need to find alternative

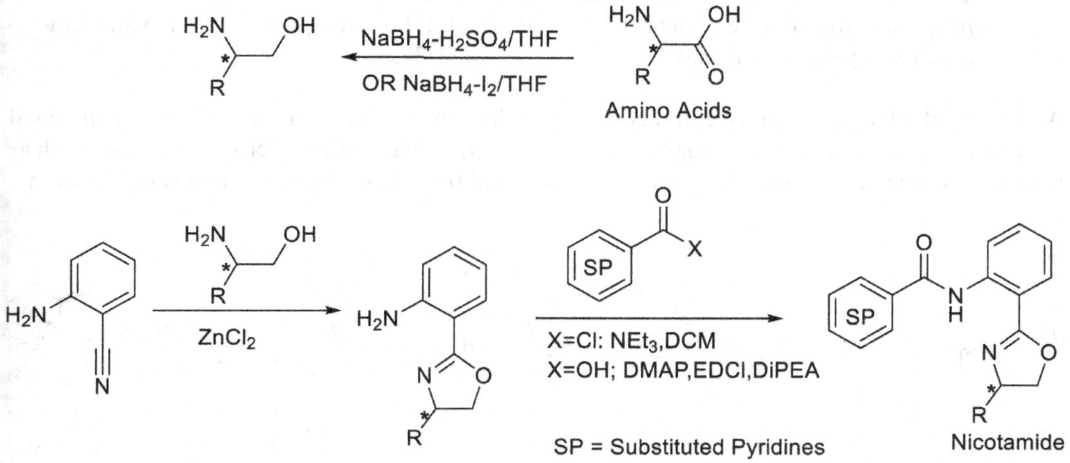

SCHEME 13.6 Synthesis of nicotamides using the chiron approach.

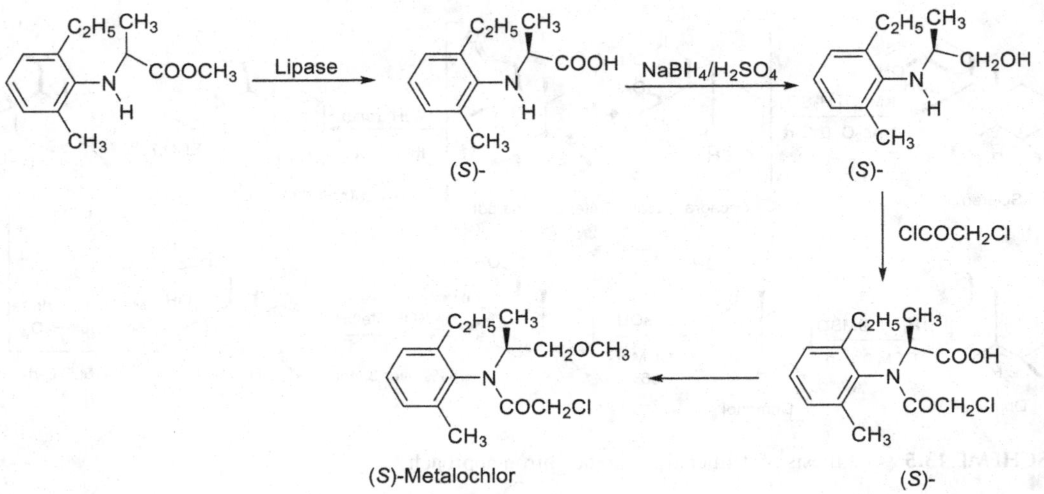

SCHEME 13.7 Synthesis of (S)-metolachlor using enzymatic transformation.

catalysts that are relatively cheap cannot be over emphasized. Consequently, the use of enzymes as biocatalyst during production of chiral pesticides is slowly gaining momentum (Cheng et al., 2018; Park and Lee, 2005; Zheng et al., 2006). For instance, Zheng et al. (2006) synthesized metolachlor using lipase as a biocatalyst (Scheme 13.7).

13.2.4 Miscellaneous Methods of Synthesizing Chiral Pesticides

Some of the chiral pesticides are amides, for example, alachlor, dimethachlor, metalaxyl, and metolachlor just to mention few, and they are all synthesized differently. For example, metalaxyl has two possible stereoisomers, namely the D- and L-isomers. It is synthesized by the reaction of 2,6-dimethyl aniline and 2-bromopropionic acid methyl ester which then produces the precursor **4**, which further reacted with methoxy acetyl chloride to produce the required metalaxyl (Scheme 13.8).

The D-configuration is known to be the most active one during pest control and attempts to separate the mixture of this isomer may be done in two different procedures:

1. The use of pure L-alanine (by substitution the configuration changes to D)
2. Separation of the mixture of the corresponding acids followed by esterification with methanol of the separated D-isomer

Another method for synthesis of chiral pesticides include the production of fipronil. This compound has a stereogenic center at the sulfur position. It is effective against a lot of crop insect pests such as bollweevils, rice insects, grasshoppers, house flies, and fruit flies. The compound is produced in a

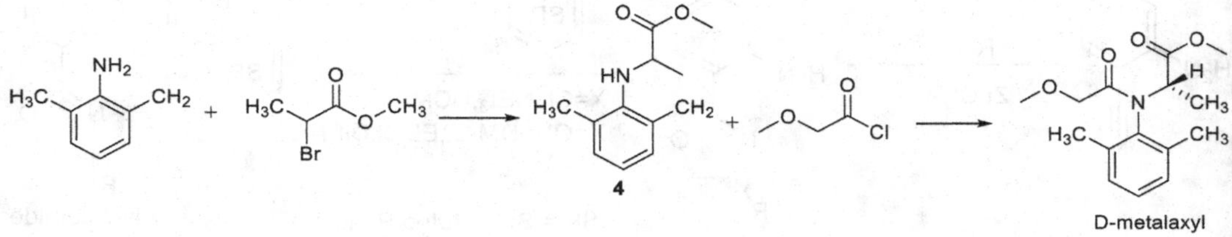

SCHEME 13.8 Synthetic scheme of metalaxyl.

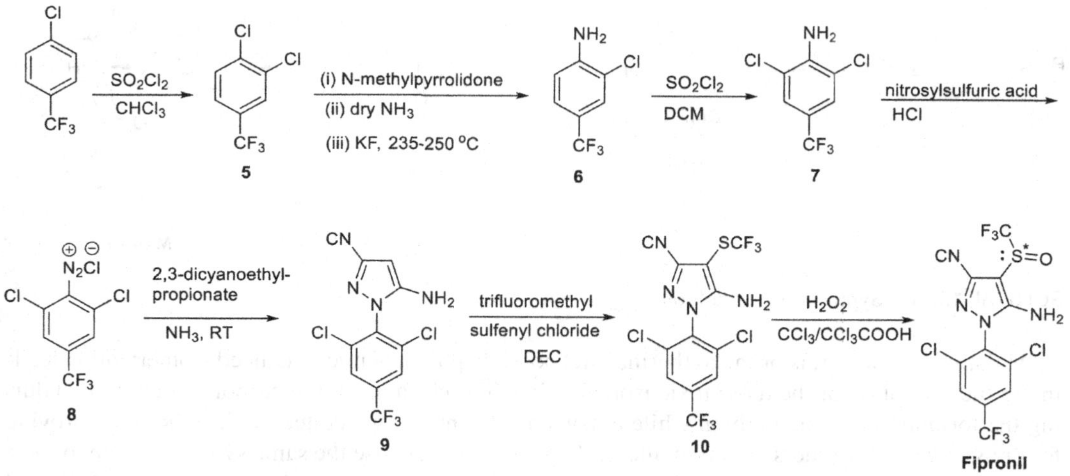

Chiral Pesticides

SCHEME 13.9 General synthesis of fipronil.

multi-step process which starts with the halogenation of 1-chloro-4-(trifluoromethyl)benzene with sulfuryl chloride in chloroform to form compound **5**. When **5** is reacting with polar groups such as N-methylpyrrolidine in dry ammonia, it produces compound **6**. The conversion of the halogenation reaction is replicated to form compound **7**. The use of nitrosylsulfuric acid with hydrochloric acid as a solvent affords the diazotization of compound **8**, which then reacts with cyanoalkyl propionate and ammonia at room temperature to obtain pyrazole compound **9**. Sulfenylation of reaction **9** by using sulfenyl chloride would then give thiopyrazole **10** which is then oxidized with peracid in trichloroacetic acid to afford the required fipronil (Scheme 13.9) (Saeed et al., 2016).

Of late, other methods are being developed: for instance, the synthesis of mecoprop was developed and proved to be the better in terms of conversion and selectivity (Yadav and Deshmukh, 2017). Mecoprop is a widely used household herbicide. A good synthetic approach/method towards the synthesis of its precursor, mecoprop methyl ester was developed using solid-liquid phase transfer catalysis (S-L PTC) with a mild base, K_2CO_3 in the presence of toluene as a solvent. A conversion of 95% together with 100% selectivity was achieved for mecoprop at 100°C. The reaction mixture of 4-chloro-2-methylphenol and methyl-2 bromopropanoate in the conditions mentioned afforded the mecoprop ester (Scheme 13.10). The ester could easily be converted to carboxylic acid in a reaction called saponification to form the chiral mecoprop.

Another widely used insecticide is an organophosphate called malathion. It has been used effectively for the control of insects on grains, fruits, nuts, cotton, and tobacco. Malathion is also a chiral pesticide, which is of the class of phosphorodithioate. It is produced by the reaction of P_2S_5 with methanol in the presence of toluene as a solvent. The dimethyl phosphorodithioic acid (DMPA) could then react with either diethyl fumarate or diethyl maleate to produce malathion (Scheme 13.11).

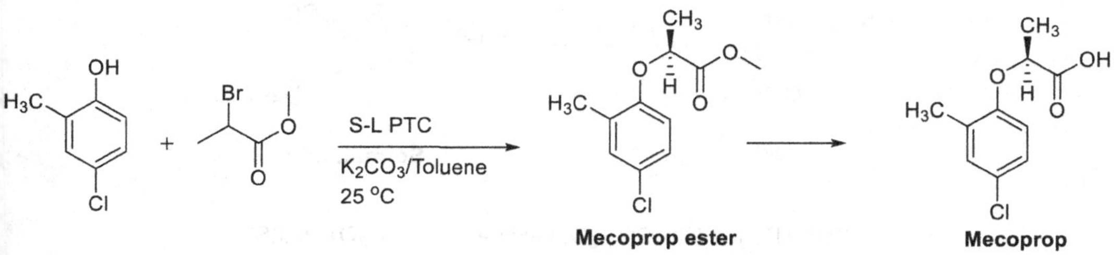

SCHEME 13.10 Synthesis of mecoprop ester using solid-liquid phase transfer catalysis.

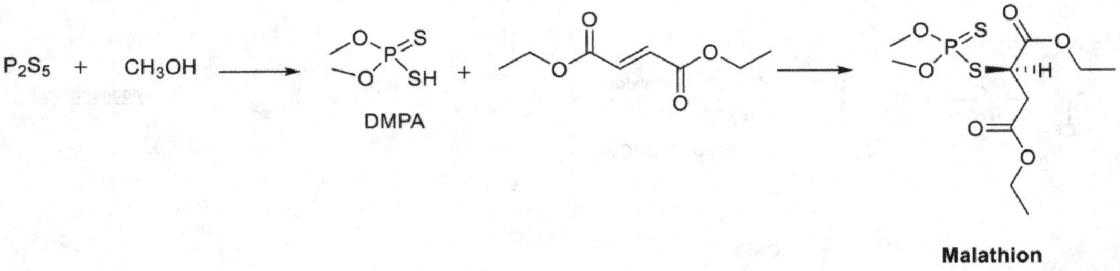

SCHEME 13.11 Synthesis of malathion.

Malathion, however, is prone to thermal and possibly photochemical induced isomerization leading to the formation of the more toxic isomalathion. The chiral center at carbon is maintained during the formation of isomalathion while a new chiral center is generated at phosphorus to provide four possible stereoisomers. Both l-malic acid and d-malic acid use the same synthetic route to form Rc- and Sc-malathion, respectively (Scheme 13.12). The acid is converted to diethylmalate followed by the hydroxyl group, which was converted to triflate in the presence of Tf_2O and lutidine. Thus, this method uses only one inversion at the carbon center and therefore reduces the opportunity for racemization (Berkman and Thompson, 1992).

13.3 MOLECULARLY IMPRINTED POLYMERS

Enantiopure chiral pesticides can be synthesized using chiral resolution techniques (Jeschke, 2018). This involves the separation of the enantiomers of racemic mixtures of chiral pesticides. Many separation techniques can be used during chiral resolution. Chromatographic methods such as chiral HPLC (Xia et al., 2019; P. Zhao et al., 2019), chiral GC (Dallegrave et al., 2016; G. Yao et al., 2019), and chiral supercritical fluid chromatography (Tao et al., 2018; Tan et al., 2017) are among the common techniques used during resolution of chiral pesticides. During these techniques, separation of the enantiomers of the chiral pesticides is achieved by passing the racemate through a chiral stationary phase (CSP). Differential retention of the enantiomers of the chiral pesticides occur on the CSP resulting in their different elution rates, causing their separation (Buerge et al., 2016; Liu et al., 2016). Materials, such as functionalized amylose and cellulose, can be used to make CSPs. These polysaccharide-based CSPs have relatively low selectivity towards the enantiomers of chiral pesticides. Selectivity during

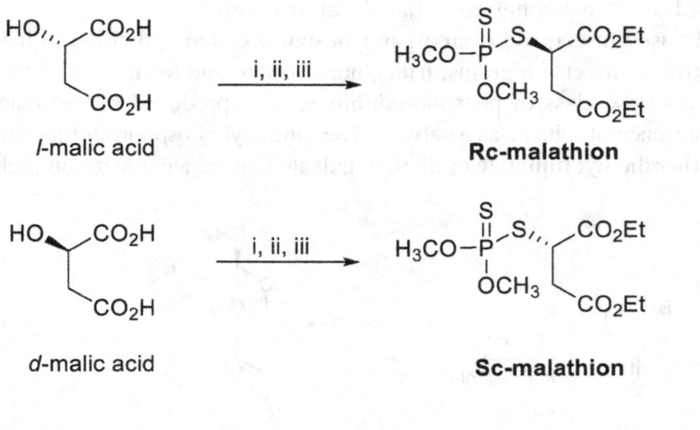

i = EtOH, H$^+$, ii = $(CF_3SO_2)_2O$, lutidine, iii = $(CH_3O)_2P(S)S^-Na^+$

SCHEME 13.12 Synthesis of Rc- and Sc-enantiomers of malathion.

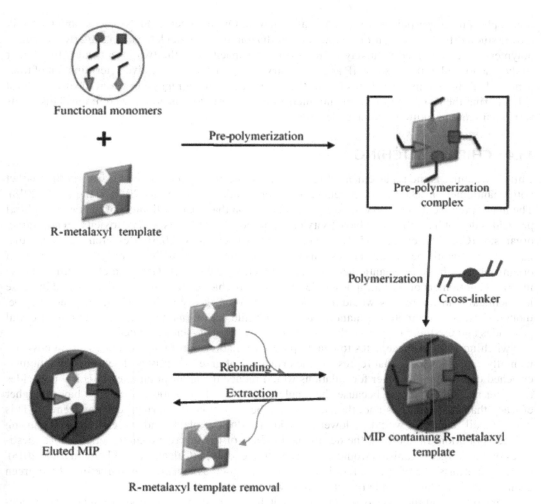

FIGURE 13.1 Preparation of chiral molecularly imprinted polymers for R-metalaxyl. (Adopted/modified from Sajini et al., 2019. Used with permission from Elsevier.)

resolution of chiral pesticides can be tremendously enhanced by using molecularly imprinted polymer-based CSPs (MIP-based CSPs) (Alotaibi et al., 2019; Rutkowska et al., 2018).

The general procedure for synthesizing chiral MIPs involves the complexation of a template, made up of the enantiomers of the chiral pesticide(s) of interest and functional monomers (Figure 13.1) (Sajini et al., 2019). This is then followed by co-polymerization in the presence of a cross-linker and an initiator in a suitable porogenic solvent. This results in the formation of a polymer that would be containing the target molecule (enantiomer). Removal of the target molecules from the polymer using a suitable eluent results in the formation of cavities in the polymer that are complementary to the shape and functional groups of the eluted chiral pesticide enantiomers (Hroboňová and Lomenova, 2018). Chiral MIPs have high chemical and thermal stability as well as greater affinity and selectivity towards specific enantiomers of chiral pesticides.

MIPs can be used as selective sorbents of chiral pesticides during sorbent-based extraction techniques or as CSPs during chromatographic techniques. For instance, Mehdipour et al. (2018) extracted chiral organophosphorus pesticides from plasma using magnet molecularly imprinted polymers. In a similar study, Boulanouar et al. (2018) used molecularly imprinted silica during solid phase extraction of organophosphorus pesticides in almond oil. Effective isolation of the organophosphorus pesticides using SPE was achieved. On the other hand, Szajnecki and Gawdzik (2019) studied the absorption of bifenthrin and diazonium insecticides on molecularly imprinted polymers. They obtained relatively high absorption capacities of the two insecticides. In another study, Cacho et al. (2008) used MIP as a stationary phase during the selective determination of thiabendazole fungicide in citrus fruits by capillary electrochromatography. However, the use of chiral MIPs during the resolution of the enantiomers of chiral pesticides is not very common. This gap is worth pursuing in future research endeavors.

13.4 CHIRAL SWITCHING

Chiral switching is the replacement of racemic chiral pesticides that are being sold on the market with enantiomer-enriched or single enantiomer formulations (Hor et al., 2017; Wang et al., 2018). The major premise for chiral switching is the realization that the enantiomers of the racemic chiral pesticides do not have the same bioactivity on the target organisms as well as toxicity on non-target organisms (Corcellas et al., 2015; Duan et al., 2018; Kaziem et al., 2020). Research has shown that usually one enantiomer would be responsible for the bioactivity of the chiral pesticide on target organisms and the other enantiomers would be inactive (Table 13.1) (Fonseca et al., 2019). These inactive enantiomers of the chiral pesticides would be discharged into the environment and increase its chemical load where they would adversely affect non-target organism. Thus, the general drive, though slow, is to promote the manufacture of only the active enantiopure formulations of chiral pesticides and use them to replace the racemates in the market—chiral switching.

Switching from the racemates to enantiopure formulations has many benefits. It is an environmentally conscious move that represents the principles of green chemistry. The use of enantiomer-enriched or single-enantiomer formulations would reduce the application rates of chiral pesticides (Yang et al., 2018). This is because the single-enantiomer formulations would be having higher efficacy than the racemates and, therefore, would be very effective during pest management. This would entail that there would be lower pesticide residues in plants and the environment causing reduced pollution. In addition, the detrimental effects of the inactive enantiomers of chiral pesticides on non-target organisms would be reduced tremendously (Calcaterra and D'Acquarica, 2018). Concerted efforts, therefore, are required to channel resources towards the marketing of the green enantiopure formulations and promote chiral switching.

There are several success stories on chiral switching of pesticides. A good example is the switching from racemic metolachlor to (S)-metolachlor that occurred in many countries during pest management. Metolachlor is an important herbicide that has two stereogenic centers and therefore four enantiomers (Yang et al., 2017). It was introduced into the market in 1976 as a racemic formulation containing all the four stereoisomers (Buser et al., 2000). Metolachlor proved to be very efficient at controlling weed growth but research revealed that its enantiomers were not equally active on target organisms. Researchers observed that about 95% of herbicidal activity of metolachlor was mainly a result of its two 1'S-enantiomers (aSS- and aRS-forms). This prompted the manufacture and commercialization of the enantiomer-enriched S-metolachlor. Currently, racemic metolachlor has been switched with S-metolachlor in many countries worldwide. For instance, Canada started to use the S-metolachlor switch in 1998, South Africa in 1998, and Austria in 1999 (Buser et al., 2000).

Another success story for chiral switching is the replacement of racemic metalaxyl on the market with R-metalaxyl. Metalaxyl is a broad-spectrum chiral pesticide that has fungicidal properties (Hamed et al., 2020). It was registered and introduced into the market in 1979 as pesticide for controlling fungi in many plants. Metalaxyl has one asymmetric carbon and therefore, it has one pair of enantiomers (Masbou et al., 2018). Bioactivity tests by researchers revealed that (R) – metalaxyl was more active than racemic and S-metalaxyl (Gu et al., 2018; Wang et al., 2018). This realization triggered the preferential commercialization of the (R) – metalaxyl formulations. Thus, there is chiral switching from racemic metalaxyl to (R) – metalaxyl in many countries in the world. The use of enantiopure (R) – metalaxyl formulations would alleviate pollution of the environment by the racemic and S-forms.

TABLE 13.1
Chiral Switching Informed by Experimental Data

Chiral Pesticide Studied	Enantiomers	Target Organism	Bioactivity of Chiral Pesticide	Activity of Racemates versus Pure Enantiomers	Chiral Switch	Reference
Fluxametamide	2	Plutella xylostella, Spodoptera exigua, Aphis gossypii, and Tetranychus cinnabarinus	Disrupts the γ-aminobutyric acid-gated chloride channels in pests	(S)-fluxametamide more active than rac- and (R)-fluxametamide	(S)-fluxametamide	Li et al. (2019)
Beflubutamid	2	Dicotyledonous weeds	Ergosterol biosynthesis inhibition	(−)-beflubutamid more active than rac- and (+)-beflubutamid	(−)-beflubutamid	Buerge et al. (2013)
Metolachlor	4	Echinochloa crusgalli	Inhibition of enzymes involved in gibberellic acid biosynthesis	(S)-metolachlor is more bioactive than rac- and R-metolachlor	(S)-metolachlor	Zhao et al. (2019)
Tebuconazole	2	Botrytis cinerea	Inihibition of fungal growth	(R)-tebuconazohas greater bioactivity than rac- and S-tebuconazole	(R)-tebuconazohas	Cui et al. (2018)
Imazamox, imazameth and imazethapyr	2	Echinochloa crusgalli and Microcysts aeruginosa	Inhibition of growth of roots and shoots	R-enantiomers more active than rac- and S-forms	R-enantiomers	Xie et al. (2018)
Metolachlor	4	Echinochloa crusgalli	Inhibition of enzymes involved in gibberellic acid biosynthesis	(S)-metolachlor is more bioactive than rac- and (R)-metolachlor	(S)-metolachlor	Zhao et al. (2019)
Napropamide	2	Scenedesmus obliquus	Inihibition of growth of weeds	(R)-napropamide more active than rac- and (S)-napropamide	(R)-napropamide	Qi et al. (2015)
Myclobutanil	2	Fusarium verticillioides	Inhibition of growth of fungi	(R)-myclobutanil more active than rac- and (S)-myclobutanil	(R)-myclobutanil	Li et al. (2018)
Lactofen	2	Echinochloa crusgali	Inhibition of weed growth	(R)-lactofen more active than rac- and (S)-lactofen	(R)-lactofen	Xie et al. (2018)
Imazapyr	2	Aradidopsis thalina	Chlorophyll synthesis inhibition	(+)-imazapyr more active than rac- and (−)-imazapyr	(+)-imazapyr	Hsiao et al. (2014)

(Continued)

TABLE 13.1 (*Continued*)
Chiral Switching Informed by Experimental Data

Chiral Pesticide Studied	Enantiomers	Target Organism	Bioactivity of Chiral Pesticide	Activity of Racemates versus Pure Enantiomers	Chiral Switch	Reference
Metolachlor	4	*Scenedesmus obliquus*	Oxidative stress in weeds	(*S*)-metolachlor more active than rac- and (*R*)-metolachlor	(*S*)-metolachlor	Liu et al. (2017)
Bitertanol	4	*Botrytis cinerea, Colletotrichum orbiculare, Alternaria solani, Fusarium graminearum, Sclerotinia sclerotiorum, Fusarium moniliforme, Phytophthora capsici,* and *Rhizoctonia solani*	Inhibition of the demethylation of sterols in fungi	(1*S*, 2*R*)-bitertanol more active than rac-bitertanol and its other enantiomers	(1*S*, 2*R*)-bitertanol	Li et al. (2020)
Flutriafol	2	*Rhizoctonia solani, Alternaria solani, Pyricularia grisea, Gibberella zeae* and *Botrytis cinerea*	Inhibition of growth of fungi	(*R*)-flutriafol has greater bioactivity than rac- and (*S*)-flutriafol	(*R*)-flutriafol	Zhang et al. (2015)
Napropamide	2	*Echinochloa crusgalli*	Inhibition of growth of weeds	(*R*)-napropamide more active than rac- and (*S*)-napropamide	(*R*)-napropamide	Xie et al. (2019)
Tetraconazole	2	*Fusarium graminearum* and *Rhizoctonia cerealis*	Ergosterol biosynthesis inhibition	(*R*)-tetraconazole more active than (*S*)-tetraconazole	(*R*)-tetraconazole	Tong et al. (2019)
Napropamide, acetochlor and propisochlor	2	*Echinochloa crusgalli*	Inhibition of growth of weeds	R-enantiomer more active than rac- and S-forms	R-enantiomer	Xie et al. (2018)
Prothioconazole	2	Various types of fungi	Ergosterol biosynthesis inhibition	(*R*)-prothioconazole is more active than rac- and (*S*)-prothioconazole	(*R*)-prothioconazole	Zhang et al. (2019)
Epoxiconazole	4	*Chlorella vulgaris*	Inhibition of growth of fungi	(*S,R*)-epoxiconazole had greater bioactivity than (*R,S*)-epoxiconazole	(*S,R*)-epoxiconazole	Kaziem et al. (2020)

TABLE 13.2
Techniques for the Determination of the Absolute Configuration of Chiral Pesticides

Class of Chiral Pesticide	Chiral Pesticide	Absolute Configuration Determination Technique	Absolute Configuration of the Chiral Pesticide	Analytical Technique	Reference
Organochlorine	o,p'-DDD	X-ray crystallography	(R)-(+)-o,p'- and (S)-(-)-o,p'-DDD	GC-ECD	Cantillana et al. (2009)
Organophosphate	Malathion	VCD spectroscopy	(R)-(+)-malathion	HPLC	Izumi et al. (2009)
Fungicide	Cyproconazole	VCD spectroscopy	(2R, 3R)-(+)-, (2R, 3S)-(+)-, (2S, 3S)-(-)- and (2S,3R)-(-)-cyproconazole	LC-MS/MS	He et al. (2019)
Fungicide	Zoxamide	ECD spectroscopy	(R)-(-)- and (S)-(+)-zoxamide	HPLC-MS/MS	Pan et al. (2017)
Fungicide	Prothioconazole	ECD spectroscopy	(R)-(-)- and (S)-(+)-prothiocanazole	HPLC-MS/MS	Zhang et al. (2017)
Fungicide	Imazalil	ECD and VCD spectroscopy	(R)- and (S)-imazalil	HPLC-MS/MS	Casas et al. (2016)
Fungicide	Triticonazole	ECD spectroscopy	(S)-(+)- and (R)-(-)-triticonazole	HPLC	Zhang et al. (2016)
Acaricide	Etoxazole	ECD spectroscopy	(R)- and (S)-etoxazole	LC-MS/MS	Yao et al. (2015)
Fungicide	Flutriafol	ECD spectroscopy	(R)-(-)- and (S)-(+)-flutriafol	HPLC	Zhang et al. (2014)
Fungicide	Mandipropamid	ECD spectroscopy	(R)- and (S)-mandipropamid	LC-MS/MS	Zhang et al. (2014)
Herbicide	Lactofen	ECD and VCD spectroscopy	(R)- and (S)-lactofen	HPLC	Xie et al. (2018)
Fungicide	Prothioconazole	VCD spectroscopy	(R)-(-)- and (S)-(+)-prothioconazole	SFC	Jiang et al. (2018)
Fungicide	Tetraconazole	ECD spectroscopy	(R)-(+)- and (S)-(-)-tetraconazole	UPLC-MS/MS	Tong et al. (2019)

Chiral switching is preceded by studies about enantioselective effects of chiral pesticides against both target and non-target organisms (Table 13.1). Toxicity and bioactivity studies are the prerequisites to assess the possibility of racemic pesticides' switch to the enantiomeric-enriched or single enantiomer formulations. Researchers extensively studied the toxicity and bioactivity of many herbicides, including napropamide (Qi et al., 2015; Xie et al., 2018; Zhao et al., 2018) (Table 13.2). It is a selective, chiral amide herbicide that is used to control a number of broad-leaved weeds. The bioactivity of napropamide, just like other chiral pesticides, is enantioselective. Most of the commercial formulations of napropamide, however, are racemic mixtures. Some researchers studied the bioactivity of racemic napropamide and its enantiomers in a bid to come up with its most effective and green chiral switch. For instance, Qi et al. (2015) studied the bioactivity of the enantiomers of napropamide as well as its racemate on the target organism, *Poa annua*. They observed that R-napropamide was more active than racemic and S-napropamide on *Poa annua*. In a similar study, Xie et al. (2019) investigated the effect of the racemate and enantiomers of napropamide on the growth of *Echinochloa crusgalli*. Their results also confirmed that (*R*)-napropamide was more active than racemic and (*S*)-napropamide as it had greater inhibitory effect on the growth of *Echinochloa crusgalli*. Thus, (*R*)-napropamide can be taken as a potential chiral switch for racemic napropamide. On the other hand, Xie et al. (2018) compared the bioactivities of the enantiomers of three herbicides and their racemates on *Echinochloa crusgalli*. The herbicides they used included napropamide, acetochlor, and propisochlor. For these three herbicides, it was found out that the *R*-enantiomers had greater activity than the S-isomers and the racemates. They also investigated the effects of the enantiomers and racemates of these three herbicides on non-target organisms. They established that the *R*-enantiomers had lower toxicity on non-target organisms than the *S*-enantiomers and the racemates. Thus, the R-enantiomers are potential chiral switches of these three herbicides.

Fluxametamide is a chiral isoxazoline insecticide that has broad-spectrum activity on various pests. It is usually developed and sold as a racemic mixture on the market. Li et al. (2019) performed studies to compare the bioactivities of racemic fluxametamide and its enantiomers. They used four target organisms, *Plutella xylostella*, *Spodoptera exigua*, *Aphis gossypii*, and *Tetranychus cinnabarinus*, during their study. Their results revealed that (*S*)-fluxametamide was more active towards target organisms than the racemic and (*R*)-fluxametamide. Thus, (*S*)-fluxametamide is a potential chiral switch for racemic fluxametamide.

13.5 ABSOLUTE CONFIGURATION OF CHIRAL PESTICIDES

The synthetic process of chiral pesticides normally ends with the determination and confirmation of the absolute configuration (R/S) of their enantiomers. Theoretically, the R/S nomenclature or Cahn-Ingold-Prelog (CIP) system is the best and most used in the chemistry community to assign absolute configuration. The CIP priority rules, which define the substituent priority, is the crucial one in this system. The criteria that are defined are as follows: (1) Higher atomic number is given higher priority; (2) when the adjacent atom of two or more of the substituents are the same, the atomic number of the next atom takes the priority; (3) multiple bonded (double and triple) atoms are equivalent to the same number of single-bonded atoms; (4) *cis* is given a higher priority than trans; (5) long pair electrons are regarded as an atom with atomic number 0; and (6) proximal groups have higher priority than distal groups. Experimentally, there are some techniques that can be used for assigning absolute configurations and these include X-ray crystallography, nuclear magnetic resonance, electronic circular dichroism (ECD), and vibrational circular dichroism (VCD) spectroscopy.

13.5.1 X-ray Crystallography

X-ray crystallography is one of the techniques that have been used for a long time for the determination of the absolute configuration of chiral compounds in general and chiral pesticides in particular after synthesis (Table 13.2). For instance, Ye et al. (2012) synthesized and determined the absolute configuration of chiral N-dichloroacetyl-2-substituted-5-methyl-1,3-oxazolidine using X-ray crystallography. The technique enabled them to assign the R- and S-configuration to the molecule. In another study, Yang et al. (2015) successfully synthesized fluorinated 2-(pyridin-2-yl) alkylamines, key intermediates for the synthesis of chiral fluorine-containing pesticides. They assigned absolute configurations to these molecules using X-ray crystallographic technique. The use of X-ray crystallography, however, is often hampered by the fact that it requires good quality crystals of the chiral pesticide in order to obtained reliable results. This is the major drawback of X-ray crystallography and this is making it to slowly lose popularity among researchers as a technique for determining

Chiral Pesticides

the absolute configuration of chiral pesticides. Circular dichroism spectroscopy is slowly replacing X-ray crystallography and is gaining a lot of popularity among researchers as an efficient technique for the determination of the absolute configuration of chiral pesticides.

13.5.2 Circular Dichroism Spectroscopy

Circular dichroism spectroscopy is a well-established and reliable technique for the determination of the absolute configuration of chiral molecules such as pesticides (Table 13.2) (Merten et al., 2019; Wang et al., 2017). It involves the use of circularly polarized light during the generation of spectra. If circularly polarized light is passed through a modulator, it splits into a counter-clockwise (left-handed) and clockwise (right hand) components (Wang et al., 2017; Burgueno-Tapia and Joseph-Nathan, 2017). If these two components are passed through a sample with chiral molecules, they are absorbed differently (Ulrih, 2017). The difference between the absorbance of the left-handed (A_L) and right-handed light (A_R) is called dichroism ($\Delta A = A_L - A_R$). Circular dichroism spectra are plotted using these differences in absorbance (ΔA). In order to assign the absolute configuration of a molecule, the spectrum obtained experimentally should be compared with the one obtained theoretically. The theoretical spectrum is usually predicted from quantum chemical data obtained using computer software (Merten et al., 2019; Polavarapu, 2016). Two types of circular dichroism spectroscopy are in common use for the determination of the absolute configuration of chiral pesticides. These are electronic circular dichroism (ECD) and vibrational circular dichroism (VCD) spectroscopy.

13.5.2.1 ECD Spectroscopy

ECD spectroscopy is a very important technique for the determination of absolute configurations of chiral pesticides (Table 13.2). It involves the use of electronic transitions that occur when circularly polarized light passes through an analyte with chiral molecules (Rogers et al., 2019). The wavelength range used during ECD spectroscopy is from visible to ultraviolet regions (Matsuo and Gekko, 2020). ECD spectroscopy has limited application during the determination of absolute configurations of chiral pesticides. The chiral pesticides whose absolute configuration can be determined using it should have electronic chromophores that absorb light in the visible spectral regions (Polavarapu, 2016). A large array of chiral pesticides do not contain electronic chromophores that absorb visible light and their absolute configuration cannot be determined using ECD spectroscopy.

There are many pesticides, however, that can be studied using ECD spectroscopy (Table 13.2). For instance, Zhang et al. (2016) used ECD spectroscopic studies to assign absolute configurations to the enantiomers of triticonazole (Figure 13.2A). They obtained a spectrum of the enantiomers of triticonazole using quantum chemical calculations, and this predicted spectrum indicated the presence of two stereoisomers of this fungicide. Another spectrum was then obtained using experimental data and the peaks on this spectrum showed a perfect match with the predicted spectrum. Thus, they were able to assign the (R)- and (S)-configurations to the enantiomers of triticonazole using the ECD data. In another study, Zhang et al. (2014) used ECD to determine the absolute configuration of flutriafol enantiomers (Figure 13.2B). The predicted and experimental spectra they obtained were mirror images, showing that flutriafol had two enantiomers. They successfully assigned (R)-(–)- and (S)-(+)-configurations to the enantiomers of flutriafol using this ECD spectral data. Zhang et al. (2017), on the other hand, used the ECD technique to assign the absolute configurations to the enantiomers of prothioconazole fungicide. With the aid of the ECD spectral data, they managed to assign the (R)-(–)- and (S)-(+)-configurations to the prothioconazole enantiomers. In another study, Tong et al. (2019) determined the absolute configurations of the enantiomers of tetraconazole using ECD spectroscopy (Figure 13.2C). Using spectral data, from the predicted and experimental spectra, they successfully assigned the (R)-(+)- and (S)-(–)-configurations to the two enantiomers of tetraconazole fungicide. Thus, ECD is a very useful technique that can be used to successfully assign absolute configurations of some chiral pesticides.

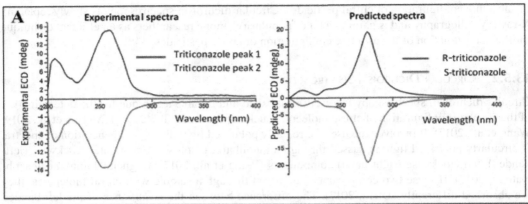
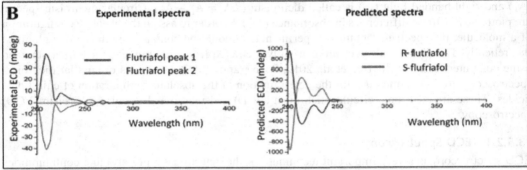
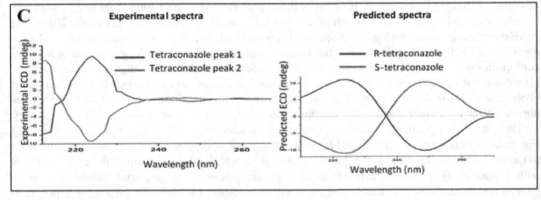

FIGURE 13.2 Experimental and predicted ECD spectra for (A) triticonazole (from Zhang et al., 2016; used with permission from Elsevier), (B) flutriafol (reprinted with permission from Zhang, Qing, Mingming Tian, Meiyun Wang, Haiyan Shi, and Minghua Wang, 2014. "Simultaneous enantioselective determination of triazole fungicide flutriafol in vegetables, fruits, wheat, soil, and water by reversed-phase high-performance liquid chromatography." *Journal of Agricultural and Food Chemistry* 62 (13): 2809–15. Copyright (2014) American Chemical Society), and (C) tetraconazole (from Tong et al., 2019; used with permission from Elsevier).

13.5.2.2 VCD Spectroscopy

VCD spectroscopy is a powerful and versatile technique for the determination of absolute configuration of chiral pesticides (Table 13.2) (Keiderling and Lakhani, 2018). This technique does not require the presence of UV/Vis chromophores and it can be carried out without any chemical derivatization of the chiral pesticide (Merten et al., 2019). VCD spectroscopy uses the vibrational transitions that occur when a chiral molecule is irradiated with circularly polarized infrared light. The technique is universally applicable to all chiral molecules (Polavarapu, 2016).

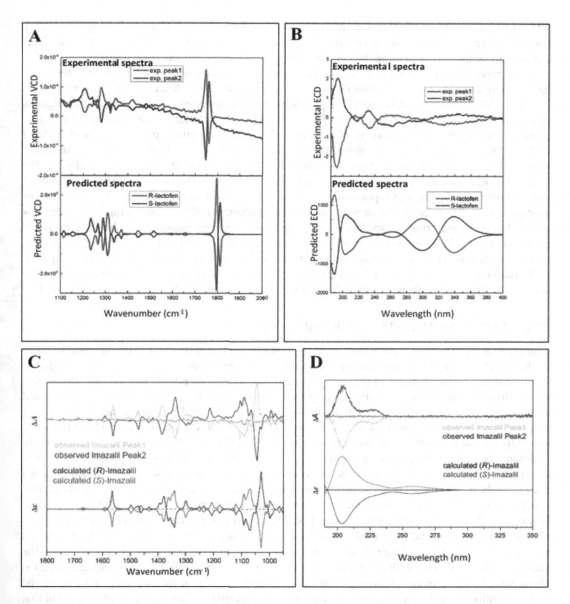

FIGURE 13.3 Experimental and predicted VCD (A) and ECD (B) spectra and for lactofen. (From Xie et al., 2018. Used with permission from Elsevier.) Experimental and predicted VCD (C) and ECD (D) spectra for imazalil. (From Casas et al., 2016. Used with permission from Elsevier.)

Many researchers have used VCD spectroscopy to assign absolute configurations to chiral pesticides (Table 13.2) (Casas et al., 2016; He et al., 2019; Izumi et al., 2009; Xie et al., 2018). For instance, He et al. (2019) used VCD spectroscopy to identify the absolute configurations of the isomers of cyproconazole fungicide. Cyproconazole has four stereoisomers and their absolute configurations were identified using VCD spectral data as (2R,3R)-(+), (2R,3S)-(+), (2S,3S)-(−), and (2S,3R). Jiang et al. (2018), on the other hand, used VCD spectroscopic studies to determine the elution order of the enantiomers of prothioconazole fungicide during chromatographic analysis. The VCD spectral pattern of the enantiomer of prothioconazole that eluted last, when using chiralpak OD column, had a perfect match with the predicted spectrum of the R-configuration. The results implied that the S-configuration of prothioconazole eluted first on the chiralpak OD column. When the chiralpak

OD column was replaced with lux cellulose-3 column during the HPLC analysis, they observed a reversal of the elution order of the enantiomers of prothioconazole. Thus, VCD spectroscopic information can be used to determine the elution order of the enantiomers of chiral pesticides.

Some researchers perform both VCD and ECD studies concurrently when determining the absolute configuration of chiral pesticides (Table 13.2) (Casas et al., 2016; Xie et al., 2018). For instance, Xie et al. (2018) used both ECD and VCD spectra for the determination of the absolute configurations of the enantiomers of lactofen (Figure 13.3A and B). The two techniques complimented each other allowing the absolute configurations of lactofen to be assigned without ambiguity. The calculated ECD spectrum showed that lactofen had two absolute configurations, R and S. This was in agreement with the experimental ECD spectrum and this was confirmation that lactofen really has the R- and S-configurations. A similar match between the predicted VCD spectrum and the experimental one was obtained during VCD spectroscopic studies. This was further confirmation of the absolute configurations of lactofen. It can be noted that the ECD spectrum of lactofen is simple with very few peaks. The VCD spectrum of lactofen, on the other hand, is more complicated and it contains many peaks. Generally, VCD spectra are more complicated than ECD spectra. In the same vein, Casas et al. (2016) determined the absolute configurations of the enantiomers of imazalil fungicide using both ECD and VCD spectra (Figure 13.3C and D). Complimentary information was also obtained from the spectra and this enabled them to assign the R- and S-configurations to the enantiomers of imazalil.

13.6 CHALLENGES AND FUTURE PROSPECTS

Chiral pesticides are among the agro-chemicals that play a pivotal role in sustainable food security in the world. As a result, many methods for synthesizing chiral pesticides have been developed to meet their increasing demand in the agro-industry. Some of the synthetic methods for chiral pesticides do not conform to the principles of green chemistry and these cause serious environmental challenges. Some challenges associated with the synthesis of chiral pesticides include the use of inefficient catalysts that would result in very low yields. The catalytic process would have low atom economy and large quantities of unreacted chemicals would be discharged into the environment. In addition, some synthetic processes use transition metal catalyst, which are very efficient but very expensive. Green chemistry requires the use of cheap and efficient catalysts. Most chiral pesticides are currently being sold as racemic mixtures. In the racemate formulations, as already alluded to, usually only one enantiomer would be active towards the target organisms and the other ones would be inactive. The inactive enantiomers would be discharged into the environment causing an exponential increase in its chemical load. Thus, synthesis and subsequent application of racemates is a serious environmental challenge. Researchers, however, are making concerted efforts to replace the racemates on the market with enantiopure formulations of chiral pesticides, a phenomenon called chiral switching. Its major premise is that the agro-industry should manufacture and market the active enantiomer of the chiral pesticides instead of the racemate. This is a noble phenomenon that upholds the principles of green chemistry but it is not free of challenges. One of the major challenges of chiral switching is racemization. The enantiopure chiral pesticide may be converted into the inactive enantiomers in the environment thereby causing adverse effects to non-target organisms. Research efforts should be directed, going forward, to the development of cheap and efficient catalysts that can be used during synthesis of chiral pesticides. Another facet of this issue on catalysis is doubling efforts of using green and biodegradable catalysts such as enzymes during synthesis of chiral pesticides. However, use of biomaterials in catalysis has its own drawbacks such as lack of mechanical strength and difficulties when recovering them due to their relatively high solubility.

13.7 CONCLUSION

The agro-industry uses a lot of chiral pesticides for pest management. Most of these chiral pesticides are still being sold as racemic mixtures. The bioactivities and toxicities of chiral pesticides on target and non-target organisms, respectively, are usually enantioselective. One enantiomer in the racemate will be more active as compared to others. Thus, the interests of researchers and environmentalists are now tilted towards the replacement of the racemates with chiral pesticides formulations containing only the active enantiomers (chiral switching). This move reduces the application rates of chiral pesticides as well as decreasing the amount of inactive enantiomers discharged into the environment. Chiral switching is, therefore, a green aspect during the manufacture and application of chiral pesticides. The absolute configurations of chiral pesticides, after synthesis, are usually determined using circular dichroism spectroscopy.

ACKNOWLEDGMENT

Authors are grateful to the Research Center at the University of Venda for financial support.

CONFLICTS OF INTEREST

The authors declare no conflict of interest.

REFERENCES

Aguiar, Alex R., Elson S. Alvarenga, Eliete MP. Silva, Elizeu S. Farias, and Marcelo C. Picanço. 2019. "Synthesis, insecticidal activity, and phytotoxicity of novel chiral amides." *Pest Management Science* 75 (6): 1689–96. doi:10.1002/ps.5289.

Alotaibi, Majdah R., M. Monier, and Nadia H. Elsayed. 2019. "Enantiomeric resolution of ephedrine racemic mixture using molecularly imprinted carboxylic acid functionalized resin." *European Polymer Journal* 121: 109309. doi:10.1016/j.eurpolymj.2019.109309.

Berkman, Clifford E., and Charles M. Thompson. 1992. "Synthesis of chiral malathion and isomalathion." *Tetrahedron Letters* 33 (11): 1415–18. doi:10.1016/S0040-4039(00)91635-6.

Boulanouar, Sara, Audrey Combès, Sakina Mezzache, and Valérie Pichon. 2018. "Synthesis and application of molecularly imprinted silica for the selective extraction of some polar organophosphorus pesticides from almond oil." *Analytica Chimica Acta* 1018: 35–44. doi:10.1016/j.aca.2018.02.069.

Buerge, Ignaz J., Astrid Bächli, Jean-Pierre De Joffrey, Markus D. Müller, Simon Spycher, and Thomas Poiger. 2013. "The chiral herbicide beflubutamid (I): Isolation of pure enantiomers by HPLC, herbicidal activity of enantiomers, and analysis by enantioselective GC-MS." *Environmental Science & Technology* 47 (13): 6806–11. doi:10.1021/es301876d.

Buerge, Ignaz J., Jürgen Krauss, Rocío López-Cabeza, Werner Siegfried, Michael Stüssi, Felix E. Wettstein, and Thomas Poiger. 2016. "Stereoselective metabolism of the sterol biosynthesis inhibitor fungicides fenpropidin, fenpropimorph, and spiroxamine in grapes, sugar beets, and wheat." *Journal of Agricultural and Food Chemistry* 64 (26): 5301–9. doi:10.1021/acs.jafc.6b00919.

Burgueno-Tapia, Eleuterio, and Pedro Joseph-Nathan. 2017. "Vibrational circular dichroism: Recent advances for the assignment of the absolute configuration of natural products (reprinted from natural product communications, Vol 10, Pgs 1785-1795, 2015)." *Natural Product Communications* 12 (5): 641–51.

Buser, Hans-Rudolf, Thomas Poiger, and Markus D. Müller. 2000. "Changed enantiomer composition of metolachlor in surface water following the introduction of the enantiomerically enriched product to the market." *Environmental Science & Technology* 34 (13): 2690–96. doi:10.1021/es0000201.

Caballo, C., M. D. Sicilia, and S. Rubio. 2013. "Stereoselective quantitation of mecoprop and dichlorprop in natural waters by supramolecular solvent-based micro-extraction, chiral liquid chromatography and tandem mass spectrometry." *Analytica Chimica Acta* 761: 102–8. doi:10.1016/j.aca.2012.11.044.

Cacho, Carmen, Leif Schweitz, Esther Turiel, and Concepción Pérez-Conde. 2008. "Molecularly imprinted capillary electrochromatography for selective determination of thiabendazole in citrus samples." *Journal of Chromatography A* 1179 (2): 216–23. doi:10.1016/j.chroma.2007.11.097.

Calcaterra, Andrea, and Ilaria D'Acquarica. 2018. "The market of chiral drugs: Chiral switches versus de novo enantiomerically pure compounds." *Journal of Pharmaceutical and Biomedical Analysis, Review issue 2017*, 147: 323–40. doi:10.1016/j.jpba.2017.07.008.

Cantillana, T., V. Lindström, L. Eriksson, I. Brandt, and Å. Bergman. 2009. "Interindividual differences in o,p'-DDD enantiomer kinetics examined in göttingen minipigs." *Chemosphere* 76 (2): 167–72. doi:10.1016/j.chemosphere.2009.03.050.

Casas, Mònica Escolà, Andreas Christopher Kretschmann, Lars Andernach, Till Opatz, and Kai Bester. 2016. "Separation, isolation and stereochemical assignment of imazalil enantiomers and their quantitation in an in vitro toxicity test." *Journal of Chromatography A* 1452: 116–20. doi:10.1016/j.chroma.2016.05.008.

Chen, Xixi, Fengshou Dong, Jun Xu, Xingang Liu, Zenglong Chen, Na Liu, and Yongquan Zheng. 2016. "Enantioseparation and determination of isofenphos-methyl enantiomers in wheat, corn, peanut and soil with supercritical fluid chromatography/tandem mass spectrometric method." *Journal of Chromatography B* 1015–1016: 13–21. doi:10.1016/j.jchromb.2016.02.003.

Cheng, Feng, Feifei Cheng, Jianyong Zheng, Guanzhong Wu, Yinjun Zhang, and Zhao Wang. 2018. "A novel esterase from *Pseudochrobactrum asaccharolyticum* WZZ003: Enzymatic properties toward model substrate and catalytic performance in chiral fungicide intermediate synthesis." *Process Biochemistry* 69: 92–98. doi:10.1016/j.procbio.2018.03.011.

Corcellas, Cayo, Ethel Eljarrat, and Damià Barceló. 2015. "Enantiomeric-selective determination of pyrethroids: Application to human samples." *Analytical and Bioanalytical Chemistry* 407 (3): 779–86. doi:10.1007/s00216-014-7905-6.

Cui, Ning, Haoyu Xu, Shijie Yao, Yiwen He, Hongchao Zhang, and Yunlong Yu. 2018. "Chiral triazole fungicide tebuconazole: Enantioselective bioaccumulation, bioactivity, acute toxicity, and dissipation in soils." *Environmental Science and Pollution Research* 25 (25): 25468–75. doi:10.1007/s11356-018-2587-9.

Dallegrave, Alexsandro, Tânia Mara Pizzolato, Fabiano Barreto, Ethel Eljarrat, and Damià Barceló. 2016. "Methodology for trace analysis of 17 pyrethroids and chlorpyrifos in foodstuff by gas chromatography–tandem mass spectrometry." *Analytical and Bioanalytical Chemistry* 408 (27): 7689–97. doi:10.1007/s00216-016-9865-5.

Duan, Jinsheng, Mingna Sun, Yang Shen, Beibei Gao, Zhaoxian Zhang, Tongchun Gao, and Minghua Wang. 2018. "Enantioselective acute toxicity and bioactivity of carfentrazone-ethyl enantiomers." *Bulletin of Environmental Contamination and Toxicology* 101 (5): 651–56. doi:10.1007/s00128-018-2474-6.

Feringa, B. L. 1988. *Chiral Synthons in Pesticide Syntheses*. Elsevier Science Publishers B.V., Amsterdam, pp. 453–99.

Fonseca, Franciele S., Daniel B. Carrão, Nayara C. P. de Albuquerque, Viviani Nardini, Luís G. Dias, Rodrigo M. da Silva, Norberto P. Lopes, and Anderson R. M. de Oliveira. 2019. "Myclobutanil enantioselective risk assessment in humans through in Vitro CYP450 Reactions: Metabolism and inhibition studies." *Food and Chemical Toxicology* 128: 202–11. doi:10.1016/j.fct.2019.04.009.

Gu, Jinping, Chenyang Ji, Siqing Yue, Dan Shu, Feng Su, Yinjun Zhang, Yuanyuan Xie, Yi Zhang, Weiping Liu, and Meirong Zhao. 2018. "Enantioselective effects of metalaxyl enantiomers in adolescent rat metabolic profiles using NMR-based metabolomics." *Environmental Science & Technology* 52 (9): 5438–47. doi:10.1021/acs.est.7b06540.

Hamed, Seham M., Sherif H. Hassan, Samy Selim, Mohammed A. M. Wadaan, Mohamed Mohany, Wael N. Hozzein, and Hamada AbdElgawad. 2020. "Differential responses of two cyanobacterial species to R-metalaxyl toxicity: Growth, photosynthesis and antioxidant analyses." *Environmental Pollution* 258: 113681. doi:10.1016/j.envpol.2019.113681.

He, Zongzhe, Fengxu Wu, Weitong Xia, Lianshan Li, Kunming Hu, Amir E. Kaziem, and Minghua Wang. 2019. "Separation and detection of cyproconazole enantiomers and its stereospecific recognition with chiral stationary phase by high-performance liquid chromatography." *Analyst* 144 (17): 5193–5200. doi:10.1039/C9AN00950G.

Hor, Yew Li, Wee Kee Phua, and Eng Huat Khoo. 2017. "Chirality switching via rotation of bilayer fourfold meta-structure." *Plasmonics* 12 (1): 83–87. doi:10.1007/s11468-016-0232-3.

Hroboňová, Katarína, and Anna Lomenova. 2018. "Molecularly imprinted polymer as stationary phase for HPLC separation of phenylalanine enantiomers." *Monatshefte Für Chemie - Chemical Monthly* 149 (5): 939–46. doi:10.1007/s00706-018-2155-5.

Hsiao, Yu-Ling, Yei-Shung Wang, and Jui-Hung Yen. 2014. "Enantioselective effects of herbicide imazapyr on *Arabidopsis thaliana*." *Journal of Environmental Science and Health, Part B* 49 (9): 646–53. doi:10.1080/03601234.2014.922404.

Izumi, Hiroshi, Atsushi Ogata, Laurence A. Nafie, and Rina K. Dukor. 2009. "Structural determination of molecular stereochemistry using VCD spectroscopy and a conformational code: Absolute configuration and solution conformation of a chiral liquid pesticide, (R)-(+)-malathion." *Chirality* 21 (1E): E172–80. doi:10.1002/chir.20793.

Jeschke, Peter. 2018. "Current status of chirality in agrochemicals." *Pest Management Science* 74 (11): 2389–2404. doi:10.1002/ps.5052.

Jiang, Ying, Jun Fan, Rujian He, Dong Guo, Tai Wang, Hui Zhang, and Weiguang Zhang. 2018. "High-fast enantioselective determination of prothioconazole in different matrices by supercritical fluid chromatography and vibrational circular dichroism spectroscopic study." *Talanta* 187: 40–46. doi:10.1016/j.talanta.2018.04.097.

Kaziem, Amir E., Beibei Gao, Lianshan Li, Zhaoxian Zhang, Zongzhe He, Yong Wen, and Ming-hua Wang. 2020. "Enantioselective bioactivity, toxicity, and degradation in different environmental mediums of chiral fungicide epoxiconazole." *Journal of Hazardous Materials* 386: 121951. doi:10.1016/j.jhazmat.2019.121951.

Keiderling, Timothy A., and Ahmed Lakhani. 2018. "Mini review: Instrumentation for vibrational circular dichroism spectroscopy, still a role for dispersive instruments." *Chirality* 30 (3): 238–53. doi:10.1002/chir.22799.

Li, Dangdang, Shasha Zhang, Zehua Song, Wei Li, Feng Zhu, Jiwen Zhang, and Shengkun Li. 2018. "Synthesis and bio-inspired optimization of drimenal: Discovery of chiral drimane fused oxazinones as promising antifungal and antibacterial candidates." *European Journal of Medicinal Chemistry* 143: 558–67. doi:10.1016/j.ejmech.2017.11.051.

Li, Lianshan, Beibei Gao, Yong Wen, Zhaoxian Zhang, Rou Chen, Zongzhe He, Amir E Kaziem, Haiyan Shi, and Minghua Wang. 2020. "Stereoselective bioactivity, toxicity and degradation of the chiral triazole fungicide bitertanol." *Pest Management Science* 76 (1): 343–49. doi:10.1002/ps.5520.

Li, Na, Luqing Deng, Jianfang Li, Zhengbing Wang, Yiye Han, and Chenglan Liu. 2018. "Selective effect of myclobutanil enantiomers on fungicidal activity and fumonisin production by fusarium verticillioides under different environmental conditions." *Pesticide Biochemistry and Physiology, Special Issue: Fungicide Toxicology in China*, 147: 102–9. doi:10.1016/j.pestbp.2017.12.010.

Li, Runan, Xinglu Pan, Qinqin Wang, Yan Tao, Zenglong Chen, Duoduo Jiang, Chi Wu, et al. 2019. "Development of S-fluxametamide for bioactivity improvement and risk reduction: systemic evaluation of the novel insecticide fluxametamide at the enantiomeric level." *Environmental Science & Technology* 53 (23): 13657–65. doi:10.1021/acs.est.9b03697.

Li, Shengkun, Dangdang Li, Taifeng Xiao, ShaSha Zhang, Zehua Song, and Hongyu Ma. 2016. "Design, synthesis, fungicidal activity, and unexpected docking model of the first chiral boscalid analogues containing oxazolines." *Journal of Agricultural and Food Chemistry* 64 (46): 8927–34. doi:10.1021/acs.jafc.6b03464.

Liu, Huijun, YiLu Xia, Weidan Cai, Yina Zhang, Xiaoqiang Zhang, and Shaoting Du. 2017. "Enantioselective oxidative stress and oxidative damage caused by rac- and S-metolachlor to *Scenedesmus obliquus*." *Chemosphere* 173: 22–30. doi:10.1016/j.chemosphere.2017.01.028.

Liu, Na, Fengshou Dong, Jun Xu, Xingang Liu, Zenglong Chen, Xinglu Pan, Xixi Chen, and Yongquan Zheng. 2016. "Enantioselective separation and pharmacokinetic dissipation of cyflumetofen in field soil by ultra-performance convergence chromatography with tandem mass spectrometry." *Journal of Separation Science* 39 (7): 1363–70. doi:10.1002/jssc.201501123.

Masbou, Jérémy, Fatima Meite, Benoît Guyot, and Gwenaël Imfeld. 2018. "Enantiomer-specific stable carbon isotope analysis (ESIA) to evaluate degradation of the chiral fungicide metalaxyl in soils." *Journal of Hazardous Materials* 353: 99–107. doi:10.1016/j.jhazmat.2018.03.047.

Matsuo, Koichi, and Kunihiko Gekko. 2020. "Vacuum ultraviolet electronic circular dichroism study of d-glucose in aqueous Solution." *The Journal of Physical Chemistry A* 124 (4): 642–51. doi:10.1021/acs.jpca.9b09210.

Mehdipour, Mohammad, Mehdi Ansari, Mostafa Pournamdari, Leila Zeidabadinejad, and Maryam Kazemipour. 2018. "Selective extraction of organophosphorous pesticides in plasma by magnetic molecularly imprinted polymers with the aid of computational design." *Analytical Methods* 10 (34): 4136–42. doi:10.1039/C8AY00955D.

Merten, Christian, Tino P. Golub, and Nora M. Kreienborg. 2019. "Absolute configurations of synthetic molecular scaffolds from vibrational CD spectroscopy." *The Journal of Organic Chemistry* 84 (14): 8797–8814. doi:10.1021/acs.joc.9b00466.

Müller, Urs. 2002. "Chemical crop protection research. methods and challenges." *Pure and Applied Chemistry* 74 (12): 2241–46. doi:10.1351/pac200274122241.

Pan, Xinglu, Fengshou Dong, Zenglong Chen, Jun Xu, Xingang Liu, Xiaohu Wu, and Yongquan Zheng. 2017. "The application of chiral ultra-high-performance liquid chromatography tandem mass spectrometry to the separation of the zoxamide enantiomers and the study of enantioselective degradation process in agricultural plants." *Journal of Chromatography A* 1525: 87–95. doi:10.1016/j.chroma.2017.10.016.

Park, Oh-Jin, and Sang-Hyun Lee. 2005. "Stereoselective lipases from *Burkholderia sp.*, cloning and their application to preparation of methyl (R)-N-(2,6-dimethylphenyl)alaninate, a key intermediate for (R)-metalaxyl." *Journal of Biotechnology* 120 (2): 174–82. doi:10.1016/j.jbiotec.2005.06.026.

Polavarapu, Prasad L. 2016. "Determination of the absolute configurations of chiral drugs using chiroptical spectroscopy." *Molecules* 21 (8): 1056–1072. doi:10.3390/molecules21081056.

Qi, Yanli, Donghui Liu, Wenting Zhao, Chang Liu, Zhiqiang Zhou, and Peng Wang. 2015. "Enantioselective phytotoxicity and bioacitivity of the enantiomers of the herbicide napropamide." *Pesticide Biochemistry and Physiology* 125: 38–44. doi:10.1016/j.pestbp.2015.06.004.

Rihs, Grety, and Hanspeter Sauter. 1982. "Der einfluß von atropisomerie und chiralem zentrum auf die biologische aktivität des metolachlor." *Zeitschrift Fur Naturforschung - Section B Journal of Chemical Sciences* 37 (4): 451–62. doi:10.1515/znb-1982-0411.

Rogers, David M., Sarah B. Jasim, Naomi T. Dyer, François Auvray, Matthieu Réfrégiers, and Jonathan D. Hirst. 2019. "Electronic circular dichroism spectroscopy of proteins." *Chem* 5 (11): 2751–74. doi:10.1016/j.chempr.2019.07.008.

Rutkowska, Małgorzata, Justyna Płotka-Wasylka, Calum Morrison, Piotr Paweł Wieczorek, Jacek Namieśnik, and Mariusz Marć. 2018. "Application of molecularly imprinted polymers in analytical chiral separations and analysis." *TrAC: Trends in Analytical Chemistry* 102: 91–102. doi:10.1016/j.trac.2018.01.011.

Saeed, Aamer, Fayaz Ali Larik, and Pervaiz Ali Channar. 2016. "Recent synthetic approaches to fipronil, a super-effective and safe pesticide." *Research on Chemical Intermediates* 42 (9). Springer Netherlands: 6805–13. doi:10.1007/s11164-016-2527-6.

Sajini, T., Renjith Thomas, and Beena Mathew. 2019. "Rational design and synthesis of photo-responsive molecularly imprinted polymers for the enantioselective intake and release of l-phenylalanine benzyl ester on multi-walled carbon nanotubes." *Polymer* 173: 127–40. doi:10.1016/j.polymer.2019.04.031.

Sasaki, Mitsuru. 2008. "Current status of organophosphorus insecticide and stereochemistry." *Phosphorus, Sulfur and Silicon and the Related Elements* 183 (2–3): 291–99. doi:10.1080/10426500701734307.

Spindler, Felix, Benoit Pugin, Hanspeter Buser, Hans-Peter Jalett, Ulrich Pittelkow, and Hans-Ulrich Blaser. 1998. "Enantioselective catalysis for agrochemicals: Synthetic routes to (S)-metolachlor, (R)-metalaxyl and (aS,3R)-clozylacon." *Pesticide Science* 54 (3): 302–4. doi:10.1002/(sici)1096-9063(1998110)54:3<302::aid-ps831>3.3.co;2-v.

Sun, Jun, Chengli Mou, Zhongyao Wang, Fangcheng He, Jian Wu, and Yonggui Robin Chi. 2018. "Carbene-catalyzed [4 + 2] cycloadditions of vinyl enolate and (in situ generated) imines for enantioselective synthesis of quaternary α-amino phosphonates." *Organic Letters* 20 (18): 5969–72. doi:10.1021/acs.orglett.8b02707.

Szajnecki, Łukasz, and Barbara Gawdzik. 2019. "Studies on sorption of bifenthrin and diazinon insecticides on molecularly imprinted polymers." *Polymers for Advanced Technologies* 30 (7): 1595–1604. doi:10.1002/pat.4590.

Tan, Qi, Jun Fan, Ruiqi Gao, Rujian He, Tai Wang, Yaomou Zhang, and Weiguang Zhang. 2017. "Stereoselective quantification of triticonazole in vegetables by supercritical fluid chromatography." *Talanta* 164: 362–67. doi:10.1016/j.talanta.2016.08.077.

Tao, Yan, Zuntao Zheng, Yang Yu, Jun Xu, Xingang Liu, Xiaohu Wu, Fengshou Dong, and Yongquan Zheng. 2018. "Supercritical fluid chromatography–tandem mass spectrometry-assisted methodology for rapid enantiomeric analysis of fenbuconazole and Its chiral metabolites in fruits, vegetables, cereals, and soil." *Food Chemistry* 241: 32–39. doi:10.1016/j.foodchem.2017.08.038.

Tong, Zhou, Xu Dong, Shasha Yang, Mingna Sun, Tongchun Gao, Jinsheng Duan, and Haiqun Cao. 2019. "Enantioselective effects of the chiral fungicide tetraconazole in wheat: fungicidal activity and degradation behaviour." *Environmental Pollution* 247: 1–8. doi:10.1016/j.envpol.2019.01.013.

Ulrih, Nataša Poklar. 2017. "Analytical techniques for the study of polyphenol–protein interactions." *Critical Reviews in Food Science and Nutrition* 57 (10): 2144–61. doi:10.1080/10408398.2015.1052040.

Wang, Cui, Zhen Yang, Lihua Tang, Ximing Wang, and Quan Zhang. 2018. "Potential risk and mechanism of microcystin induction by chiral metalaxyl." *Environmental Science & Technology Letters* 5 (11): 635–40. doi:10.1021/acs.estlett.8b00507.

Wang, Kaiqiang, Da-Wen Sun, Hongbin Pu, and Qingyi Wei. 2017. "Principles and applications of spectroscopic techniques for evaluating food protein conformational changes: A review." *Trends in Food Science & Technology* 67: 207–19. doi:10.1016/j.tifs.2017.06.015.

Wang, Ya-Li, Kai Sun, Yu-Bing Tu, Min-Long Tao, Zheng-Bo Xie, Hong-Kuan Yuan, Zu-Hong Xiong, and Jun-Zhong Wang. 2018. "Chirality switching of the self-assembled CuPc domains induced by electric field." *Physical Chemistry Chemical Physics* 20 (10): 7125–31. doi:10.1039/C7CP08279G.

Xia, Weitong, Zongzhe He, Kunming Hu, Beibei Gao, Zhaoxian Zhang, Minghua Wang, and Qiang Wang. 2019. "Simultaneous separation and detection chiral fenobucarb enantiomers using UPLC-MS/MS." *SN Applied Sciences* 1 (7): 795. doi:10.1007/s42452-019-0822-8.

Xie, Jingqian, Lijuan Zhang, Lu Zhao, Qiaozhi Tang, Kai Liu, and Weiping Liu. 2016. "Metolachlor stereoisomers: Enantioseparation, identification and chiral stability." *Journal of Chromatography A* 1463: 42–48. doi:10.1016/j.chroma.2016.07.045.

Xie, Jingqian, Lu Zhao, Kai Liu, Fangjie Guo, and Weiping Liu. 2018. "Enantioselective effects of chiral amide herbicides napropamide, acetochlor and propisochlor: The more efficient r-enantiomer and its environmental friendly." *Science of the Total Environment* 626: 860–66. doi:10.1016/j.scitotenv.2018.01.140.

Xie, Jingqian, Lu Zhao, Kai Liu, Fangjie Guo, Zunwei Chen, and Weiping Liu. 2018. "Enantiomeric characterization of herbicide lactofen: Enantioseparation, absolute configuration assignment and enantioselective activity and toxicity." *Chemosphere* 193: 351–57. doi:10.1016/j.chemosphere.2017.10.168.

Xie, Jingqian, Wei Tang, Lu Zhao, Shuren Liu, Kai Liu, and Weiping Liu. 2019. "Enantioselectivity and allelopathy both have effects on the inhibition of napropamide on *Echinochloa crusgalli*." *Science of the Total Environment* 682: 151–59. doi:10.1016/j.scitotenv.2019.05.058.

Yadav, Ganapati D., and Gunjan P. Deshmukh. 2017. "Insight into solid-liquid phase transfer catalyzed synthesis of mecoprop ester using K_2CO_3 as base and development of new kinetic model involving liquid product and two solid co-products." *Journal of Chemical Sciences* 129 (11). Springer India: 1677–85. doi:10.1007/s12039-017-1368-1.

Yang, Jialin, Jiali Zhang, Weidong Meng, and Yangen Huang. 2015. "Asymmetrical synthesis of fluorinated 2-(pyridin-2-Yl) alkylamine from fluoromethyl sulfinyl imines and 2-alkylpyridines." *Tetrahedron Letters* 56 (47): 6556–59. doi:10.1016/j.tetlet.2015.10.005.

Yang, Peng, Xiao Wang, Lin Peng, Feng Chen, Fang Tian, Chao-Zhe Tang, and Li-Xin Wang. 2017. "Optimized synthetic route for enantioselective preparation of (S)-metolachlor from commercially available (R)-propylene oxide." *Organic Process Research & Development* 21 (10): 1682–88. doi:10.1021/acs.oprd.7b00216.

Yang, Ya-Fen, Wei-Bo Hu, Lei Shi, Sheng-Gang Li, Xiao-Li Zhao, Yahu A. Liu, Jiu-Sheng Li, Biao Jiang, and Wen Ke. 2018. "Guest-regulated chirality switching of planar chiral pseudo[1]catenanes." *Organic & Biomolecular Chemistry* 16 (12): 2028–32. doi:10.1039/C8OB00156A.

Yang, Zaibo, Pei Li, Yinju He, Jun Luo, Jing Zhou, Yinghong Wu, and Liantao Chen. 2020. "Novel pyrethrin derivatives containing an 1,3,4-oxadiazole thioether moiety: Design, synthesis, and insecticidal activity." *Journal of Heterocyclic Chemistry* 57 (1): 81–88. doi:10.1002/jhet.3750.

Yao, Guojun, Jing Gao, Chuntao Zhang, Wenqi Jiang, Peng Wang, Xueke Liu, Donghui Liu, and Zhiqiang Zhou. 2019. "Enantioselective degradation of the chiral alpha-cypermethrin and detection of its metabolites in five plants." *Environmental Science and Pollution Research* 26 (2): 1558–64. doi:10.1007/s11356-018-3594-6.

Yao, Zhoulin, Zuguang Li, Shulin Zhuang, Xiaoge Li, Mingfei Xu, Mei Lin, Qiang Wang, and Hu Zhang. 2015. "Enantioselective determination of acaricide etoxazole in orange pulp, peel, and whole orange by chiral liquid chromatography with tandem mass spectrometry." *Journal of Separation Science* 38 (4): 599–604. doi:10.1002/jssc.201401065.

Ye, Fei, Shuang Gao, Ying Fu, Li-Xia Zhao, and Zhi-Yong Xing. 2012. "Synthesis and crystal structure of novel chiral N-dichloroacetyl-2-substituted-5-methyl-1,3-oxazolidines." *Heterocycles* 85 (4): 903. doi:10.3987/COM-12-12430.

Yousefi, Roozbeh, Thomas J. Struble, Jenna L. Payne, Mahesh Vishe, Nathan D. Schley, and Jeffrey N. Johnston. 2019. "Catalytic, enantioselective synthesis of cyclic carbamates from dialkyl amines by CO_2-capture: Discovery, development, and mechanism." *Journal of the American Chemical Society* 141 (1): 618–25. doi:10.1021/jacs.8b11793.

Zhang, Hu, Xiangyun Wang, Xinquan Wang, Mingrong Qian, Mingfei Xu, Hao Xu, Peipei Qi, Qiang Wang, and Shulin Zhuang. 2014. "Enantioselective determination of carboxyl acid amide fungicide mandipropamid in vegetables and fruits by chiral LC coupled with MS/MS." *Journal of Separation Science* 37 (3): 211–18. doi:10.1002/jssc.201301080.

Zhang, Qing, Beibei Gao, Mingming Tian, Haiyan Shi, Xiude Hua, and Minghua Wang. 2016. "Enantioseparation and determination of triticonazole enantiomers in fruits, vegetables, and soil using efficient extraction and clean-up methods." *Journal of Chromatography B* 1009–1010: 130–37. doi:10.1016/j.jchromb.2015.12.018.

Zhang, Qing, Mingming Tian, Meiyun Wang, Haiyan Shi, and Minghua Wang. 2014. "Simultaneous enantioselective determination of triazole fungicide flutriafol in vegetables, fruits, wheat, soil, and water by reversed-phase high-performance liquid chromatography." *Journal of Agricultural and Food Chemistry* 62 (13): 2809–15. doi:10.1021/jf405689n.

Zhang, Qing, Xiu-de Hua, Hai-yan Shi, Ji-song Liu, Ming-ming Tian, and Ming-hua Wang. 2015. "Enantioselective bioactivity, acute toxicity and dissipation in vegetables of the chiral triazole fungicide flutriafol." *Journal of Hazardous Materials* 284: 65–72. doi:10.1016/j.jhazmat.2014.10.033.

Zhang, Yinjun, Yicheng Fan, Wei Zhang, Guanzhong Wu, Jinghong Wang, Feng Cheng, Jianyong Zheng, and Zhao Wang. 2018. "Bio-preparation of (R)-DMPM using whole cells of *Pseudochrobactrum asaccharolyticum* WZZ003 and its application on kilogram-scale synthesis of fungicide (R)-metalaxyl." *Biotechnology Progress* 34 (4): 921–28. doi:10.1002/btpr.2638.

Zhang, Zhaoxian, Guizhen Du, Beibei Gao, Kunming Hu, Amir E. Kaziem, Lianshan Li, Zongzhe He, Haiyan Shi, and Minghua Wang. 2019. "Stereoselective endocrine-disrupting effects of the chiral triazole fungicide prothioconazole and its chiral metabolite." *Environmental Pollution* 251: 30–36. doi:10.1016/j.envpol.2019.04.124.

Zhang, Zhaoxian, Qing Zhang, Beibei Gao, Gaozhang Gou, Lianshan Li, Haiyan Shi, and Minghua Wang. 2017. "Simultaneous enantioselective determination of the chiral fungicide prothioconazole and its major chiral metabolite prothioconazole-desthio in food and environmental samples by ultra-performance liquid chromatography–tandem mass spectrometry." *Journal of Agricultural and Food Chemistry* 65 (37): 8241–47. doi:10.1021/acs.jafc.7b02903.

Zhao, Lu, Yue Gao, Jingqian Xie, Qiong Zhang, Fangjie Guo, Shuren Liu, and Weiping Liu. 2019. "A strategy to reduce the dose of multi-chiral agricultural Chemicals: The herbicidal activity of metolachlor against *Echinochloa crusgalli*." *Science of the Total Environment* 690: 181–88. doi:10.1016/j.scitotenv.2019.06.521.

Zhao, Pengfei, Jing Zhao, Shuo Lei, Xingjie Guo, and Longshan Zhao. 2018. "Simultaneous enantiomeric analysis of eight pesticides in soils and river sediments by chiral liquid chromatography-tandem mass spectrometry." *Chemosphere* 204: 210–19. doi:10.1016/j.chemosphere.2018.03.204.

Zhao, Pengfei, Shuang Li, Xiaoming Chen, Xingjie Guo, and Longshan Zhao. 2019. "Simultaneous enantiomeric analysis of six chiral pesticides in functional foods using magnetic solid-phase extraction based on carbon nanospheres as adsorbent and chiral liquid chromatography coupled with tandem mass spectrometry." *Journal of Pharmaceutical and Biomedical Analysis* 175: 112784. doi:10.1016/j.jpba.2019.112784.

Zheng, Liangyu, Suoqin Zhang, Fang Wang, Gui Gao, and Shugui Cao. 2006. "Chemoenzymatic synthesis of the chiral herbicide: (S)-metolachlor." *Canadian Journal of Chemistry* 84 (8): 1058–63. doi:10.1139/v06-129.

14 Chiral Pharmaceuticals
Synthesis and Chiral Switching

*Memory Zimuwandeyi, Buhlebenkosi Ndlovu,
Qaphelani Ngulube, and Edmond Sanganyado*

CONTENTS

14.1 Introduction 347
14.2 Chiral Synthetic Methods 348
    14.2.1 Chemoenzymatic Methods 348
    14.2.2 Chemoselective Drug Synthesis 350
    14.2.3 Asymmetric Organometallic Catalysis 351
    14.2.4 Asymmetric Organocatalysis 354
14.3 Trends in Chiral Drug Development 355
14.4 Conclusion and Future Research Directions 356
References 360

14.1 INTRODUCTION

The concept of chirality is a subgroup of *isomerism*, one of the most critical yet challenging concepts in organic chemistry, which describes the handedness of organic molecules. To understand chirality, one must have an appreciation of the fact that molecules occupy a three-dimensional space, where atoms have a spatial arrangement. The phenomenon arises when four distinct types of groups/atoms bonded to a sp^3 hybridized carbon center. Chirality is also observed for other heteroatoms like sulfur, silicon, and nitrogen when they have four diverse types of groups attached regardless of the hybridization. It is now widely known that the fundamental processes relevant for the action of the drug, like binding to receptors, transportation across membranes, and inhibition of enzymes, are dependent on the stereochemistry of the drug (Sanganyado, 2019). This is because biological systems are sensitive to the stereochemistry of molecules. After all, active sites have three-dimensional arrangements.

Since the efficacy of many drugs depends on the chirality of the molecules, single enantiomers should be produced where possible. Ever since the thalidomide saga of the 1960s, pharmaceutical companies have invested millions in ensuring that drugs are synthesized as single enantiomers (Kim and Scialli, 2011; Reist et al., 1998). The drug thalidomide (Figure 14.1) was administered as a racemic mixture, with adverse effects of the racemate reported in the UK, where it was approved. Other countries like the United States were spared because the FDA wanted more studies performed before approving thalidomide, although they were not aware of the problems associated with the racemate at that stage.

The drug was widely prescribed to pregnant women for the treatment of morning sickness, with devastating consequences. Because human bodies are sensitive to the chirality of molecules, these two enantiomers solicited different biological responses in pregnant women. The *R*-enantiomer was responsible for the observed therapeutic properties, while the *S*-enantiomer resulted in teratogenicity (Smith, 2009). After its banning, regulatory bodies ensured that pharmaceutical companies began to pay more attention to chiral drug synthesis and performed extensive biological tests for any drugs that were offered as racemates (Mori et al., 2018). Likewise, researchers are continuously

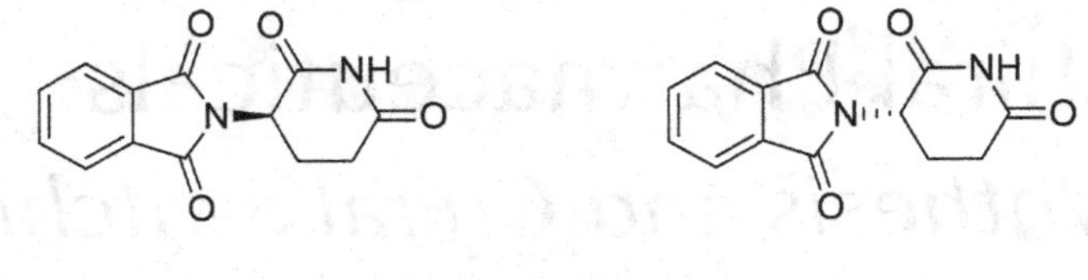

R-thalidomide   S-thalidomide

FIGURE 14.1 Structure of thalidomide.

developing novel and stereo divergent synthetic methods, ensuring that the thalidomide saga is never repeated. Despite the tremendous strides taken to avoid such tragedies, there are still several drugs being offered as racemates on the market. This creates the need for continuous development of stereoselective methods for the synthesis of drugs.

14.2 CHIRAL SYNTHETIC METHODS

The diverse nature of chiral drugs implies that there are a plethora of methods currently available to furnish these compounds. In synthesis, the desire is always to reveal the desired compound as pure as possible, employing the shortest possible route. Due to the environmental impact of chemical processes, chemists and companies alike are now more than ever incorporating the famous Sheldon's E-factor and modifications thereof in the design of their processes (Tieves et al., 2019). Syntheses of chiral drugs can be achieved by chemoenzymatic or chemical means. Each of these approaches has its own set of advantages and disadvantages; therefore, scientists try to use the best approach for each target molecule. The proceeding examples highlight some of the employed approaches to make the reader aware of the complexity and diversity of the available methods.

14.2.1 CHEMOENZYMATIC METHODS

In biological systems, many enzymes contribute to the rapid assembly of chiral products, and nature has evolved to produce only one enantiomer in most cases (Hollmann et al., 2020). This is because enzymes have an inherent ability to discriminate between enantiomers of racemic substances. The application of enzymatic processes for the synthesis of enantiopure drugs grew since the 1980s due to the high selectivity of enzymes (Margolin, 1993). In some cases, the selectivity of enzymes often significantly exceeds that of analogous chemical-synthetic methodology. The major drawbacks associated with enzyme use for syntheses include substrate specificity, stereoselectivity, stability, and reaction stability. In addition, because enzymes mostly react faster with one enantiomer, it means the maximum theoretical yield of enzymatic resolution is 50%. As a solution to this problem, dynamic kinetic resolution methods have been developed for some reactions. Dynamic kinetic resolution methods work by racemizing the unreactive enantiomer resulting in a theoretical increase in yield of 100% (Martín-Matute and Bäckvall, 2007; Nakano and Kitamura, 2014).

Despite these disadvantages, enzyme-catalyzed reactions are often highly regio- and stereoselective when optimized. Temperature and pressure conditions are optimized to avoid extreme conditions that cause isomerization, epimerization, racemization, and product rearrangements commonly observed with chemical processes (Margolin, 1993; Patel, 2001a). Immobilized enzymes can also be used for many cycles, which significantly reduce the E-factor of the processes. Recent technology has also allowed producing tailor-made enzymes (by random and site-directed mutagenesis) with modified activity and novel stereoselective applications.

As an example, Patel published a mini-review article showcasing an alternative enzymatic synthesis of the chiral intermediates for omapatrilat, an antihypertensive drug (Patel, 2001b). The same author went on to publish a review article on the use of enzymes in the synthesis of key drug intermediates (Patel, 2001a). Scheme 14.1 shows the enantioselective enzymatic reduction of ketone **1**,

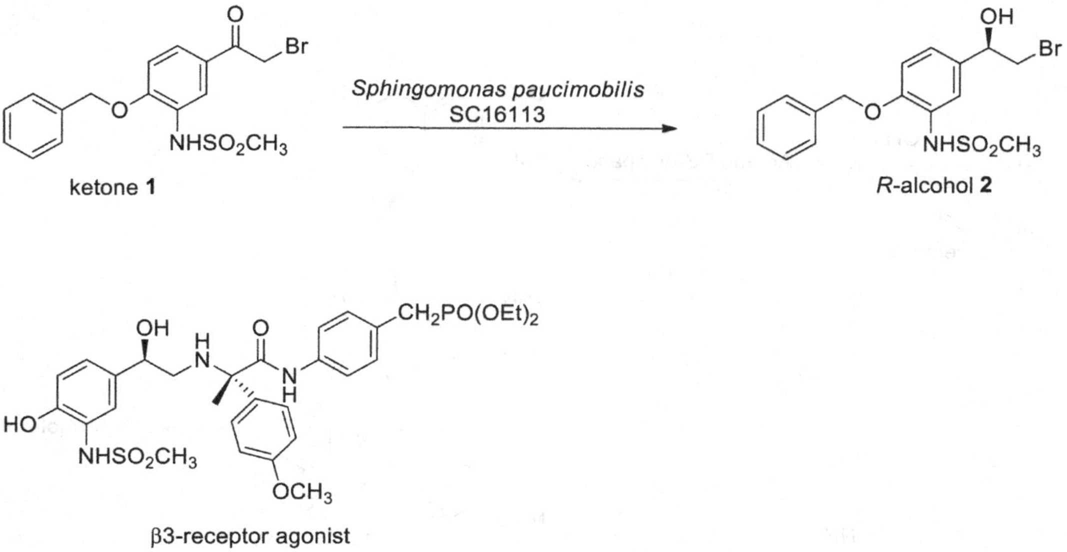

Chiral Pharmaceuticals 349

SCHEME 14.1 Enantioselective reduction of 4-benzyloxy-3-methanesulfonylamino-2′-bromoacetophenone 1 to R-alcohol 2.

which unveiled *R*-alcohol **2**, a crucial intermediate in the synthesis of β3-receptor agonist. Excellent yields of >85% were recovered, and enantiomeric excess (ee) values of >98% were obtained. In this example, the reduction of the ketone group was achieved selectively, and the recovered yields were high because the starting ketone was achiral.

Formoterol (Figure 14.2), a very potent β$_2$-agonist prescribed as a bronchodilator for patients who have asthma and chronic bronchitis, was initially synthesized by a convergent enantio- and diastereoselective method (Campos et al., 2000; Hett et al., 1998). The drug was generally offered as a racemate, but earlier studies had already revealed that the *R,R*-isomer was 1000 times more active than the *S,S*-isomer, although the activity of the *R,R*-isomer was not affected by the presence of the *S,S*-isomer (Trofast et al., 1991).

After several reports of enantiopure *R,R*-formoterol, Campos et al. (2000) reported the incorporation of a key enzymatic resolution step for two of the intermediates in the synthesis protocol (Scheme 14.2). Amano PS-lipase successfully resolved racemic alcohol **3** in the presence of vinyl acetate as the acetylating reagent furnishing the bromoacetate **4** and the alcohol **5** in 48% yield and 96% enantiomeric excess. An overall yield of 11% and enantiomeric excess of 94% was reported for the amidation of racemic amine **6**, yielding the desired amide **7** and amine **8**. Because both the starting materials were racemic, the maximum theoretical yields of the desired product were 50%, and the reactions produced an undesired side product. From these results, it is evident that enzymatic resolution reactions, although they are more environmentally friendly, they have a high E-factor because of a possible 50% waste from the unwanted enantiomer if a dynamic kinetic resolution is not employed.

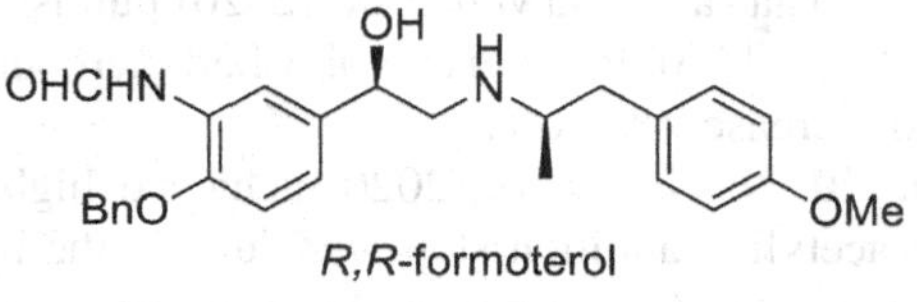

FIGURE 14.2 Structure of *R,R*-formoterol.

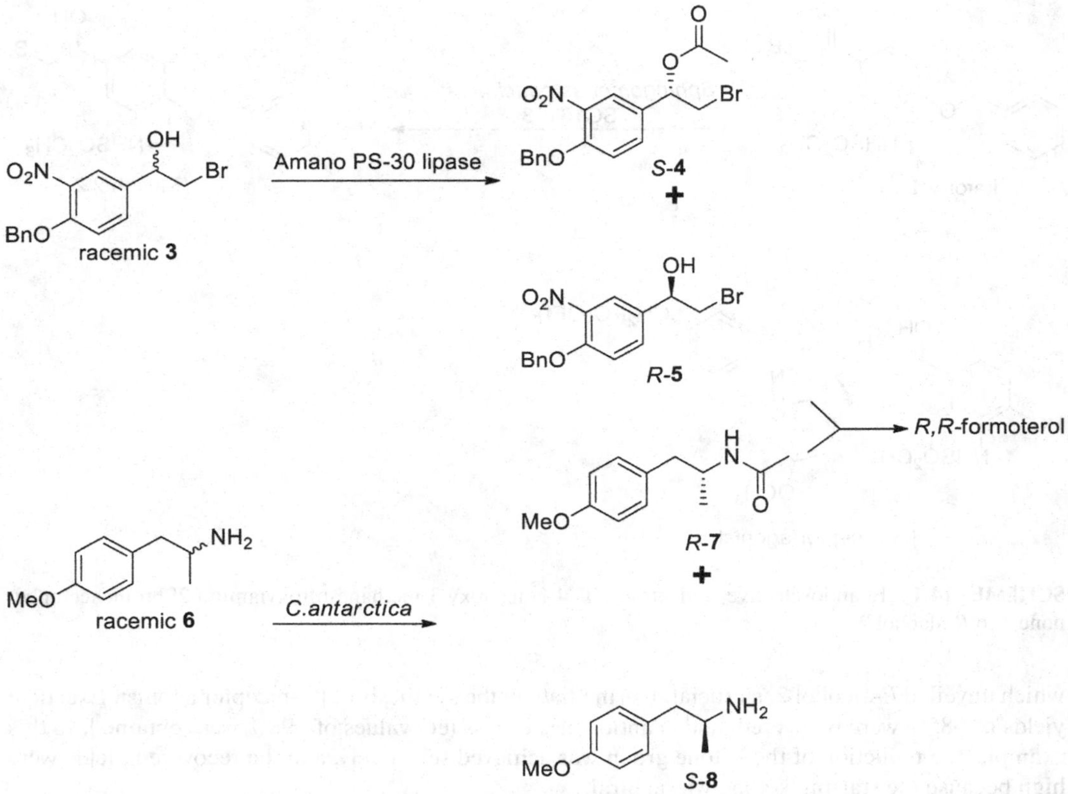

SCHEME 14.2 Enzymatic enantioselective synthesis of *R,R*-formoterol.

14.2.2 CHEMOSELECTIVE DRUG SYNTHESIS

As already alluded to, enantiopure compounds can be synthesized by chemoselective methods only, and the undesirable structure specificity of enzymes means this approach is still commonly used. The main advantage of chemoselective methods is the reactions can be designed in such a way that only the desired enantiomer is recovered, consequentially increasing the yield. Numerous approaches are used to achieve stereoselectivity, and these include use of chiral auxiliaries, using starting material from the chiral pool, and manipulation of existing synthetic routes to favor desired stereochemical outcomes (Mailyan et al., 2016; Wojaczyńska and Wojaczyński, 2020). As was the case with enzymatic methods, an example is given below to illustrate how the stereoselectivity is achieved using chemical methods.

As a result of continuous drug discovery quests for new therapeutic drugs against HIV, islatravir **9** was unveiled by Yamasa Corporation and academic collaborators in 2004, as a novel adenosine-based nucleoside reverse transcription translocator inhibitor (NRTTI) (Kageyama et al., 2011). Not only was the drug targeting a new site, but it was also the first in its category of antiretroviral drug candidates, but clinical trials are still ongoing (Schürmann et al., 2020). In a bid to improve the yield and stereoselective synthesis of islatravir, Nawrat et al. (2020) published a chemoselective method that met both these goals (Scheme 14.3). Previously published methods had reported overall yields of 17%, and they managed to increase it to 36%.

Using enantiopure ketone **10**, Nawrat et al. (2020) achieved high diastereoselectivity for the acetylide addition step. The acetylide addition step was dubbed the most critical of the synthetic route. The use of equimolar quantities of sodium trimethylethynlyaluminate **11** resulted in the highest reported diastereomeric ratio of 14:1 reported for this transformation yielding **12**, after much trial and error. Axial attack of the nucleophilic acetylide on the ketone resulted in the desired diastereomer,

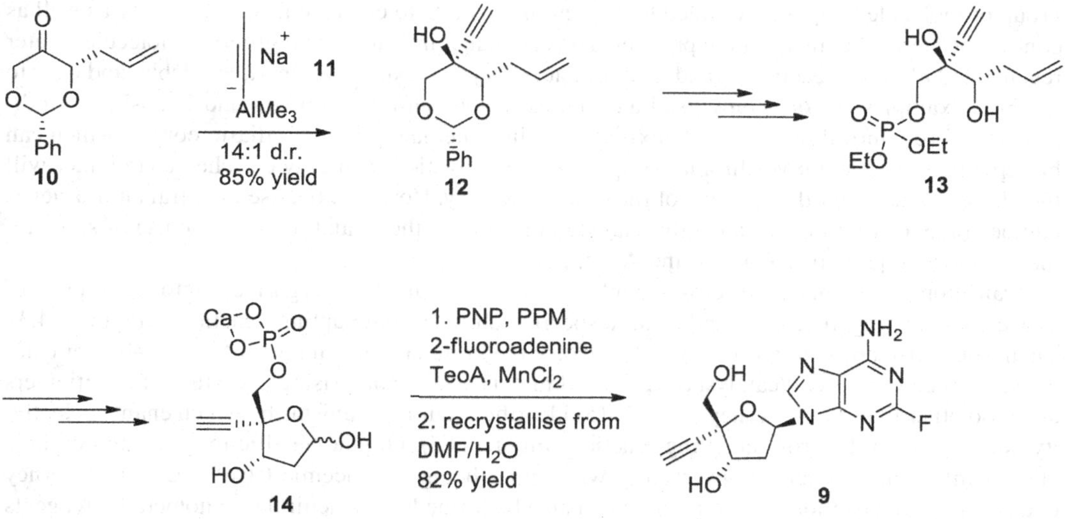

SCHEME 14.3 Synthesis of islatravir.

while equatorial attack furnished the undesired product. Deprotection of the benzylidene, followed by the regioselective phosphorylation of the primary alcohol, gave rise to **13**. This intermediate was then subjected to ozonolysis, methylation, and dealkylation yielding **14**. After reacting with 2-fluoroadenine, the desired product was recovered by recrystallization from a DMF/H$_2$O mixture.

Other examples of chemoselective drug synthesis include synthesis of reboxetine, a potent selective norepinephrine reuptake inhibitor (Shahzad et al., 2017), the antimalarial drug (+)-mefloquine (Knight et al., 2011), and antiobesity drug lorcaserin hydrochloride (Zhu et al., 2015).

In all the given examples for both the chemoselective and chemoenzymatic methods, the procedures had to be optimized to suit the substrates and reaction quantities involved. This means, when designing a stereoselective synthetic approach, the available methods serve as a starting point and compass on the steps to take to realize the expected outcome. The desired products are not guaranteed unless the exact starting materials as the method being adapted are used. Incorporation of several characterization protocols is also of paramount importance to unequivocally establish the stereochemistry of the products.

Chemoselective methods can further be grouped as asymmetric organometallic catalysis and organocatalysis. The next two subsections discuss these approaches.

14.2.3 Asymmetric Organometallic Catalysis

Enantioselective synthesis of chiral drugs using chemoselective methods is often achieved using organic or organometallic catalysis. Chiral drugs and their precursors are widely synthesized enantioselectively using organometallic complexes mediated by a metal catalyst. Most of the organometallic catalysts in use are based on metals containing a type of chiral ligand. Organometallic catalysts can be based on homogenous or heterogenous metal complexes. Homogenous metal catalysts are highly versatile and contain a chiral element comprising two coordinating atoms. The chiral ligands are typically phosphorus atoms for noble metals (i.e., Ir, Pd, Os, Rh, and Ru) and oxygen for metals (Co, Cu, Mn, and Zn) (Blaser, 2003). Nitrogen ligands are also used in complexes that are based on either of the metal groups. However, homogenous metal catalysts often produce significant amounts of waste and are difficult to separate from the reaction mixture (Ali et al., 2014). These challenges can be reduced by immobilizing the metal catalysts on insoluble solid supports such as alumina, carbon, and silica. The metal complexes can be modified using chiral auxiliaries to produce heterogeneous metal catalysts (Ali et al., 2014). A chiral auxiliary is a chiral functional group or molecule temporarily added to a reaction substrate to enhance facial selectivity as well as control the stereochemistry of the product. The original chiral functional group or molecule is later regenerated after the reaction. An ideal chiral auxiliary is inexpensive, inert, available, and easy to remove. Examples of commonly used auxiliaries include chiral sulfoxides and Evans' oxazolidinones. Before removal of the chiral auxiliary, the intermediate product is diastereomers which can be separated by preparative chromatography or crystallization. A highly enriched enantiomer will then be produced after the removal of the chiral auxiliary. However, the use of chiral auxiliaries is cumbersome as it introduces an additional step in drug synthesis and requires the use of stoichiometric reactant quantities (Gawley and Aubé, 2012).

Transition-metal-catalyzed cross-coupling reactions can be categorized into enantioselective cross-coupling reactions and enantiospecific alkyl cross-coupling reactions (Figure 14.3). Enantioselective cross-coupling reactions involve the preferential formation of an enantiomer catalyzed by a scalemic metal catalyst (i.e., a chiral metal catalyst comprising a mixture of enantiomers at a ratio other than 1:1). Cherney et al. (2015) identified four mechanisms by which enantioselectivity is achieved in the cross-coupling reaction (Figure 14.3). Organometallic reagents and organic electrophiles couple enantioconvergently with either being the racemic $C(sp^3)$ reactant (Cherney et al., 2015). An enantioenriched product can also be formed when achiral organometallic reagents and organic electrophiles couple. The enantioselective coupling can also be achieved when a prochiral organometallic reagents or organic electrophiles precursor desymmetrizes (Cherney et al., 2015). The prochiral precursor can be produced during the reaction in a stereoablative process. A stereoablative process involves the formation of a prochiral precursor through the irreversible removal of the chiral element (e.g., a chiral center) followed by the formation of a new stereogenic center through interactions with a metal catalyst (Bhat et al., 2017).

Enantiospecific alkyl cross-coupling reactions are chiral exchange reactions whereby the stereochemistry of the product is determined by that of the substrate (Lucas and Jarvo, 2017). Since alkyl-alkyl bonds are ubiquitous in drugs, alkyl cross-coupling reactions are critical in drug synthesis. In these reactions, either the organometallic reagent or the organic electrophile will be stereodefined. Enantiospecific alkyl cross-coupling reactions have been successfully used to prepare products with axial, centro, or planar chirality (Cherney et al., 2015). While palladium-based catalysts are efficient in cross-couplings reactions involving substrates with unsaturated groups, nickel-based catalysts can efficiently control the stereochemistry of electrophilic partner alkyl groups as well as nucleophilic partners (Xu and Watson, 2020).

In 2000, it was estimated that 4.3% of all commercially available drugs contained the axially chiral biaryl motif. Examples of such drugs include baricitinib, Osimertinib, pitavastatin, and valsartan (Yet, 2018). The biaryl motif can be synthesized using cross-coupling reactions catalyzed by transition metal complexes. For example, the Suzuki-Miyaura cross-coupling reaction is widely used to synthesize biaryl drugs. It involves the reaction between organohalides and an organoboron auxiliary catalyzed by palladium complexes such as $Pd(OH)_2 \cdot Fe_3O_4$ (Yet, 2018). Examples of organoboron reagents commonly used are boronic acid and boronate esters. However, few studies investigated the application of Suzuki-Miyaura cross-coupling reaction in synthesizing axially chiral biaryl motifs (Figure 14.4) (Shen et al., 2019). Previous studies successfully used a transition metal complex containing ferrocene, diene, or bishydrazones chiral ligands to synthesize an asymmetric biaryl scaffold. Application of the Suzuki-Miyaura cross-coupling reaction in chiral drug synthesis is limited since it has limited substrate scope, employs heterocyclic groups that have low selectivity, and requires bulky ortho-substituents (Shen et al., 2019). However, the strong coordination of heterocyclic groups can be suppressed by using a chiral ligand that is highly electron-donating such as N-heterocyclic carbene (Cai et al., 2019). A recent study found that the enantiocontrol ability of the chiral ligands could be enhanced using chiral ligands that are symmetric at C2 (Shen et al., 2019). N-heterocyclic carbene chiral ligands are sterically demanding, which makes it easier for the removal of the ligands in the final synthesis step (Cai et al., 2019).

Enantioselective Cross-Coupling

1. Racemic C(sp³) Organometallic Nucleophile

2. Racemic C(sp³) (Pseudo)Halide Electrophile

3. Achiral Reagents Produce a Chiral Product

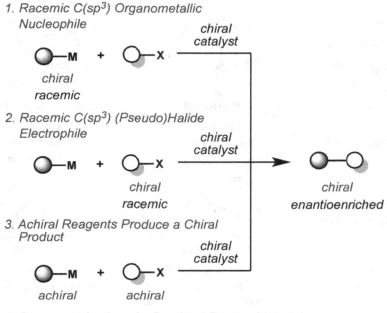

4. Desymmetrization of a Prochiral Starting Material

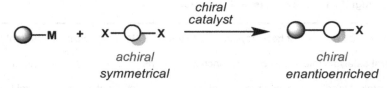

Enantiospecific Cross-Coupling

1. Stereodefined Organometallic Nucleophile

2. Stereodefined (Pseudo)Halide Electrophile

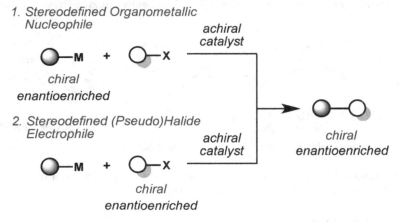

FIGURE 14.3 Mechanisms for transition-metal based cross-coupling reactions. (Used with permission from Cherney et al., 2015.)

A. Representative chiral ligands for efficient Suzuki-Miyaura couplings:

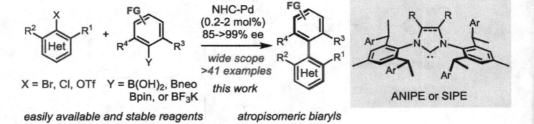

(S)-KenPhos
57-92% ee[8a]
bulky *ortho*-phosphonate

(R)-BI-DIME
90-99% ee[8j]
bulky *ortho*-substituent

resin-supported phosphine
88-99% ee[8g]
10 mol% catalyst
11 examples

PQXPhos:
helically chiral polymer

78-98% ee[8h]
bulky *ortho*-phosphonate

bis-hydrazone
70->98% ee[8e]
long reaction time
only simple substrate used

diene
48-90% ee[8k]
moderate enantioselectivity

B. Current significant challenges and limitations:

- Suppress coordination of heterocyclic substrates
- General catalysts tolerant to various functional groups
- Low catalyst loading and mild reaction conditions
- Tetra-*ortho*-substituent biaryls
- Bulky *ortho*-substituents needed
- Limited type of chiral ligands

C. A novel NHC-Pd enables a general highly enantioselective Suzuki-Miyaura coupling

X = Br, Cl, OTf   Y = B(OH)$_2$, Bneo Bpin, or BF$_3$K
easily available and stable reagents

NHC-Pd (0.2-2 mol%)
85->99% ee
wide scope
>41 examples
this work

atropisomeric biaryls

ANIPE or SIPE

FIGURE 14.4 Suzuki-Miyaura cross-coupling reaction catalyzed by palladium for the synthesis of axially chiral biaryls. (Used with permission from Shen et al., 2019.)

14.2.4 Asymmetric Organocatalysis

Organocatalysis has contributed significantly to the advancements of chiral synthesis of enantioenriched compounds, including drugs. In organocatalysis, small molecules that primarily contain carbon, hydrogen, oxygen, sulfur, phosphorus, and nitrogen atoms are used as catalysts. The rapid development in organocatalysis can be attributed to the reliability of activation modes such as enamine, iminium catalysis, hydrogen-bonding, Lewis base, and singly occupied molecular orbital catalysis. In organocatalysis, substrates can be activated through the formation of an intermediate through reversible covalent bonds or weak interactions between the organocatalyst and the substrate. Covalent bond-based activation modes are highly enantioselective due to the strength and directional nature of covalent bonds. Examples of covalent-based activation modes include chiral amines, electrophilic iminium ions, and N-heterocyclic carbene. Iminium catalysis involves enal or enone activation modes that enhance the reaction rate by lowering the energy of the lowest unoccupied molecular orbital found on a substrate (Beeson et al., 2007). Such an action enhances the enantioselective conjugation (e.g., the formation of carbon-carbon or carbon-nitrogen bonds), Friedel-Crafts alkylation, and hydrogenation (Beeson et al., 2007).

In contrast, enamine catalysis increases the energy of the highest occupied molecular orbital found on a substrate, particularly ketones and aldehydes (Beeson et al., 2007). This process has been shown to enhance the enantioselective functionalization of α-carbonyl in various electrophiles. The efficiency of iminium and enamine catalysis is often comprised of the tendency of iminium and enamine intermediate ions to interconvert. Singly occupied molecular orbital catalysis disrupts the equilibrium by oxidizing the enamine intermediates to form three π electron radicals with molecular orbitals that are singly occupied. The main challenge in oxidizing enamine intermediates lies in the identification of appropriate oxidizing agents and optimization of the reaction conditions to ensure the oxidation process is highly selective (Silvi and Melchiorre, 2018). Unlike transition-metal based catalysts, the small molecule catalysts are often insensitive to moisture and oxygen content. However, since different activation modes produce products with distinct stereochemistry, additional activation modes are required to produce new drugs. However, noncovalent bond-based approaches are also highly enantioselective since they are based on numerous weak bonds acting in concert to give directionality. Examples of noncovalent-based activation modes include hydrogen bonding, Brønsted acid, and phase-transfer catalysis.

14.3 TRENDS IN CHIRAL DRUG DEVELOPMENT

Enantiomers of chiral drugs can have different pharmacokinetic and pharmacodynamic characteristics. A eutomer is an enantiomer that has the desired therapeutic properties while a distomer does not have the desired properties (Sanganyado et al., 2020). Sometimes the enantiomers may both be therapeutically active but acting at different primary drug targets. However, in other cases, one enantiomer may be a eutomer while the other causes side effects.

Chiral switching is the process whereby a racemic drug approved for marketing is redeveloped for approval as a single enantiomer. There has been a push toward the production of single enantiomers as replacement of the racemates commercially available. Chiral switching can help improve the efficacy of a drug and its safety. Patenting a chirally switched drug is challenging. For example, the United States Food and Drug Administration does not consider single enantiomers from previously approved racemates as new molecular entities but derivatives. Interestingly, between 2002 and 2015, the number of single enantiomer drugs approved as new molecular entities nearly doubled (Figure 14.5) (Sanganyado et al., 2017).

Success in chiral switching has been mixed due to the complex nature of the behavior of chiral molecules in biological systems. Enantiomers of a chiral drug sometimes elicit their therapeutic properties through action at different primary drug targets. For example, *trans*-tramadol is an analgesic comprising a racemic mixture of (R, R)-(+)-tramadol and (S, S)-(−)-tramadol. The (+)-enantiomer inhibited serotonin reuptake due to a higher affinity for the μ-receptor, but the (−)-enantiomer inhibited noradrenaline reuptake instead (Ardakani et al., 2008). Interestingly, the (+)-enantiomer was shown to cause nausea and vomiting side effects (Grond et al., 1995). Overall, the racemic mixture has higher efficacy and better safety than the single enantiomers. Another major complication with chiral switching is some enantiomers such as profens can undergo chiral inversion *in vivo*. For example, if Enantiomer 1 is responsible for eutomer while enantiomer 2 is the distomer, it might be reasonable to switch to enantiomer 1. However, chiral switching might not be practical if

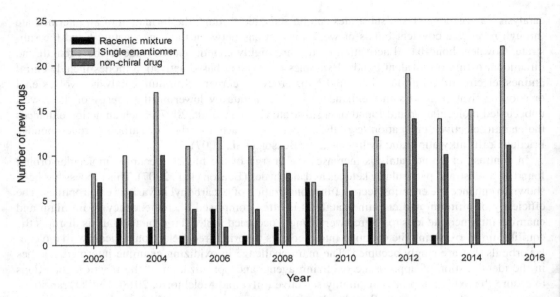

FIGURE 14.5 New molecular entities approved by the U.S. Food and Drug Administration between 2002 and 2015. (Used with permission from Sanganyado et al., 2017.)

Enantiomer 1 can convert to Enantiomer 2 *in vivo*. Table 14.1 shows a list of drugs switched from racemates to single enantiomers between 1994 and 2011.

14.4 CONCLUSION AND FUTURE RESEARCH DIRECTIONS

Chiral synthesis has played a significant role in drug discovery and development since the national regulations recommended that stereochemistry should be considered in drug development. It is estimated that there are at least 5,000 drugs in the market, of which one half are chiral compounds. There has been a trend toward the production of single enantiomers since enantiomers of a chiral drug can have different therapeutic properties. For chiral switching to succeed, there is a need for the development of sustainable, efficient, and highly selective chiral synthesis techniques.

As new drug synthesis processes are developed, there is a need for ensuring the processes adhere to green chemistry strategies. Drug production should have minimal impact on the environment. This can be achieved by following the 12 principles of green chemistry. Drug production produces the most waste per kilogram of product compared to oil refinery, bulk chemical, and fine chemical production. It is estimated that at least 80% of the waste is related to the solvent. Although most solvents used in pharmaceutical synthesis have low human toxicity, these can elicit adverse effects in the environment. Therefore, the reduction of solvent waste and the use of green solvents should be a priority in chiral drug synthesis. Future research should focus on the development of processes that require fewer amounts of solvents.

The high specificity required in chiral synthesis implies there is a need for identification of new enzymes, substrates, metal complexes, and chiral auxiliaries for the synthesis of new drugs. Computational approaches such as three-dimensional quantitative structure-activity relationship, molecular dynamics simulations, and molecular docking can help identify new drugs. *In silico* techniques can be used to identify new chiral ligands for metal complexes and new substrates. Furthermore, computational approaches are a powerful tool for catalyst design in asymmetric synthesis as they can predict the behavior of a catalyst using quantum mechanical techniques such as density functional theory and molecular dynamics simulations. Computational approaches are susceptible to human bias as they still rely on human intuition. This human bias can be reduced

Chiral Pharmaceuticals

TABLE 14.1
Drugs Switched from Racemates to Single Enantiomers between 1994 and 2011

Entry	API	Chemical Structure	Pharmacological Activity or Indications	Single Enantiomer	Launch (Country)	Company
1	Ibuprofen		Anti-inflammatory	(S)-(+)-ibuprofen (dexibuprofene)	1994 (Austria)	Spirig AG
2	Ofloxacin		Antibacterial	(S)-(−)-ofloxacin (levofloxacin)	1995 (Japan)	Aventis
3	Fenfluramine		Antiobesity	(S)-(+)-fenfluramine (dexfenfluramine)	1996 (USA)	Interneuron Pharmaceuticals
4	Ketoprofen		Anti-inflammatory	(S)-(+)-ketoprofen (dexketoprofen)	1998 (Europe)	Menarini
5	Salbutamol (Albuterol)		Antiasthmatic	(R)-(−)-albuterol (levalbuterol)	1999 (USA)	Sepracor

(*Continued*)

TABLE 14.1 (Continued)
Drugs Switched from Racemates to Single Enantiomers between 1994 and 2011

Entry	API	Chemical Structure	Pharmacological Activity or Indications	Single Enantiomer	Launch (Country)	Company
6	Bupivacaine		Local anesthetic	(S)-(−)-bupivacaine (levobupivacaine)	2000 (USA)	Purdue Pharma
7	Omeprazole		Acid reducer, proton pump inhibitor (PPI)	(S)-(−)-omeprazole (esomeprazole)	2000 (Europe); 2001 (USA)	AstraZeneca
8	Cetirizine		Antihistaminic	(R)-(−)-cetirizine (levocetirizine)	2007 (USA)	Sepracor/Sanofi-Aventis
9	Citalopram		Antidepressant	(S)-(+)-citalopram (escitalopram)	2001 (Europe); 2002 (USA)	Forest
10	Methylphenidate		Attention deficit, hyperactivity disorder	(R,R)-(+)-methylphenidate (dexmethylphenidate)	2001 (Europe)	Novartis/Celgene

Chiral Pharmaceuticals

TABLE 14.1 (Continued)
Drugs Switched from Racemates to Single Enantiomers between 1994 and 2011

Entry	API	Chemical Structure	Pharmacological Activity or Indications	Single Enantiomer	Launch (Country)	Company
11	Zopiclone		Anxiety and insomnia	(S)-(+)-zopiclone (eszopiclone)	2004 (Europe)	Sunovion/ Sepracor
12	Formoterol		Chronic obstructive pulmonary disease	(R,R)-(−)-formoterol (arformoterol)	2006 (USA)	Sunovion/ Sepracor
13	Modanafil		Narcolepsy	(R)-(−)-modanafil (armodanafil)	2007 (USA)	Cephalon
14	Leucovorin (folinic acid)		Rescue after high-dose methotrexate therapy; treatment of colorectal carcinoma in combination with 5-FU; treatment of folate deficiency	(S)-(−)-leucovorin (levoleucovorin)	2008 (USA)	Spectrum

Source: Used with permission from Calcaterra and D'Acquarica, 2018.

by employing automated techniques for analyzing the catalytic cycle. Hence, future studies should explore methods of reducing human bias in computational approaches of catalyst design.

Furthermore, the drug discovery and development processes are complex, long, and influenced by complex biological interactions. Computational techniques can be used to predict the pharmacological behavior of chiral drugs. However, current techniques, such as the prediction of ADME+T properties, overlook chirality. Molecular docking techniques are effective at predicting the behavior of different enantiomers as they determine the binding energy in a three-dimensional primary drug target. In recent years, high-throughput techniques have been used to establish the pharmacodynamics and pharmacokinetics of drugs. It has been proposed that machine learning techniques can be used to evaluate the high-throughput data, thus improving drug discovery and development.

REFERENCES

Ali, M. E., Rahman, M. M., Sarkar, S. M., Hamid, S. B. A., 2014. Heterogeneous metal catalysts for oxidation reactions. *J. Nanomater.* 2014, 1–23. https://doi.org/10.1155/2014/192038

Ardakani, Y. H., Mehvar, R., Foroumadi, A., Rouini, M.-R., 2008. Enantioselective determination of tramadol and its main phase I metabolites in human plasma by high-performance liquid chromatography. *J. Chromatogr. B. Analyt. Technol. Biomed. Life Sci.* 864, 109–15.

Beeson, T. D., Mastracchio, A., Hong, J.-B., Ashton, K., MacMillan, D. W. C., 2007. Enantioselective organocatalysis using SOMO activation. *Science (80-.).* 316, 582–585. https://doi.org/10.1126/science.1142696

Bhat, V., Welin, E. R., Guo, X., Stoltz, B. M., 2017. Advances in stereoconvergent catalysis from 2005 to 2015: Transition-metal-mediated stereoablative reactions, dynamic kinetic resolutions, and dynamic kinetic asymmetric transformations. *Chem. Rev.* 117, 4528–4561. https://doi.org/10.1021/acs.chemrev.6b00731

Blaser, H.-U., 2003. Enantioselective catalysis in fine chemicals production. *Chem. Commun.* 3, 293–296. https://doi.org/10.1039/b209968n

Cai, Y., Zhang, J. W., Li, F., Liu, J.M., Shi, S. L., 2019. Nickel/N-heterocyclic carbene complex-catalyzed enantioselective redox-neutral coupling of aenzyl alcohols and alkynes to allylic alcohols. *ACS Catal.* 9, 1–6. https://doi.org/10.1021/acscatal.8b04198

Calcaterra, A., D'Acquarica, I., 2018. The market of chiral drugs: Chiral switches versus de novo enantiomerically pure compounds. *J. Pharm. Biomed. Anal.* 147, 323–340. https://doi.org/10.1016/j.jpba.2017.07.008

Campos, F., Bosch, M. P., Guerrero, A., 2000. An efficient enantioselective synthesis of (R,R)-formoterol, a potent bronchodilator, using lipases. *Tetrahedron: Asymmetry.* 11, 2705–2717. https://doi.org/10.1016/S0957-4166(00)00238-X

Cherney, A. H., Kadunce, N. T., Reisman, S. E., 2015. Enantioselective and enantiospecific transition-metal-catalyzed cross-coupling reactions of organometallic reagents to construct C-C bonds. *Chem. Rev.* 115, 9587–9652. https://doi.org/10.1021/acs.chemrev.5b00162

Gawley, R. E., Aubé, J., 2012. Practical aspects of asymmetric synthesis, in: *Principles of Asymmetric Synthesis*. Elsevier, Oxford, U.K., pp. 63–95. https://doi.org/10.1016/B978-0-08-044860-2.00002-7

Grond, S., Meuser, T., Zech, D., Hennig, U., Lehmann, K. A., 1995. Analgesic efficacy and safety of tramadol enantiomers in comparison with the racemate: a randomised, double-blind study with gynaecological patients using intravenous patient-controlled analgesia. *Pain* 62, 313–320. https://doi.org/10.1016/0304-3959(94)00274-I

Hett, R., Fang, Q. K., Gao, Y., Wald, S. A., Senanayake, C. H., 1998. Large-scale synthesis of enantio- and diastereomerically pure (R, R)-formoterol †. *Org. Process Res. Dev.* 2, 96–99. https://doi.org/10.1021/op970116o

Hollmann, T., Berkhan, G., Wagner, L., Sung, K. H., Kolb, S., Geise, H., Hahn, F., 2020. Biocatalysts from biosynthetic pathways: Enabling stereoselective, enzymatic cycloether formation on a gram scale. *ACS Catal.* 10, 4973–4982. https://doi.org/10.1021/acscatal.9b05071

Kageyama, M., Nagasawa, T., Yoshida, M., Ohrui, H., Kuwahara, S., 2011. Enantioselective total synthesis of the potent anti-HIV nucleoside EFdA. *Org. Lett.* 13, 5264–5266. https://doi.org/10.1021/ol202116k

Kim, J. H., Scialli, A. R., 2011. Thalidomide: The tragedy of birth defects and the effective treatment of disease. *Toxicol. Sci.* 122, 1–6. https://doi.org/10.1093/toxsci/kfr088

Knight, J. D., Sauer, S. J., Coltart, D. M., 2011. Asymmetric total synthesis of the antimalarial drug (+)-mefloquine hydrochloride via chiral N-amino cyclic carbamate hydrazones. *Org. Lett.* 13, 3118–3121. https://doi.org/10.1021/ol2010193

Lucas, E. L., Jarvo, E. R., 2017. Stereospecific and stereoconvergent cross-couplings between alkyl electrophiles. *Nat. Rev. Chem.* 1, 1–8. https://doi.org/10.1038/s41570-017-0065

Mailyan, A. K., Eickhoff, J. A., Minakova, A. S., Gu, Z., Lu, P., Zakarian, A., 2016. Cutting-edge and time-honored strategies for stereoselective construction of C–N bonds in total synthesis. *Chem. Rev.* 116, 4441–4557. https://doi.org/10.1021/acs.chemrev.5b00712

Margolin, A. L., 1993. Enzymes in the synthesis of chiral drugs. *Enzyme Microb. Technol.* 15, 266–280. https://doi.org/10.1016/0141-0229(93)90149-V

Martín-Matute, B., Bäckvall, J. E., 2007. Dynamic kinetic resolution catalyzed by enzymes and metals. *Curr. Opin. Chem. Biol.* 11, 226–232. https://doi.org/10.1016/j.cbpa.2007.01.724

Mori, T., Ito, T., Liu, S., Ando, H., Sakamoto, S., Yamaguchi, Y., Tokunaga, E., Shibata, N., Handa, H., Hakoshima, T., 2018. Structural basis of thalidomide enantiomer binding to cereblon. *Sci. Rep.* 8, 1–14. https://doi.org/10.1038/s41598-018-19202-7

Nakano, K., Kitamura, M., 2014. Dynamic kinetic resolution (DKR), in: Todd, M. (Ed.), *Separation of Enantiomers.* Wiley-VCH Verlag GmbH & Co. KGaA, Weinheim, Germany, pp. 161–216. https://doi.org/10.1002/9783527650880.ch5

Nawrat, C. C., Whittaker, A. M., Huffman, M. A., McLaughlin, M., Cohen, R. D., Andreani, T., Ding, B., Li, H., Weisel, M., Tschaen, D. M., 2020. Nine-step stereoselective synthesis of islatravir from deoxyribose. *Org. Lett.* 22, 2167–2172. https://doi.org/10.1021/acs.orglett.0c00239

Patel, R. N., 2001a. Biocatalytic synthesis of intermediates for the synthesis of chiral drug substances. *Curr. Opin. Biotechnol.* 12, 587–604. https://doi.org/10.1016/S0958-1669(01)00266-X

Patel, R. N., 2001b. Enzymatic synthesis of chiral intermediates for Omapatrilat, an antihypertensive drug. *Biomol. Eng.* 17, 167–182. https://doi.org/10.1016/S1389-0344(01)00068-5

Reist, M., Carrupt, P. A., Francotte, E., Testa, B., 1998. Chiral inversion and hydrolysis of thalidomide: Mechanisms and catalysis by bases and serum albumin, and chiral stability of teratogenic metabolites. *Chem. Res. Toxicol.* 11, 1521–1528. https://doi.org/10.1021/tx9801817

Sanganyado, E., 2019. Comments on "Chiral pharmaceuticals: Environment sources, potential human health impacts, remediation technologies and future perspective." *Environ. Int.* 122, 412–415. https://doi.org/10.1016/j.envint.2018.11.032

Sanganyado, E., Lu, Z., Fu, Q., Schlenk, D., Gan, J., 2017. Chiral pharmaceuticals: A review on their environmental occurrence and fate processes. *Water Res.* https://doi.org/10.1016/j.watres.2017.08.003

Sanganyado, E., Lu, Z., Liu, W., 2020. Application of enantiomeric fractions in environmental forensics: Uncertainties and inconsistencies. *Environ. Res.* 184, 109354. https://doi.org/10.1016/j.envres.2020.109354

Schürmann, D., Rudd, D. J., Zhang, S., De Lepeleire, I., Robberechts, M., Friedman, E., Keicher, C., Hüser, A., Hofmann, J., Grobler, J. A., Stoch, S. A., Iwamoto, M., Matthews, R. P., 2020. Safety, pharmacokinetics, and antiretroviral activity of islatravir (ISL, MK-8591), a novel nucleoside reverse transcriptase translocation inhibitor, following single-dose administration to treatment-naive adults infected with HIV-1: An open-label, phase. *Lancet HIV* 7, e164–e172. https://doi.org/10.1016/S2352-3018(19)30372-8

Shahzad, D., Faisal, M., Rauf, A., Huang, J., 2017. Synthetic story of a blockbuster drug: Reboxetine, a potent selective norepinephrine reuptake inhibitor. *Org. Process Res. Dev.* 21, 1705–1731. https://doi.org/10.1021/acs.oprd.7b00265

Shen, D., Xu, Y., Shi, S.-L., 2019. A bulky chiral N-heterocyclic carbene palladium catalyst enables highly enantioselective Suzuki–Miyaura cross-coupling reactions for the synthesis of biaryl atropisomers. *J. Am. Chem. Soc.* 141, 14938–14945. https://doi.org/10.1021/jacs.9b08578

Silvi, M., Melchiorre, P., 2018. Enhancing the potential of enantioselective organocatalysis with light. *Nature.* 554, 41–49. https://doi.org/10.1038/nature25175

Smith, S.W., 2009. Chiral Toxicology: It's the Same Thing…Only Different. *Toxicol. Sci.* 110, 4–30. https://doi.org/10.1093/toxsci/kfp097

Tieves, F., Tonin, F., Fernández-Fueyo, E., Robbins, J. M., Bommarius, B., Bommarius, A. S., Alcalde, M., Hollmann, F., 2019. Energising the E-factor: The E+-factor. *Tetrahedron.* 75, 1311–1314. https://doi.org/10.1016/j.tet.2019.01.065

Trofast, J., ÖSterberg, K., Källström, B.-L., Waldeck, B., 1991. Steric aspects of agonism and antagonism at β-adrenoceptors: Synthesis of and pharmacological experiments with the enantiomers of formoterol and their diastereomers. *Chirality.* 3, 443–450. https://doi.org/10.1002/chir.530030606

Wojaczyńska, E., Wojaczyński, J., 2020. Modern stereoselective synthesis of chiral sulfinyl compounds. *Chem. Rev.* https://doi.org/10.1021/acs.chemrev.0c00002

Xu, J., Watson, M. P., 2020. Stitching two chiral centers with one catalyst. *Science (80-.).* 367, 509–510. https://doi.org/10.1126/science.aba4222

Yet, L., 2018. Biaryls, in: *Privileged Structures in Drug Discovery: Medicinal Chemistry and Synthesis.* John Wiley & Sons, Hoboken, NJ, pp. 83–154. https://doi.org/10.1002/9781118686263.ch4

Zhu, Q., Wang, J., Bian, X., Zhang, L., Wei, P., Xu, Y., 2015. Novel synthesis of antiobesity drug lorcaserin hydrochloride. *Org. Process Res. Dev.* 19, 1263–1267. https://doi.org/10.1021/acs.oprd.5b00144

Index

Note: Page numbers in *italics* indicate figures and **bold** indicate tables in the text.

A

Abbalide, *9*
Abiotic chiral inversion, 37; *see also* Inversion
Absolute configuration of chiral pesticides, **335**, 336–340; *see also* Pesticides
    circular dichroism spectroscopy, 337
    ECD spectroscopy, 337, *338*
    VCD spectroscopy, 338–340, *339*
    X-ray crystallography, 336–337
Absorption
    of bifenthrin and diazonium insecticides, 332
    chiral pesticides, 52
    dermal, 8
    longer wavelength light, 120
    pharmaceuticals, 310
    SCCPs, 171
Acaricide, **335**
Acartia tonsa, 119
Acephate, 205
Acetamiprid, **234**
Acetochlor, **58**, **238**, **334**
(R)-(-)-acetochlor, 240
(S)-(+)-acetochlor, 240
Acetofenate, **239**
(-)-acetofenate, 49
R-(-)-acetofenate, 240
(S)-(+)-acetofenate, 240
S-(+)-acetofenate, 240
(-)-acetofenate-alcohol, 49
Acetonitrile, 87
Acetylcholine, 44
Acetylcholinesterase, 44
Achiral contaminants, *4*; *see also* Contaminants
Achiral molecule, 1–2
Achiral substances, 6
Acidic amino acids, 33
Acquaviva, A., 288
Active enantiomer, 6
Adsorption, 7
Adulteration, *see* Food
Agilent 6410 triple quadruple mass spectrometer, 159
Agilent 7890 gas chromatograph, 170
Agrochemicals, 44
Akutsu, K., 163
Alaee, M., **162**
Alapyridaine (N-(1-carboxyethyl)-6-hydroxymethyl-pyridinium-3-ol inner salt), 192
(R)-(-)-alapyridaine, 192
(S)-alapyridaine, 192
Albendazole (veterinary drug), 251
Alburnus alburnus, 170, 263
Aldosterone, 5
Aldrin, 131, 154, 171, **234**
Alkylbenzene sulfonates, 35
Alkyl molecules, 11
Allylthiourea (ATU), 91
α1-acid glycoprotein (AGP), 85–86, *86*
α- and β-hexachlorocyclohexane, 11
α-carbon, 34
Alpha-cypermethrin, **58**, 59
α-diimines, 32
α-endosulfan, 169
(+)-α-HBCD, 145, 159
α-HBCD, 158, 159
(+)-α-HBCD, 143, 144, 145, 159
α-HBCDD, 159
α-HCH, 154, 156, 168
α-pinene, 216
Alternaria solani, **334**
Alvim, J., 264–265
Amfepramone, **33**
Aminium salts (R_3NH^+), 32
Amino acids, 2, 327
4-*p*-aminobenzoic acid, 119
Aminorex, 86
Ammonium acetate, 86
Amphetamine, **89**, 90, 91
R/S(±)-amphetamine, *91*
S(+)-amphetamine, 79, 90, 91
Amphetamine biodegradation, 91
Amphetamine-like compounds, 85
Amylose tris-(3,5-dimethylphenylcarbamate), 87, 205
Analgesics, 91–92
Analysis, perfluorooctane sulfonate (PFOS), 139–140
Analytical aspects of pharmaceuticals, 251–252, *252*
Andrés-Costa, M. J., 89, 93, 95
Androstenone, 116
Anguilla anguilla, 170
Anodonta woodiana, 36, 59
Antagonism, 5
Anthelmintics, 85, 265–266
Anthropogenic chiral compounds, 3
Antibiotics, 85, 93, 253, 255, 260–262
Antidepressants, 85, 88–90, **89**
Antihistamines, 85
Anti-inflammatory drugs, **89**
Aphis gossypii, **333**, 336
Aquatic biota, 8
Aquatic ecosystems, 13, 42
Aquatic organisms, 5, 6
Aquatic toxicity tests of pharmaceutical enantiomers, **94–95**
Arabidopsis thaliana, **58**, 270, **333**
Aromatic molecules, 11
Arthrobacter globiformis, 115
Aryl hydrocarbon receptor (AhR), 172
2-arylpropionic acids (2-APAs), 35, 262
Asher, B. J., 140

363

Atenolol, 79, 85, **89**, 90, **94**, **311**
(R)-atenolol, 312
R(+)-atenolol, 90
(S)-atenolol, 312
Atmospheric pressure gas chromatography (APGC), 157
Australia, 91
Avena sativa, **115**
Axial chirality, 1
Azo dyes, 13
Azole fungicides, 110–113; *see also* Fungicide
 environmental behavior, 112–113
 physicochemical properties, 110–112, **111**
 toxicity, 113, **114–115**
Azoxystrobin, **236**

B

Baccillus species, 52
Baron, E., **166**
BCPC, *43*
Beaches Environmental Assessment and Coastal Health Act, 304
Beflubutamid, **56**, **333**
Beijing, 36, 55
Benalaxyl, 42, *42*, **46**, 52, **53–54**, 203, 230, **236**
Benzimidazole derivatives, 120
Benzophenone-3, 119
Benzophenones, 120
Benzotriazoles, 120
3-benzylidene camphor, **125**
Benzylmalonate derivatives, 120
β-agonists, 85
Beta-blocker, 85, **89**, 90, 253, 268
Beta-blocker propranolol, 2, 2, 93
β-cyclodextrin, 216, 281, 282, 293
β-cyclodextrin-mediated micellar electrokinetic chromatography, 216
Beta-cypermethrin, **239**
β-endosulfan, 168
Beta hexachlorocyclohexane, 154
β-hexabromocyclododecanes, 143, 159
β-pinene, 216
β-TBCO, 145
Betaxolol, **311**
Bian, Y., 237
Biaryl motif, 352
Bichon E., **167**
Bifenthrin, *43*, **229–230**, 231, **238**
(S)-bifenthrin, 240
Bioaccumulation, 11, 13
 chiral pesticides, 44
 defined, 52
 enantioselective, 52–55
 tebuconazole, 55
Bioactive ingredients, 8
Bioactivity and toxicity, 335
Biocide, 79
Biodegradation, 5, 7–8, 50, 90
 amphetamine, 91
 chiral pesticides, 44, 52
 chiral pollutants, 31
 enantioselective, 6, 50–52
 pathways, 5
 stereoselective, 6

Biologically mediated chiral inversion, 35; *see also* Inversion
Biomagnification, 11, 12, 307
Biomagnification factor (BMF), 159
Biomass, 11
Biosolids, 13, 36, 109, 112, 254
Biota, 5, 8
Biotin–streptavidin system-based real-time immunopolymerase chain reaction assay (BA-IPCR), 161
Biotransformation, 4, 6, 35, 112; *see also* Transformation
 BDE-209, 160
 chiral pesticides, 44
 defined, 49
 enantioselective, 49
Biphenyls, **33**
Birolli, W. G., 50
Birth defects, 27
Bis-ethylhexyloxyphenol methoxyphenyl triazine, 120
Bitertanol, **229**, 231, **334**
Blanco, S. L., **164**
Blaschke, G., 306
Borit, M., 280
$(1R,2S,4R)$-(+)-borneol, 12
$(1S,2R,4S)$-(−)-borneol, 12
Botrytis cinerea, **333–334**
Boulanouar, S., 331
Brassica campestris, 312
Brassica napus, **115**
Brominated flame retardants (BFRs), 12–14, **134**, 140–145, 158, 160
 enantioselective analysis, 143–145
 environmental behavior, 143–145
 physicochemical properties, 141–143, **142**
Bromination, 132
(R)-1-bromo-1-chloromethane, 29, *29*
1-bromo-1-chloromethane, 29, *29*
Brooks, B. W., 15, 312
Buerge, I. J., 66
Bupivacaine/*(S)*-bupivacaine (levobupivacaine), 7, **358**
Bupropion, **259**
Buser, H. R., 124
Butyl methoxydibenzoyl methane, 119
Byrsonima, 281

C

Cacho, C., 332
Cadmium, 6
Caffeine, **201**, 253, 256, 283, 285
Cahn-Ingold-Prelog (CIP), 29, 336
Calderón-Preciado, D., 256
Camacho-Muñoz, D., 86
Camphor derivatives, 120
Campos, F., 349
Canada, 107, 140, 332
Canadian Food Inspection Agency, 304
Capillary electrokinetic chromatography, 205
Capillary electrophoresis (CE), 37, 252
Capillary electrophoresis immunoassay method with laser-induced fluorescence (CEIA-LIF), 260–261
Carassius auratus, 119
Carbohydrates, 2
Carbon-halogen bonds, 132

Carbon nanotubes, 6
Carbon-nitrogen ratio, 6
Carcinogenic effects, 8
Carfentrazone-ethyl, **58**
Carprofen, **258**
Carrão, D. B., 228, 240
Carson, R., 302
Carvone, 191
(R)-carvone, 191
(S)-carvone, 192, *192*
Case study, pollutants, 309–313
Cashmeran, 8, **9**, 116–118, **117**
Castrignanò, E., 87
Catalytic asymmetric synthesis, 325–326, *325–326*
Catechin, 217, 281
(+)-catechin, **200**, 281–282, 284
(−)-catechin, **200**
Cathinone, **33**
Celestolide, 117, 119
Cellobiohydrolase (CBH), 85
Cellulose tris-(3,5-dichlorophenylcarbamate), 87, 205
Cellulose-tris(3-chloro-4-methylphenylcarbamate), 87
Cellulose tris(4-methylbenzoate), 205
Ceriodaphnia dubia, 270
Cetirizine/*(R)*-cetirizine (levocetirizine), 7, **358**
Chai, T., 64
Challenges
    chiral pesticides, 67
    in exposure assessment, 309–312, *310*, **311**
    pesticides, 340
    in toxicity assessment, 312–313
Chang, G. R., 168
Charm II Chloramphenicol test, 261
Chemcatcher, 79
Chemical
    biological effects, *302*
    in food, 12–13
    stability, **135**
    transformations, 4
Chemical-specific factors influencing plant uptake of pharmaceuticals, 256–257
Chemoenzymatic methods, 348–349, *349–350*
Chemoselective drug synthesis, 350–351, *351*
Chen, F., 233
Chen, M. Y. Y., **164**
Chen, X., 240
Cherney, A. H., 352
China, 91, 107, 110, 112, 137, 217
Chinese National Environmental Protection Agency, 304
Chinese National Food and Drug Administration in Canada and China, 304
Chinese pond mussels, 36
Chiral auxiliary, 351–352
Chiral capillary electrophoresis (CCE), 281
Chiral capillary electrophoresis–ultraviolet detection, 44
Chiral chromatography mass spectrometry, 215
Chiral gas chromatography, 65–66
Chiral high-performance liquid chromatography, 44, 60–65
Chirality
    axial, 1
    chemical regulations, 304, *304*
    defined, 1, 41, 249
    environmental, 7
    in food safety, 193
    halogenated organic contaminants, 133
    helical, 1
    illustration, 42, *42*
    intrinsic, 2
    inversion and, 28–31, *28–31*
    perfluorooctane sulfonate (PFOS), 139
    planar, 1
    property of food flavors and aroma, 12
    synthetic musks, 116
    thalidomide, 27
    type, 1
Chiral liquid chromatography with tandem mass spectrometry (LC-MS/MS), 204
Chiralpak IG, 204
Chiralpak QN-AX, 139–140
Chiral pool, 7
Chiral POPs, 155–156
Chiral recognition, 2, 85
Chiral selector (CS), 157–158, **195**
Chiral separation, 60–67, 157–158, **212–213**
Chiral stationary phases (CSPs), 84–88, 158, 196
Chiral switching, 7–8, 332–336, **333–334**
Chiron approach, 326–327, *327*
Chiroptics, 1
Chloramphenicol, 86, **258**, 261
R,R(−)-chloramphenicol, 93
RR-chloramphenicol, 261
S,S(+)-chloramphenicol, 93
SS-chloramphenicol, 261
Chlordane, 5, 154, 155
Chlordecone, 154
Chlorella pyrenoidosa, 49, **58**, 59
Chlorella vulgaris, 57, **334**
Chlorfenprop-methyl, 43
Chlorfuran, 43
Chlorinated paraffins, 11
Chlorination, 132
Chlorine, 131
3-chloro-2-hydroxypropylmethacrylate, 216
Chlorobiphenyls, **33**
3-chlorobutan-2-ol, 30, *30*
Chloromidophos, 43
Chlorpheniramine, 79
Chlorpyrifos, 205, 233
Chlortetracycline, **258**
Cholinergic receptors, 44
Cinnamates, 120
Circular dichroism (CD), 1
Circular dichroism spectroscopy, 337
Cirrhinus molitorella, 159
Cis-bifenthrin, 6, **58**, 59, **207**, **239**
(S)-cis-bifenthrin, 59
Cis-chlordane, 6
Citalopram, 88, 89–90, **259**, **358**
S(+)-citalopram, 250
Citalopram/(S)-citalopram (escitalopram), 7
Clean Air Act, 304
Clean Water Act, 304
Clenbuterol, **201**, **259**, 266, 267
Climate change, 155
Climbazole, 110–112, **111**, **114–115**
Clorpyriphos, **234**
Clothianidin, **234**, 235

Clotrimazole, 110, 112
Clozylacon, *326*
CO_2, 66–67
Colletotrichum orbiculare, **334**
Comparison of enantioselective techniques, 196, **197**
Compounds, 2–3, *3*, 6, 8, 10, 13, 37
    anthropogenic, 3
    enantiomeric fraction (EF), 78
    enantiomers, *3*, 31
    inversion, **33**
    meso, 31, *31*
    organic, 7
    oxidation, 35
    pure, 5
    racemic mixtures, 5
    stereoconfiguration, 31
Configurational isomerism, 28
Constitutional isomerism, 28
Constitutional isomers, 28, *28*
Contaminants, *4*, 8, 13; see also Pollutants
Copper, 6
Coral larvae, 10
Corcellas, C., 66
Corticosteroids receptors, 5
Cortisol, 5
Couderc, M., **166**
Coulomb/electric interaction, **195**
Covaci, A., 158
COX-1 inhibitor, 263, 269
COX-2 inhibitor, 263, 269
Crassostrea talienwhanensis, 172
Crop protection, 44
Crylenes, 120
Cui, N., 54, 60
Cyamemazine, **259**
Cyazofamid, **236**
Cyclobutane, **33**
Cyclodextrin, **195**, 281
Cyclodextrin-modified micellar electrokinetic
    chromatography (CD-MEKC), 283, 285
Cyclohexane, **33**
Cyclopentane, **33**
CYD2G9 enzymes, 240
Cyflumetofen, **202**, **230**
Cyfluthrin, 34, 205
Cyhalothrin, 50, 205, **230**, 231
CYP3A enzymes, 240
Cypermethrin, 34, *43*, 205, **207**, **230**, 231, **236**
Cyprinus carpio, 161, 263
Cyproconazole, **335**
Cysteine, 35
Cytochrome P450, 95, 155

D

Dairy products, 289
Dallegrave, A., 66, 228, 233
D-amino acids, **199**
Danio rerio, **114**, 144
Daphnia magna, 10, **58**, 59, 93, **94–95**, 95, 113, **114–115**, 269, 307, 312
d-asparagine, 12
D-asparagine, 192, *192*
(−)-*d*-dinotefuran, 233

(+)-*d*-dinotefuran, 233
De Andrés, F., 312
Death toll of chemical pollutants, *106*; see also Pollutants
De Boer J., 162
Decabromodiphenyl ether (decaBDE), 154, 160–168
3-N-dechloroethylifosfamide, 86
Degree of coalescence, 36
(*R,R*)-DEHP, 125
δ-decalactone, 12
δ-octalactone, 12
δ-TBECH, 145, 159
Deltamethrin, 34, *43*, **230**, **234**
(−)-*d*-enantiomer, 233
(+)-*d*-enantiomer, 233
Deoxyribose, 2
Dermal absorption, 8
Dermatitis, 8
Desmethylcitalopram, 89, **259**, 265
Desmethylofloxacin, 87
Desmethylvenlafaxine, 90
Detection methods, POPs, 156–157
Dexibuprofen, 7
Dexketoprofen, 7
Dextramycin, **258**
D-glucose, 29
Diao, J., 59
Dias, T., 283
Diastereomer, 30, 84
(*R,S*)-diastereomer, 125
Diastereomization, 27
Dibenzoylmethane derivatives, 120
Dichlorodiphenyltrichloroethane (DDT), 5, 131–132, 154–155
Dichlorprop, **58**
Diclofenac, 256–257, 269
Diclofol, **236**
Diclofop, 38, **236**, **239**
(*R*)-(+)-diclofop, 36
(*S*)-(−)-diclofop, 36
Diclofop-methyl, 5, **236**, **238**
(*R*)-(+)-diclofop-methyl, 237
(*S*)-(−)-diclofop-methyl, 237
Dicofol, 154
Dicotyledonous weeds, **333**
Dieldrin, 154, 171
Diethylhexyl butamido triazone, 119, 120
10,11-dihydro-10-hydroxycarbamazepine, 86
Dihydroketoprofen, 86
Diketopiperazine indole alkaloids, **200**
Tris(3,5-dimethylphenylcarbamate), 87
Dimethyl phosphorodithioic acid (DMPA), 329, *330*
Ding, D., 60
Diniconazole, **234**
Dinotefuran, **202**, **229**, 233
Diode array detector (DAD), 196
2,3-di-O-methyl-6-tbutylsilyl β-cyclodextrin, 216
Dipole–dipole interaction, **195**
Dipole-induced-dipole interaction, **195**
Dipole interactions, 84
Diporeia, 140, 160
Disinfectants, 8, 107
Dispersive solid-phase extraction (dSPE), 169
Distomer, 28, 31; see also Enantiomer
Disulfoton, **234**
Disulfoton chiral organophosphates, 233

Index

D-lactic acids, **199**
DNA, 2, 5
Do Lago, A. C., 233
Dosis, I., **167**
Dreissena polymorpha, 119
Drinking water, 7
Drometrizole trisiloxane, 119
Drugs, 266–267; *see also* Illicit drugs
    development, 355–356
    expired, 7
    repurposed, 27
    switched from racemates to single enantiomers, 357–359
    unused, 7
Drug-receptor complex, 6
Dynamic chromatography, 36–37

E

ECD spectroscopy, 337, *338*
Echinochloa crusgalli, 57, **333**, **334**, 336
Econazole, 110–113, **111**
Ecotoxicity, 5, 95
Eisenia fetida, 57, 59
Electrokinetic chromatography, **212–213**
Electron capture negative ion tandem mass spectrometry (ECNI-MS/MS), 143
Electronic circular dichroism (ECD), 204, 336–337
Electronic waste (e-waste), 159
Electrophilic substitution, 34
Electrophoresis, 36
Electro-spray ionization (ESI), 88
Electrostatic interactions, 84
Enantiomer, 1
    1-bromo-1-chloromethane, 29, *29*
    active, 6
    benalaxyl pesticide, 42, *42*
    biological effects, 2
    chemical properties, 2
    composition, 5
    compounds, *3*, 31
    configurational lability, 27
    diastereomers, 84
    ethiprole, 59
    flufiprole, 49
    ineffective, 2
    interaction, 6
    inversion, *32*
    labile, **37**
    living organisms, 6
    naproxen, *92*
    pesticides, 57
    pharmacodynamics, 6
    physical properties, 2, 29
    proportions, 2
    pure, 2–3, 7
    racemic mixture, 42
    single, 6
    specific interactions, *134*
    tebuconazole pesticide, 42, *42*
    triadimefon, 55
(+)-enantiomer, 12, 78, 113, 215
(−)-enantiomer, 12, 78, 125
(R)-enantiomer, 4, 12, 36, 124, 215, 312
(R)-(+)-enantiomer, 306
S-enantiomer, 52, 57, 231
(S)-enantiomer, 4, 312
(S)-(−)-enantiomer, 306
S-(+) enantiomer, 55
(S,S)-enantiomer, 125
S(−)-enantiomer, 90
Enantiomeric fraction (EF), 52, 156
    amphetamine, 91
    atenolol, 79
    chiral compounds, 78
    defined, 3
    ibuprofen, 92
    molecular entities and, 7
    racemic, 6
    river waters, 90
    $S(-)$-propranolol, 93
    tramadol, 79
    values, 12
    variations, 4
    venlafaxine, 79
Enantiomeric ratio (ER), 156
Enantiomeric separation, 27
Enantiomerization, 27–28, 34, 55
Enantiopure pharmaceuticals, 7
Enantioresolution, 34, 36
Enantioselective food analysis, 191–217
    chirality in food safety, 193
    chiral stationary phases, 196
    comparison of enantioselective techniques, 196, **197**
    food quality control and traceability, 214–217
        enantioselective capillary electrophoresis, 217
        enantioselective electrokinetic chromatography, 216–217
        enantioselective gas chromatography, 214–216
        enantioselective liquid chromatography, 214
        multidimensional gas chromatography, 215–216
        solid phase microextraction, 215
    food safety assessment, 196–214
        enantioselective capillary electrophoresis, 205, **212–213**, 214
        enantioselective liquid chromatography, 196–204, **198–203**
        enantioselective supercritical fluid chromatography, 205, **206–211**
    mechanism of enantioseparation, *194*, 194–196, **195**
    overview, 191–193
    techniques, 196
Enantioselective multidimensional gas chromatography-mass spectrometry (enantio-MDGC-MS), 293
Enantioselective occurrence in food matrices, 257–267, **258–259**
    anthelmintic drugs, 265–266
    antibiotics, 260–262
    drugs, 266–267
    non-steroidal anti-inflammatory drugs (NSAIDs), 262–263
    psychoactive drugs, 263–265
Enantioselectivity, 6, 50
    action on target organisms, 55–57, **56**
    agonistic/antagonistic effects, 5
    analysis, 4, 12, 32, 36, 143–145
    behavior, 6
    bioaccumulation, 52–55

biodegradation, 6, 50–52
biotransformation, 49, 155
capillary electrophoresis, 8
  in food quality, 217
  in food safety, 205, **212–213**, 214
chiral pesticides, 52
degradation, *91*
electrokinetic chromatography in food quality, 216–217
enantiomer-specific interactions, 31
gas chromatography
  in food analysis, **206–211**
  in food quality, 214–216
interaction, 4
LC-MS/MS, **80–83**
liquid chromatography
  in food analysis, 196–204, **198–203**
  in food quality, 214
metabolism, 8
occurrence and fate, 139–140
of PFOS in environment, 139–140
supercritical fluid chromatography in food safety, 205, **206–211**
toxicity, 6, 57–59, **58**
transformation, 8, 92
Enantioseparation, mechanism of, *194*, 194–196, **195**
Enantiospecific alkyl cross-coupling reactions, 352
Enantiospecific toxicity, 95, 269–270
Enantiotoxicity, 5
Endocrine disruption, 240
Endogenous chemicals, 12
Endosulfan, 154
Endrin, 131, 154
Environment
  analysis of chiral pharmaceuticals in, 78–88
  behavior
    azole fungicides, 112–113
    brominated flame retardants, 143–145
  chiral inversion, 31–36
    biologically mediated, 35
    photo-induced, 34–35
    solvent-induced, 32–34
    thermal-induced, 34
  chirality, 7
  fate of pesticides, 49–55
  fate of pharmaceuticals in, 88–93
  organic UV filters in, 120, 124
  pollutants, 5
  risk assessments, 8, 32, 96
  stereoselectivity, 90
  synthetic musks in, 118
  toxicity of chiral pharmaceuticals in, 93–96, **94–95**
Environmental Protection Agency, 304, 306
Environmental quality standards (EQS), 170
Environment Canada, 304
Enzymatic transformation, 323–324, 327–328, *328*
Enzyme-mediated chiral inversion, 31
Ephedrine, *78*, **95**
(*1S,2R*)(+)-ephedrine, 93
1R,2S(−)-ephedrine, 93
Epicatechin, 281
(+)-epicatechin, **200**, 281–282
(−)-epicatechin, **200**, 281, 284
Epimerases, 35

Epimerization, 27, 35
Epinephrine, 2
ε-HBCD, 143
Epoxiconazole, **53**, **56**, **58**, **234**, **334**
(*R,S*)-epoxiconazole, 57
S,R-epoxiconazole, 59
(*S,R*)-epoxiconazole, 57
Eremias argus, 49, 64
Ergosterol biosynthesis, 45
Escitalopram, 7
Esomeprazole, 7
Essential oils, 289–291
Estimated daily intake (EDI), 169
Estrogenic effects, 8
Estrogens, 7
Ethiprole, 5, *50*, 59, **230**
(*S*)-ethiprole, 59
Ethofumesate, **230**
Ethoxylated ethyl 4-amino benzoate, **125**
Ethoxyresorufin-O-deethylase (EROD), 95, 270
Ethyl 2-methylbutanoate, **207**, 289
Ethylene dimethacrylate, 216
Ethylexyl methoxycinnamate, 119
2-ethylhexyl-2-cyano-3,3-diphenyl-2-propenoate, 121
2-ethylhexyl 2-cyano-3,3-diphenylacrylate, 120
2-ethylhexyl 4-dimethylaminobenzoate, 10, **10**, 120, 124, 125
Ethylhexyl methoxycinnamate, **10**
2-ethylhexyl salicylate, 120
Ethylhexyl triazone, 120
Etodolac, **258**
Etoxazole, **335**
Europe, 107, 141, 279, 304
European Chemical Agency (ECHA), 307
European Commission, 116, 120
European Food Safety Authority (EFSA), 13, 178–179, 308–309
European Union, 10, **10**, 107, 110, 136, 141, 172, 279, 306, **308**
Eutomer, 28, 355; *see also* Enantiomer
Evans, S., 89
Expired drugs, 7; *see also* Drugs
Exposure assessment, **14**

F

Federal Food, Drug, and Cosmetic Act, 304
Fenaxaprop-ethyl, 5
Fenbuconazole, 204
Fenfluramine, **357**
Fenoprofen, 35
Fenoxaprop, 36
Fenoxaprop-ethyl, **58**
(*R*)-fenoxaprop-ethyl, 59
Fenpropathrin, 33, *34*
Fensulfothion, *43*, 323
Fenvalerate, 34
Fernandes A., **162**
Fexofenadine, 86
Fidalgo-Used, N., 66
Finland, 280
Fiori J., 283
Fipronil, 5, 13, *43*, **46**, 49, **53–54**, 57, 233, **238**, 329, *329*
R-fipronil, 57

Index

(R)-fipronil, 36, 59
(S)-fipronil, 36, 57
Fipronil sulfone, **234**
Fixolide, **9**
Flavan-3-ols, 281
Flavanol enantiomers, **200**
Flavanol stereoisomers, **201**
Flavanone, **198**, **199**
Flavors, 291–294
(R)-flosequinan, 35
(S)-flosequinan, 35
Fluazifop-butyl, 36, 38, **47**
(R)-fluazifop-butyl, 36
(S)-fluazifop-butyl, 36
Fluconazole, 110, 112
Flufiprole, 5, 49, **202**
Flumequine, **258**
Fluorine, 131, 135
9-fluoroenylmethylchloroformate (FMOC), 216
Fluoroquinolone, 87
Fluorotelomer alcohol (FTOH), 179
Fluoxetine, 8, 79, 88–90, **89**, **94**, 95, 259, 264–265, **311**, 312
R(−)-fluoxetine, 93
S(+)-fluoxetine, 79, 88, 89, 93, 95
Flurbiprofen, 35, 87, **89**, **94**, **258**–259
Fluroxypyr methylheptyl ester, 5
Flutriafol, **56**, **203**, **334**, **335**
Fluxametamide, **333**
(R)-fluxametamide, 336
(S)-fluxametamide, 336
Fono, L. J., 90
Fonseca, F. S., 240
Fontana, A. R., **165**
Food
    additives, 13
    analysis
        chiral separations in food analysis, **212–213**
        electrokinetic chromatography, **212–213**
        miniaturized liquid chromatography, **212–213**
    authenticity and adulteration, 12, 279–294
        dairy products, 289
        defined, 279
        essential oils, 289–291
        flavors, 291–294
        fruit beverages, 288–289
        honey, 287–288
        olive oil, 286–287
        overview, 279–280
        phenols, 281–285
        seafood, 285–286
    chain, sources in, 252–257
        chemical-specific factors influencing plant uptake of pharmaceuticals, 256–257
        human consumption in food crops, 253–254
        metabolism, 257
        plant uptake processes, 256
        stereoselective sorption, 257
        stereoselective uptake, 257
        translocation, 257
        uptake and translocation in food crops, 255–257
        veterinary drugs in food crops, 254–255
    contaminants, 13, *see also* Contaminants
    packaging, 13
    processing and its impact on pesticides in food, 235–237, **236**
    products of animal origin, 233–235, **234**
    products of plant origin, 228–233, **229–230**, *232*
    quality control and traceability
        enantioselective capillary electrophoresis, 217
        enantioselective electrokinetic chromatography, 216–217
        enantioselective gas chromatography, 214–216
        enantioselective liquid chromatography, 214
        multidimensional gas chromatography, 215–216
    safety assessment, 196–214
        enantioselective capillary electrophoresis, 205, **212–213**, 214
        enantioselective liquid chromatography, 196–204, **198–203**
        enantioselective supercritical fluid chromatography, 205, **206–211**
    storage and processing, 12
Food and Agriculture Organization, 169
Food and Drug Administration, 304, 306, 347, 355, *356*
Food chain supply (FSC), 280
Food Quality Protection Act, 304
Formoterol, 349, *349–350*, **358**
Fossil fuels, 11
Fosthiazate, *43*
Fourier-transform ion cyclotron resonance mass spectrometry (LC-ESI-FTICR-MS), 284
Freshwater snail, 6
Friedel-Crafts, 116
Fruit beverages, 288–289
Fungicides, 5, **56**, 65, 229, **238**, **335**
    chiral, 57
    chiral triazole, 45
    triazole, *43*, **56**
Furalaxyl, 55
(R)-furalaxyl, 55
Fusarium graminearum, 57, **334**
Fusarium moniliforme, **334**
Fusarium verticillioides, 333

G

Galaxolide, 8, **9**, 116–119, **117**
γ-cyclodextrin, 214
γ-decalactone, 289
γ-decalatone, 193
(+)-γ-HBCD, 145
γ-HBCD, 158, 159
(−)γ-HBCD, 143, 145
γ-hexabromocyclododecanes, 159
γ-lactones, 288
γ-nonalactone, 289
γ-TBECH, 159
Gammarus pulex, 269
Gao, B., 60
Gao, J., 49, 59
Gao, Y., 55
Gas chromatography (GC), 8, 37, 84, 139, 156–157, 196, **197**
Gas chromatography coupled to ion-trap tandem mass spectrometry (GC–MS/MS), 171, 180
Gas chromatography-mass spectrometry (GC-MS), 170, 178, 180

Gas chromatography with electron capture detection (GC-ECD), 171
Gas or liquid chromatography coupled to mass spectrometry (GC–MS, LC–MS), 252
Gas or liquid chromatography-tandem mass spectrometry (GC–MS/MS, LC–MS/MS), 168, 252
Gaussian distribution, 256
Gawdzik, B., 332
GC-C-isotope ratio mass spectrometry (IRMS), 168
Geometric isomers, 28
Gibberella zeae, **334**
Gibberellic acid biosynthesis, 57
Giroud, B., 235
Giuffrida, A., 293
Glucocorticoid receptors, 5
Glufosinate, **58**
(S)-glufosinate, 59
Glutathione, 35
Glycoproteins, 2
Gomara, B., **163**
Gossypol, **33**
Great Britain, 91, 168
Green algae, 6, **114–115**
Green chemistry, 340
Ground water, 7
Guidance for Industry: Stereochemical Issues in Chiral Drug Development, 306
Guo, B., 235
Guo, H. M., 237
Gut microbiota, 13

H

Haematococcus pluvialis, 285, 286
Halogenated organic contaminants (HOCs), 11–12, 131–145
  brominated flame retardants, 140–145
    enantioselective analysis, 143–145
    environmental behavior, 143–145
    physicochemical properties, 141–143, **142**
  chirality, 133
  of emerging concern, 132–133, **134**
  future directions, 145
  overview, 131–132
  per- and polyfluoroalkyl substances (PFASs), 133–140
    enantioselective occurrence and fate, 139–140
    physicochemical properties, **135**, 135–136
    sources and transport in environment, 136–139
Halogens, 11
Haloxyfop, 38
Hashim, N. H., 92
Hazard characterization, **14**
Hazard identification, **14**
He, R., 67
Headspace solid-phase microextraction (HS-SPME), 215, 292
Headspace–solid phase microextraction stereoselective gas chromatography–mass spectrometry (HS–SPME stereoselective GC–MS), 289
Health risks of pesticides in food, 237–241, **238–239**
Helical chirality, 1; *see also* Chirality
Helvetolide, 116, **117**
Henriksson, H., 85
Henry's law constant, 256

Heptabromodiphenyl ether, 154
Heptachlor, 154
Heptakis (2,6-di-*O*-methyl)-β-cyclodextrin, 283
Herbicides, 5, 38, 41, *43*, 45, **56**, 57, **58**, 65, **238–239**, **335**
Hermansson, J., 85
Hesperetin, **198–199**
Hesperidin, 289
Heterotrophs, 91
Hewlett-Packard 6890 N gas chromatograph, 168, 177
Hexabromobiphenyl, 11, 154
Hexabromocyclododecane (HBCD), 11–12, 141–143, 155, 158–159
Hexabromocyclododecane (HBCDD), **134**, 154
Hexabromodiphenyl ether, 154
Hexachlorobenzene (HCB), 11, 154, 168
Hexachlorobutadiene (HCBD), 154, 168, 169–171
Hexaconazole, **53**, **230**
(S)-hexaconazole, 231
High-performance liquid chromatography (HPLC), 60–65, 84, 87, 96, **197**, 252, 281–283, 288, 340
High-performance liquid chromatography coupled with a diode array detector and an ion trap mass spectrometer (HPLC/DAD-MSn), 284
High-resolution gas chromatography coupled with high-resolution mass spectrometry (HRGC/HRMS), 156
High-resolution mass spectrometry (HRMS), 157
Hochmuth, D. H., 289
HOCs of emerging concern (HOC-ECs), 132–133, **134**
Homochirality, 6; *see also* Chirality
Homogenous metal catalysts, 351
Homosalate, **10**, 120, **125**
Honey, 287–288
Hormonal activities of organic UV filters, **125**
Human consumption in food crops, 253–254
Human risk of pharmaceuticals, 267–269, *268*
Human serum albumin (HSA), 85
Humic acids, 6
Huwe, J. K., **162**
Hydrogen atom transfer, 35
Hydrogen bonding, 32, 84, 85
Hydrogen bond interaction, **195**
Hydrolysis, 7, 35
Hydrophobicity, 90, **138**
1-hydroxyethylpyrene, 35
4-hydroxyflavanone, **199**
6-hydroxyflavanone, **199**
7-hydroxyflavanone, **199**
Hydroxyl radicals (OH•), 35
3-hydroxypraziquantel diastereomers, 266
4-hydroxypraziquantel metabolites, **259**
Hydroxypropyl-β-cyclodextrin (HP-β-CD), 285
2-hydroxypropyl-β-cyclodextrin, 214
Hyoscine, 2
Hyoscyamine, **33**
Hypostomus plecostomus, 159

I

Ibuprofen, 35–36, 86, **89**, 91–92, **94**, **258–259**, 263, **311**, **357**
(R)-ibuprofen, 250
R(–)-ibuprofen, 92
S(+)-ibuprofen, 92

Index

Ibuprofen/*(S)*-ibuprofen (dexibuprofen), 7
Ibuproxen, 87
Ifosfamide, 86
Illicit drugs, 3, 5, 6–7; *see also* Drugs
Imazalil, **58**, **335**
Imazameth, **333**
Imazamox, **333**
(S)-imazamox, 5
Imazapyr, **56**, **333**
Imazethapyr, 49, **333**
Imidazole, 52, 113
Indoprofen, 35
Indoxacarb, 6, **47**, **54**
(R)-indoxacarb, 6
(S)-indoxacarb, 6
Industrial halogenation, 132
Ineffective enantiomer, 2
Insecticides, 5–6, **238**; *see also* Pesticides
    in agriculture and forestry, 45
    organophosphorus, 41
    phenylpyrazole, 5
    pyrethroid, 41
Instrumental method in chiral pharmaceuticals, 84–88; *see also* Pharmaceuticals
Interim Policy for Evaluation of Stereoisomeric Pesticides, 306
International Cooperation on Harmonization of Technical Requirements for the Registration of Veterinary Medical Products, 255
Intrinsic chirality, 2; *see also* Chirality
Inversion, 27–38
    chirality and, 28–31, *28–31*
    compounds, **33**
    conformationally, **37**
    defined, 28
    determination of, 36–37
    enantiomers, *32*
    in environment, 31–36
    future directions, 37–38
    overview, 27–28
    in soil, 36
Investigation of Chiral Active Substances, 306
Iodine, 131
Ion–dipole interaction, **195**; *see also* Dipole-dipole interaction
Ion-exchangers, **195**
R-(E)-α-ionone, 292
Islatravir, 350, *351*
Isobutoxypentabromocyclododecanes (iBPBCDs), 145
Isocarbophos, **201**, **229**, *232*, **239**
(+)-isocarbophos, 241
(R)-isocarbophos, 231
Isodrin, 171
Isofenphos enantiomers, *64*
Isofenphos-methyl, **203**
Isomerism, 347; *see also* Chirality
Isopentane, *28*
Isopentyl-4-methoxycinnamate, **125**
Isophenphosmethyl, **47**
Isopropanol, 34
Isotope ratio mass spectrometry (IRMS), 291
Itraconazole, 110–112, **111**
IUPAC (International Union of Pure and Applied Chemistry), 1

J

Jacobs, M. N., **162**
Japan, 107, 217, 280
Jiang, Y., 67, 339
Jing, X., 59
Johnson-Restrepo, B., **163**
Joint FAO/WHO Expert Committee on Food Additives (JECFA), 264
Journal of AOAC INTERNATIONAL, 282
Jurgens, M. D., **166**

K

Kalachova, K., **165**, **166**
Kalathoor, R., 66
Kasprzyk-Hordern, B., 86
Kaziem, A. E., 57, 59
Ketoconazole, **53**, 110–113, **111**, **114**
Ketoprofen, 35–36, 86, **94**, **258–259**, 263, 270, **357**
S(+)-ketoprofen, 96
Ketoprofen/*(S)*-ketoprofen (dexketoprofen), 7
Kim, J., 264
Koleini, F., 233
König, W. A., 289

L

Labadie, P., **165**
Labandeira, A., **163**
Labile enantiomers, 37; *see also* Enantiomer
Lacorte, S., **164**
Lactofen, 13, **56**, **58**, **333**, **335**
(R)-lactofen, 59
(S)-lactofen, 57, 59
Lactonfen fluazifop-butyl, 5
Laizhou Bay, 268
Lake Superior Fish Tissue, 169
Lambda-cyhalothrin, *43*, **51**, **238**
(±)-lambdacyhalothrin, 50
Landfill leachate, 136, *136*, 141
Lange, C., 118
Lansoprazole, 35, 87
L-asparagine, 12, *192*, *192*
Lavandula, 290
Lead, 6
Lemna minor, **58**, 59, **114**
Lepomis gibbosus, 263
Leptophos, *43*
Leucovorin, **358**
Levalbuterol, 7
Levobupivacaine, 7
Levocetirizine, 7
Levodopa, 2
Levofloxacin, 7, 78
Levothyroxine, 2
Lewis base, 354
L-glutamide, 32
Li, D., 327
Li, Y., 65, **164**
Lidocaine, **33**
Ligand exchange capillary electrophoresis (LECE), 293
Limonene, 216

(R)-limonene, 288
(R)-limonene, 192, *192*
(S)-limonene, 192, *192*
Lin, Y., **167**, 231
Linalool, **206**, 290
(R)-linalool, 288
(R)-linalool, 215
R-(−)-linalool, 290
S-(+)-linalool, 290
Lindane, 154, 171
Lindim, C., 137
Lipophilicity, 8, 10, 11
Liquid chromatography (LC), 37, 140, 196
Liquid chromatography-mass spectrometry/mass spectrometry, 178
Liquid chromatography–tandem mass spectrometry, 179
Liquid chromatography with triple quadrupole mass spectrometry (LC-QqQ-MS), 178
Liquid extraction, 60
Liu, H., 60, **163**, 228, 240
Liu, W., 33–34, 240
L-lactic acids, **199**
London dispersion interaction, **195**
Losada, S., **164**
Loxoprofen, **258**
Lu, Y., 237
Luo, Q., **163**
Lutein, **200**
Lux Amylose-1 chiral stationary phase, 204
Lux Amylose-1 column, 214
Lux Amylose-2 column, 204
Lux Cellulose-1 column, 60
Lyoniresinol, 192
(−)-lyoniresinol, 192
(+)-lyoniresinol, 192
Lythrum salicaria, 255

M

Mackintosh, S. A., **166**
Macrocyclic antibiotics, **195**
Macrocyclic musks, 8, 116; *see also* Musks, toxicity
Macrolide glycoprotein, 84
Macrolides, 255
Mactra veneriformis, 172
Magnetic phosphatidylcholine (MPC), 168
Majoros, I., 170
Malathion, 33, **230**, 233, 329–330, *330*, **334–335**
Mandipropamid, **202, 335**
Mariaca, R. G., 289
Martinelli, T., **167**
Mass spectrometry (MS), 8, 84–85, 87, 156–157, 291
Maximum allowable limit (MAL), 52, 54
Maximum residues levels (MRL), 250, 254
Mechanism of enantioseparation, *194*, 194–196, **195**
Mecoprop, 329
Mehdipour, M., 331
Mendes, R., 228
Mentha piperita, 294
Meptyldinocap, **236**
Merino, D., 64
Meso compound, 31, *31*; *see also* Compounds
Metabolism, 8, 257
Metalaxyl, 6, *43*, **46**, **53**, **201**, **230**, *328*, 332

R-metalaxyl, 331
(R) − metalaxyl, 332
Methamidophos, 205
Methanol, 34
Method detection limit (MDL), 260
Method limits of detection (MLOD), 178
Method limits of quantification (MLOQ), 178
Method of quantitation, chiral pharmaceuticals, 88
6-methoxyflavanone, **199**
7-methoxyflavanone, **199**
Methylamphetamine, 6
4-methylbenzylidene, 120
4-methylbenzylidene camphor, 10, **10**, 119, 124, **125**
(S)-2-methylbutyl acetate, 214–215
2-methylbutyl acetate, 214
Methyl jasmonate, **200**
Methylphenidate, **358**
N-methylpiperidine, **33**
Methylxanthines, 217
Metolachlor, *43*, **56**, 57, 332, **333**, **334**
(S)-metolachlor, *328*, 332
Metoprolol, **89, 311**
Micellar electrokinetic chromatography (MEKC), 284
Micellar electrokinetic chromatography laser induced fluorescence method (MEKC-LIF), 216
Micellar electrokinetic chromatography with laser-induced fluorescence (MEKC–LIF), 293
Miconazole, **53**, 110–113, **111, 115**
Microbial mediation, 36
Microcysts aeruginosa, **58**, 59, **333**
Microelectron capture detector (μECD), 169
Micropterus salmoides, 263
Mineralocorticoid receptors, 5
Miniaturized liquid chromatography, **212–213**
Mirex, 154
Mirtazapine, 251, **259**
(R)-(−)-mirtazapine, 251
Modanafil, **358**
Moieties, 31
Molecularly imprinted polymers (MIPs), 330–332, *331*
Molecules
    achiral, 1–2
    alkyl, 11
    aromatic, 11
    chiral, 1–2, 4, 28
    defined, 28
    interaction, **195**
    natural, 4
    polycyclic musks, **9**
    structures, **9**
    without chiral carbons, 30, *30*
Molluscicides, 5
Monocrotophos, *43*
Monoterpenes, **206**
Morphine, 3
Mosandl, A., 290, 292
Mountain product, 280
MSPE, 60
Mud carp, 159
Mugil cephalus, 235
Multidimensional gas chromatography, 215–216
Multi-walled carbon nanotubes, 6
Muscone, 116, **117**
(S)-muscone, 8

Index

Musks, toxicity, 8
Mutagenic effects, 8
Mya arenaria, 172
Myclobutanil, **47**, **53**–**54**, **56**, 204, **229**, 231, **234**, **238**, 333
Mysis relicta, 140

N

Nanomaterials, 6
Nanotubes, 6
(R)-1-naphthylglycine, 263
Napropamide, **53**, **56**, 333–334, **334**
R-napropamide, 336
(R)-napropamide, 5
Naproxen, 36, 86–87, **89**, 91–92, *92*, **94**, 258–259, 263
(S(+)-naproxen), 91
(R)-naproxen, 35–36
(R)-(−)-naproxen, 36
R(−)-naproxen, 91, *92*
(S)-naproxen, 35–36
(S)-(+)-naproxen, 36
S(+)-naproxen, 78, 91, *92*, 95
Naringenin, **198**, **199**
Natural chiral molecules, 4
Navicula pelliculosa, **114**
Nawrat, C. C., 350
N-demethylation, 264
Neale, P. A., 95
Nematicides, 5; *see also* Pesticides
Neopentane, *28*
Netherlands, 178
Neurotoxicity, 57
Neurotoxins, 45
Nguyen, L. N., 92
Ni63 electron capture detector, 170
Nicotamides, *327*
Nitro musks, 8, 116, 119; *see also* Musks, toxicity
NMR, 35
N-octanol/water partition coefficient (Kow), 45
Nonfat cocoa solids (NFCS), 282
Non-racemic, 6, 156
Non-racemic OCPs, 5
Non-steroidal anti-inflammatory drugs (NSAIDs), 253, 262–263
Norfloxacin, **258**, 260
Norfluoxetine, **89**, **94**, 95, **259**
R-norfluoxetine, 264
Norway, 137, 145, 181
Nuclear magnetic resonance, 336
Nucleic acids, 31
Nucleodex β-PM, 143
Nucleophilic substitution, 35
Nucleoside reverse transcription translocator inhibitor (NRTTI), 350
Nutraceuticals, **198**

O

Oasis HLB, 79
Oasis MCX, 79, 84
Octakis (2,3-di-O-pentyl-6-O.methyl)-γ-cyclodextrin, 215
Octanol/air partition coefficient, **138**
Octocrylene, **125**
Octyl methoxycinnamate, **125**

Octyl salicylate, **125**
Ofloxacin, 7, 87, **89**, **258**, **357**
R(+)-ofloxacin, 93
R/S(±)-ofloxacin, 78
S(−)-ofloxacin, 78, 93
Oils, *see* Essential oils
Oligosaccharides, 31
Olisah, C., 235
Olive oil, 286–287
Olsen P., 280
Omeprazole, 7, 35, 87, **358**
Oncorhynchus mykiss, 263
Oomycetes, 55
O,p′-DDD, **335**
Optical isomers, 28
Optical rotatory dispersion (ORD), 1
Optional quality terms (OQT), 280
Organic carbon, 6, **138**
Organic halogenated contaminants, 11
Organic nitrogen, 6
Organic pollutants, 3, 11–12; *see also* Inversion
Organic residual compounds, 7
Organic UV filters, 8, 10, **10**, 13, 119–125
 in environment, 120, 124
 hormonal activities of, **125**
 physicochemical properties, 120, **121**–**123**
 toxicity, 124–125
Organochlorine, **239**, **335**
Organochlorine compounds (OCs), 156
Organochlorine pesticides (OCPs), 5–6, 11, 168–170
Organophosphates, *43*, 44, 229, **238**–**239**, **335**
Organophosphorus insecticides, 41
Organophosphorus pesticides, 33; *see also* Pesticides
Ort, C., 79
Overripe orange, 215
Over-the-counter (OTC) drugs, 251, 262
Oxazepam, **33**, **259**
Oxyneolignan, **198**
Oxytetracycline, 254

P

Paclobutrazol, **229**, **234**, 235, **236**
(R)-paclobutrazol, 235
(S)-paclobutrazol, 235
Palladium-based catalysts, 352
P-aminobenzoic acid, 120
Pan, X., 67
Pantoprazole, 35, 87
Papke, O., **163**
Parabens, 8
Passive sampling, 79
Patel, R. N., 348
Pažitná, A., 287
PBT models of chiral pharmaceuticals, **308**, 309–313
Pearl River Delta, 112, 159
Penconazole, *43*, **54**, **201**
(−) penconazole, 204
(+)-penconazole, 60
Pentabromodiphenyl ether, 11
Pentachlorobenzene (PeCB), 154, 168–169
Pentachlorophenol (PCP), 154, 168
Pentane, *28*
Peptides, 33

Perfluorinated compounds, 13
Perfluoroalkyl acids (PFAAs), 179
Perfluoroalkyl substances (PFASs), 133–140, 177–180
    enantioselective occurrence and fate, 139–140
    physicochemical properties, **135**, 135–136
    sources and transport in environment, 136–139
Perfluorobutanesulfonic acid (PFBS), 179
Perfluorobutanoic acid (PFBA), 179
Perfluorohexane sulfonic acid (PFHxS), 179
Perfluorooctane sulfonamidoethanols (FOSE), 136, 138
Perfluorooctane sulfonate (PFOS), 135–140, *136*, 178
    chiral analysis, 139–140
    chirality, 139
    enantioselective occurrence and fate of, 139–140
    enantioselectivity in environment, 139–140
    sources and transport in environment, 136–139
Perfluorooctane sulfonic acid, 11, 154, 179
Perfluorooctane sulfonyl fluoride, 154
Perfluorooctane sulphonamide (FOSA), 136
Perfluorooctanoic acid (PFOA), 11, 135, 154, 177–179
Permethrin, **229, 230**, 231
Persistent organic pollutants (POPs), 153–181
    chiral pollutants, 11–12
    chiral POPs, 155–156
    chiral separation, 157–158
    decabromodiphenyl ether (decaBDE), 160–168
    detection methods, 156–157
    effects, 154
    hexabromocyclododecane (HBCD), 158–159
    hexachlorobutadiene (HCBD), 169–171
    pentachlorobenzene (PeCB), 168–169
    perfluoroalkyl substances (PFASs), 177–180
    polybrominated biphenyls (PBBs), 180–181
    polychlorinated naphtalenes (PCNs), 172–177, **173–176**
    short-chain chlorinated paraffins (SCCPs), 171–172
    Stockholm Convention, 153–154
Personal care products (PCPs), 105–125
    azole fungicides, 110–113
        environmental behavior, 112–113
        physicochemical properties, 110–112, **111**
        toxicity, 113, **114–115**
    chiral pollutants, 8–10
    future trends, 125
    HOCs of emerging concern (HOC-ECs), **134**
    organic UV filters, 119–125
        in environment, 120, 124
        hormonal activities of, **125**
        physicochemical properties, 120, **121–123**
        toxicity, 124–125
    overview, 105–106
    regulatory challenges, 107
    sources and pathways of, 107–110, *110*
    synthetic musks, 113–119
        chirality, 116
        in environment, 118
        physicochemical properties, 116–118, **117**
        toxicity, 118–119
    usage, 106–107
Pesticides, 3, 5, 6, 36, 41–68, 131, 323–341
    absolute configuration, **335**, 336–340
        circular dichroism spectroscopy, 337
        ECD spectroscopy, 337, *338*
        VCD spectroscopy, 338–340, *339*
        X-ray crystallography, 336–337

absorption, 52
banned, 5
bioaccumulation, 44
biodegradation, 44, 52
biotransformation, 44
challenges, 67, 340
defined, 41
efficacy, 34
enantiomers, 57
enantioselective
    action on target organisms, 55–57, **56**
    bioaccumulation, 52–55
    biodegradation, 50–52
    biotransformation, 49
    toxicity on non-target organisms., 57–59, **58**
enantioselectivity, 6, 52
environmental fate of, 49–55
in food, 227–241
    challenges, 241
    food processing and its impact on, 235–237, **236**
    food products of animal origin, 233–235, **234**
    food products of plant origin, 228–233, **229–230**, *232*
    future prospects, 241
    health risks, 237–241, **238–239**
    occurrence, 228–235
    overview, 227–228
future prospects, 67, 340
molecularly imprinted polymers, 330–332, *331*
organophosphorus, 33
overview, 41–44
phenylpyrazole, 57
physical and chemical transformations, 44
physicochemical properties, 45–49, **46–48**
pollutants, 5–6
pre-concentration of, 60
separation techniques, 60–67
sources, 5
stereoselectivity, 55
switching, 332–336, **333–334**
synthesis, 324–330
    catalytic asymmetric synthesis, 325–326, *325–326*
    chiron approach, 326–327, *327*
    enzymatic transformation, 327–328, *328*
    synthesizing chiral pesticides, 328–330, *328–330*
types, 44–45
PH, 6, 31, 36, 52, 79, 281
Phaffia rhodozyma, 285–286
Phantolide, 8, **9**, 116–117, **117**
Pharmaceuticals, 3, 5, 6–7, 31, 77–96, 249–271, 347–360
    analysis in environment, 78–88
        chiral stationary phases, 84–88
        instrumental method, 84–88
        method of quantitation, 88
        sample preparation, 79–84
        sampling and sample storage, 78–79
        signal suppression, 88
    analytical aspects, 251–252, *252*
    chiral, 6, 7, 31
    chiral pollutants, 6–8
    described, 250
    drug development, 355–356
    enantiomers, **94–95**
    enantiopure, 7

enantioselective occurrence in food matrices, 257–267, **258–259**
  anthelmintic drugs, 265–266
  antibiotics, 260–262
  drugs, 266–267
  non-steroidal anti-inflammatory drugs (NSAIDs), 262–263
  psychoactive drugs, 263–265
enantiospecific toxicity, 269–270
estrogens and, 7
fate in environment, 88–93
  analgesics, 91–92
  antibiotics, 93
  antidepressants, 88–90
  beta-blockers, 90
  stimulants, 90–91
in freshwater, **89**
future challenges, 270–271
future perspectives, 96
future research, 356–360
human risk, 267–269, *268*
industries, 7
mammals, 35
overview, 77–78, 249–250, 347–348
polar nature, 7
sources in food chain, 252–257
  human consumption in food crops, 253–254
  metabolism, 257
  plant uptake of pharmaceuticals, 256–257
  plant uptake processes, 256
  stereoselective sorption, 257
  stereoselective uptake, 257
  translocation, 257
  uptake and translocation in food crops, 255–257
  veterinary drugs in food crops, 254–255
stereochemistry, 77
stereoselectivity, 78
synthetic methods, 348–355
  chemoenzymatic methods, 348–349, *349–350*
  chemoselective drug synthesis, 350–351, *351*
  symmetric organocatalysis, 354–355
  symmetric organometallic catalysis, 351–352, *353–354*
toxicity in environment, 93–96, **94–95**
Pharmaceuticals and personal care products (PPCPs), 251, 260; *see also* Personal care products (PCPs)
Pharmacodynamics, 6
Phencynoate, **259**
Phenols, 281–285
Phenoxyphenoxypropionate herbicides, 38
Phenthoate, 33
(−)-phenthoate, 36
(+)-phenthoate, 36
Phenylalanine, 214
Phenylbenzimidazole sulfonic acid, 119
2-phenylbutyric acid, 35
Phenylpyrazole chiral insecticides, 5
Phenylpyrazole pesticides, 57
3′-phosphoadenosine-5′-phosphosulfate, 35
Phospholipids, 31
Phosphoric acid, 44
Phosphorous, 1
Photobacterium leiognathi, **94**, 95–96
Photodegradation, 7, 44

Photo-induced chiral inversion, 34–35; *see also* Inversion
Photolysis, 34, 35, 112, 132
Physicochemical properties
  azole fungicides, 110–112, **111**
  brominated flame retardants, 141–143, **142**
  organic UV filters, 120, **121–123**
  per- and polyfluoroalkyl substances (PFASs), **135**, 135–136
  synthetic musks, 116–118, **117**
Physicochemical properties of chiral pesticides, 45–49, **46–48**
Phytophthora capsici, **334**
Pimephales promelas, 93, **94**, 269, 312
Pindolol, **311**
Piperidine, **33**
π-π interaction, **195**
Pirard C., **162**
Pirsaheb, M., 228, 231
Planar chirality, 1
Plant growth regulators, 5
Plant uptake of pharmaceuticals, 256–257; *see also* Pharmaceuticals
Plant uptake processes, 256
Plecostomus, 159
PLHC-1 cells, **94**
Plutella xylostella, **333**, 336
Poa annua, 336
Polarimetry, 1–2
PolarisQ Ion-Trap mass spectrometer, 160
Pollutants, 6–7, 301–313
  biodegradation, 31
  case study, 309–313
  challenges in exposure assessment, 309–312, *310*, **311**
  challenges in toxicity assessment, 312–313
  chiral, 6
  enantioselective analysis, 36
  environmental, 5
  future perspectives, 313
  implications of chirality on risk assessment, 303
  organic, 3
  overview, 301–303
  PBT models of chiral pharmaceuticals, 309–313
  persistent organic pollutants (POPs), 11–12
  personal care products, 8–10
  pesticides, 5–6
  pharmaceuticals, 6–8
  regulation, 303–307, *304*, **305**
  risk assessment, 13–15, **14–15**, 307–309, **308–309**
  sources, 3–5, *4*
  stability, 32
Poly- and perfluoroalkyl substances (PFASs), 302–303
Polybrominated biphenyls (PBBs), 12, **134**, 141, 180–181
Polybrominated diphenyl ethers (PBDEs), 160, 161, **162–167**, 170, 180–181
Polychlorinated biphenyls (PCBs), 11, 132, 141, 154–155, 157, 172, 255
Polychlorinated dibenzofurans (PCDFs), 154, 172
Polychlorinated dibenzo-p-dioxins (PCDDs), 154, 172
Polychlorinated naphtalenes (PCNs), 154, 168, 171–177, **173–176**
Polycyclic aromatic hydrocarbons (PAHs), 5, 11
Polycyclic musks, 8, **9**, 116, 119
Polyfluoroalkyl substances, 11
Polymers, *see* Molecularly imprinted polymers

Polysaccharides, 87, **195**
Poly-γ-glutamate, **198**
Poma, G., **166–167**
Portolés, T., **167**
Portugal, 118
Posaconazole, 110–111, **111**
Pranoprofen, **258**
Praziquantel, 86, 265–266
Pre-concentration of chiral pesticides, 60
Predicted environmental concentration (PEC)/predicted no effect concentration (PNEC) ratio, 312
6-prenylnaringenin (6PN), **198**
Pressurized liquid extraction (PLE), 180
Primary emissions, 155
Product of island farming, 280
Profen, 35, 36, 85, 262
(S)-profen, 35
Profenos, *43*
(R)-profenyl-CoA thioester, 35
(S)-profenyl-CoA thioester, 35
Programmable temperature vaporizer (PTV), 286
Promethazine, **259**
Propiconazole, 112
Propisochlor, **334**
Propranolol, **89**, 90, **94**, 268, 311
(R)-(+)-propranolol, *2–3*
R(+)-propranolol, 93
(S)-(–)-propranolol, *2–3*
S(–)-propranolol, 250
S(–)-propranolol, 93
R(+)-propranolol-d, 88
S(–)-propranolol-d, 88
Prostaglandins, 263
Protected Designation of Origin (PDO), 280
Protected Geographical Indication (PGI), 280
Proteins, 2, 31
Prothioconazole, 5, 54, **56**, **58**, 67, 205, **238**, **334**, **335**
(–)-prothioconazole, 59
(+)-prothioconazole, 59
(R)-prothioconazole, 57
S-prothioconazole, 54, 57
Prothioconazole-desthio, 49
Proton exchange, 33
Proton pump inhibitors, 35, 87
Pseudoephedrine, **95**
1R,2R(–)-pseudoephedrine, 93
1S,2S(+)-pseudoephedrine, 93
Pseudokirchneriella subcapitata, **94–95**, 95, **114**, 269
Psychoactive drugs, 263–265
Psychotropic drugs, **259**
Pulkrabova, J., 163
Pure compounds, 5
Pure enantiomers, 3, 7
Pyraclofos, **238**, 240
Pyrethrins, 45
Pyrethroids, 6, 41, *43*, 45, 49, **58**, 65, 205, 229, 231, 233, **234**, **238**, **239**
Pyricularia grisea, 334
Pyrimethanil, **234**

Q

Qi, P., 64
Qi, Y., 335
Qian, Y., 57
Qu, H., 59
Quality of life, 8; *see also* Personal care products
Quantum mechanical–molecular mechanics (QM/MM), 155
QuEChERS, 60, 169
Quizalofop-ethyl, 36

R

Rabeprazole, 35, 87
Racemases, 35
Racemic cyflumetofon, 233
Racemic drugs, 252
Racemic mixtures, 2–3, 5, 7, 29, 42, 45, 112, 156, 192, 250, 330
Racemization, 27
Rac-metolachlor, *325*, 325–326
Rac-propisochlor, 5
Ramage, S., 79
R configuration of 1-bromo-1-chloromethane, *29*
R-diclofop, 240
Recalcitrance, 11
Rectus, 2
Recycling water, 8
Redox conditions, 6
Registration, Evaluation, Authorization, and Restriction of Chemicals (REACH), 304
Regulation, pollutants, 303–307, *304*, **305**
Regulatory challenges, personal care products, 107
Reich, S., 34
Relative response factors (RRF), 169
R-enantiomers, 52, 54, 57, 59, 231
Repellents, 5
Repurposed drugs, 27
Residual organic compounds, 7
Restricted chemicals, 154
Reversed-phase liquid chromatography to gas chromatography (RPLC-GC), 286–287
Reversed phase liquid chromatography with gas chromatography/mass spectrometry (RPLC-GC-MS), 289
Rhizoctonia cerealis, 57, **334**
Rhizoctonia solani, **334**
Ribose, 2
Rice, J., 93
Risk assessments
  chiral pollutants, 13–15, **14–15**
  environment, 32
  environmental, 8
  pollutants, 303, 307–309, **308–309**
Risk characterization, **15**
River Po, 160
RNA, 2
Roach, 170
Rockenbach, I., 284
Rodenticides, 5
Roszko, M., **165**
R-triticonazole, 54
Ruditapes philippinarum, 119
Runoff, 7
Rutilus rutilus, 170
Rxi-5MS capillary column, 160

Index

S

Safe Drinking Water Act, 304
Salbutamol, 7, 79, **311**, **357**
Salicylates, 120
Salmo salar, 286
Salvia sclarea, 327
Sample preparation, chiral pharmaceuticals, 79–84
Sampling and sample storage, chiral pharmaceuticals, 78–79
Sanganyado, E., 96, 257
Sapozhnikova, Y., **166**
Scenedesmus obliquus, 333–334
Schubert, V., 290
Schurig, V., 34
(−)-sclareol, 327
Sclerotinia sclerotiorum, **334**
S-diclofop, 240
Seafood, 285–286
Secondary emissions, 155
Sedlak, D. L., 90
Sewage sludge, 8
Shaw, S. D., **163**
Shellie, R. A., 291
Shen, Q., 49
Short-chain chlorinated paraffins (SCCPs), 154, 171–172
ΣHBCD, 159
ΣHBCDD, 159
ΣPBDEΣPBDE, 160
ΣTBECH, 159
Signal suppression, chiral pharmaceuticals, 88
Silent Spring (Carson), 302
Silicon, 1
Simeconazole, **207**, **229**, *232*
Singapore, 107, **108–109**
Single enantiomers, 6
Sinister, 2
Site specific natural isotope fractionation, measured by NMR-spectroscopy (SNIF-NMR), 291
Smelcerovic, A., 290
S-metolachlor, 326
Snail, freshwater, 6
S-napropamide, 336
S-norfluoxetine, 95, 265
S-norfluoxetine, 95
Socrates, 301
Sodium azide, 79
Sodium dodecyl sulfate (SDS), 283
Soils
  Beijing, 36, 55
  chiral compounds, 6
  chiral inversion in, 36
  ecosystems, 42
  microcosms, 91
  non-racemic, 6
  pH, 6
  properties, 44
  redox conditions, 6
Solid-liquid phase transfer catalysis (S-L PTC), 329, *329*
Solid phase extraction (SPE), 79
Solid-phase microextraction (SPME), 288, 293
Solid phase microextraction and multidimensional gas chromatography (SPME–MDGC), 286
Solvent-induced chiral inversion, 32–34

Sorbus domestica, 215
Sotalol, **89**, **311**
Sotolon, **208**
Sources and pathways of personal care products, 107–110, *110*
Sources and transport in environment, PFASs, 136–139
Sources and transport of PFOS in environment, 136–139
Sources of chiral pollutants, 3–5, *4*
Spain, 137
Spirocyclicterpenoids, **200**
SPME, 60
Spodoptera exigua, **333**, 336
Sprague, M., **165**
Stanley, J. K., 15
Stereocenter, 28–30
Stereochemistry, 77
Stereoconfiguration, 31, 35
Stereogenic element, 1
Stereoisomers, 10, 29
  3-chlorobutan-2-ol, 30, *30*
  ephedrine, *78*
  galaxolide, 8
Stereoselectivity, 5
  biodegradation, 6
  chiral pesticides, 55
  chiral pharmaceuticals, 78
  environment, 90
  sorption, 257
  uptake, 257
Stereospecificity, *192*
Steric hindrance interaction, **195**
Steroid-like drugs, **259**
Stimulants, 90–91
Stir bar sorptive extraction (SBSE), 293
Stir bar sorptive extraction (SBSE)-enantio-MDGC/MS, 291
Stockholm Convention, 11, 153–154, 307
Stoschitzky, K., 312
Streptomyces venezuelae, 261
Streptomycin aureofacien, 262
S-triticonazole, 227
Sühring, R., **166**
Sulfadimethoxine, 255
Sulfamethazine, 270
Sulfonamides, 255
Sulfotransferases, 35
Sulfoxaflor, **202**
Sulfur, 1
Sun, R., 159
Supercritical fluid chromatography (SFC), 37, 44, 66–67, 87, 139, 140, 196, **197**, 252, 330
Supercritical fluid chromatography coupled to mass spectrometry (SFC-ESI-MS/MS), 205
Supercritical fluid chromatography coupled to quadrupole time-of-flight mass spectrometry (SFC-Q-TOF/MS), 205
Supercritical fluid liquid chromatography, 66–67
Suprofen, 35
Surface water, 8
Surma-Zadora, M., 171
Suzuki-Miyaura cross-coupling reaction, 352, *354*
Switzerland, 112
Symmetric organocatalysis, 354–355
Symmetric organometallic catalysis, 351–352, *353*–*354*

Synthesizing chiral pesticides, 328–330, *328–330*
Synthetic methods, pharmaceuticals, 348–355
    chemoenzymatic methods, 348–349, *349–350*
    chemoselective drug synthesis, 350–351, *351*
    symmetric organocatalysis, 354–355
    symmetric organometallic catalysis, 351–352, *353–354*
Synthetic musks, 8, 113–119, 117
    chiral, 8
    chirality, 116
    in environment, 118
    physicochemical properties, 116–118, **117**
    toxicity, 118–119
Synthons, 325
Szajnecki, Ł., 332

T

Taiwan, 169
Tan, Q., 67
Tandem mass spectrometry, 8
Tapie, N., **164**
Tebuconazole, *43*, **47**, *51*, 52, 55, **56**, 112, **229**, 231, **238**, *333*
Tebuconazole enantiomers, *64*
Tebuconazole pesticide, 42, *42*
Tebufenozide, **236**
Teicoplanin, 85
Tenebrio molitor, 55
Teratogen, 27
Teratogenicity, 347
Terpenes-4-ol, 12
(S)-α-terpineol, 288
(S)-α-terpineol, 215
Tetrabromocyclooctane (TBCO), 143, 145
Tetrabromodiphenyl ether, 11
Tetrabromoethylcyclohexane (TBECH), 12, **134**, 143, 145, 159
Tetraconazole, **56**, **334**, **335**
(-) tetraconazole, 204
(+) tetraconazole, 204
Tetracyclines, 255
Tetrahedron carbon, 1
Tetrahymena thermophila, 93, **94–95**, 95
Tetramisole, 86
Tetranychus cinnabarinus, **333**, 336
Thailand, 144
Thalidomide, 27, 33, 347, *348*
(R)-thalidomide, 27
(R)-(+)-thalidomide, 306
(S)-thalidomide, 27, 306
Theobroma cacao, 285
Theobromine, **201**
Theramenes, 301
Thermal-induced chiral inversion, 34; *see also* Inversion
Thermal stability, **135**
Thiacloprid, 205
Thiamethoxam, **234**, 235
Thioethers, 35
Thio-phosphoric acid, 44
Three-point interaction model, **195**
Tilapia, 159
Tilapia nilotica, 159
Time-of-flight (TOF), 196
Time-of-flight mass spectrometry (TOF-MS), 157
TiO$_2$, 119

Tiritan, M. E., 3–4
(+)-(3 R)-tonalid, 116
(−)-(3 S)-tonalid, 116
Tonalide, 8, **9**, 116–119, **117**
Tong, Z., 57
Toxaphene, 155
Toxaphene congers, 5
Toxicity
    azole fungicides, 113, **114–115**
    bioactivity and, 335
    chiral musks, 8
    enantioselective, 6
    enantiospecific, 269–270
    fipronil, 49
    HOCs, 133
    indoxacarb, 6
    organic UV filters, 124–125
    phenylpyrazole pesticides, 57
    synthetic musks, 118–119
Toxic Substances Control Act of 2016, 303–304
Traceability, *see* Food
Trace GC 2000, 160
Tracing, 280; *see also* Food
Traditional Specialities Guaranteed (TSG), 280
Tramadol, 79
(R,R)- (+)-tramadol, 312
(S,S)-(−)-tramadol, 312
Trans-chlordane, 6
Transformation, 8, 10
Transition-metal-catalyzed cross-coupling reactions, 352, *353*
Translocation, 257
Trans-tramadol, 355
Traseolide, 8, **9**, 116–118, **117**
Triadimefon, **47**, 49, 55, **230**
Triadimenol, 49
Triazole fungicides, 5, *43*, 45, **56**, 204, **238**
1,2,4-triazole moiety, 45
Trimethyl-β-cyclodextrin, 217
Triple quadrupole (QQQ), 196
Tris(3-chloro-5-methylphenycrbamates), 64
Triticonazole, *43*, 54, **203**, **230**, **335**
Trophic magnification factors (TMFs), 145
Tryptophan, 35
Tubifex tubifex, 311
Two-dimensional liquid chromatography, 229

U

Ultrahigh performance supercritical fluid chromatography tandem mass spectrometry (UHPSFC-MS/MS), 88
Unintentional production of chemicals, 154
United Nations, 169
United States, 107, **108–109**, 120, 137, 141
Unused drugs, 7
UPLC, 96
Uptake and translocation in food crops, 255–257
Usage of personal care products, 106–107
UVA light, 10
UVB light, 10
UVC light, 10
UV irradiation, 37
UV radiation, 10

V

Vancomycin, 84–85
Van der Waals interaction, **195**
Vapor pressure, **138**
Varian 3800 gas chromatography, 160
VCD spectroscopy, 338–340, *339*
Venlafaxine, 79, 85, 88, **89**, 89–90, **311**
Very persistent and very bioaccumulative (vPvB), 307, **308**
Veterinary drugs in food crops, 254–255
Vibrational circular dichroism (VCD), 336–337
Vibrio fischeri, 269
(−)-vinclozolin, 214
(+)-vinclozolin, 214
Voorspoels S., **162**

W

Wang, D., **164**
Wang, F., 59
Wang, J., **165**
Wang, X., 55
Wang, Z., 49, 54, 60, 266
Wastewater, sewage, 52
Wastewater effluents, 8, 79, 89, 96
Wastewater treatment plants (WWTPs), 7, 36, 77, 92–93, 109, 112, 118, 125, 251, 260, 312
Waszak, I., **166**
Water Framework Directive (WFD), 172
Water partition coefficient, 44, **138**
Waters Autospec Ultima High mass spectrometer, 168, 177
Water solubility, **138**
Western Europe, 168
Williams, M., 310
World Anti-Doping Agency, 266
World Health Organization, 169, 265

X

Xanthobacter flavus, 36
Xenobiotic compounds, 45; *see also* Pyrethroids
Xenopus laevis, **58**
Xiang, D., 59, 66
Xiao, F., 137
Xie, J., 59, 340
Xie, Y., 49
X-ray crystallography, 336–337
X-ray diffraction analysis, 143
Xu, C., 240

Y

Yamasa Corporation, 350
Yang, J., 336
Yang, Q., 237
Yao, G., 66
Ye, F., 336
Ye, X., 228, 235
Yin, J., 55
Youness, M., 52

Z

Zacs, D., **166**
Zeaxanthin, **200**
Zebrafish, 6, 8, 113, 119, 144
Zeng, H., **167**, 229
Zhang, H., 337
Zhang, P., 52, 240
Zhang, Q., 57, 60, 227, 231, 233, 266, 337
Zhang, X., 118
Zhang, Y., 155
Zhang, Z., **165**
Zhao, B., 231, 235
Zhao, M., 240
Zhao, P., 57
Zheng, L., 328
Zhou, S., 5
Zhuang, S., 240
ZnO, 119
Zopiclone, **358**
Zoxamide, **335**

Printed in the United States
By Bookmasters